Laser Treatment of Materials

Edited by
B. L. Mordike

INFORMATIONSGESELLSCHAFT · VERLAG

ISBN 3-88355-185-6

Papers presented at the European Conference on Laser Treatment of
Materials, 1992 (ECLAT '92)
Göttingen, Germany, held by
— Deutsche Gesellschaft für Materialkunde
in collaboration with
— Deutscher Verband für Schweißtechnik,
— Arbeitsgemeinschaft Wärmebehandlung und Werkstofftechnik,
— Wissenschaftlicher Arbeitskreis Laser Technik

The individual papers making up the symposium are published
as submitted by the authors.

© 1992 by DGM Informationsgesellschaft mbH
Adenauerallee 21, D-6370 Oberursel 1

Preface

Since the last ECLAT conference there has been a steady increase in the number of industrial applications. Laser cutting and laser welding are widely used and are revolutionising many production procedures, particularly in the automobile industry.

There has been only isolated, but nevertheless interesting developments in the surface treatment of materials. The problems of reproduceability, on line control, quality guarantee have in most cases still to be solved. Non destructive testing of laser treated components is another important area.

Successful application of laser technology depends on interdisciplinary cooperations. This has undoubtedly retarded progress in the past. The barriers are being broken down and the rate of development is increasing. The list of topics to be discussed at this conference is evidence of laser technology incorporating all disciplines,

> New Sources / New Processes
> Lithography
> Structuring
> Laser assisted PVD / CVD
> Thin Film Technology
> Surface Treatment
> Residual Stresses
> Applications to Non-Metallic Materials
> Laser Compatible Design
> Quality Control

The ability to predict the required laser parameters and to control the process depends on the development of suitable models. A workshop session has been devoted to modelling aspects of laser treatment in an attempt to promote this aspect of laser technology.

The organising commitee wish all participants a successful conference and hope that this years's programme will meet with the same approval as in previous years. We would like to express our thanks to all those firms and organisations who have contributed to the success of the conference whether in form of a donation or in kind.

B. L. Mordike
Chairman Programme Commitee

Contents

I. Systems and Auxiliary Apparatus

The Excimer Laser - a New Tool for Industrial Material Processing
B. Basting, Göttingen (D)   3

The High Speed Pyrometer System for Laser Welding, Cutting, Heat
Treatment and Alloying Processes Temperature Control
M. Ignatiev, A. Ermolaev, I. Titov, Moscow (CIS); S. Sturlese, Rome (I);
I. Smurov, Saint-Etienne (F)   15

Computer Controlled Laser Beam Cutting of Ceramics
H.K. Tönshoff, L. Overmeyer, F. von Alvensleben, Hannover (D)   21

High Power Laser Diodes as a Beam Source for Materials Processing
V. Krause, H.-G. Treusch, E. Beyer, P. Loosen, Aachen (D)   27

Welding with a 3 kW-Nd:YAG-Solid-State Laser
H.-G. Treusch, V. Krause, J. Berkmanns, E. Beyer, Aachen (D)   33

A Comparison of Powder- and Wire-Fed Laser Beam Cladding
F. Hensel, C. Binroth, G. Sepold, Bremen (D)   39

Dynamics of Phase Fronts in Pulse Laser Action. Influence of Pulse
Shape and Spike Structure
Smurov, C. Surry, Saint-Etienne (F); A. Lashin, Milano (I)   45

Capacitive Clearance Sensor System for High Quality Nd:YAG Laser
Cutting and Welding of Sheet Metal
S. Biermann, A. Topkaya, M. Jagiella, Gaggenau (D)   51

Kaleidoscope Beam Homogenizer for High Power CO_2 and YAG Laser
T. Ishide, O. Matsumoto, M. Mega, S. Ito, T. Mituhashi, Takasago (Japan)   57

Requirements for Beam Guiding Systems Handling CO_2 High
Power Laser Beams
M. Bea, P. Berger, A. Giesen, C. Glumann, W. Krepulat, Stuttgart (D)   63

II. Cutting and Welding

Laser Material Processing: Examples from the USA
P.E. Denney, Pennsylvania (USA)   71

High Power YAG Laser Welding and Its In-Process Monitoring
Using Optical Fibres
T. Ishide, O. Matsumoto, Y. Nagura, Tagasago (Japan); A. Yokoyama,
T. Nagashima, Kobe (Japan)　　81

Laser Beam Cutting with Solid State Lasers and Industrial Robots
M. Geiger, A. Gropp, Erlangen-Nürnberg (D)　　87

Cutting of Hard Magnetic Materials and Aluminium Alloys by High Power
Solid-State Laser
G. Spur, J. Betz, Berlin (D)　　93

Influence of Processing Parameters and Alloy Composition on the
Weld Seam in Laser Welding of Aluminium Alloys
L. Rapp, C. Glumann, F. Dausinger, H. Hügel, Stuttgart (D)　　99

Material Removal and Cutting of Carbon Fibre Reinforced Plastics
with Excimer Lasers
D. Hesse, H.K. Tönshoff, Hannover (D)　　105

Performance of Various Nozzle Designs in Laser Cutting
R. Edler, P. Berger, H. Hügel, Stuttgart (D)　　111

High Power Laser Processing of Aluminium Alloys
A.S. Bransden, T. Endres, Nürnberg (D)　　117

Welding of Different Aluminium Alloys
H. Sakamoto, K. Shibata, Yokosuka (Japan); F. Dausinger, Stuttgart (D)　　125

CO_2 Laser Welding of SiO_2-Al_2O_3 Ceramic Tubes
A. De Paris, M. Robin, I G. Fantozzi, Villeurbanne (F)　　131

Deep Penetration Welding of Steel with Pulsed CO_2-Laser Radiation
T. Wahl, F. Dausinger, H. Hügel, Stuttgart (D)　　137

Influence of the Process Parameters on the Weld Metal Hardness of Laser
Beam Welded Structural Steel Regarding the Formation of Pores
G. Kalla, K. Janhofer, E. Beyer, Aachen (D)　　143

The Laser - A New Tool for Welding Aluminium Materials
J. Berkmanns, K. Behler, E. Beyer, Aachen (D)　　151

High Power Laser Welding of Microstructure Sensitive and
Dissimilar Materials
H.W. Bergmann, T. Endres, D. Müller, Nürnberg (D)　　157

Mechanical Properties of Laser Welded Al-Alloys
A. Lang, H.W. Bergmann, Erlangen (D)   163

III. Surface Melting, Alloying and Hardening

Metallurgical Aspects of Laser Surface Treatment
B.L. Mordike, Clausthal (D)   171

Characterization and Wear Resistance of Cobalt Base Coatings
Deposited by CO_2 Laser on Steam Turbine Components
P.A. Coulon, G. Thauvin, Belfort (F); *J. Com-Nougue,*
E. Kerrand, Marcoussis (F)   181

Laser Alloyed High Speed Steels
R. Ebner, B. Kriszt, Leoben (A)   187

Load Bearing Capability of Laser Surface Coatings for
Aluminium-Silicon Alloys
E. Blank, O. Hunziker, S. Mariaux, Lausanne (CH)   193

Laser Hardening - an Effective Method for Life Time Prolongation of
Erosion Loaded Turbine Blades on an Industrial Scale
B. Brenner, G. Wiedemann, B. Winderlich, S. Schädlich,
A. Luft, D. Stephan, W.Reitzenstein, H.-T. Reiter, Dresden (D);
W. Storch, Berlin (D)   199

Microstructures and Properties of Surface Layers Produced on
Steels by Different Laser Melting Techniques
A. Luft, W. Löschau,K. Juch, R. Gassmann, V. Fux,
W. Reitzenstein, Dresden (D)   205

Improvement of Mechanical Properties of Steels by Addition of
Tungsten Carbides: Laser Cladding and Laser Welding
J.M. Pelletier, F. Fouguet, Lyon (F); *M. Robin,* Villeurbanne (F);
D. Dezert, St. Jorioz (F); *A.B. Vannes,* Ecully (F)   211

Laser Beam Surface Treatment - Is Wear No Longer the Bug Bear of Old ?
W. König, L. Rozsnoki, P. Kirner, Aachen (D)   217

Laser Cladding with Preheated Wires
A. Hinse-Stern, D. Burchards, B.L. Mordike, Clausthal (D)   223

Laser Surface Alloying of Titanium with Nickel and Cobalt
A. Weisheit, B.L. Mordike, Clausthal (D)   229

Pulse Laser Alloying Energy Spatial Distribution vs. Concentration
Fields of Alloyed Elements
I. Smurov, Saint-Etienne (F); *L. Covelli*, Milano (I); *K. Tagirov,
L. Aksenov*, Moscow (CIS)   235

The Application of Laser Technology for Improvement High Load
Friction Joints Tribotechnical Parameters
M. Ignatiev, E. Kovalev, V. Titov, A. Uglov, Moscow (CIS);
S. Sturlese, Rome (I); *I. Smurov*, Saint-Etienne (F)   241

Laser Fusion of Ceramic Particles on Aluminium Cast Alloys
H. Haddenhorst, E. Hornbogen, Bochum (D)   245

Synthesis of Nitride and Carbide Compounds of Titanium by Means
of a Solid State Laser Source
I. *Smurov*, Saint-Etienne (F); *L. Covelli*, Milano (I);   251

Thermochemical Nitridation of TA6V Titanium Alloys by CO_2 Laser
P. Laurens, D. Kechemair, L. Sabatier, T. Puig, L. Beylat, Arcueil (F);
P. Jolys, P. Lagain, Suresnes (F)   257

Hardness Improvement in Laser Surface Melted Tool Steel by
Cryogenic Quenching
A.R. Costa, M.A. Anjos, Ouro Preto (Brasil); *R. Vilar*, Lisboa (P)   263

Solidification in Laser Surface Alloying of Al and AlSi10Mg with Ni and Cr
E.W. Kreutz, N. Pirch, M. Rozsnoki, Aachen (D)   269

Alloying and Dispersion of TiC-Base Hard Particles into Tool Steel
Surfaces Using a High Power Laser
W. Löschau, K. Juch, W. Reitzenstein, Dresden (D)   281

Microstructure and Properties of Laser Processed Composite Layers
A. Schüßler, Karlsruhe (D)   287

Laser Carburizing of a Low Carbon Steel
D. Müller, H.W. Bergmann, T. Heider, T. Endres, Erlangen (D)   293

The Micro-Characterization of Laser Gas Nitrided Titanium
A. Bharti, M. Roy, G. Sunderarajan, Hyderabad (India)   299

Laser Heating and Structure of Steel
*V.M. Schastlivtsev, V.D. Sadovsky, T.I. Tabatchikova,
I.L. Yakovleva*, Yekaterinburg (CIS)   307

Effects of Transformation-Hardening without Absorptive Coatings
Using CO_2- and Nd:YAG-Lasers
T. Rudlaff, F. Dausinger, Stuttgart (D)   313

High Intensity Arc Lamps for Transformation Hardening and Remelting
of Metals
H.K. Tönshoff, M. Rund, Hannover (D)   319

Possibilities for the Combinations of Thermochemical Diffusion
Treatments with Thermal Laser Treatments
D. Müller, M. Amon, T. Endres, H.W. Bergmann, Erlangen (D)   325

Treatment of Overlapping Runs by cw-CO_2 Laser
F. Nizery, G. Deshors, Arcueil (F); J. Mongis, J.-P. Peyre, Senlis (F)   331

Fatigue Strength of Laser Hardened Plain Carbon and Alloyed Steels
S. Schädlich, B. Winderlich, B. Brenner, Dresden (D)   337

Influence of Laser Hardening on the Fatigue Strength of Notched
Specimens from X20Cr13
B. Winderlich, B. Brenner, C. Holste, H.-T. Reiter, Dresden (D)   343

Optimization of the Laser Hardened Track Geometry for Different
Kinds of Power Density Distribution
D. Lepski, W. Reitzenstein, Dresden (D)   349

Characteristic Features of Laser-Induced Hardening in Maraging Steels
I.L. Yakovleva, T.I. Tabatchikova, Yekaterinburg (CIS)   355

Laser Material Processing with 18kW Using a Variable Beam Profile
Achieved with a Deformable Optic
H.W. Bergmann, T. Endres, R. Anders, D. Müller, Erlangen (D);
B. Freisleben, Röthenbach (D)   363

Laser Cladding of Paste Bound Carbides
E. Lugscheider, H. Bolender, Aachen (D); H. Krappitz, Hanau (D)   369

Power Absorption during the Laser Cladding Process
C.F. Marsden, J.-D. Wagnière, A. Frenk, Lausanne (CH)   375

Laser Cladding of Ni-Al-Bronze on Cast Al Alloy AA333
Y. Liu, J. Mazumder, Urbana (USA); K. Shibata, Yokosuka (Japan)   381

Laser Cladding with Amorphous Hardfacing Alloys
V. Fux, S. Nowotny, D. Pollack, A. Luft, W. Schwarz, Dresden (D)   387

CAD/CAM Supported Laser Surface Treatment of Complex Geometry
W. König, D. Scheller, P. Kirner, Aachen (D)   393

Producing Hardfacing Layers on Aluminium Alloys by High Power
CO$_2$-Laser
R. Volz, Stuttgart (D)   399

Adhesion and Microstructure of Laser Cladded Coatings
R. Gassmann, A. Luft, S. Schädlich, A. Techel, Dresden (D)   405

Laser Cladding with a Heterogeneous Powder Mixture of WC/CO
and NiCrBSi
B. Grünenwald, J. Shen, F. Dausinger, Stuttgart (D);
S. Nowotny, Dresden (D)   411

IV. Laser Application in the Automotive Industry

Anwendung der Lasertechnik im Automobilbau
W. Kirmse, Stuttgart (D)   419

Recent Trends in Laser Material Processing in the Japanese
Automotive Industry
K. Shibata, Yokosuka (Japan)   433

V. Thin Film Technology

Laser Ablation and Arc Evaporation
W. Pompe, B. Schultrich H.-J. Scheibe, P. Siemroth,
H.-J. Weiß, Dresden (D)   445

Characterization and Applications of PLD Oxide Ceramic Films
G. Erkens, O. Lemmer, T. Leyendecker, M. Alunovic, J. Funken,
E.W. Kreutz, H. Sung, A. Voss, Aachen (D)   ·   451

Preparation of Thin Films by Laser Induced Deposition (Laser-PVD)
H. Haferkamp, C. Möhlmann, Hannover (D)   461

In Situ Laser Assisted PVD and Characterization of Y-Ba-Cu-O Layers
inside a Laser Coupled Scanning Electron Microscope
K. Wetzig, S. Menzel, Dresden (D)   467

Growth Kinetics of Thin Films Deposited by Laser Ablation
A. Richter, Emden (D); G. Keßler, Dresden (D); B. Militzer, Berlin (D)   473

Laser Pulse Vapour Deposition of Metal-Carbon Superlattices for
Soft X-Ray Mirror Applications
R. Dietsch, H. Mai, W. Pompe, B. Schöneich, S. Völlmar, Dresden (D);
S. Hopfe, R. Scholz, Halle (D); B. Wehner, Dresden (D);
P. Weißbrodt, Jena (D)   479

Laser-CVD on High-Tensile Ceramic Fibres
V. Hopfe, A. Tehel, S. Böhm, Chemnitz (D)   485

Laser Induced Chemical Vapour Deposition of Conductive and
Insulating Thin Films
G. Reisse, R. Ebert, U. Illmann, F. Gaensicke, A. Fischer, Mittweida (D) 491

Film Thickness Variation of Pulsed Laser Deposited Carbon
G. Reisse, B. Keiper, A. Fischer, Mittweida (D)   497

Deposition of Hard Coatings by Laser Induced PVD Technique
F. Müller, K. Mann, Göttingen (D)   503

Photophysical and Photochemical Processes in Excimer Laser
Depostion of Titanium from $TiCl4$
P. Kubát, P. Engst, Prague (CSFR)   509

Formation and Spectroscopy of Thin Films Produced by Laser
Vaporization of Ceramics
F.W. Froben, M. Ritz, H. Yu, Berlin (D)   515

Polishing of Lead-Crystal-Glass Using cw-CO_2-Lasers
A. Geith, C. Buerhop,H.W. Bergmann, R. Weißmann,
A. Helebrant, Erlangen (D); R. Jaschek, Vilseck (D)   521

Fundamental Aspects of the Damage Free Removal of Hard Coatings
Using Excimer Laser
R. Queitsch, K. Schutte, H.W. Bergmann, J. Naser, Erlangen (D);
H. Kukla, Nürnberg (D)   527

Paint Stripping With Short Pulsed Lasers
H.W. Bergmann, R. Jaschek, G. Herrmann, Erlangen (D)   533

Fretting Wear of PVD-TiN Coatings Oxidized with Excimer Laser Radiation
J.P. Celis, M. Franck, J.R. Roos, B. Blanpain, Leuven (B); E.W. Kreutz,
M. Wehner, K. Wissenbach, Aachen (D)   539

VI. Precision Machining and Structuring

Precision Drilling With Short Pulsed Copper Vapor Lasers
R. Kupfer, H.W. Bergmann, Erlangen (D)   547

Surface Microstructural Changes on Alumina and Zirconia after Excimer
Laser Treatment
A. Tsetsekou, T. Zambetakis, C.J. Stournaras, Chalkida (GR);
G. Hourdakis, E. Hontzopoulos, Iraklion (GR)   553

Integration of Materials Processing with YAG-Lasers in a Turning Center
M. Wiedmaier, E. Meiners, F. Dausinger, H. Hügel,
T. Rudlaff, Stuttgart (D)   559

Laser-Plasma Deposition of TiN Coatings
R. Becker, S. Burmester, S. Metev, Bremen (D)   565

Material Removal Processing with High-Power-CO_2- and Nd:YAG-Lasers
M. Stürmer, Hannover (D)   571

Influence of Process and Material Parameters on Ablation Phenomena
and Mechanical Properties of Ceramics and Composites for Excimer
Laser Treatment
M. Geiger, N. Lutz, T. Rebhan, M. Goller, Erlangen (D)   577

Contoured Material Removal Using cw-Q-switch Nd:YAG-Laser Radiation
B. Läßiger, M. Nießen, P. Ott, H.-G. Treusch, E. Beyer, Aachen (D)   585

Vapourization Cutting of Metal Sheets With Copper Vapour Lasers
H.W. Bergmann, J. Hofmann, Erlangen (D); H. von Bergmann,
W. Klopper, Pretoria (South Africa); R. Kupfer, Sulzbach-Rosenberg (D) 591

Process of Generating Three-Dimensional Microstructures with
Excimer Lasers
H.K. Tönshoff, J. Mommsen, Hannover (D)   597

Marking of Silicate Glasses with Excimer Laser Radiation
- Influence of Glass Properties and Laser Parameters -
C. Buerhop, N. Lutz, R. Weißmann, M. Geiger, Erlangen (D)   603

High Resolution Excimer Laser Based Micromachining
B. Burghardt, D. Basting, Göttingen (D); H.-J. Kahlert, Jena (D)   609

Microstructuring of Glass with Excimer Lasers
B. Wolff-Rottke, J. Ihlemann, H. Schmidt, Göttingen (D)   615

Laser Texturing R-Ba-Cu-O Superconductors
D. Dubé, B. Arsenault, C. Gélinas, P. Lambert, Boucherville (Canada)   621

VII. Residual Stresses

Residual Stresses and Microstructures in the Surface Layers of Different
Laser Treated Steels
K.-D. Schwager, B. Scholtes, E. Macherauch, Karlsruhe (D);
B.L. Mordike, Clausthal (D)   629

Prediction of Thermal, Phase Transformation and Stress Evolutions
during Laser Hardening of Steel Pieces
M. Boufoussi, S. Denis, J.Ch. Chevrier, A. Simon, Nancy (F);
A. Bignonnet, Bievres (F); J. Merlin, Lyon (F)   635

VIII. Beam-Material-Interaction

Laser Target Interaction: a Comparison between CO and CO_2-Lasers
M. Stöhr, E. Zeyfang, Stuttgart (D)   643

Characterization of Excimer Laser Treatment of Metals Using Short
Time Diagnostics; Influence on Possible Industrial Applications
K. Schutte, R. Queitsch, H.W. Bergmann, Erlangen (D)   649

Two-Dimensional Model for Materials Removal for Laser Drilling
A. Kar, J. Mazumder, Urbana (USA); T. Rockstroh, Evandale (USA)   655

Laser Induced Thermal Shock Cracks in High Temperature Materials
inside a Scanning Electron Microscope
S. Menzel, K. Wetzig, Dresden (D); J. Linke, Jülich (D)   661

Effect of Ambient Pressure and Shielding Gas on Penetration Depth
and CO_2 Laser-Induced Plasma Behavior
M. Kabasawa, M. Ono, K. Nakada, S. Kosuge, Kawasaki (Japan)   667

Materials Processing Using a Combination of a TEA-CO_2-Laser and
a cw-CO_2-Laser
R. Jaschek, R. Taube, K. Schutte, A. Lang, H.W. Bergmann, Erlangen (D)   673

Modelling CO_2-Laser Welding of Metals
G. Simon, U. Gratzke, J. Kroos, B. Specht, M. Vicanek, Braunschweig (D)   681

IX. Modelling and Simulation

Cladding with Laser Radiation: Properties and Analysis
H. Schlüter, E.W. Kreutz, B. Ollier, N. Pirch, A. Gasser,
K. Wissenbach, Aachen (D)   687

Modelling of Keyhole/Melt Interaction at Laser Deep Penetration Welding
M. Beck, P. Berger, H. Hügel, Stuttgart (D)   693

Melting-Solidification Phenomena in Pulsed-Laser Treatment of Metallic
Materials. Two Dimensional Simulations
A. Lashin, M. Poli, Milano (I); I. Smurov, Saint-Etienne (F)   699

Approches in Modelling of Evaporative and Melt Removal Processes in
Micro Machining
E. Meiners, A. Kessler, F. Dausinger, H. Hügel, Stuttgart (D)   705

Knowledge Based System for Laser Hardening
D. Schlebeck, M. Bachmann, Ilmenau (D)   711

Modelling of Laser Heating of the Fine Ceramic Particles
A.I. Bushma, I.V. Krivtsun, Kiev (CIS)   719

List of Authors   725

Subject Index   731

I.
Systems and
Auxiliary Apparatus

The Excimer Laser - a New Tool for Industrial Material Processing

D. Basting
Lambda Physik GmbH, Göttingen/FRG

Introduction

Applications of lasers are widespread in modern industry. Many production processes have been made possible or became more efficient by the use of CO_2 or Nd:YAG lasers which count for 3/4 of all industrial laser sales. With the introduction of excimer lasers a totally different new laser tool has been made available at the beginning of the eighties.

Initial applications were hampered by the absence of an industrially qualified laser and a misunderstanding of this different kind of laser. People tried to use the excimer laser as they used the CO_2 or Nd:YAG laser. Difficulties arose from the fact that the excimer laser emits in the ultraviolet between 157 nm (VUV) and 351 nm (UV) and that it is a repetitively pulsed laser with a pulse duration as short as several 10 nanoseconds (10^{-8} s) and emits a large cross section symmetric beam with a pulse energy of up to several Joules.

Laser	Pulse duration	Peak power	Repetition rate	Average power
CO_2 laser	cw	10 kW	cw	10 kW
CO_2 laser	10^{-8} s	100 MW	100 Hz	100 W
Nd:YAG laser	cw	2 kW	cw	2 kW
Nd:YAG laser	10^{-8} s	100 MW	20 Hz	20 W
Excimer laser	10^{-8} s	40 MW	500 Hz	200 W

Fig. 1: Comparative data of commercial excimer lasers for material processing.

Pulse energy		100 - 4000 mJ
Pulse duration		10 - 40 ns
Pulse stability		3 - 5 %
Repetition rate		< 500 Hz
Average power		200 W
Wavelenghts	F_2	157 nm
	ArF	193 nm
	KrF	248 nm
	XeCl	308 nm
	XeF	351 nm
Efficiency	ca.	2 %

Fig. 2: Typical data of excimer laser

The Excimer Laser - a Research Tool

More than a few thousand excimer lasers are being used worldwide in research laboratories. They were designed primarily for flexibility and outstanding specifications such as high energy per pulse and multiple wavelength operation. Already in 1989, a research excimer laser set the world record of 750 Watt at 308 nm.

The Excimer Laser Technology for Industrial Use

Industrial applications started after the unique features of this UV laser source were fully recognized. Intense development activities of the laser industry have been carried out to satisfy the requirements of the industrial user. Design criteria are:

- adequate power at a given wavelength for high throughput
- good beam quality, e.g. high beam intensity uniformity and low divergence
- long operation time with little or no user interaction
- reliable operation with predictable maintenance intervals
- long lifetime of the laser and its major components
- economic operating costs.

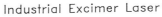

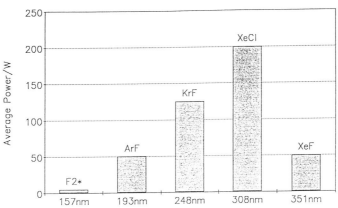

* research result

<u>Fig. 3:</u> Maximum average power for industrial excimer lasers (Lambda 4000).

A variety of industrial excimer lasers is available to date with an average output power of up to 200 Watt. These lasers are optimized to operate at a single wavelength only. Pulse energies exceed 1/2 J at repetition rates of up to 400 Hz. While the first excimer lasers still suffered from a short lifetime of a gas fill, nowadays 10^7 - 10^8 pulses with one gasfill are common. This is accomplished by the use of a new gas purification technique which removes gas impurities by a cryogenic process. This process is virtually maintenance-free, the cryogenic unit is operating maintenance-free for more than 10,000 hours. If a gasfill is used-up, which means that the power of the laser can no more be maintained, a refill cycle starts. It first pumps out the used-up gas and then adds fresh gas fully automatic, controlled by the laser's microprocessor controller. In order to reduce operating cost, small fractions of the halogen compound of the excimer laser gas are injected before a total replacement of the gas takes place.

Only when this gas injection does not improve the performance any more, a total gas refill is necessary. A typical excimer laser gas mix consists of 120 - 180 liter and costs about DM 50 - 100 (KrF), DM 100 - 180 (XeCl).

The operation of the excimer laser is surveyed continuously by sensors such as energy monitor, pulse counter, gas pressure sensor, temperature sensor, and high voltage controller. Also, the cooling water flow and safety interlocks are monitored.

Interfaces such as RS 232, RS 422 or IEEE 488 are common. Most up-to-date excimer lasers even allow monitoring of all operational parameters online by a telephone link (modem) by a remote computer controller.

Especially in the industrial production environment it is important to guarantee a high uptime. As an example, the most widely used industrial excimer laser, the LAMBDA 3000, reaches uptimes of typically 90 - 95 %.

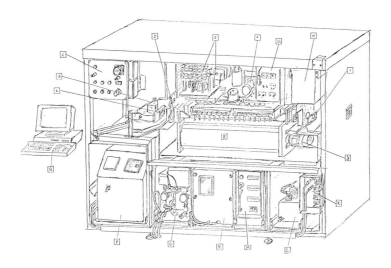

A	Rear mirror	K	Gas supply valves
B	Energy monitor	L	Vacuum pump
C	Main electrical module energy monitor	M	Electrical module
D	Pneumatically controlled laser window	N	High voltage power supply
E	High voltage discharge circuit	O	Water-oil-temperature exch. module
F	Thyratron	P	Gas purifier
G	Thyratron supply	Q	Personal computer
H	Gas circulating driver	R	Laser tube
I	Front mirror		
J	Motor gas circulation		

Fig. 4: Building blocks of a typical industrial excimer laser (Lambda 3000).

The measure for the reliability is the MTBF (mean time between failures). Carefully designed excimer lasers show MTBFs of more than 1,000 hours as a typical value. The maintenance of an excimer laser mostly includes cleaning of the optics after several 10 million pulses. Exchange of components is necessary after every 100 hours, thyratron and gas filter after 1,000 hours. One of the more recent innovations in excimer laser design is the long life of the laser tube. In case of the LAMBDA 3000, at 308 nm several 1,000 of hours have been demonstrated before it must be replaced. As far as consumption costs are concerned, the following components must be considered: gas cost, cooling water cost, electrical power, and components. For industrial 150 Watt lasers, this amounts to DM 250/hour.

The beam of an excimer laser has a rectangular symmetry with a typical dimension of 10 mm times 20 - 40 mm. The beam profile is an important parameter and is determined by the discharge conditions and the laser resonator. However, under continuous operation, these parameters naturally change leading to an alternation of the beam profile. Initial applications suffered from this behaviour. More modern designs do not rely on the original beam quality, but use beam delivery systems and homogenizers to create a symmetrical flat top homogeneous profile, which will be discussed later.

Application

The short wavelenght of the excimer laser makes it possible to create structures in the sub-micron range. Very soon, it was discovered that the excimer laser is an ideal tool for micromaterial processing. In addition to the small structure the short wavelength leads to so-called cold ablation which means that organic materials can be ablated without heat diffusion to the substrate involved in Nd:YAG or CO_2 laser processing. This invention by Prof. Srinivasan made possible a whole new world of machining applications of polymers such as polyimide. The applications are so numerous that only a few important applications can be mentioned.

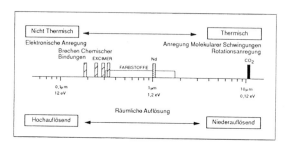

Fig. 5: Photoenergy and wavelength of different lasers.

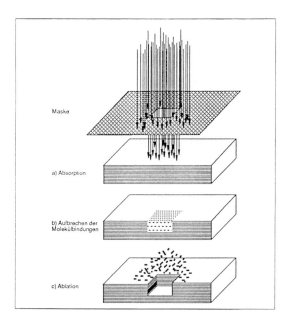

Fig. 6: Schematic diagram of photochemical ablation.

Micromachining

The removal of polymers using excimer lasers has been studied intensely during the past eight years. Remarkable feature is the almost complete absence of heat diffusion into the substrate. In contrast to normal laser processing, where the beam is focused and which relies on highest possible intensity, high-resolution high-quality processing is achieved with an excimer laser at remarkably low energy densities. Polymers like polyimide can be processed with extreme precision at energy densities of several hundred mJ per cm^2 only, materials are removed at a processing rate of typically less than 1 µm per pulse. This micromachining process has found its application in industry in a variety of processes, most of them in the innovate semiconductor packaging and electronic industry. The most prominent application of this process is the production of so-called multichip modules. These are multilayer printed circuit boards where it is important to bring as many chips as possible on the least volume. Part of the design process of these modules is the necessity to drill many holes from one to the other layer through an isolating polyimide film to produce vias with a diameter of <100 µm. Conventional methods as mechanical drilling are not possible any more. Only photolithographical or indirect processes are possible. Hybrid circuits for PC's, main frames and super computers require the ability to remove various sorts of polyimide insolating materials and create circuit patterns on a microscopic scale that is considerably beyond the limitations of traditional mechanical punch and drill techniques. With excimer lasers one can micromachine polyimide down to a feature size of 10 µm. An additional advantage of the direct etching process is that it avoids additional steps such as chemical etching.

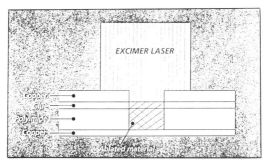

Fig. 7: Principle of selective polyimide ablation at 500 mJ/m². The excimer laser does not damage the copper mask, but removes the polyimide to create a clean wall and base suitable for metallisation to form vias.

The flexibility and the possibility of 3-dimensional structuring of plastic materials by the excimer laser can be demonstrated by the picture of a small plastic gear. It has been designed by direct writing with the focused excimer laser beam.

Fig. 8: Scanning electron micrograph of a gear obtained by excimer laser dry etching of polyimide (courtesy of Laser Laboratorium Göttingen).

Micromachining of Metal and Ceramics

The number of materials which can be processed by the excimer laser is huge, as practically all materials absorb very well in the ultraviolet. This includes diamonds, ceramics or glass. Especially the 193 nm radiation is used at energy densities of more than 40 - 50 J/cm^2. With a short pulse duration and the small absorption depth heat influences and melt zones are very thin. This essentially prevents thermal loading of peripheral zones which could lead to "micro-cracks".

Material Processing

Ultraviolet laser light holds particular promise for its ablility to modify surface structures at the pre-ablation threshold and to lay down ultrathin layers of film. Some applications such as those in the semiconductor industry are already established. Others like excimer based thin film transistor (TFT) processing show promise in the feasability stage. They include surface treatment (annealing, doping, planarization), physical vapor deposition and chemical vapor deposition.

Systems for Excimer Laser Material Processing

Almost all excimer laser applications make use of its rectangular flat top beam profile. The beam is projected onto the target with the best possible uniformity and at an energy density appropriate for the intended use. To prevent radiation damage of the mask the image is reduced by a factor of 2 - 20 typically. The illumination of the mask is performed through a homogenizer which is based on the principle of superimposing the different beam segments. The result is a superb beam with a flatness of +/- 5 % using more than 90 % of the original beam.

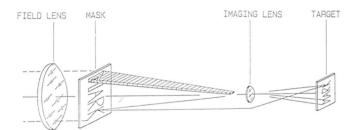

FIELD LENS   MASK   IMAGING LENS   TARGET

Fig. 9: Schematics of excimer laser beam delivery system.

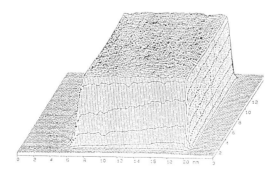

Fig. 10: Homogenized excimer laser beam.

Complete machines are now available on the market which include online monitoring of the process and the beam characteristic. Energy densities can be adjusted within a few percent. The substrate is automatically positioned with an accuracy of 1 μm or better. Depending on the degree of precision, the number of axes to be controlled and the degree of automatization, the price for the workstation easily exceeds the costs of the excimer laser by far. The more the excimer laser enters the industrial production floor, the more we will see specialized machines doing the work.

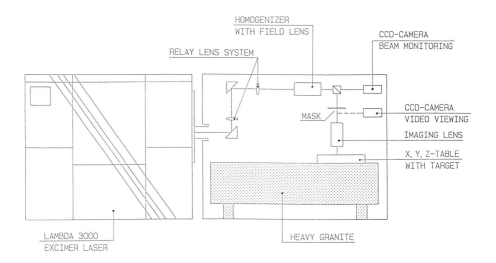

Fig. 11: Excimer laser micromachining workstation (Courtesy: MicroLas, Jena)

Outlook

The development of excimer lasers will continue. Besides higher power up to 500 W operating costs will be the area where one can expect the most progress. While the market is still dominated by one company (Lambda Physik), it is expected that more companies will enter that market.

Japanese companies, so far, do not play an important role. However, the second largest excimer laser company, the Canadian Lumonics, has been acquired by a large Japanese firm. In addition, seven large corporations such as NEC, Toshiba, Mitsubishi and a lot of smaller companies in Japan are developing excimer laser technology demonstrating the strategic point of view.

In contrast, in Europe in addition to Lambda Physik only one little company, SOPRA in France, is active in the development of a very high energy excimer laser with up to 15 J per pulse. Siemens gave up its own development, to market the excimer laser of a U.S. company, Cymer, which has several Japanese shareholders such as Nikon, Canon and Mitsubishi. The only other excimer laser company in the U.S., Questek, has recently been purchased by VISX, a medical firm, and will concentrate on that market.

Not only is the development effort of excimer lasers shifted towards Japan, also many applications are expected to be persued most rigorously in Japan. Such important technologies as flat panel displays which may use excimer lasers during the manufacturing process are almost exclusively produced by Japanese firms.

The excimer laser will find its application mostly in the electronic industry. The challenge for excimer laser manufacturers is to follow their customers' requirements "technically" and "geographically".

The High Speed Pyrometer System for Laser Welding, Cutting, Heat Treatment and Alloying Processes Temperature Control

M. Ignatiev, A. Ermolaev, I. Titov, Institute of Metallurgy, Moscow, Russia
I. Smurov, Ecole Nationale d'Ingenieurs, Sent-Etienne, France
S. Sturlese, Centro Sviluppo Materiali, Rome, Italy

Introduction

The thermal state of material in zone of exposure to laser radiation is one of the main integrated parameters describing the evolution of physicochemical processes and structural phase transformations in the surface layer of a material. In view of this, information on the dynamics of surface temperature of materials during irradiation are of major practical interest. It is frequently difficult to calculate the temperature fields, rates of heating and cooling, and the characteristics of the thermal cycles because of the lack of sufficient information on the thermal changes of the optical and thermophysical properties of materials, and also because of need to allow for thermochemical reactions and losses of heat due to convection and radiation.

Appreciable difficulties are encountered in the experimental determination of the surface temperature because of the specific measurement conditions: the small size of the irradiation zone (1-5 mm), the wide range of temperature variation (300-6000°K), and the high rates of variation (1-1000 kK/s).

Under these conditions, noncontact technique of optical pyrometry is the most suitable.

Two-channel high speed optical pyrometer system (PS) was developed to measure surface temperature versus time profiles of metals and alloys irradiated with pulse-periodic or continuous wave laser.

The thermocycles during processes of laser welding, cutting, heat treatment, alloying and cladding were investigated.

Experimental methods

Worked out PS consists from optical pyrometer, interface board and IBM PC/AT with special program set for signals registration and treatment. Optical pyrometer permit to measure the brightness bodies temperature on two wavelengths (that can be switched over) and corresponding color temperature.

The brightness temperatures may be measured at one or two fixed wavelengths by using of known values of monochromatic emissivities (1). The true temperature is defined from Wien formula.

The color registration regime allows to measure the temperature practically equal to true temperature under condition of greybody or small monochromatic emissivities variation at effective wavelengths.

The range of temperature measurements is 1200- 5000°K, with the space resolution 200 μ and response time 20 μs. The range of wavelengths overtuning is 0,5 - 1,0 μ.

The effective wavelengths were formed in a zones of relative transmission of the laser plasma radiation for each concrete laser processes.

The measurements of brightness and color temperatures are based on spectral disassembling with diffraction grating of light flow radiated by studied area, conversion of quasi monochromatic radiation on two wavelengths into electric signals with photodiodes and getting results in the form of exit electric signals proportional to measured temperatures by method of analogous calculations by Wien formula.

A signals were recorded by means of interface board directly into IBM PC/AT memory. The special program set permits to obtain on monitor screen the dynamic of temperature variation.

The pyrometer was calibrated using a standard pyrometric tungsten lamp up to temperature $2500^\circ K$ and at higher temperatures the calibration dependence was extrapolated using the Wien approximation (1).

The temperature measurement methods include recording the spectra from the plasma and surface, determining the spectral transparency windows for the surface laser plasma, reflectometry with the auxiliary laser in the region of relative transparency, determining the monochromatic emissivities $\varepsilon(\lambda=const)$, measuring the brightness temperatures and (or) color temperature and recovering the true temperature (2,3).

Spectral diagnostic of the laser plasma jet at the treated surface was carried out using diffraction spectroscope. For example, the relative transparency windows for laser plasma jet for stainless steel in air lie around 0.60, 0.57, 0.55 and 0.49 μ. In this case the monochromatic emissivities $\varepsilon(\lambda=const)$ of the materials were determined by reflectometry (3) with He-Ne laser operating at 0.63 μ together with an Ulbricht integrated sphere. Fig.1 shows the quasistationary emissivity of stainless steel (curve 1) and mild steel (curve 2) in air for continuous wave laser welding with various power densities. Up to 3 kW/cm2, there is extensive oxidation in the irradiated parts. The $\varepsilon(\lambda=0.63~\mu)$ starts at 0.6 approximately for stainless steel or 0.5 for mild steel and correspondingly attains the maximum values of 0.95 and 0.83. Up to 13 kW/cm2, there are reductions to 0.65 (stainless steel) and 0.74 (mild steel) probably because of rising temperature, with hydrodynamics activation in the liquid and sublimation from the oxides. The monochromatic emissivities experimental data were used to determine the true surface temperature from brightness temperature for the effective wavelength $\lambda=0.63~\mu$.

A 2.5 kW and 6.0 kW carbon dioxide lasers and YAG pulse-periodic laser with a pulse energy of 1-50 J and pulse duration 1-5 ms were used in our experiments. The stainless steel, mild steel and refractory metals specimens without absorption coating were mounted on numerical control (CNC) x-y table and were irradiated with laser. During laser treatment helium was used to protect a metal surface from oxidation.

We investigated the thermocycles during processes of laser welding, cutting, heat treatment, alloying and cladding. The main tasks of our experiment were: to test the PS for wide range of laser treatment processes; to study some features of surface heating and cooling at treated and heat affected zones (HAZ).

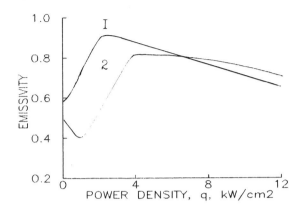

Fig. 1: Monochromatic emissivity of:
1-stainless steel; 2-mild steel.

Welding, cutting

We tested the PS for flash welding of plate mild steel with continuous wave laser. Fig.2 shows the thermocycles for laser welding measured by PS at various distance from center of weld (laser power - 1.9 kW; traverse speed - 300 mm/min). The first parts of these curves (dashed) are the surface plasma radiation. The seconds parts (solid) - the true surface temperature. The special methods are needed to eliminate the influence of laser plasma, but the peak temperature, cooling stage and size of HAZ may be easy measured.

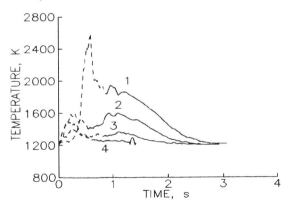

Fig. 2: Thermocycles for flash laser welding of plate mild steel.
Distance from the center: 1 - 1 mm;
2 - 2 mm; 3 - 2.5 mm; 4 - 3 mm.

The task of monitoring the HAZ for laser cutting is similar the same one for laser welding and may be decided with PS.

Heat treatment

We investigated the thermocycles during process of continuous wave laser hardening of mild steel for the traverse speed (TS) from 400 mm/min to 4000 mm/min.
The threshold of sharp change of thermocycles character for mild steel was discovered. When the TS was more than 3100 mm/min (laser power 1.7 kW) the PS did not registry the thermocycles because of the true steel surface temperature did not reach lower level of instrument sensitivity. If the TS value was 15-25 mm/min less we observed two situation. In the first case, we saw only upper part of the heating curve from 1200 to $1350^{\circ}K$ (Fig.3, curve 1). But in the second case, insignificant variation of beam scanning speed (nearly 10 mm/min) leads to fast heating of the steel surface up to temperature $1700^{\circ}K$ (Fig.3, curve 2). We could not see the fluent transition between this two situation. When we decreased the beam scanning speed more than 100 mm/min we observed the intensive melting of steel surface (Fig.3, curve 3, there is shelve on the mild steel melting temperature level $1810^{\circ}K$).
On our opinion, such type of threshold thermocycles change is explained by influence of surface oxidation on absorption coefficient. It is known, that absorption coefficient increase into 3-5 times due to surface oxidation. We really have the same situation in our experiments (and also in most industry laser hardening processes) because of the turbulence of shielding gas jet at low value of gas expenditure.

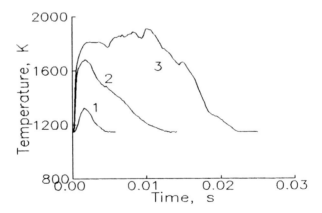

Fig. 3: The heat treatment thermocycles. Traverse speed:
1 - 3085 mm/min; 2 - 3075 mm/min;
3 - 2900 mm/min.

Alloying

The specimens of mild steel with Mo coatings were irradiated by pulse-periodic YAG laser with pulse duration 4 ms. The thickness of coating was 35 μ.
Fig.4 shows the temperature in the center of the focus spot. The laser action (power density - 28.0 kW/$cm2$) on the specimen with Mo coating led to sharp temperature growth up to 3100 K (more than Mo melting temperature). The Mo coating destruction was observed in the central part of the irradiated zone. It led to rapid surface temperature decrease to temperature of naked steel substrate.
The heating and melting of naked steel substrate was the second stage of alloying process. The maximum surface temperature was 2500 K. Thus the melting as Mo coating as the steel substrate led to effective alloying of surface steel layer with Mo.

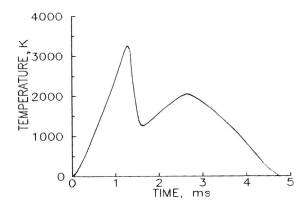

Fig. 4: The temperature in the center of the treated spot for laser alloying from Mo coating.

Cladding

The PS has been used for the investigations of maximum surface temperature and cooling rate of cladded stellite-6 layer on mild steel substrate. This parameters are very important for production the cladded layer with extremely high microhardness. Fig.5 illustrated the typical thermocycle for laser cladding. The maximum surface temperature of cladded layer reached stellite-6 melting point with the small overheating and the cooling rate was nearly 1 kK/s at the following laser treatment parameters: power 3.0 kW, traverse speed 300 mm/min and spot size 3 mm.

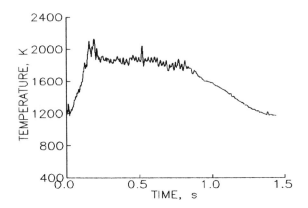

Fig. 5: Laser cladding thermocycle

Conclusions

The developed PS may be successfully used for monitoring and control the surface temperature for continuous wave and pulse-periodic laser treatment processes.

Acknowledgments

The authors would like to thank Dr. J.C. Ion and staff of Laser Processing Laboratory Lappeenranta University of Technology for experimental assistance and helpful discussion.

References

(1)   Radiative Properties of Solid, Energy, Moscow (1974) 452.
(2)   A. Uglov, A. Ermolaev, V. Zavidei: Sov. J. Quant. Electr. 20, N 4 (1990) 453.
(3)   A. Uglov, A. Ermolaev, V. Zavidei: High Temp. Thermo Phys. 28, N 4 (1990) 601.

Computer Controlled Laser Beam Cutting of Ceramics

H.K. Tönshoff, Laser Zentrum Hannover e.V., FRG
L. Overmeyer, Laser Zentrum Hannover e.V., FRG
F. von Alvensleben, Laser Zentrum Hannover e.V., FRG

Introduction

In industrial applications, ceramics are being increasingly used, due to their outstandig physical and chemical properties. However, brittleness and hardness lead to difficulties in machining. Besides conventional methods such as grinding with diamond tools, ultrasonic and electrical discharge laser cutting is a promising alternative (1,2,3).

The most critical point in using lasers to cut ceramics in industry is crack damage, which is caused by thermal overload of the material during the cutting process. Investigations have shown that the bending strength of laser cut samples is about 20 - 40 % lower than of ground samples (4). Several strategies have been introduced to reduce crack damage, either by using special components, i. e. a Q-switch for Nd:YAG-lasers, or by applying special process environments, such as reactive gases (5,6). Another method is computer controlled laser beam cutting combined with a UV-sensing technique which will be described in this paper.

Effects on Laser Cutting of Ceramics

Material properties, laser parameters and workpiece geometry have a main effect on the result of the laser cutting process, and especially on crack formation. Material removal processing by vaporization is desirable, because a melting process generates a wavy and rough cut surface and large heat affected zones. This requires high laser beam power and intensity. Whereas high beam power is useful for vaporization, low crack damage can be achieved with minimum average beam power which leads to small heat affected zones. A low volatilization temperature and vaporization heat allow high rates of vaporization. This rate can be increased by high absorption coefficients. In order to minimize temperature gradients, the amplitude and speed of the temperature cycle should be low and constant. Also, reduced heat transfer and no further chemical reactions should be ensured, i. e. by using suitable process gases. In order to prevent overheating, enough cooling zones should be available, based on the size and shape of the workpiece which will be processed (7).

Furthermore, material properties, regarding the thermal stress factors, should be considered. The crack critical tensile stress caused during the cooling phase can be reduced by low elasticity and shear moduli, resulting in a low Poisson's ratio, a low thermal expansion coefficient, and a high breaking strength (1). Although a low thermal conductivity

increases the temperature gradient in the material, it is
suitable for laser processing in terms of concentrating the
absorbed energy for ablation and small heat affected zones.

Closed loop control for process optimization

A controlling system was developed for cutting ceramics with
a minimum of crack damage. For the realization of well adapted controllers, it is advantageous to describe the process
mathematically. A mathematical model for the process 'cutting
of ceramics', useful for a theoretical conception of a controller, is not available. Therefore, a process identification had to be found empirically. Input variables of the
cutting process are the beam characteristics such as power,
mode, pulse frequency and duration. Moreover, the process gas
attributes such as pressure, sort and flow are relevant. Very
important for a good cutting result is the cutting speed. The
output variable of the cutting process is in general the
cutting quality. A main characteristic of a high quality
laser cut is provided by a minimum of cracks in the cutting
edge. The on-line measurement of the laser-induced cracks,
e.g. by acoustic emission analysation brought no satisfying
results (8).

A prooved method to get real-time information about a laser
material treatment process is the detection of the UV-plasma
radiation. During laser welding, for example, this method
provides information about defects (9). The investigation of
the UV-plasma radiation while cutting ceramics brought out,
that the signal contains information concerning cutting quality.

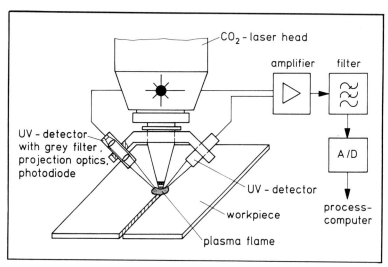

Fig. 1: Sensing system for UV-radiation

To detect the UV-radiation, a lens system was used to project the plasma flame onto a UV-sensitive photodiode, as shown in fig 1. A grey filter was placed in front of the projection lens system to reduce the radiation intensity fitting for the operative range of the photodiode. The control system was applied to a 750 Watt high frequency excited CO_2 laser. The laser control was equiped with two external analog interfaces, one for pulse frequency and the other for pulse duration. Thus, the laser power remains constant while the pulse energy can be changed externally. The schematics of the system is shown in fig 2.

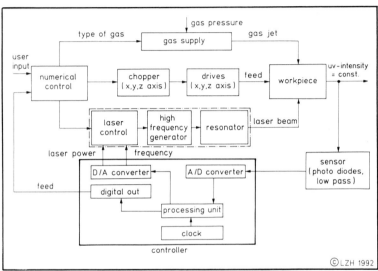

Fig. 2: Schematics of closed-loop system

Research results

The described results were achieved at cutting 4 mm thick Aluminiumoxide ceramic. During the time the laser reacts with the ceramic, the UV-signal rises linearly. During the quiescent period, the UV-signal falls exponentially towards zero. This means the plasma flame quenches. If the quiescent period is too short to cause a quenched plasma, the crack rate rises immediately. Thus the real-time UV-signal can be used to choose the pulse frequency. A condition for the choice of the pulse frequency is that the plasma flame is quenched. As a result, the pulse frequency has to be chosen at about 100 Hz. The real-time sawtooth-like UV-signal has to be prepared for use in the control circuit. A low pass filter system with a variable cutoff frequency between 10 and 50 Hz serves as an integrating element for the signal.

In fig. 3 the filtered signal is documented for different treatment parameters along a cut. The diagramm shows the UV-signals from six cuts with equal treatment conditions, but a rising pulse duration which is equivalent to a rising average power. With a rising average laser power the UV-signal rises too. Moreover the oscillation amplitude of the signal is amplified. Not only the average laser power, but also a changed cutting speed influences the UV-radiation. If the cutting speed increases linearly, the UV-radiation intensity increases nearly inversely parabolic. The physical background of these phenomenon will be investigated.

Comparison of the samples by flourescent penetration tests brought out that a minimum of cracks was reached if the oscillation amplitude of UV-signals was held to a minimum. Thus, the controller should be used to reduce the oscillation of the filtered signal around the empirically found operation point.

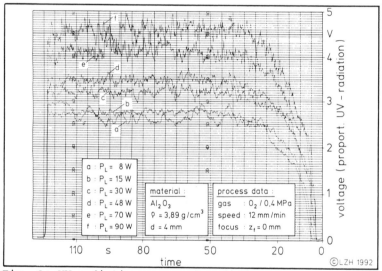

Fig. 3: UV-radiation curves for different laser parameters

Round the operation point, the cutting process has a linear characteristic as far as the interdependence between the UV-signal and the laser pulse duration is considered. Therefore, it is possible to use a parameter optimized proportional controller with a provided set point. The used PID-algorithm was implemented to a computer equiped with the necessary A/D- and D/A- devices, and with an external clock (10). The sampling period of the controller had to be the same as the pulse frequency of the laser. According to that, the controller was able to change the duration of each pulse. In fig. 4 the result of the computer-controlled cut is shown. The controller was able to reduce the oscillation amplitude of the UV-signal during the cut. The flourescent penetration test of the controlled cut samples brought out that the crack length was also reduced.

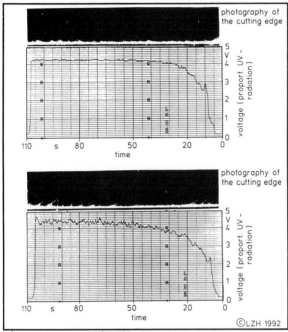

Fig. 4: UV-radiation during controlled (a) and conventional (b) laser cutting

Discussion

Controlling the UV- plasma radiation during laser cutting of ceramics is a method to assure treatment quality. Concerning other laser material treatment processes, eg. cutting or welding of metals, the possibility to control the UV-plasma in order to assure optimized quality has to be investigated.

The analysation of the UV-signal emitted while laser cutting of ceramics will open the chance to realize computer- controlled adjustment of optimized treatment parameters. The investigation has shown that an optimized operation point for cutting ceramics is reached at a minimum of average laser power sufficient for complete separation. Adaptive control algorithms implemented on the controlling computer will allow the recognition of an optimized operation point during the cutting process, and at the same time control the UV- radiation at this point.

References

(1)   H.K. Tönshoff, C. Emmelmann: "Laser Cutting of Advanced Ceramics", Annals of the CIRP, Vol. 38/1/(1989)

(2)   G. Spur, I. Sabotka, W. Tio: "Überblick über trennende Fertiungsverfahren zur Hartbearbeitung von Keramiken", Fachblattfür Materialbearbeitung, Vol. 64, Nr. 4 (1987)

(3)   C. Emmelmann, M. Gonschior, O. Gedrat, F. von Alvensleben and C. W. Byun: "Machining of ceramic with Laser Radiation", paper presented at ISEM X, Magdeburg (May 1992)

(4)   B. Wielage, J. Drozak: "Lasertrennen von Ingenieurkeramik und Werkstofverbunden", VDI Berichte Nr. 797 (1990)

(5)   N. Morita, H. Sakamoto, T. Watanabe, Y. Yoshiba: "Crack- free Processing of hot-pressed Silicon Nitride Ceramics using pulsed YAG-Laser, Machining Characteristics of Advanced Materials; Vol. 16, ASME (1989)

(6)   H. K. Tönshoff, W. Pompe, et. al.: "Chemisch unterstützte Mikrostrukturierung keramischer Werkstoffe mit UV-Strahlung", DFG-Antrag im SPP "Strahl- Stoff-Wechselwirkung bei der Laserstrahlbearbeitung (1991)

(7)   C. Emmelmann: Trennen von Keramik mit Laserstrahlung, Dr.-Ing. Dissertation, University of Hannover (1991)

(8)   C. Emmelmann et al.:"Qualitätskontrolle für die Laserstrahlbearbeitung von Keramik", Neue Werkstoffe, 8/91

(9)   R. Beck, M. Jurca: "UV- Sensor zur Überwachung von Laserstrahlschweißungen", DVS Bericht 113 (1988)

(10) H.C. Houpis, G.B. Lamont: "Digital Control Systems, Theory, Hardware, Software", McGraw-Hill Book Company (1985)

High Power Laser Diodes as a beam source for Materials Processing

V. Krause, H.-G. Treusch, E. Beyer, P. Loosen, Fraunhofer Institut für Lasertechnik, Aachen (FRG)

Introduction

The progress of high power laser diode systems in the power range between 20 and 50 Watt during the last years has increased the awareness to use these devices for materials processing. Foreseeable progress will result in higher laser power and devices up to the 1 kW level will be available within the next years. Within this paper a general description and evaluation of the different configurations of laser diode systems as beam sources for material processing applications is given.

Besides the principal configuration for these laser diode systems, the characteristic data (beam quality, power efficiency, etc.) of already and future existing systems are compared. Existing and future directions of development are described. The comparison of the charakteristic data of laser diode systems with the required characteristic data for a specific material treatment process gives the possible field of application for laser diode systems. Competing applications are compared with these systems for their industrial usability. With the example of soldering of microelectronic components (surface mounted devices) the application of laser diode systems is described. The soldering behaviour up to an average power of 20 Watt is described.

Laser Diodes

Since the development of the first laser diode with an operation temperature of - 200°C in 1962 the development of the following 20 years made devices operating at room temperature but moderate power levels(<100 mW) available. During the last 8 years the average power of these devices was increased by a yearly factor of 2 up to an average power of 20 Watt in 1992. At the same time the price development of laser diodes was characterized by a price reduction of a factor 1.6 every year, which is typical for semiconductor devices (5).

In comparison to YAG and CO_2 lasers diode lasers are different in terms of efficiency, lifetime, maintainance and lasing volume (see table 1).

	Laser Diodes	Nd : YAG	CO_2
Efficiency	30 - 60 %	1 - 3 %	5 - 10 %
Wavelength	780 - 830 nm	1064 nm	10.6 µm
Power	multi kW (with coupled laser diodes)	up to 3 kW	up to 20 kW
Lifetime	20.000 - 100.000 h	10.000 h	10.000 h
Maintainance	maintainance free	each 200 h (lamps)	each 500 h
Price/Watt	200 - 1000 DM/Watt	300 - 1000 DM/Watt	200 - 400 DM/Watt
Fiber Delivery	possible	possible	not possible
Voltage	up to 100 V	up to 1000 V	up to 10 kV
Watts per lasing Volume	1000 Watt/cm^3	50 Watt/cm^3	1 Watt/cm^3

Table 1 : Comparison of laser diodes with Nd:YAG and CO2 lasers.

The size of a laser diode is much smaller compared to other laser systems. One laser diode bar (10 mm x 0.6 mm x 0.1 mm) with a string of 800 individual emitters, see figure 1, emitts up to 50 Watt. The beam quality of the individual laser diode emitter is diffraction limited. Because of the small beam emitter area of 1 x 3 µm^2 the laser radiation is spread in a large divergence angle. Common divergence angles are in the range of 1000 mrad in the plane perpendicular and 200 mrad in the plane parallel to the emitter string. The overall beam quality of a laser bar is conserved only in the plane with the large divergence angle (1000 mrad). In the other plane the string of the not coherent emitters interleaved by non radiant spacings results in a beam quality of 3000 times diffraction limited for a bar of 10 mm length.

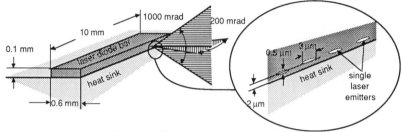

Fig.1 : Laser diode bar with laser emitters mounted on a heatsink.

The maximum achievable power of the individual emitter is limitted to about 60 mW. With an emitting size of arround 1 μm x 3 μm the power density of the emitter facets is $2 \cdot 10^6$ W/cm². The power of a diode laser system has to be scaled up by increasing the number of emitters per system. The combination of the single laser emitters can be realized in three different ways, see figures 2, 3,4.

1. stacking to a close packed 2-dim. array
2. lens multiplexing
3. fiber multiplexing

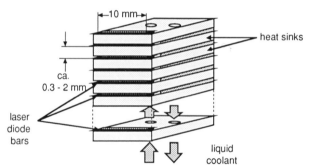

Fig.2 : Stacking of laser diodes to a close packed 2-dimensional array.

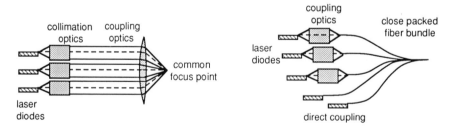

Fig.3 : lens multiplexing for high power.   Fig. 4 : fiber multiplexing for high power.

Each combination technique poses limits to the achievable power density by non radiant spaces between the laser emitters and optical losses (reflection, optics abberations, etc.). The achievable power density for the different technical aproaches is shown in table 2.

1. For the stacking technique a packaging density up to 25.000 emitters/cm^2 is the current state of the art. The resulting power density depends on the heat sink technology and reaches 120 W/cm^2 for high average power stacks and $1.5 \cdot 10^3$ W/cm^2 for high peak power stacks. The stacking approach is free of optics losses. (1,2)

2. The lens multiplexing technique uses microlenses in front of the laser emitters to generate one collimated beam. The effective power density is determined by the ratio of the emitter surface to the non radiant area between the emitters, which is ~0.3 for the laser diode bar described in figure 1. It is further reduced by the ratio of micro lens area to system area, which is 0.1 to 0.9. Optical losses of 10% - 30% exist due to cut off by the pupil function of the lenses because of the high numerical aperture of the laser diodes. Abberations of these fast lenses reduces the usable power even further. This results in power densities of $2 \cdot 10^4$ W/cm^2 to $5 \cdot 10^5$ W/cm^2 (3,4,6,20.)

3. In a fiber multiplexed laser device each emitter (or a group of up to 10 emitters) is coupled directly or by optics into a single fiber of round or rectangular cross section. Fiber coupling generates additional non radiant spaces on the output side of the fiber bundle(fiber claddings and holes generated by close packed round fibers), which reduce the effective power density by a factor of 1.5. Further reduction of the power density by non radiant spaces is generated by the different cross section of the fiber and the laser emitters. The resulting power density is reduced by a factor of 15 - 100 for fiber coupling of single emitters and by a factor of 250 - 10^3 for coupling of several laser emitters (up to 10). This results in power densities of 10^4 W/cm^2 to $8 \cdot 10^4$ W/cm^2 for fiber multiplexing of single emitters and 10^3 W/cm^2 to $2 \cdot 10^4$ W/cm^2 for several emitters. Fiber multiplexing offers to arrange the laser diodes independent from each other, which differs from the approaches 1+2, where the laser diodes have to be stacked close together or have to be arranged in a certain order to fulfill the requirements for lens multiplexing(16,18,19)

	2- Dimensional Stack	Lens Multiplexing	Fiber Multiplexing (Single Emitters per Fiber ~10 μm)	Fiber Multiplexing (Multiple Emitters per Fiber > 100 μm)
Power Density at Laser Diode	$2 \cdot 10^6$ W/cm^2	$2 \cdot 10^6$ W/cm^2	$2 \cdot 10^6$ W/cm^2	$2 \cdot 10^6$ W/cm^2
Optical Losses	0 %	10 - 40 %	10 - 50 %	20 - 50 %
Reduction factor (radiant area/ total area)	$5 \cdot 10^2$ - 10^4	$4 \cdot 10^2$	15 - 100	100 - 10^3
Power Density of Laser System	10^2 - $4 \cdot 10^3$ W/cm^2	10^4 - $5 \cdot 10^5$ W/cm^2	10^4 - 10^5 W/cm^2	10^3 - $2 \cdot 10^4$ W/cm^2
Overall Laser System Efficiency	30 - 60 %	20 - 55 %	10 - 50 %	10 - 50 %

Table 2 : Achievable power densities, efficiency and losses for different combination techniques for laser diodes.

At this time first devices are available up to average laser powers of 50 Watts. Prototype versions of powers of up to 1 kW average power exist (1). The devices in stacked construction form are the most advanced with the highest pulsed and cw laser power and a maximum power density of $1.5 \cdot 10^3$ W/cm^2. Fiber multiplexed systems with up to 10 emitters per fiber achieve already power densities of $8 \cdot 10^3$ W/cm^2. Fiber multiplexed devices with one emitter per fiber have not been developed, although the technique is well established in the communication sector. The highest power density of up to $5 \cdot 10^5$ W/cm^2 can be accomplished by lens multiplexing, which has already been reported with $2 \cdot 10^4$ W/cm^2 (11). 'Theoretical improvement' factors of 15 for lens multiplexing techniques and not even realized fiber multiplexing of single emitters per fiber document the possible improvements in this area. An overview of the current state of the art is given in table 3 :

	Achieved Power	Achieved Power Density	Theoretical Improvement	Achieved Efficiency	
2-Dimensional Stack (high average power)	$4 \cdot 10^3$ W (pulsed) $1.4 \cdot 10^3$ W (cw)	$1.2 \cdot 10^2$ W/cm^2 (cw)	max. 6	100 % (no optics)	(1),(2)
2-Dimensional Stack (high peak power)	$1.5 \cdot 10^3$ W (pulsed) 60 W (cw)	$1.5 \cdot 10^3$ W/cm^2(pulsed)	max. 3	100 % (no optics)	(1), (14), (15)
Lens Multiplexing	50 W (cw)	$2 \cdot 10^4$ W/cm^2	max. 25	60 %	(4), (11), (12), (20)
Fiber Multiplexing (Single Emitters per Fiber)	no realisation known	/	/	/	/
Fiber Multiplexing (Multiple Emitters per Fiber)	20 W (cw)	$8 \cdot 10^3$ W/cm^2	max. 2.5	50 %	(10), (13), (15), (16), (18), (19)

Table 3 : Currently implemented laser diode systems.

Materials Processing

Materials processing applications with laser radiation are currently dominated by CO_2 - and Nd:YAG lasers. The applications differ in terms of power density and beam interaction time, as outlined in figure 4. With the development of high power laser diodes the semiconductor based beam sources will be introduced for certain material processing applications in the near future.

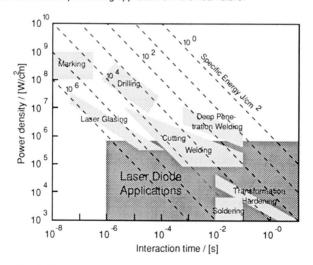

Fig. 4 : Power density required for materials processing applications.

The power density of laser diode systems is limited to about $5 \cdot 10^5$ Watt/cm^2 (see table 2), the power level of the devices is not limited however and devices in the range of several watts to multi kilowatts can be realized by adding more laser emitters to the system. Potential areas of applications for laser diode systems are (see fig. 4):

1. Soldering
2. Transformation hardening
3. Welding
4. Laser glasing
5. Cutting

Soldering, the application which requires the lowest power densities and power levels, has already been realized (7), (8), (9) with fiber and lens multiplexed devices and is also demonstrated in this paper in the last section.

Currently, no applications in the fields of transformation hardening, welding and laser glasing have been reported. As laser diode systems with the required power densities above 10^3 W/cm^2 already exist the foreseeable increase to an average power of several hundred Watt will make transformation hardening applications possible. For welding and glasing techniques power densities above 10^5 W/cm^2 are required. The development of laser diode systems is currently advanced to $2 \cdot 10^4$ W/cm^2 with lens multiplexing, which is at least 5 times below the required density. Devices above this limit can only be realized by lens multiplexing or coupling of individual laser emitters to single fibers. Since this fiber coupling technique needs new assembly techniques, it is expected that the lens multiplexing approach will reach these power density levels above 10^5 W/cm^2 first. The even liability of laser diode systems with electrical/optical efficiencies of 20 - 40 % will bring significant advantages to the field of micro and macro processing.

Soldering

Soldering of electronic devices with the laser as a heat source which has no mechanical contact to the electronic device is already described in several publications (7), (8), (9). The laser beam is focussed onto the soldering pad, where the solder and connector are heated above the melting point by the absorbed laser radiation. The advantages of laser soldering compared to conventional techniques like reflow or vapor phase soldering are the short process times, some ms to 1 s, which reduce the risk of thermal damage of the devices. The non mechanical contact and small focus diameter (20 ... 1000 μm) offers the possibility to solder with individual real time control very small connectors (< 200μm), which is not possible with conventional techniques. The temperatur of the connector and solder can be used to control the soldering process. Diode laser systems open the regime of a fast power modulation (~ up to 100 kHz).

Soldering investigations are carried out with a fiber coupled laser device with a maximum average power of 20 Watt. The laser system consists of 50 fibers of 200 μm diameter. The fibers are combined to a bundle of 1.6 mm diameter and the radiation is focused to a spot of 0.8 mm diameter. The whole set up with diode laser system, fiber, optics is shown in figure 5.

The investigations are carried out with SO-28 devices, which are placed on the solder pads coated with solder paste (~1 mg per pad, SnPbAg 62/36/2). The laser power is varied between 3 and 13 Watts at soldering durations of 30 ms to 10 s. Process times between 3s (3 Watt)and 80ms (13 Watt) and process energies between 1 J and 10 J give 'good' soldering results. The minimum process energy for good solder joints is 1 J , with a 30% increase up to 1.3 J at 3 Watt. This can be explained by the losses of heat conduction at low power input. The process efficiency including reflection and heat conduction losses is 0.5 which can be determined by the required energy (400 - 600 mJ) to heat the pad, connector and solder onto the melting temperature of the solder (180°C) and the real process energy of 1 J to 1.3 J.

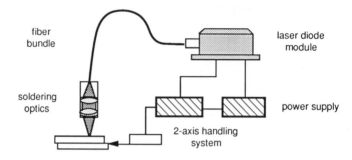

fiber
bundle

laser diode
module

soldering
optics

power supply

2-axis handling
system

Fig. 5 : Fiber coupled laser diode soldering system

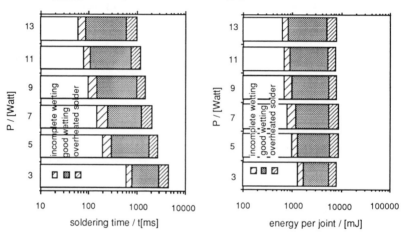

Fig 6 : Soldering results with a fiber coupled laser diode system

Reference

(1) R. Beach   IEEE J-QE, Vol. 28, No. 4, pp. 966-976, (1992)
(2) J.G. Endriz   IEEE J-QE, Vol. 28, No. 4, pp. 952-965, (1992)
(3) J. Snyder   Appl. Opt., Vol. 30, No. 19, pp. 2743-2747, (1991)
(4) T.Y. Fan   Appl. Opt., Vol. 30, No. 6, pp. 630-632, (1991)
(5) K. Cheo   Handbook of solid state lasers, Dekker, (1989)
(6) Y. Asahara   Appl. Opt., Vol. 25, No. 19, pp. 3384-3387, (1986)
(7) V. Krause   DVS Berichte, Bd. 141., pp. 39-42, (1991)
(8) L. Musiejovsky   VTE 2/91, pp. 50-55, (1991)
(9) A.V. Polijanczuk   IEE advances in interconnection technology, No. 1991/082 , (1991)
(10)Sony Corp.   Technical description,Tokyo (J), (1991)
(11) Diomed   Technical description, Cambridge (GB), (1991)
(12) Mc Don. Douglas   Technical description, St. Louis (USA), (1992)
(13) ADLAS GmbH   Technical description, Lübeck (FRG), (1991)
(14) Heimann GmbH   Technical description, Wiesbaden (FRG), (1992)
(15) Spectra Diode   Technical description, San Jose (USA), (1991)
(16) D. Scifres   US-patent, No. 4.688.884, (1987)
(18) D. Scifres   US-patent, No. 4.763.975, (1988)
(19) R. Jacobs   US-patent, No. 5.022.043, (1991)
(20) T.Y. Fan   US-patent, No. 5.081.637, (1991)

Welding with a 3 kW-Nd:YAG-Solid-State Laser

H.-G. Treusch, V. Krause, M. Nießen, J. Berkmanns, E. Beyer; Fraunhofer-Institut für Lasertechnik, Aachen

Introduction

Significant advantages are expected by the use of solid state lasers in the multi-kW-range instead of CO_2-lasers for materials processing, as soon as high average power could be combined with sufficient beam quality for beam delivery via optical fiber for the following reasons:
- superior absorptivity on metal surfaces
- less influence by laser induced plasmas

Currently high power solid state lasers are using either several laser rods in a common resonator or a spatial overlap of several laser beams in the focal plane of a lens. In the first case more than 6 laser rods are needed for an average power of 3 kW and the thermal lensing effect of each rod contributes a source of instability to the laser system (1). The second case accepts the disadvantages of a decreasing beam quality with the square root of the number of single beams and a beam delivery via multiple of fibers, which therefore demands an increase in positioning tolerances for parts or optics (2).

In this paper an alternative concept of a high power solid state laser using temporal multiplexing for 3 kW average power is presented. Applications in welding are discussed.

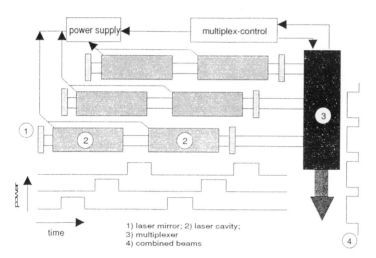

1) laser mirror; 2) laser cavity;
3) multiplexer
4) combined beams

Fig. 1: Scheme of temporal multiplexing method for three solid state laser

1. Laser system configuration

The temporal multiplexing method renders possible a combination of a multiple of laser beams of medium average power into one beam of high average power without any degradation of beam quality. The method of multiplexing is shown in figure 1 schematically.

The multiplexing method is realized and tested with a prototyp laser system, consisting of six laser rods, two each in a common cavity, which deliver an average power of 520 W each in a pulsed mode. The maximum average power of the laser system was measured to 3.1 kW at a beam quality of 17 mm*mrad (radius*half angle of divergence). Two rotating segmented and one stationary 90 degree bending mirrors form multiplexer. The axes of rotation are perpendicular to the surface of the mirrors for constant bending angle during the puls duration. The individual laser is triggered by encoders on the rotating axes to lock the emision to the rotating mirror in the output path.

2. System parameter

The maximium average output power of a pulsed solid state laser at constant average pump power is depending on the pulse duration, pulsshape and average peak power. The average peak power is determined by the ratio of pulse energy and pulse duration. A characterization of the system with respect to the peak power by changing the flash lamp current is done at a constant pulse duration of 3 ms. The average pump power is kept constant by adapting the pulse repetition rate to its maximum value at the specified lamp current. The maximum average output power of the laser system in an oszillator-amplifier-configuration is only achievable at an average peak power above 10 kW and pulse duration above 1 ms. The output power behind the focussing optics is shown in figure 2 as a function of average peak power.

Applications in materials processing and the usability of fiber optics as a beam delivery system are determined by the beam quality besides the output power of the laser system. The caustic of a focussed beam is measured as a function of average pump power by a beam diagnostic system based on a CCD-camera with a motor driven optic to determine the energy distribution in the beam cross section. The beam quality is then calculated by fitting a gaussian beam propagation to the measured caustic. The achieved beam quality as a function of average pump power per laser cavity is compared with data calculated by resonator theory in figure 3. The maximum value of the beam quality (beam radius x half divergenc angle) is 20 mm*mrad. This value limits the usage of fiber optics for beam delivery to a minimum fiber core diameter of 0.4

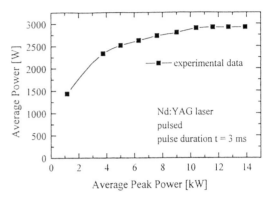

Fig. 2: Average output power as a function of average peak power of a pulsed Nd:YAG laser

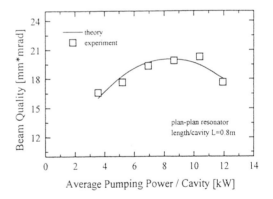

Fig. 3: Beam quality as a function of average pump power per laser cavity compared to resoator theory

mm at a numerical apperture of N.A.= 0.2 (40 mm*mrad). The maximum beam diameter and coupling angle at the fiber entrance should both never exceed 70% of the specific fiber data.

First trials in coupling a 3 kW beam to a 0.6 mm fiber show another limitation of the minimum core diameter due to the high average power. Above a threshold of 2.3 kW average power the fiber entrance starts to glow in the cross section of the laser beam. The threshold is determined to 1.8 MW/cm², which is in a good agreement to data from literature (1).

3. Welding with a pulsed solid state laser at high average power

Due to the limitation of the fiber optics to an average power at < 2.3 kW the welding results discussed below are carried out with fixed optics. The beam is expanded by a factor of 2 and then focussed with a lens (f=150 mm, F-number 7.5) to a spot diameter of 0.75 mm. Welding is performed in stainless steel (1.4301), aluminum (AlMgSi1 and AlFe1.5Mn) and zinc coated steel. For the first investigations in welding an oscillator-amplifier setup is used instead of the configuration shown in figure 1.

3.1 Stainless Steel

The welding depth is investigated in a 10 mm plate as a function of laser parameters and processing speed. The shielding gas is delivered by a dragging nozzle of 6 mm diameter with an inclination of 30 degree. At a flow rate of 15 l/min no influence of the kind of gas (nitrogen, helium and argon) on the welding process can be detected. Without shielding gas the width of the seam is decreased due to the reduction of the surface tension by oxidation without increasing the welding depth at low welding speed (high line energy 150 Ws/mm). The surface tension forces the hot liquid metal on the surface to flow in a region of lower temperatur. In figure 4 the welding depth is shown as a function of welding speed for different peak power and for different focus position with respect to the surface. At a peak power of 5 kW the melt expulsion (ablation) increases strongly at welding speeds below 2 m/min. The expulsion can be reduced by decreasing the peak power or defocussing the beam to reduce the pulse power density at the surface below 1 MW/cm². But the reduction of peak power is also leading to a reduction of average power as shown in figure 2. At peak power density levels below 1 MW/cm² welding melt expulsion is absent down to welding speeds of 0.5 m/min. Cross sections of penetration welds in figure 5 demonstrate the obtained weld geometry and welding depth at different welding speeds from 1 - 6 m/min. Only at low welding speed of 1 m/min and below the

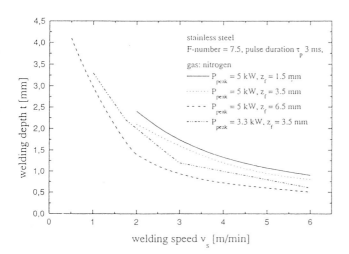

Fig. 4: Welding depth as a function of welding speed in stainless steel 1.4301 at different peak power and focus position with respect to the metal surface

depth of welding with pulsed solid state lasers becomes comparable to those using a cw solid state laser or CO_2-laser at the same power level. At higher welding speed the losses due to rapid cooling between the single laser pulses and the high vaporization rate, which can be detected by a vapour plume length of several cm above the metal surface, are leading to a smaller penetration depths. In addition the low power density of 1 MW/cm² is not sufficient to reopen the keyhole for deep penetration welding at high welding speeds.

3.2 Aluminum

The welding geometry as a function of welding speed is investigated for two different aluminum alloys at full penetration of the plate thickness of 1.25 mm AlFe1.5Mn and 2.5 mm AlMgSi1.0. The shielding gas is

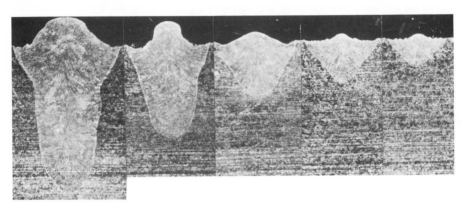

Fig. 5: Cross sections of penetration welds in stainless steel (1.4301) at 1,2,3,4 and 6 m/min, average power 2.2 kW, peak power 3.3 kW, focus position 3.5 mm, pulse duration 3 ms.

delivered by a dragging nozzle at a flow rate of 15 l/min. Again, as for stainless steel, the influence of different gases like helium, argon or nitrogen is negligible. But at full penetration there is a need of an additional shielding gas at the back of the plate to prevent oxidization. Figure 6 shows cross sections of seam welds in a 1.25 mm AlFe1.5Mn plate with argon as a shielding gas from above and below the plate. Full penetration is achieved up to a welding speed of 8 m/min. No cracks are observed, but above a welding speed of 3 m/min the micro structure of the solidified material shows the characteristic marks of the pulsed laser operation.

Fig. 6: Cross sections of through welds in 1.25 mm AlFe1.5Mn at welding speed of 2, 3, 4, 6 and 8 m/min; average power 2.5 kW, peak power 5 kW, pulse duration 5 ms, focus position $z_f=0$

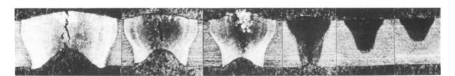

Fig. 7: Cross sections of through weld in 2.5 mm AlMgSi1.0 at welding speed of 0.5, 0.7, 1, 2, 3 and 4 m/min, average power 2.5 kW, peak power 5 kW, pulse duration 3 ms, focus position $z_f=0$

The influence of oxidization on the seam geometry and the micro structure is demonstrated in figure 7 in a 2.5 mm AlMgSi1.0 plate. Helium is used as shielding gas to the top and no shielding gas is applied to the bottom of the plate. At full penetration of the plate, achieved at welding speeds below 2 m/min, the surface tension of the molten material at the back is reduced by oxidization. Due to this reduction and to the pressure of melt expulsion, the surface curvature of the molten pool is forced into the material. Also the micro structure of the solidified material is changed, if full penetration is achieved and oxidization is not prevented. The color of the solidified material in figure 7 is changed from black to white. This identifies the change from a fine micro structure to a coarse one, which results from slower cooling. The oxidization prozess mixes oxids into the material leading to a crack in the center line of the seam at low welding speed The capability of high average power pulsed solid state lasers in welding of aluminum alloys is demonstrated in figure 8 and 9 with two butt welds of the alloys mentioned above. No cracks are detected in the cross section and a smooth surface is achieved on both sides.

Fig. 8: cross section of a butt weld of 1.25 mm Al-Fe1.5Mn, welding speed 4 m/min, laser parameter as in figure 6, argon at both sides.

Fig. 9: Cross section of a butt weld in 2.5 mm AlMgSi1.0, welding speed 2 m/min, laser parameter as in figure 7, argon at both sides.

3.3 Zinc coated steel

One of among several possibilities (3) to overcome the problem of lap welding of zinc coated material is to keep the diameter of vapor capillary as large as possible and the volume of the molten material at the cappilary wall as small as possible. This could be achieved in a pulsed laser operation mode with a pulse duration below 1 ms and a beam diameter comparable to the thickness of one sheet metal. Figure 10 shows two cross sections of lap welds of 1 mm zinc coated steel. The laser and process parameters are given in the table below. With a pulse duration of 3 ms there is a need of a 0.1 mm gap between the plates for the escape of the zinc vapor, but with a pulse duration of 1 ms and twice the peak power welding with zero gap is possible at a welding speed of 2 m/min. The high peak power opens a keyhole with a diameter in the range of the beam diameter, through which the zinc vapor can escape easily.

parame-ter set	average power	peak power	pulse duration	repetition rate	focus position	shielding gas	welding speed
A	2.1 kW	3.3 kW	3 ms	200 Hz	-1 mm	argon	1 m/min
B	2.7 kW	6.9 kW	1 ms	390 Hz	-1.5 mm	argon	2 m/min

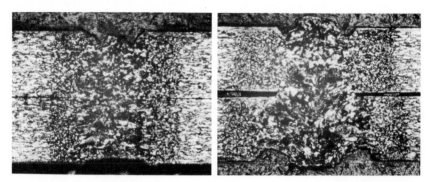

Fig.10: Cross sections of lap welds of two 1 mm zinc coated steel sheets, zinc layer 0.007 mm on both sides, left picture: paramter set A, 0.1 mm gap, right picture: parameter set B, zero gap

Conclusions

The temporal multiplexing method is a useful tool for scaling up solid state laser into the high power multi-kW-range at acceptable beam quality. The oszillator-amplifier configuration offers some significant disadvantages at low peak power with respect to efficiency and maximum average output power. Due to this reason the configuration is now changed into the two-cavity-oszillator setup, shown in figure 1, which enables the maximum output power of 3 kW already at a peak power of 4 kW, while every cavity can run at a peak power of 4 kW. The pulsed operation mode of solid state laser leads to a higher penetration depth in laser welding of aluminum alloys compared to the continuous wave mode. At an average power above 2 kW crack free welds are possible. Also for lap welding of zinc coated sheets the pulsed mode offers significant advantages in adapting the pulse duration to the expansion of the zinc vapour. This enables gap free welding of coated sheets. Due to the high absorptivity and a low thermal diffusivity of steel welding in the continuous wave mode is preferable.

References

(1) N. Hodgson, "High power solid state laser in rod-, slab- and tube geometry"
    Laseroptoelektronik H.23, March 1991
(2) K. Tönshoff, "New possibilities in material processing with kW-solid state lasers"
    Optical Science and Engineering, The Hague, 12.-16. March 1990
(3) R. Imhof, "Laser welding of zinc-coated sheets and quality control"
    ISATA Proceedings, 1.-5. June 1992, Florence

Acknowledgements

The basic principles of the multiplexing method is investigated in the frame of an EUREKA-project, which is supported by the BMFT (Bundesministerium für Forschung und Technologie), the VDI-TZ, Düsseldorf and Rofin Sinar Laser, Hamburg.

A Comparison of Powder- and Wire-Fed Laser Beam Cladding

F. Hensel, C. Binroth, G. Sepold
BIAS, Bremer Institut für Angewandte Strahltechnik, Bremen, FRG

Introduction

During the last few years laser cladding techniques have been further developed, so that now a days they can be used for industrial applications (1 - 8). Especially the Dynamic Material Feeding processes (DMF), where filler material is directly inserted into the laser affected zone on the substrat surface, have many advantages such as processing of limited local areas, low heat input into the base material, good adhesion strength of clads due to metallic bonding, low dilution of filler material and substrat material, refined structures due to high cooling rates.
Filler materials in DMF are predominantly wire or powder, resulting in modified processes which lead to essential differences like

- process efficiency,
- dilution and
- workpiece geometry.

After a short explanation of two modified processes the above mentioned differences are demonstrated and some general application aspects are shown. Not included are technical demands on powders and wires as to their metallurgical capabilities, and their geometrical sizes or the influence of a line shaped heat source.

Powder feed process

The principle of laser powder feed cladding is shown in fig. 1. A powder feeder

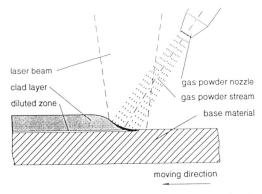

Fig. 1: Principle of laser beam cladding with powder-feed

transports the powder into the laser affected zone through a nozzle by means of an inert gas (usually argon). The intensity of the laser beam in the affected zone must be below the plasma formation level at intensities of $I<2*10^6$ W/cm^2, (9).

The use of different optical elements for beam shaping such as beam integrators, facet mirrors or scanners can influence the geometry of the clad with respect to its relation of height to width ratio. In case of a line shaped illumination of the surface the inert gas stream with powder has to be flattened using appropriate nozzles so that a constant amount of material is fed into the affected zone.

This technique leads to plated surfaces when creating a series of rows next to oneanother without the need to change the principle arrangement.

Commercially available powders for conventional surfacing technologies can be used for laser beam cladding.

Wire feed process

The principle of laser wire feed cladding is shown in fig. 2. The wire and the laser beam are directed under an oblique angle to the surface. The intensity of the laser beam must be higher than the plasma formation level at intensities $I>2*10^6$ W/cm^2, (9). The laser plasma is induced between wire and surface. The main part of the laser energy is absorbed by the wire and the remaining part by the surface. The use of working gas e.g. CO_2 for plasmacontroll as well as a shielding gas is necessary. The gas-nozzle encloses the wire-nozzle in this arrangement.

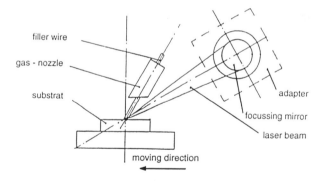

filler wire

gas - nozzle

substrat

adapter

focussing mirror

laser beam

moving direction

<u>Fig. 2:</u> Principle of laser beam cladding with wire feed

If a cladding on a surface is to be achieved a number of single traces with some overlapping have to be performed. In this case the set-up has to be changed so that the wire-laser plane is distorted to the moving direction under an angle of 35°. The surface of the layer is rippled so that it is not possible to position one cladding

on the next one. Slight inaccuracies in postioning of beam, wire, and workpiece surface can lead to ripples and slag inclusions in the "valleys".
Solid wires as well as flux-cored ones are suitable for the process.

Comparison and applications

- Process efficiency

During laser powder feed cladding the encoupling of energy into the material occurs by absorption of the laser beam and conduction of heat. Therfore only slow processing speeds are available (v_t= .5 m/min). The process efficiency is approximately 60 to 70 % concerning of the absorptivity of the powder material and the surface.
Further the powder is melted to some percentage in the range of between 50 and 80 % (weight) depending on the parameters.

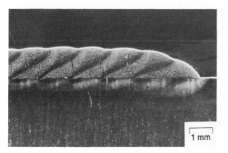

(P_L= 3.5 kW; F= 4.3; Δf= +20 mm v_t= .5 m/min)

(P_L= 3.5 kW; F= 4.3; Δf= 0 mm v_t= 2.1 m/min)

Fig. 3: Cross- section of a clad of Stellit 21 produced by powder feed cladding on mild steel

Fig. 4: Cross- section of a clad of Stellit 6 produced by wire feed cladding on mild steel

First experiments for wire feed cladding have shown, that due to lower absorptivity at lower intensities melting occurs at decreased wire feed speeds.
At higher intensities of I>2*10^6 W/cm^2 a plasma is formed, which leads to better conditions for encoupling of the laser energy, e.g. 90% of the laser beam energy. Due to the presence of the plasma much higher processing speeds (v_t= 2.1 m/min) can be reached.
It can be summarized that the efficiency of the wire feed cladding process is superior as to the efficiency of the powder feed process with respect to encoupled energy, enabling higher processing speeds or a higher amount of filler material.

- Dilution

During laser powder feed cladding in many cases the formation of a plasma must be avoided in order to avoid deep penetration causing a higher dilution of the powder material. If the powder stream is positioned towards the laser beam in such a way that the powder primarily melts and the substrat is mainly heated by molten material, very small dilution rates of less than 5% become possible, see fig. 3.
Laser beam wire feed cladding enables dilution rates of less than 20% when using the wire-laser arrangement as described above, see fig. 4.
The melting line within the carrier material is wavy when using wire cladding. The reason is the deeper penetration and higher dilution of the clad material in the middle of the layer.

- Workpiece geometry

The explained particularities of the two processes lead to different applications.
A smooth first surface layer enables the deposition of a second smooth surface layer etc.. Even after some layers the surface of the clad remains even and smooth, see fig. 5. This leads to smaller machining operations after cladding. It is also possible to clad more complicated geometries such as curves and bends with the relativly simple arrangement. Automated cladding with robots is possible because the slight track deviations - they occur in connection with teach-in programming and discrepancies in data and contour in CNC programming - do not affect the relative unsensitive process.

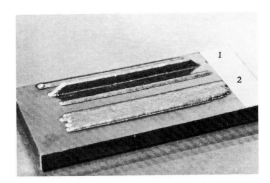

Fig. 5: Laser powder feed clads   (1. cladding -10 track multi layer
2. cladding - 6 layers side by side)

Due to its charactristics laser beam powder cladding is especially suitable for :
- clad thicknesses in the range of minimum .2 mm,
- localized multi-layer claddings of >10mm,
  (higher thicknesses, fine structure, high purity, graduated claddings),
- cladding of relative complicated geometries,
- automated cladding.

In comparison to powder feed cladding the laser wire feed cladding process is characterized by the demand on precise adjustment of laser beam towards wire and workpiece surface. If there are little deviations in adjustment, the wire is molten in a none uniform way, leading to interruptions in the process. A slight rippling of the produced cladding is typical which may cause higher expenditures in the machining operations of the work piece.

Even though it is easily possible to lay clads side by side, experience has shown that it is difficult to create them layer on layer. The reason for these difficulties are disturbances of process due to an uneven surface of the preceeding layer.

It is very important that the parameters v_W and v_t should be chosen in such a way that an contineous melting of the wire is guaranteed.

Deviations in geometry of the workpiece make necessary a tracking of the working head resulting in a noncontinuous melting of the wire which could lead to process instabilities. Due to a complicated set-up of the processing head, all movements should be carried out by the workpiece itself.

The set-up (inclination of beam and wire nozzle, see fig. 2) promotes a cladding of parts with difficult access.

Under technical aspects wire cladding is suitable for:

- one-layer technology,
- thicknesses in the range of approx. .3 mm to 1.5 mm,
- simple geometries (circular or linear),
- surfaces with difficult access (such as grooves).

	laser beam	
	wire cladding	powder cladding
arrangement	inclined	more or less verticaly
cladding accomplishment	.5- 2 kg/h [**]	.3- 2 kg/h [*][**] (10)
efficiency of energy	over 90%	60- 70%
utilizing of filler material	fully	up to 80 %
dilution	< 20 %	< 5 %
sensitivity of process	high	low
surface geometry	rippled	even
attainable height of claddings	app. .3- 1.5 mm (one layer)	app. .2- >10 mm (single or multilayer)
suited workpieces for cladding	- cylindrcal or even - difficult to access surf.	- 3-D contours - easy to access surf.

Table 1: Comparison of processes   [*] additionally dependent on powder utilisation
[**] at power ranges of 3-4 kW

44

Summary

The particularities of the cladding processes with wire or powder feeding are to be examined in connection with the foreseen applications. So the processes must be considered e.g. with regard to efficiency of process, required thickness of the clad, required dilution rate, geometry of the workpiece as well as expenditures for follow-up-machining, see table 1.
Although the laser beam powder feed process has a low efficiency in comparison with the wire feed process some advantages have to be stressed:

- the thickness of the cladding is controllable in all ranges,
- the volume of machining after cladding is low due to a smooth and even surface and
- application of robots is possible.

Whereas at the wire feed cladding

- the thickness of the cladding is limited,
- the volume of machining after cladding is higher and
- application of robots is not recommendable.

Besides generell known laserspecific advantages like locally limited processing and low heat input etc. the user gets some new .possibilities to achieve clads of high quality as to structures and construction of multy layers. These unique features cannot be produced by conventional thermal processes.

References

(1) V.M. Weerasinghe, W.M. Steen: "Laser in Materials Processing" L.A. 1983, ASM Metals Park, Ohio 44073
(2) R. Becker, C. Binroth, G. Sepold: "Laser`87", Optoelektronik, Springer Verlag, 1987
(3) R. Becker, H. Kohn, G. Sepold, P. Seyffarth: Schiffs-Ingenieur-Journal, 36 (1990) 210
(4) F. Hensel, C, Binroth, G. Sepold: "High- Tech- Finishing`92, Praxis- Forum Oberflächentechnik"
(5) C. F. Marsden, A. Frenk, J. D. Wagnière, R. Dekumbis: "ECLAT`90"
(6) R. Becker, C. Binroth, G. Sepold: "Laser`89, Laser- Optoelectronics in Engineering", Springer Verlag, 1990
(7) R. Becker, G. Sepold, K. Shibata, H. Matsuyama: "ECLAT`90"
(8) D. Burchards, A. Hinse: "ECLAT`90"
(9) E. Beyer, K. Wissenbach, G. Herziger: Feinwerktechnik & Meßtechnik 92 (1984) 3
(10) G. Sepold, R. Becker: "SPIE International School, Laser Surface Microprocessing", Tashkent 1989

Dynamics of Phase Fronts in Pulse Laser Action. Influence of Pulse
Shape and Spike Structure

Smurov I.Yu., Surry C., ENISE, Saint-Etienne, France
Lashin A.M., IMU, C.N.R., Milano, Italy

Mathematical model

For the modelling of heat processes of pulse laser action on metals (that correspond to comparatively high energy density flow), it is necessary to join both processes - melting and surface evaporation in the frames of one mathematical model: this leads to the situation when two moving (in a common case with different velocities) phase boundaries are tracked [1-3].
The mathematical model proposed includes the process of heating, melting, evaporation, cooling and solidification under the irradiation of arbitrary energy flow on a metal slab [1,2,]. It is assumed that the energy flow is absorbed on the irradiated surface; convection and radiation mechanisms of heat losses from both the sides of the slab are considered; melting (solidification) is determined by a classical Stefan boundary condition.
The used mathematical model can be written in the following form:

$$\frac{\partial^2 T_1}{\partial x^2} = \frac{1}{a_1}\frac{\partial T_1}{\partial t} \qquad S_1(t) \leq x \leq S_2(t), \quad t_m \leq t \leq \infty$$

$$q_0(t)-\alpha_g[T_1(x,t)-T_g]-\sigma[\varepsilon_1 T_1^4(x,t)-\varepsilon_g T_g^4]=-\lambda_1\frac{\partial T_1}{\partial x} + \rho_1 L_v\frac{dS_1}{dt}, \quad x=S_1(t)$$

$$\frac{dS_1}{dt} = \frac{V_*}{\sqrt{T_1(S_1(t),t)}} \exp\left[-\frac{T_*}{T_1(S_1(t),t)}\right]$$

$$\lambda_1\frac{\partial T_1}{\partial x} = \lambda_2\frac{\partial T_2}{\partial x} - \rho_2 L_m\frac{dS_2}{dt}, \qquad T_1 = T_2 = T_m, \quad x=S_2(t); \quad S_2(t_m)=0$$

$$\frac{\partial^2 T_2}{\partial x^2} = \frac{1}{a_2}\frac{\partial T_2}{\partial t} \qquad S_2(t) \leq x \leq L$$

$$-\lambda_2\frac{\partial T_2}{\partial x} = \alpha_f(T_2-T_f) + \sigma(\varepsilon_2 T_2^4-\varepsilon_f T_f^4), \quad x = L$$

$$T(x, t=0) = T_0$$

where: $T_1(x,t)$ temperature of liquid phase; $T_2(x,t)$ temperature of solid phase; x, t distance and time, respectively; $S_1(t)$, $S_2(t)$ positions of evaporation and melting phase boundaries, respectively; $q_0(t)$ absorbed energy density flux; $a_{1,2}$ thermal diffusivity of liquid and solid phase, respectively; $\lambda_{1,2}$ thermal conductivity of liquid and solid phase, respectively; $\rho_{1,2}$ densities of liquid and solid phase, respectively; L_v specific heat of evaporation; L_m latent heat of melting; $\alpha_{g,f}$ convection heat losses coefficient of irradiated and rear surfaces of the slab; $\varepsilon_{1,2}$ emissivities of irradiated and rear surfaces of the slab; $\varepsilon_{g,f}$ emissivities of the environment near the irradiated and rear; surfaces of the slab T_m temperature of melting; t_m starting time for melting; L thickness of the slab; T_0 initial temperature.

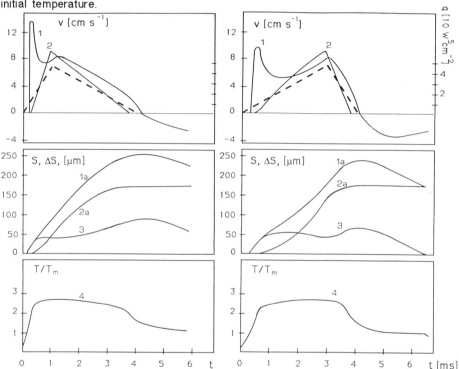

Fig.1. Time dependence of: melting (1) and evaporation (2) front velocities; positions of melting (1a) and evaporation (2a) fronts; melt thickness (3), surface temperature (4). Shape of energy pulse - dashed curve. Material - Ti.

Constants V_*, T_* are determined by Herz-Knudsen's law of evaporation:

$$V_* = \frac{P_v}{2\rho_1(2\pi k/m)^{1/2}} \exp\frac{L_v}{T_v(k/m)} \ , \ T_* = \frac{L_v}{(k/m)}$$

where: $k = 1.38 \cdot 10^{-23}$ J/K, Boltzmann constant; m atomic mass of slab material; T_v boiling temperature corresponding to the pressure P_v.

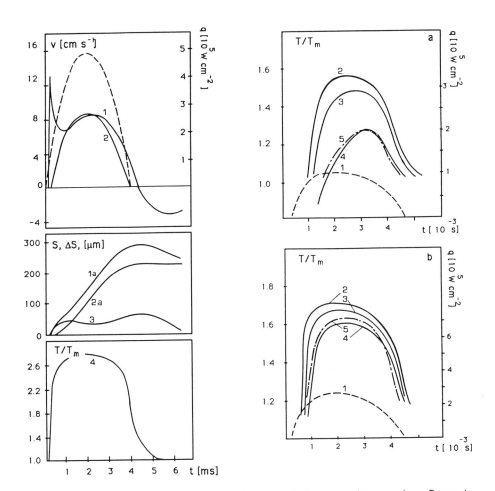

Fig.2. Evolution of surface temperature of Ti slab during pulse laser action. Curve 1 - shape of laser pulse; 2-4 results of numerical simulation correspond to the absorptivity values 1.0; 0.8; 0.45 respectively; 5 - experimental results; a - laser pulse energy 1.2 J; b - 3.0 J.

The phase change fronts are tracked continuously and the latent heat release is treated as a moving boundary condition. In both regions of liquid and solid phases the moving uniform grids consist of a fixed number of points. Each grid point moves with a different velocity. The Crank-Nicolson's technique of various derivatives is used [4].

The calculations were made for titanium and steel slabs of 1 mm thickness. In all the cases examined below the slab can be considered as a semi-infinite body.

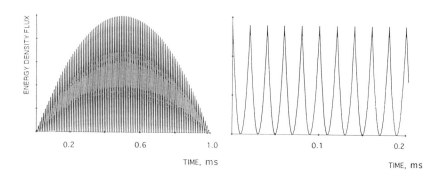

<u>Fig.3.</u> Model spike structures of laser pulses used in simulation: (a) - free generation, (b) - ordered generation.

Results and discussion

The influence of the shape of energy pulse on heat processes dynamics is considered on the example of triangle pulses with the same energy input, duration and maximum value of energy density flux (fig.1). The only difference is the position of extreme value of energy density flow q_{max}, that leads to different rates of increase of energy flow. In this numerical experiment pulse duration is equal to 4 ms, $q_{max}=5*10^5$ W cm^{-2}.

The shape of the curve corresponding to the dependence of the evaporation front velocity on time is close to the shape of energy pulse. This is the result that the transient period of the velocity is comparatively short $t_{st}(q_{max}=5*10^5$ W cm$^{-2})=0.1$ ms; that is why the velocity change follows a comparatively slowly variations of energy density flow.

The velocity of melting is a nonmonotone function of time with two peak values. The first one is caused by the increase of surface temperature on one hand, and by the thickening of melt layer - from the other, that increase its thermal resistance. The second peak is a direct result of interaction of two phase boundaries: velocity of motion of the evaporation front increases up to the values exceeding the melting velocity, that causes the melt layer to become thinner. This is the reason for an increase of heat flux reaching the melting phase boundary and therefore increase of melting velocity.

It must be noted that such parameters as depth of the propagation of the melting front and thickness of the evaporated layer only weakly depend on the shape of energy pulse provided that it's duration, q_{max} and energy per pulse remain the same. But the difference can sharply increase with the decreasing of q_{max} up to the threshold of melting.

As a rule the real shape of laser pulse corresponding to the case of chaotic (free) generation is approximated by the smooth curve (fig.2). But in practice a radiation pulse with duration about 10^{-3} s consists of a sequence of spikes with duration about 10^{-6} s and the duty ratio (pulse period to pulse duration ratio) about 2-5. In free generation pulse the spike's amplitudes are different; in, so-called, ordered generation - the amplitude of the individual spikes and the time interval between successive spikes remain constant during the considerable part of the pulse (fig.3). In practice it is important to understand in which case it is necessary to consider this real spike structure and when it appears possible to approximate its shape by a smooth curve. In other words this is the question of the result's reproducibility of laser pulse action with chaotic generation.

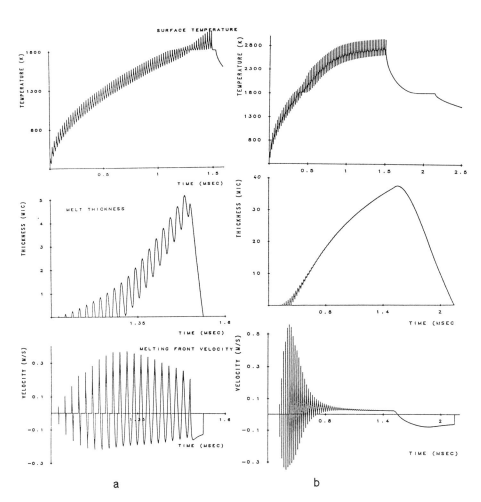

Fig.4. The action of laser pulse (duration 1.5 ms) with regular (ordered) generation (from fig.3b); (a) - $q_{max}=4.6*10^4$, (b) - $q_{max}=1.2*10^5$ W cm^{-2}. Respectively, from up to down: surface temperature (K); melt thickness (μm); melting front velocity (m/s). Material - steel.

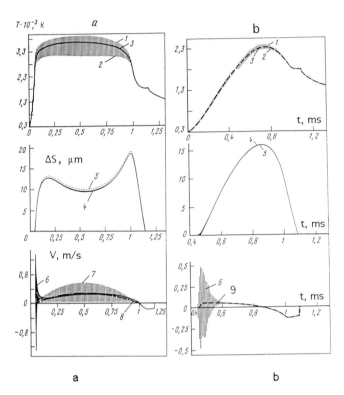

Fig.4. The action of laser pulse (duration 1 ms) with chaotic generation (fig.3a); (a) - $q_{max}=4*10^6$, (b) - $q_{max}=2*10^5$ W cm^{-2}; curves: 1,2 - upper and low boundaries of the temperature's oscillations, 4 - melt thickness, 6 - melting, 7 - evaporation front velocity. Curves: 3 - temperature, 5 - melt thickness, 8 - evaporation, 9 - melt front velocity correspond to the action of the similar pulse but with the smooth parabolic shape and equal energy input.

References

(1) I.Smurov, A.Lashin: Thermal Physics Processes Modelling of Pulse Energy Flow Action on Metal Plates. In the book: 'Physics - Chemical Processes in Materials Treatment by Concentrated Energy Flows' (1989) "Nauka'", Moscow, 160.
(2) I.Smurov, A.Uglov, A.Lashin, P.Matteazzi , L.Covelli , V.Tagliaferri: Int. J. Heat Mass Transfer (1991) v. 34, No. 4/5, 961.
(3) A.Kar, J. Mazumder: J. Appl. Phys. (1990) 68, 3884.
(4) M.A.Hastaoglu: Int .J. Heat Mass Transfer (1986) 29, 495.

Capacitive Clearance Sensor System for High Quality Nd:YAG Laser Cutting and Welding of Sheet Metals

S.Biermann, A.Topkaya, M. Jagiella, Weidmüller Interface GmbH, Gaggenau

Introduction

With the availibility of high power Nd:YAG laser systems the interest in robot systems with glass fibre beam guiding is rapidly increasing. For a continuous high quality laser processing it is necessary to ensure a constant distance between the laser working head and the workpiece. Both, the position of the focal plane of the laser beam and the gas flow conditions are important to be kept constant. For laser processing purposes it is necessary that the laser working head is positioned at an optimum clearance to the workpiece with a tolerance of not more than +- 0.1 mm. These tolerances are clearly exceeded by deviations from the programmed 3D contour during the operation caused by workpiece tolerances, thermal effects (warping), and inertness of the robot arm . The method of capacitive clearance control has shown to be appropriate to compensate these deviations to keep the distance between working head and workpiece within the required tolerances.

Principle of Capacitive Clearance Control

The capacity C_p depends on the area of the plates (a), the permittivity (ϵ_r) of the material between the plates and the clearance (d) between the plates. The nonlinear 1/d response, Fig. 1, is a disadvantage for closed loop clearance control applications. If the capacitor is connected to an AC voltage, it has a capacitive reactance X_{Cp} depending on the frequency of the AC-Voltage and the

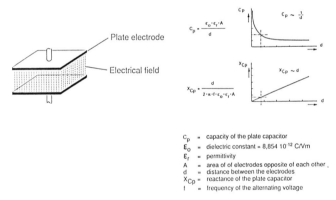

Plate electrode

Electrical field

$$C_p = \frac{\epsilon_o \cdot \epsilon_r \cdot A}{d}$$

$$C_p \sim \frac{1}{d}$$

$$X_{Cp} = \frac{d}{2 \cdot \pi \cdot f \cdot \epsilon_o \cdot \epsilon_r \cdot A}$$

$$X_{Cp} \sim d$$

C_p = capacity of the plate capacitor
ϵ_o = dielectric constant = $8{,}854 \ 10^{-12}$ C/Vm
ϵ_r = permittivity
A = area of of electrodes opposite of each other .
d = distance between the electrodes
X_{Cp} = reactance of the plate capacitor
f = frequency of the alternating voltage

Abb. 1 Principle of Capacitive Measurement Technique for Clearance Control

capacity . Therefore X_{Cp} is proportional to d. The new developed Lasermatic®[1] II system for capacitive clearance control uses this advantage[1]. The copper nozzle of the cutting head acts as the electrode of the clearance sensor. As the contour of a cutting nozzle is rather different from a ideal plate capacitor, inhomogeneous electric fields occur. This results in a nonlinear clearance characteristic depending from the real nozzle contour and additionally in a lateral spreading of the electrical field. For this reason objects close to the nozzle influence the measured signal. Using an "active shielding potential" and special designed cone nuts, the lateral sensitivity can be adapted by compression of the field lines, Fig. 2. The active measurement technique gives a drastical increase of the sensitivity of the sensor. Therefore an extrem slim contour of the nozzle tip has been realized, which is best suited for 3D applications.

---

[1]. Lasermatic® is a registered trademark of Weidmüller Interface GmbH

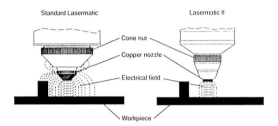

Standard Lasermatic   Lasermatic II

Cone nut

Copper nozzle

Electrical field

Workpiece

Abb. 2 Compression of the Electrical Field Lines using Active Shielding Technique

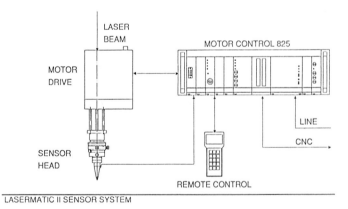

LASER
BEAM

MOTOR CONTROL 825

MOTOR
DRIVE

LINE

CNC

SENSOR
HEAD

REMOTE CONTROL

LASERMATIC II SENSOR SYSTEM

Abb. 3 Scheme of Lasermatic® Clearance Control System

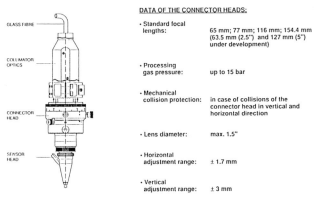

DATA OF THE CONNECTOR HEADS:

• Standard focal lengths:	65 mm; 77 mm; 116 mm; 154.4 mm (63.5 mm (2.5") and 127 mm (5") under development)
• Processing gas pressure:	up to 15 bar
• Mechanical collision protection:	in case of collisions of the connector head in vertical and horizontal direction
• Lens diameter:	max. 1.5"
• Horizontal adjustment range:	± 1.7 mm
• Vertical adjustment range:	± 3 mm

Abb. 4 Scheme of the Lasermatic® Cutting Head for High Power Nd:YAG Laser

Lasermatic® II Sensor System

Fig. 3 shows the complete Lasermatic® system configuration for Nd:YAG welding and cutting applications. The system consists of the following components:
* the cutting head
* the control unit containing sensor measurement electronic, sensor control, motor drive control and motor amplifier
* control panel
* fast motor drive
Experiences have proved that the sensor control system should be independend from the robot control system to garantee minimal reaction time in the closed loop.[2] Fig. 4 shows the scheme of the Lasermatic® cutting head for Nd:YAG Laser. The laser beam has to be collimated the glass fibre to generate a parallel beam. The focussing optic is integrated in the connector head together with the vertical and horizontal adjustment. Lens holder and the used focal length differ from stan-

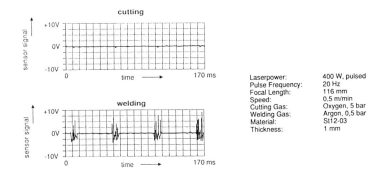

Laserpower:	400 W, pulsed
Pulse Frequency:	20 Hz
Focal Length:	116 mm
Speed:	0,5 m/min
Cutting Gas:	Oxygen, 5 bar
Welding Gas:	Argon, 0,5 bar
Material:	St12-03
Thickness:	1 mm

Abb. 5 Signals of the Capacitive Lasermatic® Sensor while Cutting and Welding with a pulsed Nd:YAG Laser

dard CO_2 - laser equipment. The sensor and the signal preprocessing and transportation is integrated in to the sensor head. The contour of the sensor head has to be adapted to the focal length to garantee an optimum slim construction of the cutting head.

Results of Welding and Cutting Test

Several welding and cutting tests with pulsed and continuous wave Nd:YAG laser have been performed. Fig. 5 shows the signals of the capacitive sensor for cutting (a) and welding (b) of a flat steel sheet using a pulsed Nd:YAG laser. The sensor signals for the cutting process are rather smooth. The laser pulses can be hardly recognized. In comparision the signals during welding have some strong disturbances during the laser pulses. The experiments have shown that it is possible to eliminate these disturbances using a signal filtering technique without loosing the distance information. The hypothesis that only the pulse pauses were used for measurement could be disproved by using a cw Nd:YAG Laser, Fig. 6. Again the signal during the cutting process is very smooth, while during the welding strong disturbances occur. The signals are slightly bended because in this experiment a steel sheet with a radius was processed. After low pass filtering the sensor signal for welding the disturbances are suppressed and the clearance control works without any problems. The tests were performed for a variety of laser and welding parameters. In the carried out tests parameters like laser power, welding or cutting speed, focal position, material and welding geometry have no influence on the sensor signal.

As it is well known, that the capacitive sensor technique is well suited for laser cutting applications, this investigations prove that capacitive clearance control works as well for Nd:YAG laser welding processes. In spite of this, former investigation in CO_2- laser welding proves it rather difficult to get usable signals for clearance control using standard Lasermatic® sensors. This difference between Nd:YAG and CO_2-laser welding is assumed to be due to the different absorption cross section[3] of the metal vapour at 10,6 and 1,06 μm. Therefore very different ionisation levels of the welding plasma plume occur. This results in a short circuit in the case of CO_2- laser welding and limited signal disturbance in Nd:YAG- laser welding, which can be easily compensated. It can be assumed, that the capacitive clearance control is limited by a still unknown intensity level of the Nd:YAG laser beam or for special materials with high ionisation rate due to a higher ionisation level of the welding plasma.

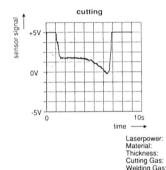
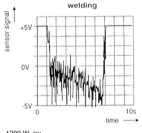

Laserpower:	1200 W, cw
Material:	St12-03
Thickness:	0,75 mm
Cutting Gas:	Oxygen, 5 bar
Welding Gas:	Argon, 0,5 bar

Abb. 6 Signals of the Capacitive Lasermatic® Sensor while Cutting and Welding with cw Nd:YAG Laser

Applications

The capacitive clearance control using standard Lasermatic® devices has big advantages especially in 3D applications using robots with glass fibre guided Nd:YAG laser beam, Fig. 7. The slim contour of the Lasermatic® nozzle allows a sensor controlled welding process even with difficult workpiece geometries.

Special advantages for the manufacturing technology can be expected, when the combination of Nd:YAG laser cutting and welding processing can be used. Taken care only for the Lasermatic® sensor control system the only difference between both applications are the different diameter of the nozzle orifice and the different distance between nozzle and workpiece. The Lasermatic II system allows any control distance between 0.3 and 9.5 mm. Therefore an easy change between cutting and welding is possible.

Acknowledgement

The results concerning sensor assisted welding have been supported in BRITE research project No. 3257 'Sheet-metal welding with kW Nd:YAG-laser for the automotive industrie'

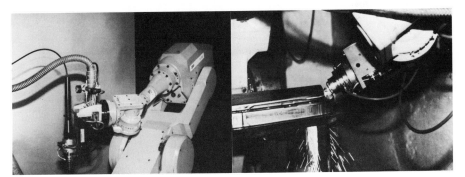

Abb. 7 3D Cutting of metal sheets using high power Nd:YAG laser in job shop production (BKLT, Bad Aibling)

References

1. M. Jagiella, S. Biermann, H.Spörl, A. Topkaya: Sensorik setzt neue Maßstäbe. Laser 5, S.288-291 (1991)

2. C. Olaineck: Nd:YAG-Laser und Roboter als Team. Laser Praxis. München: Carl Hanser Verlag, 1992, S. LS25-LS28

3. A.Treusch, private communication

Kaleidoscope Beam Homogenizer for High Power CO2 and YAG Laser

T. Ishide, M. Mega, O. Matsumoto, Takasago Research & Development Center, Mitsubishi Heavy Industries, Ltd., Takasago, Japan
S. Ito, T. Mitsuhashi, Takasago Machinery Works, Mitsubishi Heavy Industries, Ltd., Takasago, Japan

1. Introduction

For surface treatment with the CO_2 and the YAG laser, the beam of high uniformity can be obtained stably for a long time with the kaleidoscope which can obtain a uniform beam through multiple beam reflection (1-2). However, for its practical application, it is necessary to obtain the beam uniforming conditions and characteristics of the kaleidoscope for the laser used and further the beam absorptivity and so on for the wavelength.

Therefore, the authors have designed and manufactured the kaleidoscopes for the CO_2 laser and the YAG laser (utilizing optical fiber) on the basis of the beam intensity analysis, and also have compared the beam absorptivities of the two obtained beams which have the same uniform distribution shape but different wavelengths, and applied practically to laser hardening of the blower.

2. Test equipment and samples

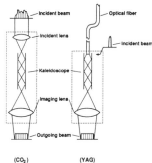

Fig. 1: Composition of the kaleidoscope

Fig. 1 shows the basic compositions of the CO_2 and YAG laser kaleidoscopes. In the CO_2 laser (5kW) the oscillation beam is condensed to the port of the kaleidoscope by the incidence lens, uniformed into a rectangle, and then imaged on the work surface in the required magnification through the imaging lens. On the other hand, in the YAG laser (1.2kW) the incidence lens portion is composed of optical fibers and the beam is led directly into the kaleidoscope. Its imaging lens is of a combined one for the purpose of aberration correction. The materials to be hardened are pure iron, SK-3 tool steel and SF55A.

3. Beam intensity analysis in kaleidoscope

The ray tracing method has been used to obtain the beam intensity uniforming conditions for the kaleidoscope incidence beam conditions and kaleidoscope dimensions. In addition, for the CO_2 laser of the high coherency beam, ray superposition has been carried out in consideration of beam interference. For the kaleidoscope the rectangular straight tube shape has been studied.

3.1 Analysis method

(1) Ray tracing

For ray tracing in the kaleidoscope the incidence beam has been divided into minute elements and the element beams outcoming from their respective points have been traced. Fig. 3 shows the beam tracing method in the kaleidoscope. The reflected beam C_i, which starts from the Point P_i, runs in the space enclosed by the kaleidoscope surface π and is reflected in i times, and the reflection plane π_j are expressed as follows:

$$C_i = \left(\mathbf{P}_i, \mathbf{v}_i, a_i\right), \quad \pi_j = \left(\mathbf{Q}_j, \mathbf{n}_j\right)$$

where P_i: Position vector of point P_i, $\mathbf{v}_i$: Direction vector of beam C_i, a_i: Beam intensity, $\mathbf{Q}_j$: Position vector of point Q_j, $\mathbf{n}_j$: Normal vector of reflection plane π_j

The following formulas are obtained between C_i and the next reflected beam $C_{i+1} = (P_{i+1},\ v_{i+1},\ a_{i+1})$.

$$P_{i+1} = P_i + \frac{(Q_i - P_i) \cdot n_j}{v_i \cdot n_j} v_i \qquad \text{Equ. 1}$$

$$v_{i+1} = -2(v_i \cdot n_i) \cdot n_j + v_i \qquad \text{Equ. 2}$$

$$a_{i+1} = R a_i \qquad \text{Equ. 3}$$

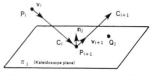

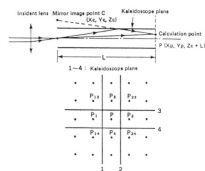

Fig. 2: Ray tracing method

With the above Equ. 1 through Equ. 3 the reflected ray C_{i+1} has been calculated to obtain the beam intensity at the outgoing surface of the kaleidoscope. In the calculation the reflectivity $R=1$, neglecting beam scattering and absorption in the kaleidoscope.

Fig. 3: Beam superimposition from minor image points

(2) Beam superposition

Fig. 3 shows the beam intensity analysis method that considers the beam interference. Superposition has been carried out for the beams from all mirror image points that are in the incident beam range of the certain point P on the kaleidoscope outgoing surface. That is to say, when the beam is to be in $u = s e^{i\alpha}$ (s: amplitude, α: phase), the superposition of n beams are expressed as follows:

$$\sum_{i=1}^{n} u_i = \sum_{i=1}^{n} S_i e^{i\alpha_i} \qquad \text{Equ. 4}$$

Then the beam intensity I is as follows:

$$I = \left| \sum_{i=1}^{n} S_i e^{i\alpha_i} \right|^2 = \left(\sum_{i=1}^{n} S_i \cos \alpha_i \right)^2 + \left(\sum_{i=1}^{n} S_i \sin \alpha_i \right)^2 \qquad \text{Equ. 5}$$

Each α_i has been obtained by calculating its beam passage length from the mirror image point to the calculation point P.

3.2 Outgoing beam intensity of CO_2 laser

Two kinds of the incident beam intensity distribution of the CO_2 laser have been considered, on the basis of the actual oscillation beam measured with a beam analyzer.

In both cases the intensity distributions are assumed to be symmetrical to the optical axes. The incident beam has been divided into minute elements, the element beam from each point sent into and condensed by the incident lens, and then the beams traced in the kaleidoscope to obtain the beam intensity at the outgoing port of the kaleidoscope.

Among the beam intensity analysis results, the beam uniformed state for the unit kaleidoscope length l is shown in Fig. 4. The following value has been used to evaluate the beam intensity uniformness for various conditions: $\eta = (I_{max} - I_{min})/I_{max} \times 100(\%)$, where, I_{max}: Maximum value of beam intensity, I_{min}: Minimum value of beam intensity.

In Fig. 5, the η values for various focal distances f of the incident lens in the single and multiple modes are arranged for the l/a values. The η values become more uniform as the value l increases and as the values a and f decreases. With Fig. 5 the l and a values to obtain the uniform beam can be obtained easily for the given f value.

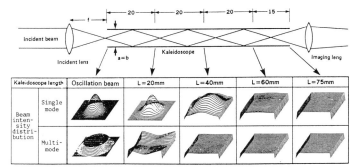

Fig. 4: Beam uniforming states for kaleidoscope length (CO_2 laser)

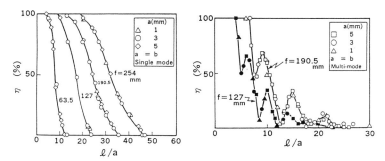

Fig. 5: Beam uniforming condition for kaleidoscope shape

Fig. 6 shows an example of the beam intensity distribution considering the interference at the outgoing port of the kaleidoscope. Since this figure shows the beam intensity with contours, interference fringes of two directions intersect orthogonally to show the state where minute high-density beams gather regularly. Fig. 7 shows variations of the interference pattern due to the kaleidoscope port dimension at the beam intensity cross-section.

Thus, in order to obtain the uniform beam, it is required to select the interference pattern suitable to the working purpose.

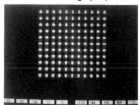

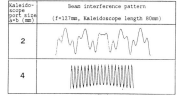

Fig. 6: Example of calculated beam interference pattern

Fig. 7: Effect of kaleidoscope port size on
beam interference (Single mode)

3.3 Outgoing beam intensity of YAG laser

For the high-power YAG laser, since the oscillated beam after being transmitted through optical fibers can be treated as the incoherent light source, its intensity analysis has been carried out in ray tracing only.

Fig. 8 shows the intensity distribution of the optical fibers. (GI & SI types) In ray tracing, this intensity distribution has been divided into minute elements, and the beam having the beam intensity outcoming in accordance with NA of the optical fiber traced (3).

For the purpose of compacting the kaleidoscope head, its incident port dimensions have been selected as small as possible within the range where the optical fibers can be coupled, and determined to be 2×2mm. The kaleidoscope length required to make the beam uniform with the above dimensions has been calculated through the beam intensity analysis and the results are shown in Fig. 9. The beam becomes uniform at the kaleidoscope length l=25mm.

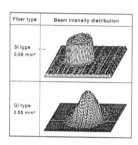

Fig. 8: Measured beam intensity distributions of SI and GI type fiber

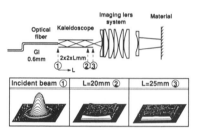

Fig. 9: Beam uniforming states for kaleidoscope length (YAG laser)

4. Evaluation of kaleidoscope performances

On the basis of the analyses shown in the preceding sections the kaleidoscopes for the CO_2 and YAG lasers have been manufactured. The dimensions are 3×3×75mm and 4×2×90mm for the CO_2 laser and 2×2×25mm for the YAG laser. All are made mainly of copper and their inside surfaces are coated with gold (Figs. 10 & 11). The kaleidoscope head for the YAG laser is as compact as 60mm in diameter ×150mm in length, and its imaging magnification can be selected freely between 1 through 5.

Fig. 12 shows the acryl burn patterns of the beams adjusted to the same beam dimensions of 4×4 mm. Fig. 13 shows the acryl burn pattern cross-section and the calculation results through beam tracing for the CO_2 laser. Fig. 14 shows the measurement results of the YAG laser outgoing beam with the beam profiler. The beam intensity obtained through beam tracing agrees with the measured value comparatively well.

Besides the laser power losses in the kaleidoscopes (including lenses) are as low as approximately 15% for the CO_2 laser and approximately 10% for the YAG laser.

Fig. 10: Appearance of CO_2 laser kaleidoscope head

Fig. 11: Appearance of YAG laser kaleidoscope head

Fig. 12: Uniformed beam shapes by kaleidoscope (Acryl pattern)

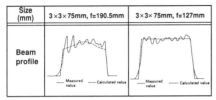

Size (mm)	3×3×75mm, f=190.5mm	3×3×75mm, f=127mm
Beam profile		

Measured value ——— Calculated value   Measured value ——— Calculated value

Fig. 13: Example of uniform beam shape (acryl pattern) and beam intensity analysis results (sectional shapes at beam centers) (CO_2 laser)

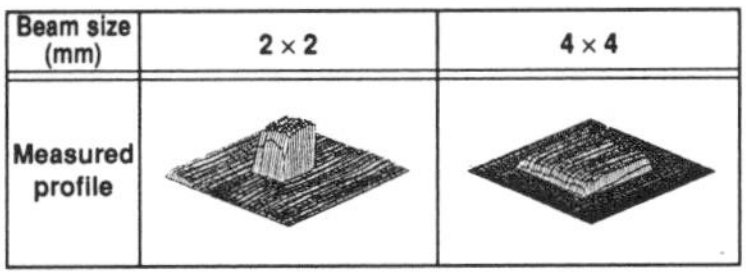

Beam size (mm)	2 × 2	4 × 4
Measured profile		

Fig. 14: Kaleidoscope beam intensity distribution

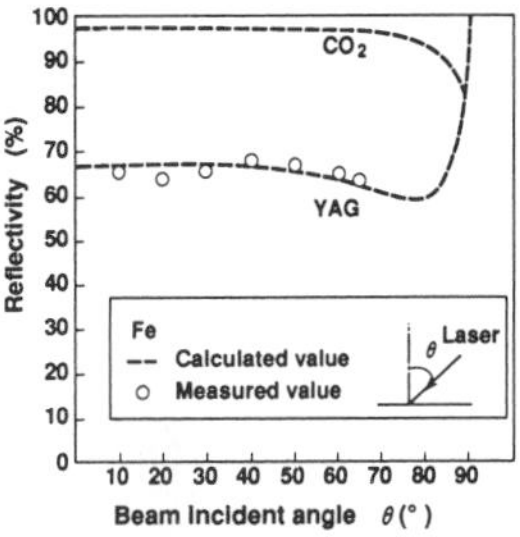

5. Application to surface hardening

5.1 Comparison of beam absorptivities

The relation of the calculated and measured values of the beam incident angle and reflectivity to pure iron are shown in Fig. 15 for the CO_2 and YAG lasers. In addition, Fig. 16 shows the beam irradiation results of the SK-3 material shaped to the same beam dimensions of 4×4mm. In the figure are shown the measured results of the beam absorptivity for the cases with and without the beam absorbent. For the case with the beam absorbent both the CO_2 and YAG lasers secure absorptivities as high as 60 to 70%, though the YAG laser has a little higher value. On the other hand, without the absorbent the CO_2 laser has a constant value of 6% irrespective of the treating speed, but for the YAG laser the absorptivity increases with the decreasing treatment speed and becomes nearly 70%. The hardened depth at the hardened section is shown in Fig. 17. With the figure it is found in hardening without the absorbent that the CO_2 laser has obtained no hardened section but the YAG laser can have the hardened depth nearly to that without the absorbent by treating at a low speed.

Fig. 15: CO_2 and YAG laser beam reflectivity

In order to clarify the cause the temperature and reflectivity of the irradiated section have been measured on the material irradiated by the beam and also the reflectivity and oxidized layer depth after beam irradiation have been measured at room temperature. These results

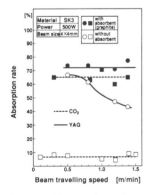

Fig. 16: Beam absorption rate by CO_2 and YAG lasers

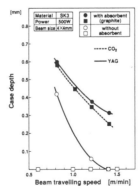

Fig. 17: Hardened depth by CO_2 and YAG lasers

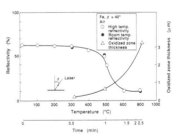

Fig. 18: Effect of the surface oxidation to the beam reflectivity

are shown in Fig. 18. In this figure the reflectivity decreases rapidly when the material temperature is 400°C and higher. However, since the reflectivities of the same samples at room temperature have quite the same tendencies and the oxidized surface layer depth of each sample increases with the decreasing reflectivity, it becomes clear that the reflectivity decrease is caused by surface oxidation due to the material temperature rise.

Thus, for hardening with the YAG laser narrow places can be treated with optical fibers and also this hardening can be applied to places where the beam absorbent difficult to be applied or hardening without the absorbent can be available.

5.2 Application to products

The CO_2 laser hardening with the kaleidoscope has been carried out for a blower whose material is SF55A. Fig. 19 shows the appearance and the hardened state. Fig. 20 shows the cross-sectional macro photograph of the hardened section.

The YAG laser hardening state of a rotating/sliding section with the kaleidoscope coupled with the optical fibers is shown in Fig. 21.

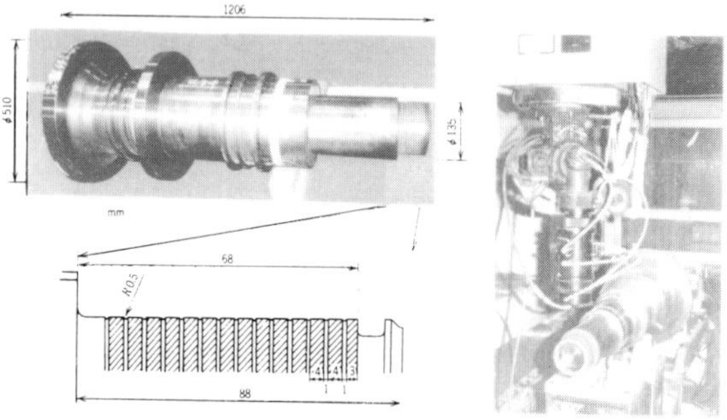

Fig. 19: Real application to fan (CO_2 laser)

6. Conclusion

In CO_2 and YAG laser hardening with the kaleidoscopes the authors have the following conclusions:

(1) By carrying out the beam intensity analysis, the relation of the outgoing beam intensity uniformity to the kaleidoscope dimensions and beam incoming conditions has been clarified.

Fig. 20: Macro sections of laser hardening parts

Fig. 21: Laser hardening condition (YAG laser)

(2) On the basis of the beam intensity analysis results the kaleidoscope heads of 4×2×90mm and 3×3×75mm for the CO_2 laser and the optical-fiber directly-coupled type YAG laser head of 2×2×25mm have been manufactured and it has been confirmed that uniform beams can be obtained with them.

(3) The beam absorptivity in laser hardening without the absorbent has been approximately 6% with the CO_2 laser, but as high as 60% or more for low-speed hardening with the YAG laser. The laser hardening with the kaleidoscope has been applied to a blower practically.

References

(1) S. Shono, T. Ishide, et al.: "Uniforming of Laser Beam Distribution and its Application to Surface Treatment," International Institute of Welding, Doc.IV-450-88, July (1988).
(2) I. Miyamoto, H. Maruo, et al.: "Novel Shaping Optics of CO_2 Laser Beam; LSV Optics-Principles and Application", SPIE Vol.1276, CO_2 Laser Applications II, March (1990).
(3) T. Ishide, O. Matsumoto, et al.: "Optical Fiber Transmission of 2kW CW YAG Laser and Its Practical Application to Welding," SPIE Vol.1277, High-Power Solid State Lasers and Applications, p.188-198. March (1990).

Requirements for Beam Guiding Systems Handling CO_2 High Power Laser Beams

M. Bea, P. Berger, A. Giesen, Ch. Glumann, W. Krepulat, Institut für Strahlwerkzeuge (IFSW), Universität Stuttgart, Pfaffenwaldring 43
D-7000 Stuttgart 80, Germany

Abstract

The quality of laser material treatment processes depends to a large extend on an adequate power density profile on the workpiece. Stable and predictable beam propagation properties are necessary conditions for successful beam shaping.

Optical components (beam guiding and shaping), apertures and the atmosphere the beam propagates through influence the transmitted beam quality during the beam propagation. Therefore the transient behaviour of ZnSe components and of reflecting mirrors has been investigated under the irradiation of high power CO_2 laser beams. The different characteristics are discussed in detail. To compensate the residual thermal induced effects a low cost adaptive copper mirror system has been developed and investigated.

To enlarge the range of possible distances between laser and multiple processing units the use of enhanced coated mirrors and mirror equipped beam expanders inside a multiplexing beam guiding system (MBGS) is discussed. The propagation properties of laser sources with different beam qualities and output power levels up to 5 kW have been measured, both inside the MBGS as well as in special experimental setups (up to 30 m). In particular, the influence of apertures and of pollution on the focusability and on the transported laser beam power is shown. Polluted air inside the MBGS could destroy beam quality and cause degradation of the optical elements. To prevent this risk, the atmosphere in the MBGS should be protected against pollution. Two alternative concepts (special "aerodynamic windows" at the openings and flushing the whole system with clean dry air) have been investigated.

Introduction

Recent investigations (1) show, that an excellent beam quality is a prior condition to achieve better results and higher process efficiency in laser material treatment especially for difficult materials such as aluminium and its alloys. Figure 1 illustrates this showing the welding depth and seam width versus welding speed of AlMgSi 1 for three different high power CO_2 lasers. The laser output power used was 3.6 kW for the lasers with stable resonators and 2.7 kW for the laser with an instable resonator. The power density distribution on the workpiece is quite different: a broad nearly flat profile from the stable resonators with the low beam propagation factor K and the typical power density distribution with a narrow and very high peak power density for the unstable resonator. It is obvious to see that the peak power density determines the efficiency of the process.

Besides the beam quality also the effective f-number (focal length / beam diameter at the focussing element) influences the laser material process. As smaller the effective f-num-

ber as smaller the focus diameter on the workpiece. Figure 2
shows the different seam geometry (seam depth and seam
width) for welding of stainless
steel for different effective f-numbers.

Resuming these effects some
requirements for beam guiding
systems handling CO_2 high
power laser beams can be derived:
In general all the elements of a
beam delivery should be able to
preserve the beam quality independent of the number of the
optical elements and independent of the distance between
lasers and working stations. For
multiplexing beam guiding sys-

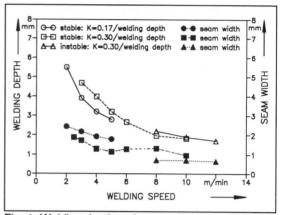

Fig. 1: Welding depth and seam width versus welding speed behaviour of AlMgSi 1 using three lasers diffe-ring in their output power level and their power den-sity profile.

tems (MBGS) which are able to connect various lasers with numerous working stations the beam diameter of all the lasers should be equal at all the working stations.

Influence of Optical Components on the Beam Quality

Since the number of the optical
components used in a beam
guiding system is large the influence of these components on
the laser beam quality was studied in detail.

Any optical component irradiated
by a laser absorbs a certain part
of the energy or power depending on the component and the
coatings, the angle of incidence
as well as the state and the
orientation of the polarization.
Typical absorptance values for
uncoated copper mirrors are

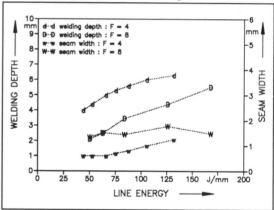

Fig. 2: Welding depth and seam width versus line energy behaviour of stainless steel depending on the realized effective f-number.

between 0.5 % and 1.0 %, depending on the angle of incidence and the state of the polarization. Copper mirrors which are provided with an enhanced reflecting coating show an absorptance of about 0.1 % up to 0.35 %. Transmissive optical elements such as ZnSe windows (both sides AR coated) absorb about 0.15 % up to 0.2 % (2).

The yielded increase in temperature will result in expansion and mechanical distortion and in the case of transmissive components in a change of the index of refraction, as well. This will change the optical path length through the component which is called "optical deformation".

Using the measured values of the absorptance, time dependent calculations of the deformation of the wavefront have been carried out. In a first step, the temperature distribution is calculated and then the mechanical distortion is evaluated. Finally a ray tracing algorithm is used to calculate the optical distortion from the temperature distribution and the mechanical deformation by evaluating the optical path length.

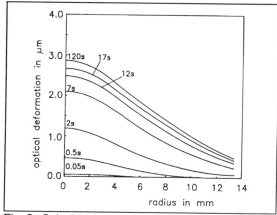

Fig. 3: Calculated optical path length difference vs. radius of a ZnSe window.

Figure 3 shows the calculated optical deformation for a ZnSe AR/AR plano/plano component under an irradiation of 1.5 kW and an average power density of 1.0 kW/cm².

Figure 4 shows a comparison between calculated and measured mechanical deformation of an uncoated copper mirror on the optical axis under an irradiation of 0.8 kW but an average power density of 8.3 kW/cm².

These results show that the phase front of a reflected or transmitted beam will be affected and distorted. A distorted phase front will degrade the beam quality and the focusability.

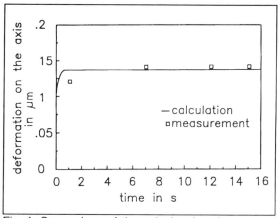

Fig. 4: Comparison of the calculated and measured mechanical deformation on the optical axis of an uncoated Cu mirror.

Comparing the two investigated materials in detail obviously it turns out that firstly AR/AR coated ZnSe components (lenses, windows) affect considerably more the transmitted beam as a copper mirror would do. Secondly, for ZnSe the time constant turns out to be about 10 s meanwhile the deformation of copper has completely finished after less than

1 s. Regarding applications in materials processing, where typical working periods are within 1 to 30 s, Cu mirrors can be regarded as static whereas ZnSe shows its largest dynamic change within this time (2,3).

Design of a Multiplexing Beam Guiding System (MBGS)

Figure 5 shows schematically the concept of a MBGS such as realized at the IFSW (3,4). In its final state it will connect 4 high power CO_2 lasers (1 kW - 5 kW) with 8 working stations and each laser can be guided to each station.

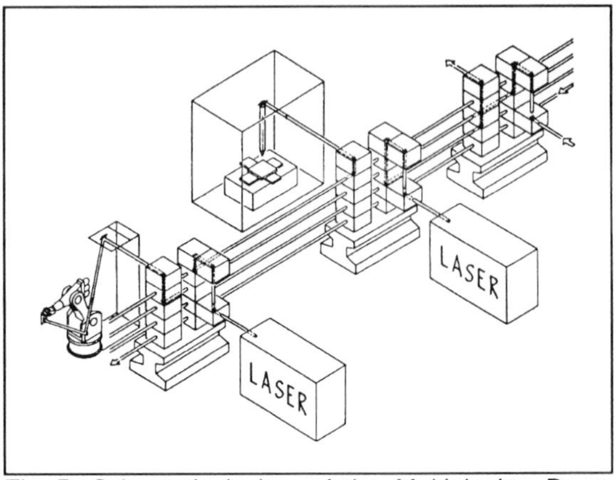

Special beam expanders were designed for transforming all the laser beams to the desired beam diameters. Choosing an anastigmatic design spherical mirrors with good optical properties can be used. These beam expanders are capable of maintaining the laser beam diameter nearly constant at 35 mm within a defined distance without any negative influence on the beams.

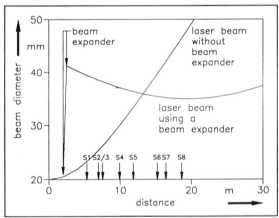

Fig. 5: Schematical view of the Multiplexing Beam Guiding System (MBGS) at the IFSW.

Figure 6 shows the beam diameter of one of the 5 kW lasers as function of the distance from the laser. The locations of the working stations are marked. For comparison the free propagation of the same laser beam is shown demonstrating the strong influence of the beam expander used.

The installation of the MBGS consists of a modular system of standardized boxes and tubes connecting lasers and processing stations. This self-contained design ensures the safe use in any kind of factory. Thus the highest safety standard is guaranteed.

Fig. 6: Measured laser beam diameter as function of the distance from the laser.

The boxes contain either fixed mirrors, moving mirrors or beam expanders. These stacked boxes (Fig. 5) can be exchanged arbitrarily, are reuseable and allow to change or to extend the design of the system at any time to an optimized layout.

The mirrors used in the MBGS are optimized with respect to a minimum surface deformation and provided with a special enhanced coating with an absorption of less than 0.2 % so that the total losses remain below 2 % even after 10 deflections. The beam pointing stability of the total system is much better than 0.1 mrad measured after 2000 movements of the beam deflectors.

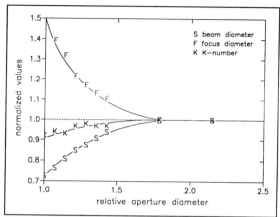

Fig. 7: Influence of apertures on the beam quality (K), the focus diameter (F) and the remaining beam diameter (S).

For preserving the laser beam quality all the apertures of the system have to be large enough. Figure 7 demonstrates for a low power laser with excellent beam quality (K ≈ 1) the influence of an aperture on the beam quality (beam propagation factor K), focus diameter and beam diameter. For a given beam width a TEM_{00} mode remains unaffected as long as the diameter of the smallest aperture will be larger than 1.8 times the beam diameter.

Aerodynamic Flows as Protection against Pollution

To protect the atmosphere inside the MBGS against a polluted environment material windows such as to seal laser resonators could be used. As discussed this would be a disadvantageous solution due to the limitation of power density, the thermal deformation and degradation of the windows. Therefore two technical protection concepts basing on aerodynamical flows were developed and investigated experimentally (5).

The first concept - called air curtain - consists of a nozzle providing a free jet (velocity: 20 m/s) which flows close to any unavoidable opening (laser - MBGS, MBGS - station) and seals the MBGS. The other concept - called open end - is simply realized by scavenging the MBGS, the outflowing air (velocity: 2 m/s) prevents pollution. For reducing the gas consumption all the unused openings are shut automatically. Both concepts use dry clean air (air blower). Flow visualization and tracer gas showed a perfect sealing effect. Optical measurements (HeNe interferometry) did not show any influence on the transmitted beam quality.

Low Cost Adaptive Optical Element

In many cases where transmissive elements have to be used the degradated laser beam can be "repaired" during its further propagation to the workpiece. Regarding this the IFSW developed a low cost adaptive mirror system based on a conventional water cooled

68

copper mirror (6). Using the pressure of the coolant (water) to deform the mirror plate a changing surface shape can be produced depending on the chosen pressure. Figure 8 shows a comparison between a measured surface shape of the adaptive mirror and a measured optical deformation of a ZnSe window under irradiation. The widespread range of complete conformity of the two curves indicates the possibility to compensate thermal induced effects with a satisfying efficiency.

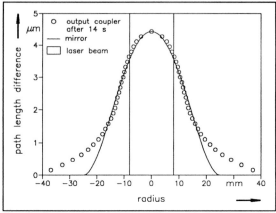

Fig. 8: Comparison between the optical deformation of a ZnSe window under irradiation and a choosable surface of the adaptive mirror.

References

(1)   J.Rapp, C. Glumann, F. Dausinger, H. Hügel: "Influence of Processing Parameters and Alloy Composition on the Weld Seam in Laser Welding of Aluminium Alloys", ECLAT (1992).

(2)   S. Borik, A. Giesen: "Finite Element Analysis of the Transient Behaviour of Optical Components under Irradiation", SPIE (1991) 420.

(3)   A. Giesen: "Optics and Beam Delivery for High Power Lasers for Materials Processing" LAMP'92 (1992).

(4)   U. Zoske, M. Bea, D. Fritz, A. Giesen: "A Beam Guiding System Combining Several Lasers with Various Working Stations", ICALEO (1991) 52.

(5)   W. Krepulat, P. Berger, H. Hügel: "Aerodynamisches Fenster zum Schutz optischer Komponenten", interim report: EUROLASER (EU 194) BMFT-No.: 13 EU 00620 (1991) 2.

(6)   M. Bea, S. Borik, A. Giesen, U. Zoske: "Transient Behaviour of Optical Compnents and Their Correction by Adaptiv Optical Elements", ICALEO (1991) 83.

Acknowledgements

This report is partially based on projects supported by the Bundesministerium für Forschung und Technologie (BMFT) of the Federal Republic of Germany which are recorded under code number 13 EU 00620 and 13 N 5809 respectively.

II.
Cutting and Welding

Laser Material Processing: Examples From The USA

P. E. Denney, Applied Research Laboratory-The Pennsylvania State University, State College, PA., USA

Introduction

Utilization of lasers by industries in the United States has been primarily limited to low power systems accomplishing high volume production (1-5). This is reflected in the use of lasers to fabricate window spacers, juice cans, and razor blades. For these and other products, lasers accomplish the tasks which would not be possible by conventional welding at very high production rates.

Recent developments in laser technology, processing capabilities, and new approaches to production, have resulted in new applications for lasers. Many of these processes are "niche" oriented processes that , because of its specific characteristics, can only be accomplished using laser technology.

This paper is a review of a number of past and present programs that are being conducted at the Applied Research Laboratory The Pennsylvania State University (ARL Penn State) for the U.S. Navy's Manufacturing Technology (MANTECH) office and other government and private sponsors. For the U. S. Navy and U.S. industries in general, the change in global politics, economics, and budgetary constraints has increased emphasis on decreasing the total life-cycle cost of existing or new systems. To this end, most of these laser materials processing programs are designed to develop and demonstrate laser processes for specific components or applications. This paper will cover a number of these programs dealing primarily with welding, cladding, cutting, and innovations in laser systems and delivery techniques that meet program goals.

Laser Welding

To decrease the weight of combat structures, the U.S. Navy is investigating the use of corrugated core, lightweight structures which have been laser welded (LASCOR)(Fig. 1), these structures have the load bearing capabilities of structures weighing nearly twice as much. A number of programs are under way to develop, fabricate, and test LASCOR panels with the results being tabulated into a design manual for new construction or retrofit or for use during overhaul. In addition to the weight savings, the material has excellent thermal properties (Fig. 2). Possible applications include decks, bulkheads, large doors, and ramps. LASCOR is also being investigated for possible application in double hull structures.

Another possible application for the laser in the fabrication of structural members is in the welding of innovative structural elements for ships. One example is the laser welding of high strength, thin section "Tee" stiffeners. The use of thin (< 10 mm (0.39 in)), high strength (>550 MPa (80 ksi) YS) materials, such as HSLA steels, results in a reduction of weight, which in combatants greatly influences operating costs and performance. Normally these structural elements can be hot rolled from lower strength steel alloys but for HSLA alloys, the members must be welded.

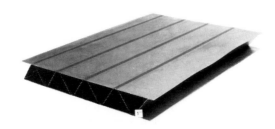

Fig. 1: Example of LASCOR panel. Approximate size; 690 mm X 760 mm (24 in X 30 in). Panel was fabricated from an High Strength Low Alloy (HSLA) steel alloy with sheet material thicknesses of 2.0 mm (0.080 in) for the face and 1.1 mm (0.045 in) for the core.

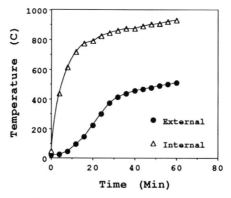

Fig. 2: Results of standard ASTM E119 fire box test for LASCOR panel.

Normal welding techniques such as Gas Metal Arc (GMA) or Submerged
Metal Arc (SMA) could be used but in the thickness of materials
required (3 mm-7mm)(0.12-0.26 inches) the resulting distortion
would not be acceptable. Laser welding has been used to produce
production runs of the "lightweight tees" (over 5,181 meters
(17,000 feet))(Fig. 3). Likewise, welding of 11 mm thick HSLA
crucifix structures that would be used in the construction of
double hull ships is possible by GMA or SMA, however severe
distortion would result. These materials were laser welded in four
passes with no measurable distortion using filler material to
achieve acceptable fillet configuration and size.

Fig. 3: Example of laser welded HSLA "Tee"
section. Web thickness 3mm (0.118 in) and flange
thickness 5 mm (0.197 in).

The use of the lasers in the modification and refurbishment of
used components has been demonstrated. A component in the steam
catapult launch system for U.S. Navy aircraft carriers was
previously a two part structure that was bolted together.
Corrosion and alignment of the part was costly maintenance item.
A solution was the joining of the track, a high alloy steel, to
the support structure by laser. This was possible without
distortion of the components. The depth of the required weld
(32 mm (1.25 inches)) was greater than could be accomplished
with a single pass with the available laser. Also, to meet the
mechanical properties for the weldment, the chemistry of the
weld had to be altered. To meet both of these requirements, a
multi-pass, filler added laser welding technique was developed
and qualified. The procedure is now being implemented as a
production practice.

Laser Cladding

The use of laser cladding technology for repair or to improve
the corrosion and/or wear resistance of a structure is a major
area of research at ARL Penn State. Other processes are capable
of depositing material at higher rates and lower cost during the
fabrication of Navy components but most conventional processes
such as Gas Tungsten Arc (GTA), GMA, and SMA have high heat
input and high dilution rates which limit their applications. A
number of past and present programs at ARL Penn State deal with
the deposition of difficult to clad materials on existing
structures.

One laser cladding program that has been completed, was related
to components of the previously mentioned catapult system.
Process procedures for the deposition of wear and corrosion
resistant materials by laser have been demonstrated, and in some
cases are being used to fabricate or repair actual components.
Most of this work has been accomplished with high powered CO2
lasers (>10 kW) using defocused and oscillated optics (Fig. 4-
5). Components that have been repaired or fabricated and have
been used in actual service include catapult tracks and support
structures. The laser cladding process was also used to
fabricate tracks for a helicopter retrieval system that operates
during adverse weather conditions.

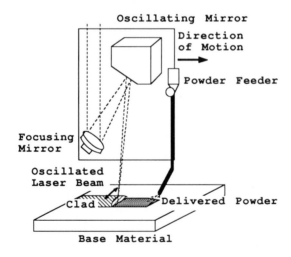

Fig. 4: Diagram of 14 kW CO_2 laser cladding
system that is used at ARL Penn State for process
development.

Fig. 5: Example of laser cladding using 14 kW
laser at ARL Penn State.

Refurbishment of valves and shafts are major problems for U.S.
Navy shipyards during a major overhaul of a combatant. Under
normal operating conditions, valve seats and wear surfaces
experience material loss that may ultimately exceed design and
performance criteria. Replacing the valve or shaft is not
always possible due to time constraints associated with the
overhaul schedule. Therefore the cost for refurbishment by
conventional GTA welding may exceed the cost of a new component.

Penn State has developed procedures that meet the Navy's
requirements for hardfacing and corrosion resistant cladding for
a number of material combinations (Fig. 6). These procedures
have been demonstrated on actual components with some returning
to service (Fig. 7).

Laser Cutting
The cutting of materials by focused laser beams is one of the
most common applications for lasers in industries (6-7).
Presently, laser systems are being used in the U.S. for cutting
everything from automotive body parts to plastic components. A
number of programs have been conducted to examine innovative
ways to use the laser for cutting advanced or unique materials.

A number of programs have been conducted at ARL Penn State in
the area of laser cutting of advanced materials especially metal
matrix composites. Aluminum (Al-Al$_2$O$_3$, Al-SiC, Al-Graphite),
titanium (Ti-Alminides), and ceramics have successfully been cut
with low power CO2 lasers (<1.5 kW). Polymeric composites have
also been successfully cut as well.

Fig. 6: Top; cross section of Co-W laser clad
(dark region) on austenitic stainless steel. Clad
width is approximately 32 mm (1.25 in) and was
clad at 11 kW and 2.54 mm/sec (6 ipm). Bottom;
microhardness profile for same laser clad.

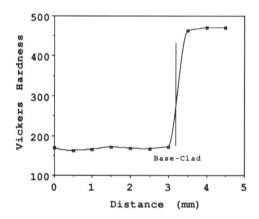

Fig. 7: Example of Co-W laser cladding on actual
steam valve. Cladding was accomplished with no
pre-heat and without removing previous clad.

Another major program that is being investigated is the use of
lasers for cutting of materials that have contaminated by
hazardous materials. A number of military structures have been
asbestos and PCBs which must be removed prior to flame cutting
the structure for disposal. This removal is a manual operation
which is costly and results in hazardous exposure for the
workers and the production of secondary waste that must be
disposed of properly. Because of the low heat input associated
with the laser cutting process, procedures are being developed
to determine if contaminated materials can be cut automatically
without the production of toxic by-products. Results to date
have been favorable for asbestos and the removal of leaded
paint. A study on the byproducts that are produced during laser
cutting of PCB contaminated materials is to occur during the
summer of 1992.

Laser Innovations
In the maintenance of equipment for the U.S. Navy and some
industries, large components must often be moved to welding
shops to properly be refurbished. This may mean the removal of
other components or the cutting of an access way through
structural components to allow for the removal of the part to be
refurbished. This removal of components or structural
alterations is expensive and time consuming. Laser processing
of these components has been limited to those items that could
be taken to a laser work cell for processing. This is not
significantly more economical when compared to the conventional
method of refurbishment. Portable lasers have been used in the
repair of boilers in nuclear power plants by U.S. and Japanese
companies, but these repairs have been limited to lap type welds
(8).

A major effort is underway at ARL Penn State to develop, demonstrate, and deliver a portable, high power (>1.5 kW) Nd:YAG laser system for use by the U.S. Navy in the refurbishment of "shipboard" components. This is to be accomplished using a containerized laser optic system integrated with uses a 150 meter fiber optic cable to reach a work area. The first goal is to develop cladding methods for valves and hull penetrations. Two new lasers have been developed an (1.8 kW and 2.4 kW), and demonstrated for cladding as well as cutting and welding (Fig. 8). The laser light from these systems have been delivered through a single fiber 150 meters in length with only 12% loss in power. Laser procedures have been developed for a number of materials and processes including welding, cladding (Fig. 9-10), and cutting. The system has been used for cladding of Co-W based hardfacing materials on Ni-Cu alloy shafts (Fig. 11) and on a Mn-Mo cast steel valve component both of which met U.S. Navy standards.

Summary
Innovations in lasers and laser materials processing techniques have resulted in advanced applications which, in the near future, could result in new applications of laser materials processing. These include the use of lasers in the fabrication of thick section structures and lightweight construction materials, cladding for the refurbishment and repair of components, cutting of contaminated and advanced materials, and the use of portable lasers to process materials insitu instead of in a workcell.

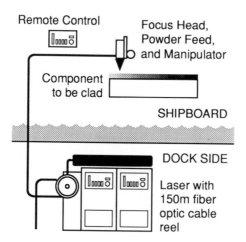

Fig. 8: Diagram of portable 1.8 kW Nd:YAG system for Navy applications.

Fig. 9: Cross section Co-W laser clad made with 1.8 kW Nd:YAG on austenitic stainless steel.

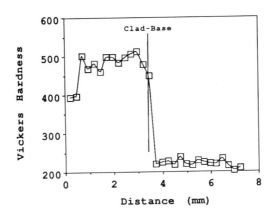

Fig. 10: Microhardness profile for Co-W clad on austenitic stainless steel produced by 1.8 Nd:YAG laser and shown in cross section in Fig. 9.

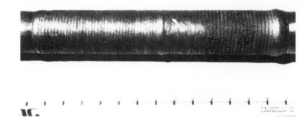

Fig. 11: Example of Co-W hardfacing clad on Ni-Cu alloy shaft. Clad was made using 1.8 kW Nd:YAG laser system.

References
(1) D. Belforte: "ICALEO '89 Proceedings, Volume 69-Laser Material Processing", LIA (1989) 1.
(2) J. A. Miller, J. Chevalier: Welding J. 62 (1983) 50.
(3) D. Belforte: Laser Focus World 26 (1991) 69.
(4) Y. Iwai, N. Okumura, O. Miyata: "ICALEO '89 Proceedings, Volume 69-Laser Material Processing", LIA (1989) Supplemental.
(5) T. Araya, T. Umino: J. Mater. Eng. 13 (1991 299.
(6) D. Belforte: "The Industrial Laser Handbook; 1992-1993 Ed.", (1992) 21.
(7) U. Schreiner-Mohr, F. Dausinger, H. Hugel: "ICALEO '91 Proceedings, Volume 69-Laser Material Processing", LIA (1989) 263.
(8) "Technology Report", ILR 4 (1990) 12.

High Power YAG Laser Welding and Its In-process Monitoring Using Optical Fibers

T. Ishide, Y. Nagura, O. Matsumoto, Takasago Research & Development Center, Mitsubishi Heavy Industries, Ltd., Takasago, Japan
T. Nagashima, A. Yokoyama, Kobe Shipyard & Machinery Works, Mitsubishi Heavy Industries, Ltd., Kobe, Japan

1. Introduction

In-process monitoring of laser welding has been studied in laboratories variously (1-2). However, because of the following problems, any example of its practical use is not seen. That is to say, the relation between welding phenomena and monitored data is not comprehended correctly. The present technology is not so high to make the monitored data reflect on control due to the laser welding speed higher than those of the conventional welding process, and so on.

Then, from the viewpoint of preventing welding defects previously, the authors have studied the system to judge welding conditions through information processing of monitored data, as well as have developed the in-process monitoring method of laser welding. In this paper, we will describe the results of this study.

2. Objects to be welded

For YAG laser welded sleeving of nuclear power plant steam generator tubes (3) the 2kW class YAG laser is loaded on a container installed outside the reactor containment vessel and the beam is transmitted to the steam generator through an optical fiber (hereinafter called a "fiber") of 200m long, as shown in Fig. 1. Since the beam is to be branched to several welding heads, the above fiber is connected to fibers of each 25m long. The welding head, as shown in Fig. 2, is inserted into the damaged heat exchanger tube. Then 3-pass lap welding is carried out from the inside of the previously-inserted sleeve through it as shown in Fig. 3.

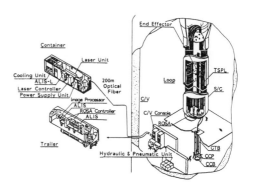

Fig. 1: S/G Laser welding sleeve repair system

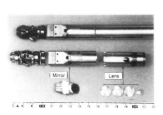

Fig. 2: Laser welding head

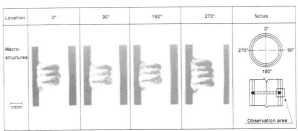

Fig. 3: Cross sectional macrostructures by laser welded sleeving facility

For this welding pulse welding of 650W in laser mean power and 50% in duty is adopted.

Since this welding is to be carried out in a narrow space of about 16.4mm in inside diameter under the radioactive atmosphere and welding penetration should be controlled correctly, it is necessary and essential to monitor welding conditions remotely. The most important detected object is variation of the welding power. Optical parts may be degraded by spatter, fume, etc. and this becomes the main cause to decrease the welding power. Then, we have studied monitoring of welding power variation mainly.

3. Monitoring by sounds

The expert laser welding worker judges the quality of welding inside the sleeve by means of sounds generated during welding. Therefore, we carried out welding with the fiber-transmitted beam and detected the welding sounds with a microphone for the proper laser power and decreased one, as shown in Fig. 4. After frequency analysis of these data, we have obtained a prospect to judge the welding quality by keeping our eyes upon the specified frequency.

Then, we monitored the welding sounds in a long tube (10m long) practically. However, since the welding sounds were absorbed by a rubber fixing function of the welding tool and disturbance noises also existed, correct judgement was difficult to attain.

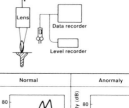

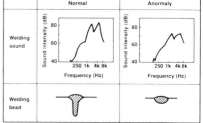

Fig. 4: Monitoring by welding sounds

4. Monitoring by light

4.1 Observation of welding phenomenon

For the purpose to study monitoring by means of light we observed radiation conditions of welding sections with a high-speed camera. The results are shown in Fig. 5. At the pulse-on time in the beam waveform laser plume of very high radiation intensity was formed, and at the pulse-off time radiation from the weld pool was recognized clearly. In addition, the laser plume extinguishing time agreed well with the rising-up time of the beam pulse waveform.

From the above observation and the fact that the laser plume is easily affected by the assist gas

Fig. 5: Laser plume formation

flown to protect the optical system, we intended to study mainly detection of radiation from the weld pool, which is considered to have the most direct relation to penetration.

4.2 Selection of monitoring wavelength

Fig. 6 shows the results measured with a photodiode for the radiation intensity of the welding section at the pulse-off time. Any correlation is not recognized between the laser power increase and radiation intensity. We consider one of the causes is that, in spite of the increase of the average radiation intensity of the welding section with the increasing welding power, the spectral sensitivity characteristics of the sensor used, $I_s(\lambda)$, was not constant for the wavelength, and so sufficient correlation was not obtained between the radiation intensity and laser power.

Therefore, on the basis of the Planck's law of radiation, we studied to have a good correlation between the radiation intensity of the welding section and welding power even with any sensor by paying attention to the specified wavelength, that is to say, to obtain such wavelength that can be used to detect the average temperature change of the welding section most sensitively.

The filter characteristics for the wavelength is to be $I_f(\lambda)$ and spectral sensitivity of the sensor $I_s(\lambda)$. The appropriate monitoring wavelength for this case was studied with the following equations.

$$E = \int_{T_1}^{T_2}\int_\lambda I_s(\lambda)I_f(\lambda)\varepsilon(\lambda,T)E(\lambda,T)d\lambda dT \quad \text{Equ. 1}$$

$$E(\lambda,T) = C_1\lambda^{-5}\{\exp(C_2/\lambda T)-1\}^{-1} \quad \text{Equ. 2}$$

Equ. 2 shows the Planck's law of radiation. In Equ. 1, T_1 and T_2 are the temperatures of the monitored object, λ is the wavelength and ε is the emissivity. Fig. 7 shows the calculation results with Equ. 1 for 1000 to 1600°C in average temperature of the monitored region. From these results monitoring sensitive to temperature change, namely welding power change, proves to be attainable by paying attention to the wavelength of λ=0.9 to 1.0 micron.

4.3 Study of monitoring area

To detect the welding section radiation, the technique shown in Fig. 8 is simple. This uses the power transmission fiber directly and so makes the equipment compact. In this method the light from the area of about 0.6mm in diameter around the optical axis comes in and is led to the sensor, as shown by the spot diagram in Fig. 9. Fig. 10 shows the detection results of radiation from the welding section with attention paid to the appropriate wavelength. The results show that the radiation intensity will not change even with welding power changed and that monitoring sensitive to welding power change cannot be expected.

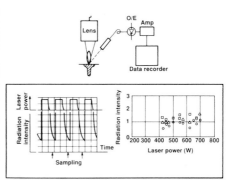

Fig. 6: Monitoring the radiation from the welding part

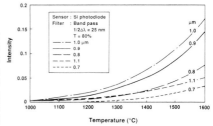

Fig. 7: Monitoring intensity and wavelength against the temperature of welding part (Calculation)

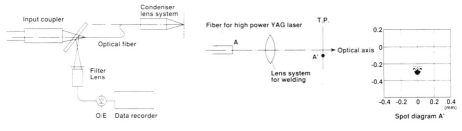

Fig. 8: Monitoring method by the fiber for high power transmission

Fig. 9: Monitoring area by the fiber for high power transmission

Then, we arranged seventeen small-diameter fibers (monitoring fibers) of 0.2mm in core diameter around the beam transmitting fiber to detect radiation from a wide area in the beam irradiated point (Fig. 11). Fig. 12 shows the spot diagram of the light from the end faces of the monitoring fibers. With them it is clear that the radiation intensity is detected at a large area of 2.4mm in diameter around the optical axis.

With this equipment detection was carried out at various wavelengths and their correlation to the welding power was obtained. The results are shown in Fig. 13. The higher the welding power is, the higher the radiation intensity is, and the slope at λ=0.9µm is especially large. Therefore, it is found that monitoring sensitive to welding power change can be attained in this wavelength. In addition, it is confirmed that the appropriate monitoring wavelength studied in Fig. 7 is proper roughly.

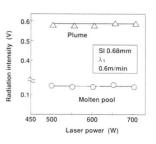

Fig. 10: Welding power vs. radiation intensity

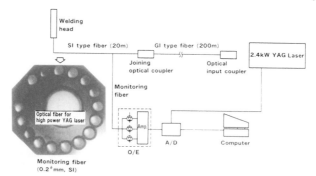

Fig. 11: Monitoring system of welding condition

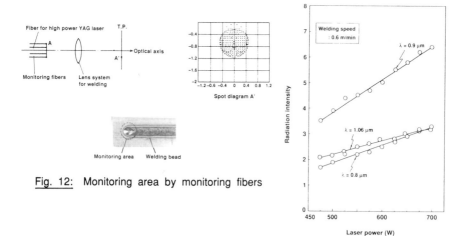

Fig. 12: Monitoring area by monitoring fibers

Fig. 13: Welding power vs. radiation intensity

5. Monitoring results

Fig. 14 shows timing in practical-monitoring. At the pulse-on time the reflected beam was monitored and at the pulse-off time the radiation of the welding section was monitored. Fig. 15 shows the monitoring results during 3-pass welding carried out in a tube. During this monitoring the laser power (Laser), radiation of welding section (λ_1 & λ_2), laser reflected beam (λ_3), motor running speed (Speed), and load ratio of motor (Duty) were also detected and shown in the figure.

With them it is clear that the more the pass number is, the higher the temperature is and the stable the welding is.

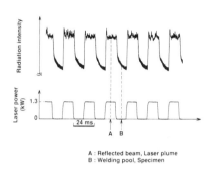

A : Reflected beam, Laser plume
B : Welding pool, Specimen

Fig. 14: Monitoring timing

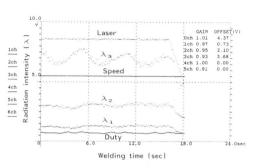

Welding time (sec)

Fig. 15: In-process monitoring data of laser sleeving (Normal)

6. Intelligent monitoring of welding

Some degree of skill is necessary to estimate the cause with the above-mentioned monitoring results in case of any welding anomaly as well as to judge the quality of welding. In order to save this labor, we have studied intelligent monitoring system of the monitoring data with a computer.

The monitoring objects and anomalies are summarized in Fig. 16. Processing of the monitoring data is divided into judgement of welding to normal/anomaly through multi-variate analysis, and in case of anomaly, its discrimination through the neutral network.

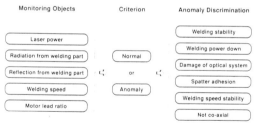

Fig. 16: Monitoring objects and judgement of welded condition

Fig. 17 shows the processing flow chart. The monitoring data are read as sets of about 200 data per channel and compressed into 38 data through mean value processing. These data are compared with the data previously-registered as normal welding data. When the data are in normal distribution, they are processed through "judgement with Makaranovis distance" to judge the welding quality. If not, they are processed through "standard value comparison judgement." From now, we will judge welding to normal/anomaly.

The neural network system is used for anomaly judgement. The information of 5 channels together with other welding information is divided into 29 discriminating sections shown in Fig. 18, and compared with 6 kinds of anomaly standard patterns. This pattern comparison utilizes the neural network shown in Fig. 18. For the quantity of these neurons, 29 neurons in the input layer are provided the same as those of the discriminating sections, 6 neurons in the output layer the same as those of anomalies, and 5 neurons in the middle layer. The error back propagation method is used in learning anomaly standards.

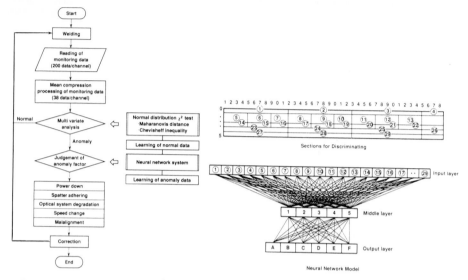

Fig. 17: Flow chart of intelligent monitoring system

Fig. 18: Neural network model for anomaly discrimination

6. Conclusion

In the YAG laser sleeve welding technology for nuclear power plant steam generator tubes we have put to practical use the remote in-process monitoring of welding in narrow spaces and its intelligent monitoring system by monitoring with attention paid to the specified wavelength, utilizing monitoring fibers arranged around the power transmitting fiber.

References

(1) L. Li, et al.: "On-line Laser Weld Sensing for Quality Control," ICALEO (1990), Proceedings, p.411~421.
(2) D. D. Voelkel, et al.: "Visualization and Dimensional Measurement of the Laser Weld Pool," ICALEO (1990), Proceedings, p.422~429.
(3) T. Ishide, et al.: "Optical Fiber Transmission of 2kW CW YAG Laser and Its Practical Application to Welding," SPIE Vol.1277, p.188~198 (1990).

Laser Beam Cutting with Solid State Lasers and Industrial Robots

M. Geiger, A. Gropp
Forschungsverbund Lasertechnologie Erlangen - Lehrstuhl für
Fertigungstechnologie, Universität Erlangen-Nürnberg, Germany

1. Introduction

Nowadays three-dimensional laser beam processing is at the border of industrial use, about some applications is reported elsewhere [1, 2]. But to overcome such questions like process control or system technology and to gain a broad knowledge about the process itself, especially for cutting with solid-state-lasers, many research institutes and R&D-departements of industrial companies are working also in this area of laser material processing.

Because of the main advantages like low investment costs and easy beam guidance in comparison to the combination CO_2-laser/-gantry robot [3] and the development of beam powers of cw and pulsed Nd:YAG-Lasers in the kW-range in recent times the combination Nd:YAG-Laser and industrial robot is expected to have a high growth in the very next future. Therefore here is reported about some experiences gained with such a system in two- and three-dimensional laser beam cutting of mild steel.

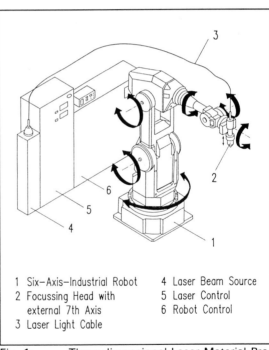

1 Six–Axis–Industrial Robot
2 Focussing Head with external 7th Axis
3 Laser Light Cable
4 Laser Beam Source
5 Laser Control
6 Robot Control

Fig. 1: Three-dimensional Laser Material Processing with a six-axis-industrial robot with revolute coordinates and a pulsed Nd:YAG-laser

2. Experimental Setup

The investigations have been carried out with a six-axis industrial robot (IR), that has a horizontal reach of about 1800 mm, a maximum path speed of 1500 mm/s, a maximum load capacity of 30 kg and a producer-specified repetition accuracy of less than ±0.2 mm. It is combined with a pulsed Nd:YAG-laser with a typical mean power of about 150 W. Beam guidance is done by a step index fibre with a core diameter of 0.6 mm. The distance nozzle-workpiece and with it the focal position is controlled by a capacitive sensor, integrated as an external 7th axis (**fig. 1**).

3. Diagnostics

The deviations of the robot from a straight line, given by a pilote HeNe-laser beam, have been measured with the help of a position sensitive detector (PSD), when moving along this line. This signal is presented in **fig. 2** for 3 different speeds which are about the expected cutting speeds for the available laser system. The deviations are all along the investigated distance at most about 0.2 mm. The deviations with a higher frequency,

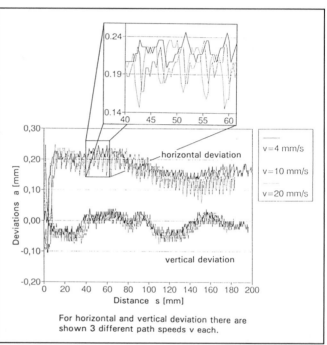

For horizontal and vertical deviation there are shown 3 different path speeds v each.

Fig. 2: Guiding behaviour of a six-axis-industrial robot: Deviation from a main axis (here: x-axis) in there horizontal and vertical direction

which could be put down to the position control loop of the robot, are less than 0.1 mm, independent of the regarded path speed. Vertical deviations could be balanced by using a control system for the distance between nozzle and workpiece, unlike horizontal deviations. These are expected to influence directly shape and measures of the workpiece. The presented deviations however are only valid for movements where there is no reorientation of the focussing head necessary. Cutting experiments, where a rotation of the focussing head is required to keep the beam axis perpendicular to the workpiece's surface, e.g. cutting over an edge with a narrow bending radius, have shown, that the deviations from the programmed path get higher. Additionally the maximum path speed is limited according to the radius of the path and the location of the rotation axis in the work room of the robot [3].

4. Technological Results

In **fig. 3** surface profiles of laser cut edges with different sheet thickness and therefore different cutting speeds are presented, measured according the German Standard DIN 2310 [4]. With decreasing sheet thickness and therefore increasing cutting speed there is only a slight change in the cut surface. The peak to valley difference is always in the range of 50 to 80 microns and therefore smaller than the mentioned deviations of the IR. The results show

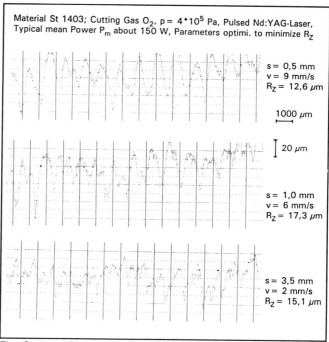

Material St 1403; Cutting Gas O_2, p = $4*10^5$ Pa, Pulsed Nd:YAG-Laser, Typical mean Power P_m about 150 W, Parameters optimi. to minimize R_Z

s = 0,5 mm
v = 9 mm/s
R_Z = 12,6 µm

1000 µm

20 µm

s = 1,0 mm
v = 6 mm/s
R_Z = 17,3 µm

s = 3,5 mm
v = 2 mm/s
R_Z = 15,1 µm

Fig. 3: Surface Profiles of different Sheet Thicknesses s when cutting Mild Steel with a pulsed Nd:YAG-Laser (cutting speed v) [measuring length 15 mm; measuring speed 0.5 mm/s]

that the robot and not the laser process is responsible for the deviations. A correlation between surface peak frequency and pulse frequency of the laser could not be found.

Beam guiding for Nd:YAG-lasers is usually done by glass fibres, which will have some influence on the beam: The intensity distribution is homogenized, the focussability gets worse and there are transmission losses of about 10 % [5]. With the given fibre and the focussing head with its optical system a minimum spot size of 0.49 mm should be theoretically possible. The real width of the cut kerf in dependence of the focal position is shown in **fig. 4** at the beam entry - (top) and exit side and for the middle of the sheet. The beam entry side of the kerf is rather wide and of nearly the same width all over the exa- mined range with a slight minimum if the focal point is on the work- piece's sur- face. The theoretical spot size is only reached at the beam exit side, the kerf is V-sha- ped. But this means, that much more of the working gas is incou- pled into the kerf, which therefore acts like a nozzle,

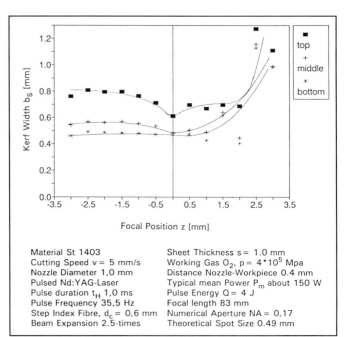

Material St 1403   Sheet Thickness s = 1.0 mm
Cutting Speed v = 5 mm/s   Working Gas O_2, p = $4 \cdot 10^5$ Mpa
Nozzle Diameter 1,0 mm   Distance Nozzle-Workpiece 0.4 mm
Pulsed Nd:YAG-Laser   Typical mean Power P_m about 150 W
Pulse duration t_H 1,0 ms   Pulse Energy Q = 4 J
Pulse Frequency 35,5 Hz   Focal length 83 mm
Step Index Fibre, d_c = 0,6 mm   Numerical Aperture NA = 0,17
Beam Expansion 2.5-times   Theoretical Spot Size 0.49 mm

Fig. 4:   Kerf Width in dependence of Focal Position

thus impeding the adherence of dross. The kerf formation doesn't seem to be influenced by the focal position significantly. Only for focal positions relatively high above the surface there is a clear increase in kerf width and a tendency to parallel kerf walls.

The variation of the cutting speed when working with different sheet thicknesses is shown in **fig 5**. The maximum cutting speed v_{max} is thereby defined as the highest speed where the separation is complete, regardless to cut quality. The second speed of the diagram $v(R_{Z, min})$ is the "high-quality-speed" where the roughness of the cut surface is minimal. With increasing sheet thickness the cutting speed decreases, as in all thermal cutting methods. The cut surface roughness at optimized parameters is thereby rather constant in the examined thickness range.

Out of the experience got by systematic experiments in three-dimensional laser cutting a proposal for a model part for this application was developed, **fig. 6**. A semi-finished product was chosen as the workpiece to enable changing of material and sca- ling very easily. The tube is cut in three different positions, each parallel to one of the main axis of the robots's system of world-coordinates. Each position is marked by the cut letter of the corresponding main axis, to find out if there are preference positions or not. With the cut contours it is possible to judge all important features in laser cutting, for example cut surface quality, corner - and kerf formation, and the guiding behaviour of the IR. Cutting a programmed plane circle in a curved surface will show the function of a distance control and give a chance to measure circle irregularity

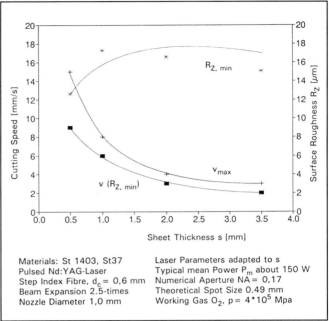

Materials: St 1403, St37
Pulsed Nd:YAG-Laser
Step Index Fibre, $d_c = 0,6$ mm
Beam Expansion 2.5-times
Nozzle Diameter 1,0 mm

Laser Parameters adapted to s
Typical mean Power P_m about 150 W
Numerical Aperture NA = 0,17
Theoretical Spot Size 0.49 mm
Working Gas O_2, $p = 4*10^5$ Mpa

Fig. 5: Laser beam cutting of mild steel with a pulsed solid state laser: cutting speed and cut surface quality

and to rate a inclined cut that is very common for some applications, e.g. when cutting over a crimp.

5. Summary and Outlook

The combination of a Nd:YAG-laser with an industrial robot is a powerful tool that enables a lot of applications especially in three-dimensional laser material processing. The accuracy of the parts is determined by the guiding behaviour of the robot, in particular at complex tasks of motion requesting a reorientation of the focussing head. Cutting speed, the cutable sheet thickness and also the cut surface quality are quite good with regard to the relatively low mean laser power. Some new investigations that are dealing with an optimization of the beam guiding system, aiming at a smaller spot size, are expected to effect an increase in cutting speed and quality. Systematic experiments of cutting contours resulted in a proposal for a model part for three-dimensional laser cutting.

6. Acknowledgements

The research project, which is the basis of the here reported investigations, has been supported by the German Ministry of Re-

92

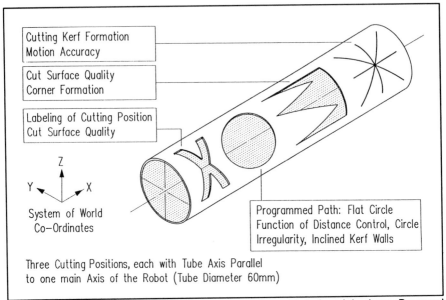

Cutting Kerf Formation
Motion Accuracy

Cut Surface Quality
Corner Formation

Labeling of Cutting Position
Cut Surface Quality

Z
Y ← ↑ → X

System of World
Co-Ordinates

Programmed Path: Flat Circle
Function of Distance Control, Circle
Irregularity, Inclined Kerf Walls

Three Cutting Positions, each with Tube Axis Parallel
to one main Axis of the Robot (Tube Diameter 60mm)

<u>Fig. 6:</u> Three-dimensional laser beam cutting with industrial robots: Proposal
for a model part

search and Technology under the registration number 13 N 5777
and several industrial companies. The responsibility for the
content of this paper is taken by the authors.

7. Literature

[1] Geiger, M.; Hoffmann P.: Laser Applications in the Automo-
tive Industry. In: Laser Applications in the Automotive
Industries. Proceedings of the 25th ISATA. Florence 1992.
ISBN 0 947719 48 2, pp. 69-83

[2] Wollrab, P.M.; Billinger, A.: Laser-Roboter im Praxis-Test.
In: Roboter, August 1990, Heft 4, pp. 22-26

[3] Geiger, M.; Gropp, A.; Hoffmann, P.: Technological Comparison of CO_2- and Nd:YAG-Laser Manufacturing Systems. In:
Laser Applications in the Automotive Industries. Proceedings of the 25th ISATA. Florence 1992. ISBN 0 947719 48 2,
pp. 511-518

[4] DIN 2310: Thermal Cutting, part 5: Laser Beam Cutting of
metallic materials, principles of process, quality, dimensional tolerances. Berlin: Beuth, 1990

[5] Homburg, A.; Meyer, C.: Strahlanalyse eines kW-Festkörper-
lasers mit Lichtleitfaser. In: Laser Magazin 1989, Nr. 4,
pp. 19-25

Cutting of Hard Magnetic Materials and Aluminum Alloys by High Power Solid-State Laser

G. Spur, J. Betz, Institut für Werkzeugmaschinen und Fertigungstechnik der Technischen Universität Berlin, Berlin, FRG

Introduction

In recent years, laser beam cutting of steel and aluminum sheets has been established as a reliable manufacturing technology. Besides high power CO_2-lasers the neodymium lasers are gaining acceptance in industrial applications. The small wavelength which allows the transmission of the beam to the workpiece by fibre optic devices, a better absorption of the radiation by most metals and the significant smaller size of the laser system favour the Nd:YAG laser over other types.

The paper focusses on technological investigations into laser beam cutting of aluminum alloys and hard magnetic materials with a fibre optical beam transmission.

Experiments

The laser cutting of the specimens were carried out in the Laser Material Processing Laboratory of the Festkörper-Laser Institut, Berlin. A high power solid-state laser of Nd:YAG type was employed: a neodymium laser from Haas, type LAY 600, with a maximum power output of 600 W (cw) in pulsed mode conditions. The resonator consists of two cavities and operates in the oscillation mode. It provides a circular beam with 7.0 mm in diameter. The beam quality is better than 30 mm millirad over the entire range of selected parameters and the intensity distribution is close to a Gaussian mode as verified by the abel inversion method of knife edge data (1).

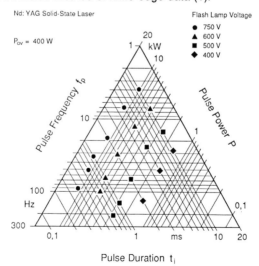

Fig. 1: Working domain of the Nd:YAG solid-state laser for an average laser beam power of 400 W.

The most important laser parameters, the pulse duration, pulse frequency and flash lamp voltage, and can be adjusted between 0.1 and 10 ms, 1 and 1000 Hz and 400 and 750 V, respectively. Different parameter setups which all yield an average power of P_{av} = 400 W are shown in figure 1 (2). The results show a linear dependance of the pulse power on the flash lamp voltage and a maximum pulse power of 12 kW at 750 V. The average laser power of 400 W can be generated with a minimum pulse duration of 0.4 and 2.0 ms for 750 and 400 V, respectively. Based on past engineering experiments on laser beam cutting highest cutting speed can be expected at short pulse duration and high pulse frequency.

The work station consists of a CN-controlled crosstable, a gas nozzle with a gas mixing unit and the optical beam delivery. The beam was focussed to a spot size at the focal point of 0.3 mm by a 100 mm focal length lens. The Rayleigh length of the focussed beam was z_R = 1.0 mm.

The transmission of the laser beam to the work station is realised by a step index light-fibre with a 600 μm diameter and a numerical aperture of NA = 0.18. The beam is guided over a distance of 20 metres. The total loss of laser power due to the transmission and optical systems lies between 18 and 21 %.

The substrate materials investigated in this paper were commercially pure aluminum, Al99,5, the aluminum alloys AlMg3 and AlMgSi0.5 and the hard magnetic compounds NdFeB and SmCo with a material thickness of 0.5 to 3.0 mm and 1.0 to 5.0 mm, respectively. A summary of material properties of the hard magnetic materials are given in table 1.

material	number	$(BH)_{max}$	JH_c	T_c
NdFeB 160/80		160 kJ/m^3	800 kA/m	310 °C
SmCo 112/100	2.4135	112 kJ/m^3	1000 kA/m	727 °C

Table 1: Material properties of the investigated magnetic materials (3).

Both materials were cut in an unmagnetic state but contained a preferable orientation in the direction of the sheet thickness. The surface of the materials demonstrated a grey diffuse appearance and had an average peak-to-valley height R_z = 30 μm.

The framework of the present investigation is concerned with laser beam cutting of two dimensional shaped substrates. Low angle corners, thin bars and small circles are of particular interest in industrial applications.

Results

In figure 2, a schematic sketch of laser beam cutting with micrographs of the machined surface and different cross sections are given. The pulse power density of the incident beam was 8.5x10^6 W/cm^2 and the workpiece was AlMg3 (s = 1.5 mm). Investigations with different focal point positions yielded the maximum cutting speed at z_f = -0.5 mm which was chosen in this experiment. With regard to the cut quality (burr height, surface roughness) a focal point position near the lower side of the workpiece (z_f = -1.5 mm) or above (z_f = 1.0 mm) the surface achieved better results. But in this case, the amount of lateral heating of the workpiece was so high that bending and movement of thin metal sheets occured.

The process gas, that was delivered coaxially to the laser beam, was compressed air with a pressure between 0.3 and 0.4 MPa. At the erosion front, the formation of a 10 μm thick resolidified layer is achieved (Fig. 2a). Its thickness is quite uniform over the entire height. The inclination of the erosion front is 82° relative to the surface. The figure 2b shows the microstructure of the cut surface. From this picture it can be concluded that a mixture of liquid and solid material, plasma and vapor forms within the kerf during the cutting process and resolidification occurs. Even a maximum gas pressure of 1 MPa could not suppress this formation of melt zones and burr. It also comes out, that the melt

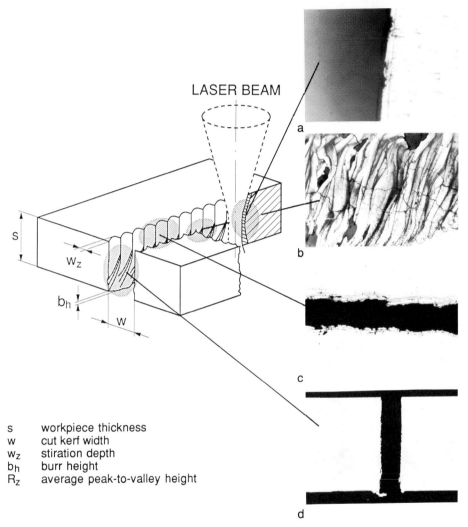

LASER BEAM

a

b

c

s	workpiece thickness
w	cut kerf width
w_z	stiration depth
b_h	burr height
R_z	average peak-to-valley height

d

Fig. 2: Schematic arrangement and micrographs of laser beam cutting.

reacts with the process gas by forming thin Al-oxide layers with pores and a network of cracks. Figures 2c and 2d show a top view and the cross section of the kerf. A sinusoidal profile which can be interpreted as a reproduction of regular pulses can be observed.

96

The width of the kerf is about 0.35 mm and the thickness of the lateral resolidified layer is 0.1 mm at its maximum. The investigations demonstrated an increase of the kerf width with increasing cutting speed. The kerf geometry varied with the laser- and gas parameters, where the gas guidance and pressure turned out to be most important.

Of importance for the productivity of the process is the maximum cutting speed. In rough cutting the cut geometry and surface roughness of the kerf play an inferior role. Therefore the maximum cutting speed can be defined as the feed rate at which the part is separated. In figure 3a the maximum traverse speed for the aluminum alloy AlMg3 as a function of the thickness is given. The experiments were carried out with different process gases.

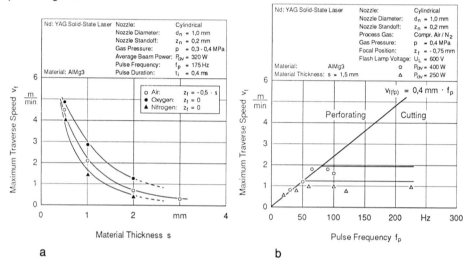

Fig.3: Maximum cutting speed as a function of material thickness (a) and pulse frequency (b).

An estimation of the maximum cutting speed at different laser parameters can be obtained by using the following easy consideration. As the study of cutting experiments demonstrated the pulse energy (Q_i) of the used laser parameters was high enough to penetrate the material thickness in one pulse. Therefore the pulse energy was at least equal to the melt energy (Q_s) for the volume $w^2 \pi/4 \, s$ ($Q_i \geq Q_s$). According to that, the important parameters to achieve maximum cutting speed are the pulse frequency and the average laser beam power or rather the pulse energy. In figure 3b the maximum traverse speed as a function of pulse frequency is shown. The plane is divided by the line $v_{f(fp)} = 0,4$ mm f_p into the two manufacturing processes cutting and drilling. The value 0.4 mm is the experimental average kerf width. In case of cutting along the border line means cutting without an overlap of the single laser pulses. If the laser parameter yield in less pulse energy ($Q_i < Q_s$) the range of cutting (fig.) will be reduced by a line of constant traverse speeds. This line is calculated by multiplying the geometrical maximum cutting speed with the quotient Q_i / Q_s. The whole estimation was carried out under the premise of negligible heat radiation and heat conductivity.

In figure 3b, the experimental results shows a good comparison for cutting of 1.5 mm thick AlMg3 with 250 and 400 W.

The cutting experiments on the hardmagnetic materials showed good results for cutting the compound NdFeB. This material seems to be less sensitive to heat induced stresses. The best results were achieved with an average beam power of 230 W and a pulse duration of 0.3 ms. In figure 4a and b the cut kerf width, the burr height and the surface roughness as a function of cutting speed are shown.

Because the rare earths are chemically very reactive (4), the process gas was argon or nitrogen at relative low pressure of 0.2 to 0.3 MPa. The magnetic properties of the laser cut samples are the most important quality criterion for the use of this cutting technique. As the above reported results show laser cutting will always thermally affect the substrate in the region near to the kerf. To quantify this influence the magnetic properties were measured.

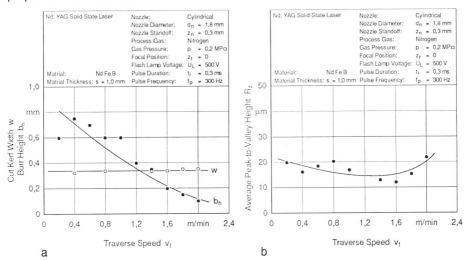

a   b

Fig.4: Cut kerf width, burr height (a) and surface roughness (b) as a function of cutting speed.

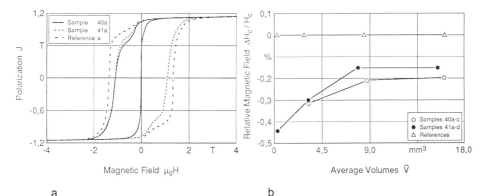

a   b

Fig.5: Room temperature hysterese loops of laser cut samples (a), relative magnetic field as a funktion of specimen volume.

The measurements were taken with a Foner-Vibrationsmagnetometer at the Max-Planck-Institute of Materials Science, Stuttgart. The measured hysterese loops of the laser cutted samples and of one reference sample are given in figure 5a. The measurements were carried out at room temperature until a magnetic induction of ± 5 Tesla, which is high enough to completely magnetize this material. The laser samples were cut with an average beam power of 230 W and a pulse duration of 0.3 ms. The samples 40 and 41 were machined at a pulse power of 1.5 and 3.8 kW and the cutting speed 1.8 m/min and 0.8 m/min, respectively.

To demonstrate the effect of laser cutting on the magnetic properties the relative magnetic field strength $\Delta H_c/H_c$ was calculated and plotted over the volume of the cutted samples (figure 5b). Its shows, that the relative change of $\Delta H_c/H_c$ increase with decreasing specimen volume.

Conclusions

The investigated materials behave different with regard to laser cutting. Aluminum and its alloys are well known to demonstrate a relatively poor absorbtion and also have a high thermal conductivity. The raving formation of thin metal oxide layers at elevated temperatures results in some problems to achieve high quality cuts. The hardmagnetic materials demonstrates a good absorption of the laser light but is very sensitve to thermal induced stresses. Also the highly reactive rare earth elements tend to make the process uncontrollable.

The cutting experiments of aluminum result in a maximum cutting speed of 4.8 m/min for the alloy AlMg3 with a 0.5 mm thickness . The cut kerf varied between 0,3 and 0.5 mm and the surface roughness lies between 20 and 40 μm. The most importand parameters as far as surface quality is concerned are the gas pressure and the cutting speed. The best results of surface roughness, R_z = 10 μm could be achieved with cutting at high pressure (1MPa) and a large nozzle diameter (2.4 mm).

For cutting of the 1 mm thick magnetic material the maximum feed rate was 2.0 m/min. The cut kerf width was about 0.4 mm and the maximum peak-to-valley height has been R_z between 10 and 20 μm. The profiles of the surface roughness via the feed rate shows slight minimum at 60 or 70 % of the maximum cutting speed. The change of the magnetic properties $\Delta H_c/H_c$ were 15 and 21 %.

Support for this project is given by the Federal Ministry for Research and Technology (BMFT) and the companies, Siemens AG (Erlangen) and Preussag AG (Berlin).

References

(1)   R.M O`Connell, R.A. Vogel: Abel inversion of knife-edge data from radially symmetric pulsed laser beams. Applied Optics, Vol. 26, No. 13, 1, July 1987.

(2)   M.H.H. van Dijk: Pulsed Nd:YAG Laser Cutting. The Industrial Laser Annual Handbook 1987, S. 52 - 64.

(3)   N.N.: DIN 17410 (5. 1977) Dauermagnetwerkstoffe. Technische Lieferbedingungen. Beuth Verlag, Berlin 1977.

(4)   G. Schneider, E.-Th. Henig, G. Petzow, H.H. Stadelmaier: Phase Relations in the System Fe-Nd-B. Z. Metallkunde, Bd. 77 (1986) H. 11, S. 755 - 761.

Influence of Processing Parameters and Alloy Composition on the Weld Seam in Laser Welding of Aluminium Alloys

J.Rapp [1,2], C.Glumann [1], F.Dausinger [1], H.Hügel [1]
[1] Institut für Strahlwerkzeuge (IFSW), Universität Stuttgart
Pfaffenwaldring 43, D-7000 Stuttgart 80 (FRG)
[2] Max-Planck-Institut für Metallforschung, Institut für Werkstoffwissenschaft
Seestr. 92, D-7000 Stuttgart 1 (FRG)

1. Introduction

Welding of aluminium alloys with a laser offers distinct advantages over conventional techniques, such as TIG- or MIG-welding. In the keyhole welding mode deep weld seams with narrow heat-affected-zones at high processing speeds can be produced. Main problems associated with laser welding of aluminium alloys in general are the high surface reflectivity and the high thermal conductivity. This means that the coupling of the laser beam into the work piece is impeded and a considerable amount of the input power is lost by heat conduction. Therefore the threshold intensity for the development of a keyhole is much higher for aluminium than for steel. In addition metallurgical defects, such as hot cracking, pores or evaporation loss of volatile alloying elements may affect the laser weldability of the different aluminium alloys. Conclusively the welding process and the resulting weld seam react more sensitive on changing process parameters than in the case of steel.

The purpose of these studies was to determine the influence of the workpiece, i. e. alloy composition and heat conduction, as well as laser beam parameters, i. e. beam quality, power and processing speed, on the development of the weld seam. This contribution summarises some of the main results of these investigations.

2. Experiments

In the first part of the investigation overlap welds on AlMgSi1- and AlMg5Mn-plates, with a single sheet thickness of 1.25 mm, as well as bead-on-plate welds on 2.5 mm thick AlCuMg2-plates were produced. The alloy compositions are listed in tab. 1. The weldings were done with a 1.5 kW CO_2-laser TRUMPF TLF 1500. In the second part of the study, bead-on-plate welds were made on 5 mm thick AlMgSi1-alloy plates with the 5 kW CO_2-lasers TRUMPF TLF 5000 and TLF 5000 TURBO. For the beam parameters of the lasers see tab. 2. Helium at a gas flow rate of 10 to 15 l/min was employed as a plasma-jet. The nozzle with 5 mm in diameter was placed at an angle of approximately 10 to 20 degrees to the plate and perpendicular to the welding direction. The processing speed was variied from 0.8 to 10 m/min. Prior to welding the alloy plates were properly cleaned and degreased. After the metallographic preparation the weld seam shape of cross and longitudinal sectioned specimen were investigated by optical microscopy and correlated to the different material and laser parameters. Specimen sectioned parallel to the plate surface were cut out of the

Element	Mg	Fe	Si	Mn	Ti	Cu
AlMgSi1 (AA 6082)	0.82	0.28	0.92	0.68	0.05	0.01
AlCuMg2 (AA 2024)	1.32	0.22	0.26	0.61	0.05	4.34
AlMg5Mn (AA 5182)	4.55	0.32	0.16	0.12	0.04	0.02

Tab. 1   Composition of the three Al-alloys in wt-%.

AlMgSi1-welds in order to reveal the microstructure in the fusion- and heat-affected-zone (HAZ).

3. Results and Discussion

3.1 Influence of the magnesium-content on the penetration depth and pore formation

Comparing the three aluminium alloys containing variing amounts of magnesium, the penetration depth increases with rising Mg-content (fig. 1). For the same value of the heat input per unit length a magnesium content of 4.55 wt-% in the AlMg5Mn-alloy provided an approximately 30 per cent deeper penetration than for AlMgSi 1 (0.82 wt-%) and 22 per cent deeper than for AlCuMg2 (1.32 wt-%). Sakamoto [1] showed that the threshold intensity of aluminium alloys decreases with increasing content of volatile alloying elements like magnesium and zink. As he found no correlation between the threshold intensity and the thermal properties of the distinct alloys, he suggested that the ease of evaporation of aluminium alloys is a dominant factor governing the threshold intensity.

In laser deep welding the vapor pressure of the evaporating metal sustains the keyhole against surface tension forces and the hydrostatic pressure of the surrounding liquid. Hereby the vapor pressure of the alloy composes of the partial vapor pressures of the alloying elements. Since the partial vapor pressure of

laser	TLF 1500	TLF 5000 TURBO	TLF 5000
K-number	0.7	0.3	0.18
F-number	8	4	4
d_f [µm]	160	200	300
P [kW]	1.1	3.6	3.6

Tab. 2   Beam parameters of the CO_2-lasers (K ... beam-quality-number; F ... focusing-number; r_f ... focus diameter; P ... beam power on the work piece surface).

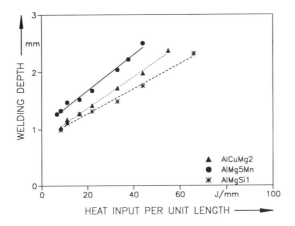

Fig. 1   Welding depth as a function of the heat input per unit length for three
Al-alloys: P = 1.1 kW; K = 0.7; F = 8; 10 l/min He.

magnesium is much higher than that of aluminium, the presence of a small content of
magnesium results in a sufficient vapor pressure to produce a stable cavity at a lower
temperature than for pure aluminium. Consequently this leads to a dependence of the
vaporization temperature on the chemical composition of the alloy [2]. Further
investigations [3] showed that the vaporization temperature of Al-Mg-alloys decreases
with increasing magnesium-content resulting in a considerable rise in penetration.

Summing up, it may be said that in laser deep welding the evaporation of volatile
alloying elements like magnesium plays a dominant role for the development of the
weld seam by influencing the vapor pressure in the keyhole.

Longitudinal cuts revealed cavities with an irregular shape in the weld centerline,
which were be attributed to keyhole instabilities (see ref. [3] and [4]). They clearly
could be distinguished from hydrogen induced pores of spherical shape. No influence
of the magnesium content on the frequency of these cavities have been detected.
However in the investigated parameter range a strong dependence of the number and
size of the cavities and pores on the processing speed could be determined. Even for
the highly Mg-alloyed AlMg5Mn-plates cavity- and pore-free welds could be achieved
beyond a processing speed of about 6 to 8 m/min.

3.2.   Influence of the beam quality and power on the penetration depth

The experimentally observed influence of the beam parameters (see tab. 2) on the
welding depth and the averaged seam width is shown for AlMgSi1-welds in fig. 2 and
fig. 3. Comparing the graphs for the two 5 kW-lasers with different K-numbers in
fig. 2, it can be seen that for a fixed focusing device (F = const.) and beam power
level the penetration depth can be enhanced by using a laser with a higher beam

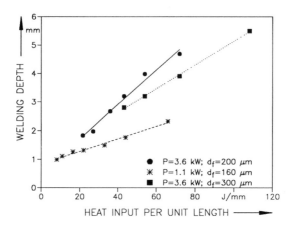

Fig. 2   Welding depth in AlMgSi1-alloy plates as a function of the heat input per unit length for different laser beam parameters.

quality. In addition narrower welds can be produced (see fig. 3), thus reducing the thermal load of the work piece.

When welding with a beam power of 1.1 kW, at values of heat input per unit length below 20 J/mm deep penetration welding with good weld quality is still possible due to the high beam quality of the laser, which has been used. At values beyond 20 J/mm, which can be achieved with a laser beam power of 1.1 kW only by low welding speed values (< 3.5 m/min), a lower penetration depth is observed despite of a smaller focus diameter than for a beam power of 3.6 kW (fig. 2 and tab. 2). At a low

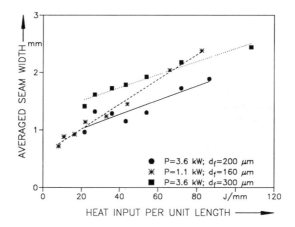

Fig. 3   Averaged seam width of AlMgSi1-welds as a function of the heat input per unit length for different laser beam parameters.

power level and low welding speeds the thermal efficiency is lower and thus the heat conduction losses related to the absorbed beam power are higher than for the high power level. This is confirmed when comparing the corresponding plots of the average seam width in fig. 3. Due to the higher fraction of heat conduction losses beyond 20 J/mm the weld seam gets broader for a beam power of 1.1 kW than for 3.6 kW (K=0.3).

Conclusively in order to get weld seams as narrow as possible and to minimize the thermal load of the workpiece a laser with a high beam quality should be used. In addition the power of the incident laser beam should be adjusted to the desired penetration depth.

3.3   Influence of processing speed on hot cracking in AlMgSi1-welds

Hot cracking in aluminium alloys is strongly dependent on the chemical composition. Recent investigations [3] revealed that in laser welding of the low crack sensitive AlMg5Mn-alloy crack-free welds can be achieved. AlMgSi1, which is an alloy with a large solidification interval, is highly susceptible for the formation of hot cracks. Metallographic investigations showed that in the experiments intercrystalline transverse weld hot cracks occured beyond a welding speed of about 4 m/min (see fig. 4). Within the investigated parameter range the appearance of cracks was found to be independent of beam power and plate thickness.

The cracking takes place in the last stages of the solidification process because of highly exceeding strains. These are imposed by solidification shrinkage stresses and localized in regions, where the residual melt is distributed as thin liquid films along the grain boundarys of the already solidified grains. The amount of strain and its accommodation is strongly dependent on the solidification conditions and the fusion

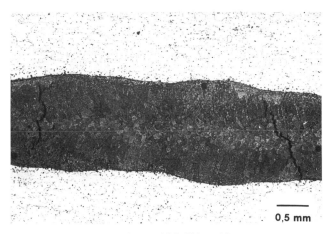

0,5 mm

Fig. 4   Transverse hot cracks in an AlMgSi1-weld:
P = 3.6 kW; K = 0.18; F = 4; v = 6 m/min; 15 l/min He.

zone grain structure. Therefore hot cracking in highly sensitive AlMgSi1-welds is dependent on the laser process parameters. The weld pool and thus the solidification front elongates with rising welding speed. The formation of columnar crystals during solidification is promoted and the hot crack sensitivity of the resulting fusion zone microstructure rises (see ref. [4]). Below a welding speed of 4 m/min crack-free welds with a finegrained fusion zone microstructure was obtained.

4. Conclusion

It was shown for AlMgSi1-, AlMg5Mn- and AlCuMg2-alloy plates that the base material composition has a great influence on the development of the laser-weld seam. The investigations confirmed that a small content of magnesium with its high vapor pressure strongly affects keyhole-formation and -stability, thus providing deeper penetration by lowering the vaporization temperature of the alloys. No correlation between pore and cavity frequency and magnesium content could be detected. Choosing a processing speed beyond 8 m/min was found to be sufficient for producing cavity- and pore-free welds even for the AlMg5Mn-alloy plates.

The comparison of different laser beam parameters in welding AlMgSi1-alloy sheets showed that by choosing a laser with a higher beam quality narrower weld seams can be produced and therefore the thermal load of the workpiece can be reduced. The investigation also revealed that applying a laser with a sufficiently high beam quality even a beam power of 1.1 kW was found to be enough up to a penetration depth of about 1.5 mm.

Hot cracking in AlMgSi1-welds was found to be strongly dependent on the processing speed. Below 4 m/min the solidification conditions promote the development of a less crack sensitive microstructure in the fusion zone allowing to achieve crack-free welds.

This report is based on the EUREKA project 194 funded by the Bundesminister für Forschung und Technologie under the contract number 13 EU 0062. The author is responsible for the content of the paper.

5. References

[1]   H. Sakamoto, K. Shibata, F. Dausinger: Proc. of the 4th Europ. Conf. ECLAT'92 Göttingen (1992) to be published

[2]   M. J. Cieslak, P. W. Fuerschbach: Metallurg. Trans. B 19 B (1988) 319

[3]   J. Rapp, C. Glumann, F. Dausinger, H. Hügel: Proc. of the 25th Int. Symp. ISATA'92 Florence (1992) 461

[4]   J. Rapp: diploma thesis University of Stuttgart (1991)

Material Removal and Cutting of Carbon Fibre Reinforced
Plastics with Excimer Lasers

H.K. Tönshoff, D. Hesse, Laser Zentrum Hannover e.V.,
Hannover, FRG

1. Introduction

Excimer lasers are an ideal tool for fine- and micromachining.
They are especially suitable for the removal of thermally
sensitive materials demanding a minimal heat affected zone
(1,2). Carbon fibre reinforced plastics with epoxy resin
matrix call for high control since matrix and fibre material
have very different decomposition temperatures and heat
conductivities. Fundamental research has already pointed out
the removal behaviour of the matrix and fibre material. Based
on these investigations, the cutting process of composite
materials was optimized.

2. Removal of epoxy resin

The wavelength and the energy density of the laser beam have a
decisive influence on the removal behaviour. Because of the
high absorption coefficient at a wavelenth of λ = 193 nm, a
homogeneous removal up to 0.3 m/pulse can be achieved
(Fig. 1).

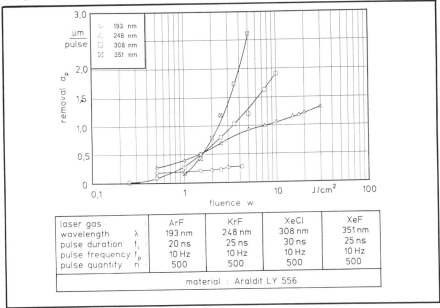

Fig. 1: Removal rates of epoxy resin irradiated with excimer
lasers

With an increasing wavelength, the decreasing absorption
coefficient enables greater penetration depth, and thus
removal rates up to 2.0 m/pulse, but with a highly
inhomogeneous character (Fig. 2).

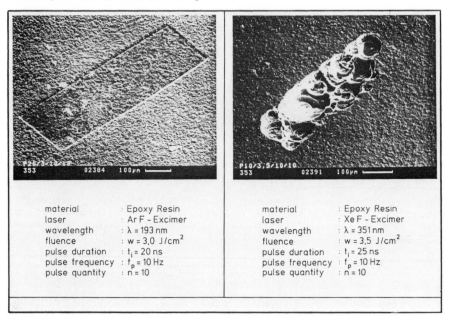

material	: Epoxy Resin	material	: Epoxy Resin
laser	: Ar F - Excimer	laser	: Xe F - Excimer
wavelength	: $\lambda = 193\,nm$	wavelength	: $\lambda = 351\,nm$
fluence	: $w = 3,0$ J/cm^2	fluence	: $w = 3,5$ J/cm^2
pulse duration	: $t_i = 20\,ns$	pulse duration	: $t_i = 25\,ns$
pulse frequency	: $f_p = 10\,Hz$	pulse frequency	: $f_p = 10\,Hz$
pulse quantity	: $n = 10$	pulse quantity	: $n = 10$

Fig. 2: Ablation of epoxy resin with different wavelengths

A compromise of high removal rates and good contour sharpness
is offered by the wavelengths $\lambda = 248$ nm and $\lambda = 308$ nm, which
do not show saturation up to energy densities of 30 J/cm .

High removal volumes per time t_v require, besides high energy
densities, high pulse frequencies that may cause thermal side
effects. In the range of frequencies from 10-150 Hz, the
removal rates a_p increase at an energy density of w = 1.5
J/cm , from 0.5 m/pulse up to 1.1 m/pulse (Fig. 3). At an
energy density of w = 5 J/cm , the removal a_p increases in the
same frequency range, from 1.2 m/pulse to 1.7 m/pulse. By
supplying air as a process gas at a preasure of $p_g = 0.5$ MPa
coaxial to the laser beam, the removal rates a_p stay constant
at an energy density of w = 1.5 J/cm and pulse frequencies f_p
between 10 Hz and 150 Hz. At an energy density of w = 5 J/cm ,
the increase in the removal rate a_p is clearly less. The
higher contour sharpness, by supplying a process gas, and the
constant removal rates lead to the conclusion that thermal
side effects are avoided or at least decreased.

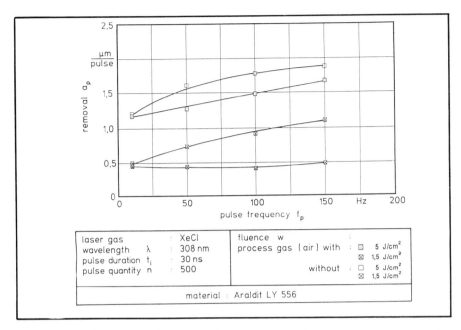

Fig. 3: Influence of pulse frequency and process gas on the removal behaviour of epoxy resin

3. Removal of carbon fibres

Carbon fibres can be cut with excimer-lasers without melting the cutingedges. Analogous to the removal behaviour of the epoxy resin, very homogeneous geometries and removal rates up to 0.2 m/pulse can be achieved at a laser wavelength of λ = 193 nm. At wavelengths of λ = 248 nm and λ = 308 nm, an optimal compromise of high removal rates and contour sharpness can be obtained.

At energy densities between w = 5 J/cm and w = 30 J/cm and a wavelength of λ = 248 nm, the removal rates a_p of the carbon fibre and the epoxy resin are almost the same. At low energy densities of w < 5 J/cm down to the removal threshold, the carbon fibre removal is less than the epoxy resin removal. At energy densities of w > 30 J/cm, the epoxy resin removed overproportionally. Based on this konwledge, the parameter area for the cutting of carbon fibre compounds can be derived.

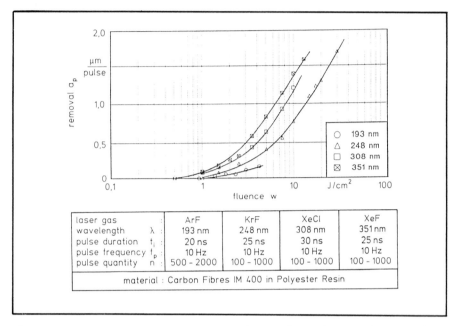

Fig. 4: Removal rates of carbon fibres irradiated with excimer lasers

4. Cutting of carbon fibre reinforced epoxy resins

Carbon fibres show a decomposition temperature of $T_Z = 3600°C$, whereas the epoxy resins melt at $T_Z = 350°C$. This difference in the decomposition temperatures leads to several millimeter deep matrix destruction along the cutting edge when cutting with thermal interacting lasers (3). When cutting with excimer lasers, the cutting edges are not thermally damaged, so that component behaviour is not influenced.

The excimer lasers used, with a mean power of $P_L = 160$ W, enable cutting of carbon reinforced epoxy resins with a feedrate of $v_f = 2.5$ mm/min. The material thickness is limited, due to narrowing the cut width clearance to s = 1.0 mm. Due to the circular form of the laser beam on the workpiece, any contour can be produced (Fig. 5).

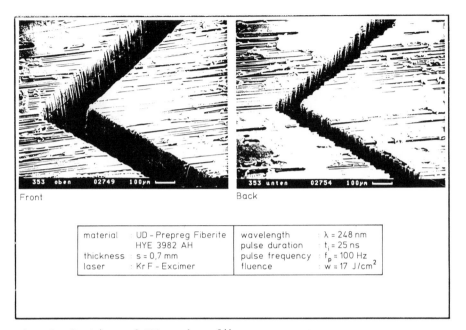

Front   Back

material	: UD - Prepreg Fiberite HYE 3982 AH	wavelength	: $\lambda = 248$ nm
thickness	: s = 0,7 mm	pulse duration	: $t_i = 25$ ns
laser	: Kr F - Excimer	pulse frequency	: $f_p = 100$ Hz
		fluence	: $w = 17$ J/cm^2

Fig. 5: Cutting of UD carbon fibre prepregs

While cutting, a carbon deposit develops on the workpiece surface that can not be avoided by variation of the process parameters. By supplying a process gas via a nozzle, coaxially to the laser beam towards the workpiece surface, reaction products can be combined and removed from the interacting zone.

By using inert gases, nitrogen, helium or argon, no deposit decrease can be achieved. While using oxygen, nearly all reaction products can be combined and removed (Fig. 6). Additionally, a higher contour sharpness can be obtained.

These results can be transmitted to other modifications of epoxy resins and carbon fibres by small variations in the laser process parameters. The machining of aramide fibre compounds with equally advantageous and characteristic features is principally possible, by using special process parameters.

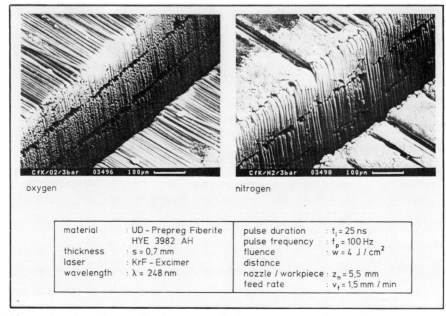

oxygen

nitrogen

material	: UD - Prepreg Fiberite	pulse duration	: $t_i = 25$ ns
	HYE 3982 AH	pulse frequency	: $f_p = 100$ Hz
thickness	: $s = 0,7$ mm	fluence	: $w = 4$ J / cm^2
laser	: KrF - Excimer	distance	
wavelength	: $\lambda = 248$ nm	nozzle / workpiece	: $z_n = 5,5$ mm
		feed rate	: $v_f = 1,5$ mm / min

Fig. 6: Cutting of fibre reinforced plastics with process gases

5. Literature

(1) H.K. Tönshoff, R. Bütje, Material Processing with Excimer Lasers, 5th Int. Conf. Lasers in Manufacturing, Stuttgart 1988

(2) H.K. Tönshoff, R. Bütje, W. König, Excimer Laser in material processing, Annals of CIRP, Vol. 37/2/1988

(3) R. Bütje, Möglichkeiten und Grenzen beim Laserschneiden von Kunststoffen, Werkstoff und Innovation, 2 (1991)

Performance of various nozzle designs in laser cutting

R. Edler, P. Berger, H. Hügel, Institut für Strahlwerkzeuge (IFSW), Universität Stuttgart, Pfaffenwaldring 43, Stuttgart, Federal Republic of Germany

Abstract

In this paper the laser cutting process is regarded under the aspect of melt ejection. Therefore cutting experiments were done in the combination of high pressure nozzles and high power lasers to show the influence of the gas pressure and the nozzle type. The used nozzles are a nozzle with conical-cylindrical inner contour, a single Laval nozzle and a new twin-jet-Laval-nozzle. The gas pressures were variied up to 3 MPa. Cutting results are presented for CO_2 lasers with output powers up to 12 kW. Ferrous materials up to a thickness of 35 mm were cut.

The comparison of the conical-cylindrical nozzle and the twin-jet-Laval-nozzle showed, that in the case of cutting thin steel plates the twin-jet-Laval-nozzle allows higher cutting velocities, larger distances and larger distance variations. Detailed investigations showed, that the flow from the twin-jet-Laval-nozzle and the single Laval nozzle deliver higher stagnation pressures over a larger distance range than those from the conical-cylindrical nozzle. The advantage in reaching higher stagnation pressures results in higher cut velocities for thin materials.

In cutting thick materials the melt ejection gets more difficult. In the cut kerf the gas seperates from the cutting front at the lower part of the kerf and a backward flowing hinders the melt ejection.

Introduction

Not only the focussed laser beam parameters influence the cutting process, but also the cutting gas plays an important role. Just the gas flow enables the cutting process: on the one hand, the gas forces the melt to move out of the kerf and on the other hand, if cutting with oxygen is performed, an additional heat input is given. Other authors have investigated the influence of the gas on the cutting process (1,2,3,4), but not all phenomena are understood.

Nozzles

Three different nozzles have been used for our laser cutting experiments: a nozzle with conical-cylindrical inner contour, a Laval nozzle, both mounted coaxially to the laser beam axis and the twin-jet-Laval-nozzle with gas exits excentrically mounted to the laser beam (5).

To qualify laser cutting nozzles, three ways have been developed. To check the free-jet behaviour, Schlieren pictures (or shadowgraphs and interferograms) are useful. One can prove, how regular the flow is, and detect the position and strength of disturbancies possibly occuring. The measurement of the stagnation pressure gives a qualified insight in the ability of the flow impinging on the workpiece. The third method is the proof of the reliability of the nozzle during laser cutting.

Figure 1 shows the Schlieren pictures of the free jets with 0.4 MPa chamber pressure. The conical-cylindrical nozzle shows the behaviour of an underexpanded jet: after the exit of the orifice the flow expands to reach the ambient pressure and when he has reached it, the forming compression waves and shocks force the jet to contract and so on. The shocks change energy useful for the ejection of the melt in heat useless for the process. The flow of the single Laval nozzle shows only weak disturbancies and is nearly perfect. The combined flow of the twin-jet-Laval-nozzle is marked by deviation of the two single jets merging to one common jet.

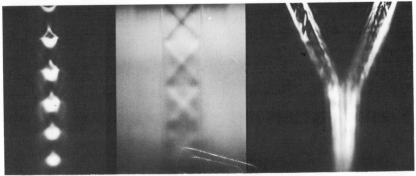

Fig.1: Schlieren pictures of (left) a conical-cylindrical nozzle, (middle) a single Laval nozzle and (right) a twin-jet-Laval-nozzle. 0.4 MPa.

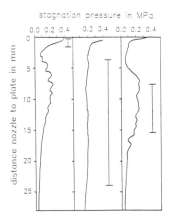

Fig.2: Stagnation pressures of the conical-cylindrical nozzle (left), the single Laval nozzle (middle), and the twin-jet-Laval-nozzle (right). 0.4 MPa.

The stagnation pressure distributions along the flow center lines of the three nozzles are given in figure 2, marked are the working distances. The example of the single Laval nozzle shows, that a high pressure niveau over a long distance is achieved. For comparison the conical-cylindrical nozzle shows strong fluctuations resulting from a chamber pressure higher than the critical pressure. The weak fluctuation in the case of the twin-jet-Laval-nozzle results from the conjuction of the two single Laval jets and the entrainement of ambient gas in the deviation region. But nevertheless the pressure niveau is higher than that of the conical-cylindrical nozzle, and also the niveau remains high over a larger distance.

Cutting experiments

The cutting experiments were done with a 1 and a 5 kW CO_2 laser at the IFSW (TRUMPF lasers with stable resonators, low multimode, f = 150 mm, F = 4 and F = 3.6) and a 14 kW CO_2 laser at MAUSER/Oberndorf (UTIL, instable resonator, transverse flow, f = 300 mm , F = 5) (6). The cut material was mild steel with thicknesses of 1, 4 and 8 mm and high alloyed steel, 10 and 35 mm thick. The used optics were mirror optics. A sealing window was necessary using coaxial nozzles. Cutting experiments, done with the 1 and 5 kW lasers, were performed with a sealing window even in the case of the twin-jet-Laval-nozzle for the sens of comparison. Only in the case of the 14 kW laser all cuts were done without any transmitting window. Due to the window, in the case of using the conical-cylindrical and the single Laval nozzle, the highest pressure was limited to 0.5 MPa. The polarization of the 1 and 5 kW laser is linear and it was cut with parallel polarization. The polarization of the 14 kW laser was undefined.

Cutting results

Melt cuts
Figure 3 shows cutting results, obtained with the conical-cylindrical and the single Laval nozzle on 1 mm thick mild steel using the 1 kW laser. The maximal cutting velocity is given as a function of the laser power. The nozzles have nearly the same cutting results concerning the maximum velocity. Concerning the quality there is also no significant difference.

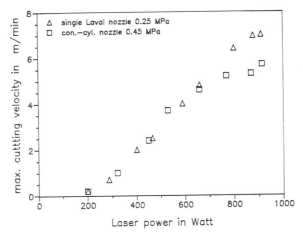

__Fig.3__: Maximum cutting velocity as a function of the laser power for 1 mm thick mild steel (1 kW laser, N_2).

In the next figure, the maximum cutting velocity is given as a function of the laser power for the results obtained with the 5 kW laser. Again, for all three nozzles the maximum velocities are nearly the same.

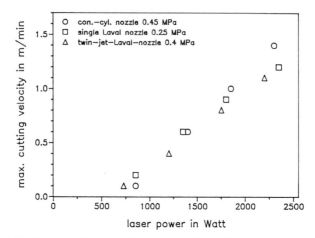

Fig.4: Maximum cutting velocity as a function of the laser power for 4 mm thick mild steel (5 kW laser, N_2).

How expected from the pressure diagrams, there is no striking difference between the various nozzles, if they are used within their optimal working distances. In varying the working distances the Laval nozzles show an advantage, see figure 5. If it is impossible to use a sealing window, because of such high pressures, which would crack a window, or because of such high laser powers, which would deform thermally the window in an unacceptable extent, the use of a twin-jet-Laval-nozzle would be advised. As an example for this, the figure 6 shows the cut surface of a mild steel sheet 8 mm thick, cut with 7000 W and 2 MPa nitrogen.

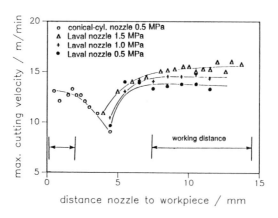

Fig.5: Maximum cutting velocity as a function of the distance between nozzle and workpiece for 1 mm thick mild steel (3.6 kW laser, N_2, conical-cylindrical nozzle and twin-jet-Laval-nozzles)

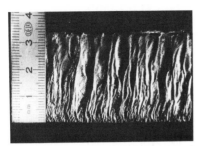

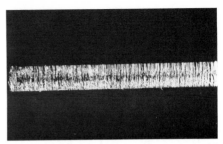

Fig.6: Cut surface of mild steel, 8 mm thick, cut with the 14 kW UTIL at MAUSER/ Oberndorf. (7 kW, 2 MPa N_2, 1.6 m/min).

Fig.7: Cut surface of a 34 mm thick alloyed steel, cut with the 14 kW UTIL at MAUSER/ Oberndorf (7 kW, 2 MPa O_2, 0.5 m/min).

Oxygen cuts

High alloyed steel with a thickness of 34 mm was cut with oxygen using the 14 kW laser (figure 7). Here only the twin-jet-Laval-nozzle was used because of the high pressure and the high laser power.

Investigations inside the cut kerf

The comparison of own cut surfaces affirmed earlier investigations (7), which showed the influence of the nozzle-beam alignement on the cutting results.

We observed in some cases, that the cut surface was slighty oxidized while the cut front was shiny. The striations had an other incliniation as the cut front: the cut front was steeper. The inclination of the striations changed inside the cut kerf. First, the melt is driven down nearly vertically, then forced aside.

A reason for this phenomena is a seperation of the gas flow from the cutting front. A detachement of the gas flow has been observed regarding a cut kerf simulated with two plates (8). There occured the detachement stochastically, once on one side then on the other side of the kerf. Here the detachement seems to occur on the cut front.

To prove this idea, a gas jet was blown on a plate with a laser cut kerf. Varying the position of the gas jet axis to the cut front, there was a sucking of ambient air inside to the cut kerf at the bottom: was the axis of the gas jet behind the cut front - that means during the cutting process a after-running of the gas jet - there was no sucking, was the jet axis before it or even on it, there was a big sucking force towards the cut kerf. Even water pearls on the bottom of the plate nearby the kerf were forced to move to the kerf and spread inside it (see also figure 8). Olsen (9) already found out, that running after gas jets improve the cut quality. (The effect of the displacement of the laser beam to the cutting front has investigated theoretically in (10)). So, during the cutting process the gas-flow seperation acts on the molten layer. This effects a thickening of the melt and even a backward flow, so that the melt ejection is hindered.

116

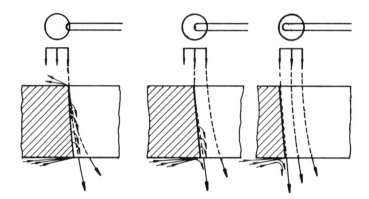

Fig.8: Schematic view of the behaviour of a gas jet at the cutting front (6).

Conclusion

Investigations of the flow out of various nozzles showed, that Laval nozzles deliver higher pressures over a larger distance range. Cutting experiments confirm this behaviour. If it is necessary to renounce on transmitting optics: the twin-jet-Laval-nozzle allows the realization of high pressures and the use of high laser powers. The melt ejection efficiency depends on the effectiveness of the gas jet inside of the kerf. The presented investigations give an better insight of the fact, that running after gas jets increase the cut quality, leading gas jets lead to detoriations.

References

(1)   J. Fieret, M. Terry, B.A. Ward: SPIE Vol. 801 High Power Lasers, 1987, p. 243.
(2)   P. Berger, M. Herrmann, H. Hügel: Proceedings LASER 89, München 1989, Springer Verlag, Berlin 1990, p. 630-634.
(3)   E. Beyer, D. Petring: Proceedings ICALEO 90 (1990), p. 199.
(4)   Nonhof, C.J.: Material Processing with Nd-Lasers, Ayr, Scotland, Electrochemical Publications, 1988.
(5)   R. Edler, P. Berger: Proceedings ICALEO 1991, to be published.
(6)   T. Beck: Universität Stuttgart, Diploma thesis, 1992 (Institut für Strahlwerkzeuge IFSW-1992).
(7)   J.T. Gabzdyl, W.M. Steen, M. Cantello: Proceedings of ICALEO 87, 1987.
(8)   H. Zefferer, D. Petring, E. Beyer: DVS 135, (1991), p.210-214.
(9)   H.-O. Ketting, F.O. Olsen: Proceedings LASER 91, München 1991, Springer Verlag, to be published.
(10)   U. Schreiner-Mohr: Proceedings of ICALEO 91, (1991), to be published.

HIGH POWER LASER PROCESSING
OF ALUMINIUM ALLOYS

A.S.Bransden[+], T.Endres[*]
[+] GEAT-Industrielaser GmbH, Nürnberg
[*] Department of Materials Science II, University Erlangen-Nürnberg

ABSTRACT

The emergence of reliable, high power laser systems in excess of 10kW demonstrates an acceptance by industry of the laser as a competitive machine tool. However, little work has been published on applications at these higher power levels, particularly on aluminium alloys. The use of high power overcomes the reflectivity and conductivity problems associated with these materials and allows high speed and efficient processing. This paper gives a brief summary of recent work in the areas of cutting, welding and surface treatment. In cutting, a number of alloys are assessed including pure aluminium while in welding, work has concentrated on AlCu and AlMg including the welding of laser cut edges. In surface treatment, results of alloying silicon into AlSi12 to achieve hypereutectic alloys are described. In all areas, process parameters are given and microstructural aspects discussed.

1. INTRODUCTION

The CO_2 laser treatment of aluminium has traditionally been difficult owing to the high reflectivity of the metal to 10.6µm radiation and its high thermal diffusivity. For surface heat treatment the absorbtion can be increased by the use of special coatings or treatments, these often providing a second function namely supplying the additive in the case of alloying or cladding. In welding and cutting however, coatings are not employed and a reliance has to be placed on the capability of being able to deliver a high enough energy flux to create localized melting and/or evaporation and the establishment of a keyhole which effectively traps the laser beam and increases process efficiency. For lasers of low to medium power, high power densities can be achieved by tight focusing or by the use of pulsed or Q-switched lasers but the average powers are low and the treatment depths are limited to a few mm. With the increased availability of lasers with high average power (>10kW) focusing is less critical and cw operation allows efficient cutting and welding of aluminium of 10mm thick and more. No data exists on high power laser cutting of aluminium. Some data at lower (up to 2kW) laser powers [1,2,3,4] is available. 10kW laser welding of Mg and Li aluminium alloys [5] are reported, while [6] compares the weldability of pure aluminium, AlCu, AlZn and AlMg alloys. A review of laser welding of aluminium [7] shows only results up to 9kW laser power. In surface treatment, the use of high powers results in high coverage rates making viable the treatment of production components. Many examples exist of silicon alloying of aluminium alloy but generally only with laser powers up to 5kW [8,9]. This paper gives a brief summary of recent results obtained in the cutting, welding and silicon alloying of aluminium using a high laser power.

2. MATERIALS

Four grades of aluminium were investigated in this work, the processes and the compositions being given below in Table 1. Thicknesses were up to 25mm.

Aluminium grade	Composition				Processes investigated
	Mg	Si	Cu	Mn	
Al99.5	-	-	-	-	Cutting
AlCu	0.4-1.0	0.2-0.8	3.5-4.5	0.4-1.0	Cutting Welding
AlMg4.5Mn	4.0-4.9	4.0-4.9	0.1	0.4-1.0	Cutting Welding
AlSi	-	9.0	-	-	Alloying

Table 1. Details of the aluminium grades used.

3. THE EQUIPMENT

The laser used in this work was an 18kW CO_2 cw laser model SM41-18 manufactured by United Technologies Industrial Lasers (UTIL), USA. The laser is of fast transverse flow configuration, DC excited, of modular design and employs an aerodynamic window and all-mirror optics. The laser is coupled to a 5-axis gantry system manufactured by Messer Griesheim, the optics moving in four axes (Y,Z,Θ,ϕ) and the workpiece in one (X). The facility is situated at the University of Erlangen-Nürnberg. For the cutting work a 300mm focal length parabolic focusing mirror was used with an off-axis cutting assist jet using gas pressures of 5bar. For the welding and alloying work a 500mm focal length parabolic mirror was used with appropriate gas shielding of helium. Silicon additive material for the alloying was supplied by a powder blowing technique using argon as powder transport.

4. RESULTS AND DISCUSSIONS

4.1 CUTTING

Trials were performed on various grades of aluminium as shown in Table 1 and using various gases (He, N_2, CO_2, O_2) as cutting assist. The threshold cutting speed was defined as the maximum speed which produced a clear cut free of internal dross but not necessarily dross-free on the underside. Full optimization of the cutting process has not been carried out and the use of an off-axis cutting nozzle is probably not ideal. Nevertheless, indications are given as to the potential of high power lasers in the cutting of aluminium.

Figure 1 shows the effect of laser power on the cutting speed using helium as assist gas. At higher speeds the cutting efficiency drops off due to the inability of the off-axis gas jet to supply sufficient pressure to enable good clearing of the kerf of molten material. The same applies in the case of thicker sections where gas pressure towards the bottom of the kerf is at a minimum and blocking of the kerf by molten material becomes the limiting factor in obtaining maximum speed. Higher speeds are obtained with the AlMg alloy, and it is likely that evaporating magnesium is assisting in keeping the kerf open. The differences become less as thickness increases. Large thicknesses (eg. 25mm can be cut), Figure 2 but at relatively slow speeds determined by the rate at which molten metal can be ejected from the kerf.

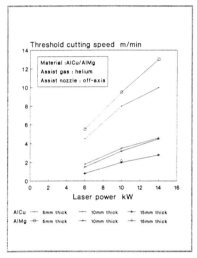

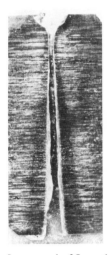

Figure 1   Influence of laser power and material thickness on threshold cutting speed.

Figure 2.   Laser cut in 25mm thick AlCu produced at 0.4m/min with 10kW.

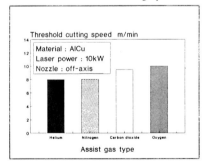

Figure 3.   Effect of assist gas on threshold cutting speed for 5mm thick material.

The type of assist gas also determines the maximum cutting speed as indicated in Figure 3 for the cutting of 5mm thick AlCu. Maximum speeds are obtained with oxygen as expected while helium and nitrogen (inert gases) produce the lowest speeds.

In terms of cut profile, helium produces the best result showing near-parallel kerfs (with a slight waisting at mid-height) in the 5mm thick material, Figure 4. Cuts become more unstable as speed increases particularly in the lower regions of the kerf. The use of CO_2 and O_2 both produce a noticeable increase in the width of the kerf due to the burning effect of the oxygen. The increase in width of the kerf allows easier clearing of the molten material which is partly responsible for the higher cutting speeds.

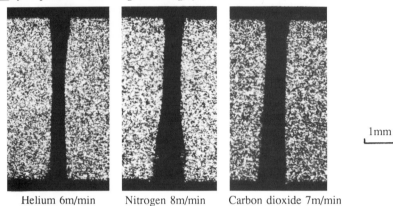

1mm

Helium 6m/min   Nitrogen 8m/min   Carbon dioxide 7m/min

Figure 4.   Effect of cutting assist gas and speed on kerf profile.

Surface roughness measurements show little difference between the gases helium, nitrogen and carbon dioxide but oxygen is noticeably rougher. The kerfs tend to be smoother at the centre, Figure 5 showing roughness results (Rz) taken at kerf top, centre and bottom positions and an the appearance of the associated kerf (SEM) of a cut produced with nitrogen. Typical roughness values of 30-50µm should permit the rewelding of laser-cut edges particularly if filler wire is employed.

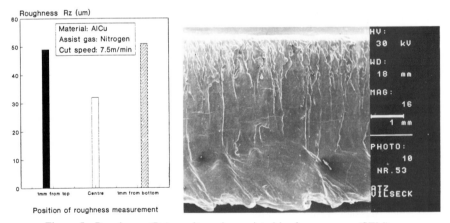

Figure 5. Roughness (Rz) results and associated kerf appearance (SEM).

4.2 WELDING

Welding of 10mm thick AlCu and AlMg has been performed at 10kW. The AlMg material welds faster than the AlCu for similar penetration depths, respective speeds being 3.5 and 3m/min. Full penetration results in some drop-through but this can be controlled by the use of backing bars or side-hand welding. Figure 6a shows a partially penetrating bead-on-plate weld in AlCu and little porosity is seen. Laser-cut edges (10kW, helium) have been rewelded in the as-cut condition, Figure 6b, some drop-through occurring with a little top surface undercut resulting from the imperfect fit-up of the edges (Roughness Rz 50μm). The porosity probably originates from the adherent oxide remaining on the laser-cut edges.

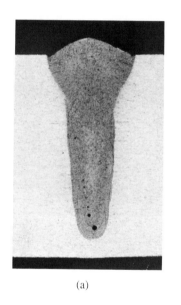   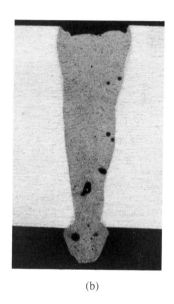

(a)   (b)

Figure 6.   Laser welds in 10mm thick AlCu alloy showing (a) a bead-on-plate weld and (b) a weld of two laser-cut edges.

For the AlCu alloy, some softening (90HV) occurs in the weld zone due to dissolution of the copper, but artifical ageing at 200°C for 1hr has been shown to restore the hardness to the parent value of 120HV. For the AlMg alloy, no significant loss of magnesium occurs and hardness in the weld zone is similar to the parent value of about 100HV.

4.3 ALLOYING

Silicon enhancement of the cast alloy AlSi9 has been carried out at up to 10kW. Figure 7 shows a cross section of track produced at 8kW, 1m/min and detail of the resulting microstructure. Good uniformity of the silicon distribution is seen. The microstructure contains primary silicon particles in a matrix of aluminium/aluminium eutectic with little porosity and no cracking. In excess of 40wt% Si can be readily acheived and resulting eutectics have been measured to contain 18%Si which is far in excess of the thermal equilibrium value of 11.5%.

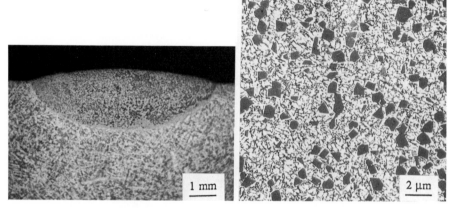

Figure 7.   Details of track profile and microstructure of laser alloyed aluminium silicon alloy.

Figure 8 shows the effect of treatment speed on track case depth and hardness for a given silicon addition. Maximum depths and hardness values are respectively 3mm and 160HV. The higher dilution occurring at the lower speeds reduces hardness. Layers 2mm thick can be deposited at 100cm^2/min with 14kW.

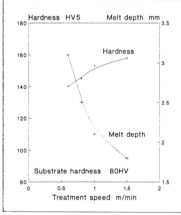

Figure 8.   Effect of process speed on hardness and depth.

5. CONCLUSIONS

This work has shown some of the potentials of the use of high power CO_2 lasers in the areas of cutting, welding and alloying of aluminium. Cutting of aluminium in thicknesses greater than 10mm shows the ability in producing reasonably narrow, parallel high speed cuts. The welding of the laser-cut edges shows no real problems and the small weld undercut can probably be overcome (eg.by side-hand welding or the use of a wire filler technique). The use of lasers in silicon alloying allows localised treatment of components either for direct wear improvement or as a means of strengthening surfaces to increase load-bearing capacity.

6. ACKNOWLEDGMENTS

The authors wish to thank Dr R.Nuss (Geat-Industrielaser GmbH) for his encouragement and R.Zeller and B.Dill (University Erlangen-Nürnberg) for their help in the analysis and the preparation of this paper.

7. REFERENCES

1.   Bergmann H.W., Geiger M., Nuss R., Juckenath B., Biermann S., Laser cutting of aluminium and titanium alloys for aerospace applications. Opto Elektronik Magazin Vol.5, No.2, 1989, pp196-200.

2.   Kupfer R., Biermann B., Bergmann H.W., Cutting of Al-alloys using different lasers.Proc. ECLAT'90, 3rd European Conference on Laser Treatment of Materials, Erlangen, Sept.1990.

3.   Giere R., Decker I., Ruge J., Influence of laser beam cutting on the mechanical-technological properties of grain refined steel and aluminium alloys. Proc. Conf ECO3, The Hague, 12-14 March 1990.

4.   Michaelis A., Schäfer J-H., Uhlenbusch J., Viöl W., Cutting and welding of aluminium with a high repetition rate pulsed CO_2 laser, Ibid 3.

5.   Marsico T.A., Kossowsky R., Laser beam welding of aluminium alloy plates 7039, 5083 and 2090. LIA Vol.69 ICALEO (1989),pp61-71.

6.   Bransden A.S., Megaw J.H.P.C., The laser welding of aluminium and its alloys, Laser Users Club Report, CLM-RR-LUC6, June 1984, Culham Laboratory, UK.

7.   Thorstensen B., Laser welding of aluminium. Industrial Laser Handbook, Pennwell Books, 1989.

8.   Bransden A.S., Funnell J.G., Megaw J.H.P.C., Investigation of laser alloying of aluminium alloy. Proc.Conf. ISATA 20th International Symposium on Automotive Technology and Automation, Florence 29 May-2 June 1989.

9.   Lindner H., Bergmann H.W., Endres T., Remelting and alloying of Al-Si alloys. Ibid 3.

Laser Welding of Different Aluminum Alloys

H. Sakamoto and K. Shibata, Nissan Motor Co.,Ltd., Nissan Research Center, Yokosuka, Japan
F. Dausinger, Universität Stuttgart, Institut für Strahlwerkzeuge, Stuttgart, FRG

1. Introduction

Laser welding is one of the most common methods for joining steel. Its application to the welding of aluminum alloys has been studied in recent years (Ref. 1-7), but a practical technique for use in production operations has yet to be devised. The main reason for this is the low degree of energy efficiency that is obtained because of the high reflectivity and high thermal conductivity of aluminum alloys. Compared with steel, it is reported that the welding of aluminum alloys requires twice the laser power density. In addition, the additives contained in aluminum alloys strongly affect their weldability. For these reasons, it is difficult to optimize laser welding conditions.
In this study, Al-Mg, Al-Mg-Si and Al-Zn alloys and industrial pure aluminum were welded using a carbon dioxide laser. Welding was performed under various laser power intensities and traverse velocities. The threshold power intensity required to accomplish keyhole welding was investigated by measuring the welded depth. The relationship between threshold intensities and the properties of the aluminum alloys was investigated.

2. Procedure

2.1 Aluminum alloys

The materials varied in the welding experiment included an Al-Mg alloy, two Al-Mg-Si alloys, an Al-Zn alloy and industrial pure aluminum. Table 1 shows the additive contents of the five different aluminum alloys. The Al-Mg alloy, Al-Mg-Si_1 alloy and industrial pure aluminum were rolled plates and the Al-Mg-Si_2 alloy and Al-Zn alloy were extrusions.

mol%

	I.P.Al	Al-Mg	Al-Mg-Si_1	Al-Mg-Si_2	Al-Zn
Si	0.019	0.048	0.558	0.433	0.040
Fe	0.140	0.053	0.058	0.111	0.085
Cu	N.D	0.182	0.013	0.026	0.031
Mn	N.D	0.020	N.D	0.030	0.010
Mg	N.D	4.867	0.367	0.845	0.828
Cr	——	0.026	0.031	0.016	N.D
Zn	N.D	0.004	N.D	0.004	2.350
Ti	0.006	0.006	0.011	0.011	0.017
Zr	——	——	——	——	0.049
V	0.005	——	——	——	——

Table 1 Composition of alminum alloys

2.2 Experimental setup

Figure 1 shows the experimental setup. A carbon dioxide laser beam in the TEM01* mode was focused by a parabolic mirror having a focal length of 150 mm. The beam diameter on the workpiece was about 360 µm. Laser power intensity was varied from 1.5 to 6.0 e+4 W/mm². Helium gas was used for shielding the molten pool. Traverse velocity was varied from 3.0 to 7.0 m/min.

3. Results

Cross sections of the Al-Mg alloy welded at two different power intensities are shown in Fig.2(a) and (b). Traverse velocity was 3.0 m/min. and the shielding gas flow rate was 20.0 l/min.
The small welded depth seen in Fig. 2 (a) indicates that keyhole welding was not accomplished because the power intensity was not high enough.

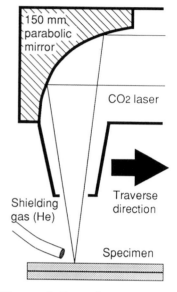

Figure 1 Experimental setup

When the intensity was above the "threshold level", keyhole welding was accomplished and large welded depth was obtained as seen in Fig.2 (b).

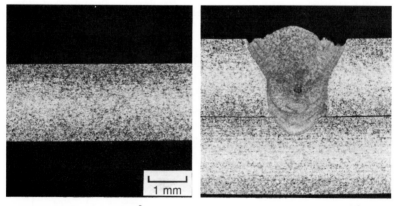

(a) 1.81e+4 W/mm²   (b) 2.96e+4 W/mm²

Figure 2 Cross sections of welded seam

Figure 3 shows the welded depth obtained under various intensities. Each alloy had a threshold level at which the welded depth increased significantly. The threshold level varied among the alloys. The Al-Mg alloy had the lowest threshold level and industrial pure aluminum had the highest. At the same intensity, the welded depth increased with a lower threshold level.

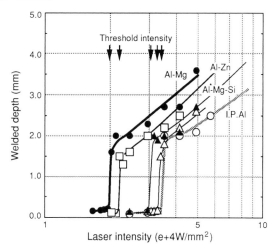

Figure 3 Welded depth vs. laser intensity

Figure 4 shows the threshold intensities of each alloy under various traverse velocities. Decreasing the traverse velocity had only a slight effect on lowering the threshold intensity.

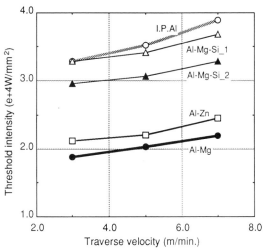

Figure 4 Threshold intensity vs. traverse velocity

4. Discussion

It is assumed that the threshold intensity depends on the properties of the alloy. Figure 5 shows the thermal conductivity of each alloy at various temperatures. Average values were found in order to compare the threshold intensity of each alloy.

Figure 6 shows the correlation between the average thermal conductivity and threshold intensity of each alloy. It was found that an alloy having high thermal conductivity did not always have high threshold intensity.

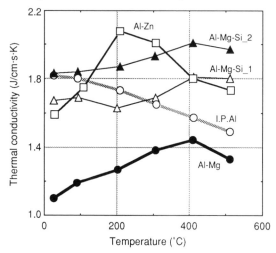

Figure 5   Thermal conductivity of aluminum alloys

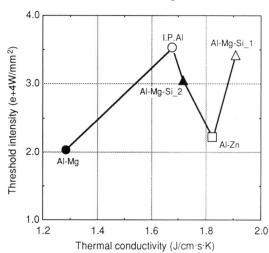

Figure 6   Threshold intensity vs. thermal conductivity

The threshold intensity of each alloy is plotted against the
magnesium and zinc content in Figure 7. Magnesium and zinc
additives have a lower evaporating temperature than pure
aluminum (Fig. 8).

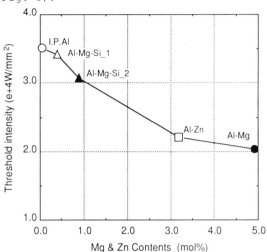

Figure 7   Threshold intensity vs. Mg & Zn contents

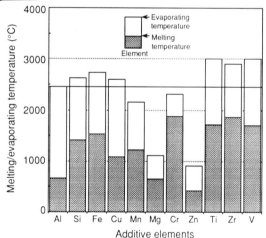

Figure 8   Melting/evaporating temperature of additive elements

It is seen that the threshold intensity decreased with a higher
content of magnesium and zinc. It is thought that elements a
having low evaporating temperature help to accomplish keyhole
welding because a keyhole is induced by the vapor pressure of
evaporated elements. The ease of evaporation of aluminum
alloys, rather than their thermal conductivity, is a dominant
factor governing the threshold intensity.

5. Conclusion

Al-Mg, Al-Mg-Si and Al-Zn alloys and industrial pure aluminum were welded with a carbon dioxide laser.
(1) Alloys having a higher content of magnesium and zinc showed a lower threshold intensity in keyhole welding.
(2) There was no correlation between thermal conductivity and threshold intensity.
(3) The ease of evaporation of an aluminum alloy is a dominant factor governing the threshold intensity.

References

(1) M. Bazan et al: Welding Journal 6 (1985) 27.
(2) B. Thorstensen and J. Mazumder: Proc. 4th Int. Conf. on Lasers in Manufacturing (1987) 203.
(3) M.J. Cieslak and P.W. Fuerschbach: Metallugical Transaction B Vol. 19B (1988) 319.
(4) G.Sepolt, T.C. Zuo and Ch. Binroth: Proc. of ECLAT'90 (1990) 671.
(5) K. Behler, E. Beyer and R. Schäfer: SPIE Vol. 1020 High Power CO2 Laser Systems and Applications (1988) 164.
(6) T.A. Marsico and R. Kossowsky: L.I.A. Vol 69 ICAREO (1989) 61.
(7) S. Katayama and C.D. Lundin: Journal of Light Metal Welding & Construction Vol. 29 No. 8 (1991) 349.

CO_2 Laser Welding of SiO_2-Al_2O_3 Ceramic Tubes

A. De Paris, GEMPPM-CALFETMAT, INSA, Villeurbanne, F.
M. Robin, UCB-IUTB-CALFETMAT, Villeurbanne, F.
G. Fantozzi, GEMPPM, INSA, Villeurbanne, F.

Introduction

Because of their weaks points such as difficult processing and small resistance to thermal shock, application of ceramics has been practically restricted. Methods for processing complicated shape parts and products as well as use of ceramics in severe environnement will make a new mark in expanding advanced ceramic application (1).

Some joining ceramic techniques were developed in the last years like mechanical joining, brazing (2), diffusion welding (3), etc, but they cannot provide vacuum-tight, structurally strong, heat resistant joint.

A good technique is the direct fusion joining of ceramic materials to themselves and to metals. In a pioneering work, Hokanson et al (4) welded with a electron beam pieces of Al_2O_3 together and to W, Mo, Co and Kovar. Preheating and controlled cooling rates were required to eliminate cracking. Similar research to evaluate the potentialities of this joining method was carried out by Rice (5). Generally, alumina-to-alumina joints could be welded fairly well, although problems in joint reproducibility and cracking were encountered.

One method very promising is to use a CO_2 laser in fusion welding of ceramics. The potential advantages of the CO_2 laser beam stem from the fact that it can be used in air, can heat dieletric materials without any problems with a high absorptivity at 10.6 µm radiation (6).

Experimental procedure

The laser used throughout our experiments was a CO_2, 10.6 micrometer, continuous wave laser CILAS CI-4000 with a maximum power output of 3.8 kW.

As the weld specimens, tubes Ø 8x5 mm of SiO_2-Al_2O_3 ceramic with 60% wt alumina were used (Degussa 60 E). The microstructure of the parent ceramic material is needles like mullite and crystals of alumina in an amorphous silica matrix, figure 1.

Table 1 summarize some typical physical properties of ceramic tubes 60 E.

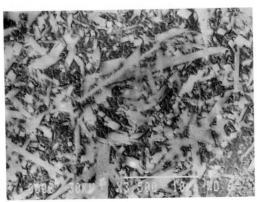

Fig.1: Microstructure of parent material. Needles like mullite in an amorphous silica with crystals of alumina.

Principal constituent	%	60% Al_2O_3
Alkalis	%	3.0
Maximum working temperature	°C	1600
Apparent density	g/cm^3	2.6
Open porosity	%	0
Bending strengh	MPa	125

Table 1: physical properties of ceramic 60 E.

One of the major problems encountered in fusion welding of ceramic materials is the control of cracking caused by thermal stresses. If a section of ceramic material is melted, a very high temperature gradient is set up in the adjacent unmelted material with proportionally high thermal stresses (5). Cracks are likely to appear during or after welding as this process produces tensile stresses both in the unmelted area beneath the bead during heating and in the shrinking bead during cooling and because tensile strength in ceramics is known to be fairly low. By providing a supplementary heating in a broader zone around the weld, the net thermal gradient has benn kept so low that no thermal stresses, high enough to cause cracking are reached (7). This supplementary heating also allows the part to be heated and cooled slowly enough to avoid thermal shock. In order to prevent weld cracking, ceramic specimens were preheated at 900°C with radiant energy produced by halogen lamps and collected by reflectors, figure 2.

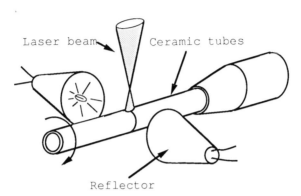

Fig. 2: Service heating system used to reach 900°C before laser treatment.

Results and discussion

Welding technique was established by an investigation of the effects of the processing parameters on weld characteristics. Several interactions can be set up between power beam, travelling speed and focal position. Figure 3 shows how penetration depth and weld bead cross-section vary for different welding speeds and focusing conditions, under a constant (140W) laser beam powers.

One problem found when welding SiO_2-Al_2O_3 ceramic material is the porosity. Since ceramic material is very porous, porosities tend to be produced in the weld bead. A lot of round small pores not connected to each other were found along fusion boundary. These pores seem to originate in coalescence of porosities of the parent material or of residual vapors due to silica vaporization at high temperature. The greater concentration of pores in the interface is obtained with an out of focus distance from the workpiece d=+10 mm and a welding speed V_T=50 cm/min. Lowring the welding speed decreases the porosities but a depression appears at the top of the bead. Consequently our main experiments are performed

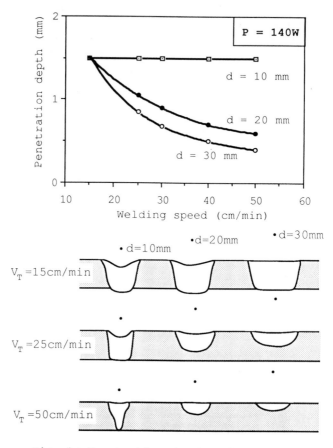

Fig. 3: Penetration depth and weld bead cross-section for a 140W laser beam power.

out of focus with d=+30 mm and a welding speed V_T=15 cm/min. With this parameters we minimize the porosities in the weld bead. Figure 4 shows a tube welded by laser beam.

Figure 5 shows a cross-section of a weld bead polished and etched with HF(10%) in H_2O. The structure of the weld zone unmodified with respect to the substrate: we note the presence of stoichiometric mullite crystals in silica glass, and the absence of alumina crystals.

Fig. 4: Exemple of a tube welded by
laser beam with the parameters presets.

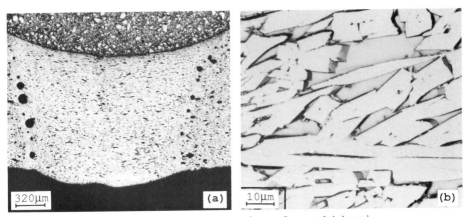

Fig. 5: (a) Cross-section of a weld bead
with porosities at interface, (b)
microstructure of the weld bead, mullite
needles in an amorphous silica matrix.

The study of the relationship between welding parameters and weld
zone characteristics was supported by several evaluation techniques.
Determination of cracking was based on visual examination at high
magnification. No crack was found. Hermeticity of the joints was
confirmed by helium leak checking. Flexural bend testing of welded

ceramics indicated that good joint strengths can be obtained. Under simple beam bending the parent ceramic achieved a surface stress of approximately σ_f=125MPa. Laser beam welded joints withstood σ_f=125MPa, the fracture beginning in the parent material zone under tension, and propagating then to the weld bead.

Vickers indentation technique was used for measuring the hardness. The welds exhibits an increase in hardness of about 50% with regard to parent material due to both microstructural and densification changes, figure 6.

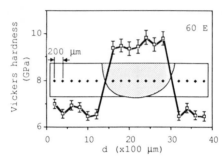

Fig. 6: Vickers indentation (Hv 300).

Conclusion

Utilization of a CO_2 laser beam for joining ceramics allows reliable and efficient joints to be made. Full-penetrated, ceramic to ceramics, sound welds can be obtained and exhibit the same mechanical strength as the base material. This new joining technique is likely to expand the use of ceramic materials in several industrial fields.

References

(1) M. Ikeda: Taikabutsu Overseas. 5(1985)27.
(2) H. Mizuhara, K. Mally: Weld. J. 64(1985)27.
(3) B. J. Dalgleish, M. Lu, A. G. Evans: Acta Metall. 36(1988)2029
(4) H. A. Hokanson, S. L. Rogers, W. I. Kern: Ceram. Ind. 81(1963)44
(5) R. W. Rice: Naval Research Laboratory (NRL) n° 7085(1970).
(6) H. Maruo, I. Miyamoto, Y. Arata:Proc. 1st. Int. Laser Processing Conference. LIA, Anaheim, CA (1981).
(7) A. De Paris, M. Robin, G. Fantozzi: J. Phys. 1(1992)127.

Deep Penetration Welding of Steel with Pulsed CO_2-Laser Radiation

T. Wahl[*], F. Dausinger, H. Hügel, Institut für Strahlwerkzeuge, Universität Stuttgart, Stuttgart, FRG
[*]Recent adress: Lasertechnik Salem, Salem, FRG

1   Introduction

When welding with cw-mode CO_2-lasers, rising line-energy causes negative effects on the weld bead. The resulting characteristic "wine-glass" shape is due to the heating of the bead surface by the laser-induced plasma occuring at low welding velocities, absorbing fractions of the laser power and thus screening the weld. The absorbed energy is partially conducted to the weld bead surface, leading to a relatively broad and shallow trace of heat conduction welding shape superposed to the deep penetration weld.

This behaviour is overcome when welding in a pulsed mode, switching off the beam and letting the plasma decay in the pulse pauses (1,2). Scope of this presentation is to present recent work on this field, using modern CO_2-lasers with good beam quality, in the average power range of 250 W to 2.600 W.

2   Characteristic Figures and Variables to Describe Pulsed Mode

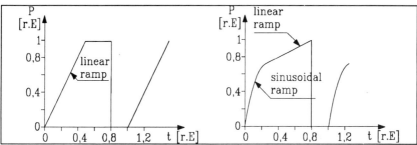

Fig. 1:   Principal sketches of the puls shapes used in the experiments: a) linear ramp, used with the 5-kW-laser, 1b) sinusoidal ramp used with the 1.5-kW-laser. (P [r.E] ... beam power relative units, t [r.E] ... time relative units)

In addition to the well known parameters of pulsed laser-mode such as frequency f, pulse length l_{pu}, and pause length l_{pa}, others as well such as pulse shape play an important role in pulsed laser materials processing. In fig. 1 the pulse shapes used in this work with different ramps and variable holding time resulting in duty cycles between 50 and 80 % are shown.

3   Experimental Setup

The experiments were carried out using three rf-exited, fast longitudinal flow CO_2-lasers TRUMPF TLF 1500 with 1.110 W, TRUMPF TLF 5000 with 4.100 W and TRUMPF TLF

138

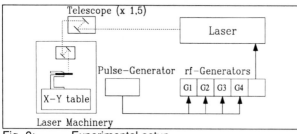

Fig. 2: Experimental setup.

5000 turbo with 4.400 W respectively at the workpiece. The pulse shaping was done by a programmable variable pulse form generator, feeding directly the lasers' rf-control. Fig. 2 shows the experimental setup.

The laser beams were focused on the workpieces with an off-axis parabolic mirror focusing optics with an aperture of 42 mm and a focal length of 150 mm, the F-numbers being about 4 for the 5-kW-lasers and 7 for the 1.5-kW-laser. Helium was used as shielding gas, fed from an off-axial nozzle with 4 mm inner diameter in welding direction at a flow rate of 5 to 15 l/min.

4   Experimental Results

4.1   Welding without Pulse Shaping

First experiments were carried out with the TLF 5000. The focus diameter was approximately 0.3 mm yielding values of the cw-intensity (and also the maximum pulse intensity) of 6 to $7*10^6$ W/cm². The laser beam was pulsed within a frequency range of 5 to 10.000 Hz with the lasers built-in pulsing facility, allowing to produce rectangular pulses with variable duty-cycle and frequency.

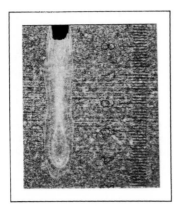

Fig. 3:   Typical weld with cut-in resulting from sputter found without puls shaping.

The investigated pulse parameters split into two regimes. A large regime was found producing results characterized by extreme sputter and cut-in at frequencies below 6 kHz, fig. 3. The penetration depth is deeper than in the corresponding cw-welds.

A smaller regime producing sound welds was found at frequencies between 6 and 8.5 kHz and duty-cycles of 75% to 80%. The sound welds are characterized by a welding depth approximately equal to that of cw-welds of the same mean laser power, fig. 4. This could mean that the weld, i.e. the capillary, is not affected by the high-frequency interruptions of the beam, so that the weld might be regarded as a quasi-cw-weld.

As a consequence of this, it was tried to reduce the sputtering in the low-frequency-regime, where, as predicted (1,2), higher penetration was found than in cw-mode, by the means of a pulse-shaping facility. The scope was to reduce the slope of laser power at

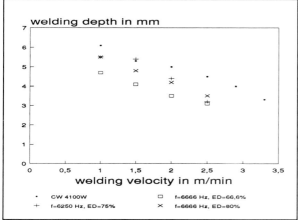

the beginning of the pulse, being very steep for rf-exited lasers, to approximate the pulse form in (1,2) from dc-exited lasers, known to be considerably smoother in general.

4.2 Welding with Pulse-Shaping

4.2.1 Medium Laser Power

Experiments were carried out with the TLF 1500 with 1.1 kW of cw- and maximal pulse power at the work-piece. The focus diameter being

Fig. 4: Welding depth over welding velocity for sound welds received from the TLF 5000 without puls shaping. The average power is 3.400 W for 66.6%, 3.600 W for 75%, 3.800 W for 80% duty cycle and 4.100 W for cw-operation.

approximately 0.13 mm, this power range leads to a cw- (and maximum pulse-) intensity of about $8*10^6$ W/cm^2. The pulse shapes used consisted of a sinusoidal ramp-in of 25 % of the total pulse length. The holding time was varied, leading to a variation of the duty-cycle of 50% to 80 %, see fig. 1a). The pulse frequency was varied from 100 to 900 Hz, the welding velocity from 0.25 to 5 m/min. The resulting values of average power were in the range of 250 to 660 W.

It was found that sound weld beads and little sputter result from pulse overlap rates (regarding the propagation while the pulse is switched on) of 60 % to 80 %. This means that the frequency has to be choosen with respect to the welding velocity.

On the other hand, throughout the whole pulse parameter range, it was found that high frequencies lead, at the same average power level, to lower penetration depths. A maximum is found at 200 to 400 Hz. Fig. 5 a) and b) shows the welding depth and width versus welding velocity for the cases of pulsed mode, 400 Hz and duty cycle 80 % resulting in 660 W of average power, cw-mode with 660 W and cw-mode with 1.100 W. It can be seen that pulsing leads to approximately the same penetration depth as cw-welding at full power, while significantly higher penetration is found compared to welding in the cw-mode with the pulse average power. Also, a better bead shape without "wine-glass" is observed.

4.2.2 High Laser Power

The experiments of chapter 4.2.1 were complemented with the TLF 5000 turbo delivering 4.400 W. A focal diameter of approximately 0.2 mm led to an intensity of $1.4*10^7$ W/cm^2.

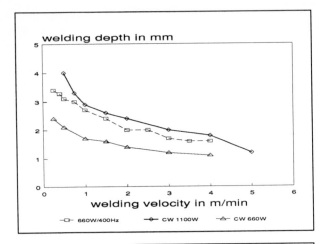

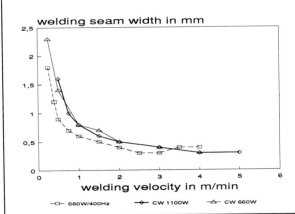

Fig 5a),b): Welding depth and seam width for welds produced with the TLF 1500 with puls shaping.

The pulse shape was slightly changed to a linear ramp-in of 40 % of the total pulse length that was found as the optimal pulse shape at this higher power level. The duty cycle varying between 53% to 80% resulted in average powers of 1.250 to 2.600 W.

The welds carried out with high average powers of above 1.500 W showed a strong tendency to sputtering and a low seam quality, while the penetration depth was comparable to that of cw-mode. At average power values below 1.500 W sound welds were found.

To understand this behaviour, high-speed video movies were taken from the welding process. These movies showed that the molten pool turbulences were the more severe, the larger the molten pool was. From these observations a maximum line energy can be estimated above which instabilities preferentially occur.

Low values of line energy can be due to low power or high welding velocity. So it was seen that, because of the otherwise resulting large molten pool, a relatively high average power can only be used at high velocities. Regarding the overlap-rate having to exceed 60 %, this leads to high frequencies and thus to high power slopes, which tend to sputtering, see chapter 4.1. All these factors limit the pulsed laser welding to a power level of less than 1.500 W and a velocity range of less than 5 m/min when using the equipement regarded in this work.

To demonstrate the influence of beam quality, i.e. the achievable focus diameter, the cw-

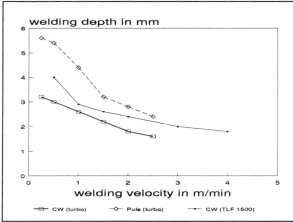

Fig. 6: Influence of the beam quality on the welding performance: Welding depth versus welding velocity for the TLF 5000 turbo, cw with 1.250 W and pulsed with 1.250 W, and the TLF 1500 cw with 1.100 W.

welds with this laser at 1.250 W and cw-welds with the TLF 1500 at 1.100 W were compared, see fig. 6. The TLF 1500, although delivering less power to the workpiece, shows a 10 to 25 % higher penetration depth than the TLF 5000 turbo.

The results of pulsed welding with the TLF 5000 turbo, which also are shown in fig. 6, show a penetration depth 50 to 80 % higher than the cw-mode weldings, reaching up to 5.5 mm at 0.5 m/min. From these findings it can be estimated that with a pulsed laser of 1.5 kW average power (with a beam quality comparable to that of the TLF 1500) yielding pulse maximum power values in the range of 4 to 5 kW, a still considerably higher welding depth would be achievable.

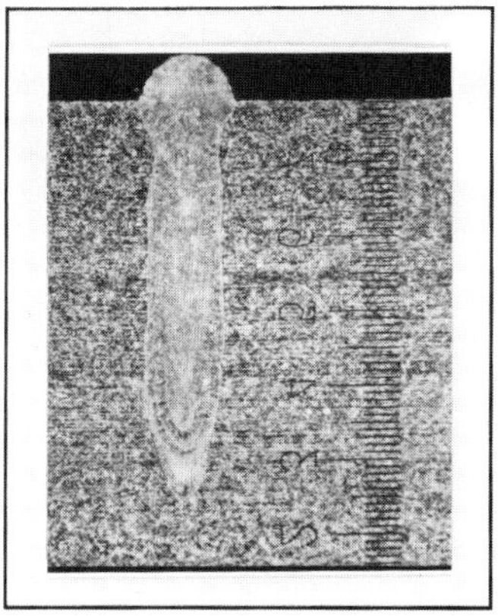

In this pulsing regime, sound welds without "wineglass"-shape are found, see fig. 7, when the laser was used with the pulse shaping facilitiy. With standard built-in rectangular pulse-shapes, strong sputter can not be avoided.

5 Conclusion

Pulsed weldings were carried out using three different rf-exited fast axial flow CO_2-lasers driven in a switching mode. It was observed that in only a frequency range below 1 kHz a rise in penetration depth versus cw-weldings at the same average power occurs. In a frequency range above 6 kHz, quasi-cw-type behaviour was established.

It was noticed that even at welding velocities as low as 0.3 m/min, no "wine-glass" shape was found.

Fig. 7: Typical pulsed welding from the TLF 5000 turbo with puls shaping. The parameters are: P = 1.250 W, f = 100 Hz, v = 1.5 m/min, d = 3.5 mm.

Hence it can be estimated that the laser-induced screening plasma plays no role in this mode of welding, as was expected from the experiments of (1,2).

It was found that sound welds can be produced, exceeding the penetration depth of cw-welds at the same average power by a factor of up to 3 for average powers as low as 250 to 300 W and a factor of up to 1.8 for average powers of about 1.250 W. In contrast to (1,2), where a dc-exited laser was applied, the use of a pulse shaping facility was turned out to be absolutely necessary. The reason lies in the fact that rf-exited CO_2-lasers exhibit by a strict response of laser pulse shapes to exitation pulse shapes, leading to steep slopes when exciting with a rectangular pulse.

A maximum power slope limits the pulse frequency and, assuring an overlap-rate of at least 60 %, the welding velocity to about 5 m/min. The size of the molten pool resulting in these relatively low velocities limits the pulsed mode welding to laser average powers of about 1.500 W.

A maximum attainable penetration depth appears to be determined by the maximum pulse intensity. Finally, from the results the conclusion can be drawn that maximum benefit from pulsed welding could be taken from applying a laser type providing time shaped higher power pulses at relatively low average power level. A conceivable availability of such a laser at the price of a standard low power system could increase the potential of laser welding as an efficient production technology.

6   Literature

(1)   S. Kimura, S. Sugiyama, M Mizutame: Proc. of the Int. Conf. ICALEO '86, ed. by C. M. Banas and G. L. Whitney, Arlington VA (1986) 89

(2)   T. Ishide, S. Shono, T. Ohmae, H. Yoshida, A. Shinmi: Proc. of the Int. Conf. LAMP '87, Osaka (1987) 187

7   Acknowledgement

The authors wish to thank Mr. M. Sprenger and Dipl. Ings. J. Scholz and H. Kunce for their work contributing to this paper.

This work was sponsored by the BMFT under contract number 13 N 5661. The authors are responsible for the contents of this contribution.

Influence of the Process Parameters on the Weld Metal Hardness of Laser Beam Welded Structural Steel Regarding the Formation of Pores

G. Kalla[*], K. Janhofer[**] and E. Beyer[*]

[*] Fraunhofer-Institut für Lasertechnik, Aachen
[**] Thyssen Stahl AG, Duisburg

Introduction

Laser beam welding is a high energy density, low heat input process. Metal sheets up to 25 mm thickness can be joined in a single pass by the deep penetration welding. Compared to e.g. submerged arc welding, the 5 to 8 times lower heat input /1,2/ affects the microstructure of the base material only to a small extend. A narrow heat affected zone without a coarse grain zone is obtained. On the other hand, the short heat cycle with rapid cooling causes an increase in hardness in the fusion zone of ferritic steels /3-7/. It has to be taken into consideration that the maximum hardness should fall below hardness limits recommended by welding codes /8/.

There are several possibilities to reduce the hardening effect e.g. by preheating / post weld heat treatment or an appropriate chemical composition of the base material / filler material. These methods often cause additional operating or material expenses. Apart from these methods, it is the aim to investigate the possibility to reduce the weld metal hardness by the adaption of the welding parameters.

It is of interest to study the influence of the line energy to steer the cooling rate of the fusion zone. This possibility can either be realized by an increased laser beam power or by slower travel speeds. This alternative is investigated within the following taking account on the efficiency of the process and the formation of pores in the weld seam.

Weldments

The investigations are carried out using a 22 kW continuous wave CO_2 laser. The structural steel Fe 510 (St 52-3) at 10, 15, and 20 mm thickness is autogeneously laser beam welded. The chemical composition and the base metal hardness is given in table 1.

Plate Thickness	Chemical Composition of Fe 510 (Analysis of the Workpiece in %)										Base Metal Hardness
	C	Si	Mn	P	S	Cr	Ni	Cu	Al	Nb	
10 mm	0.169	0.222	1.220	0.018	0.008	0.032	0.034	0.012	0.012	0.012	165 HV 10
15 mm	0.163	0.268	1.330	0.021	0.011	0.025	0.023	0.010	0.051	0.001	165 HV 10
20 mm	0.176	0.306	1.470	0.037	0.024	0.027	0.019	0.040	0.006	0.001	165 HV 10

Table 1: Chemical Composition of the Welded Plates

The welding is carried out in the H/V-position to prevent drop throughof the melt at very high heat input. Plasma suppression is accomplished with 30 l/min Helium flow out of a gas nozzle with 4 mm inner diameter. An increased line energy reduces the temperature gradient in the weld, which results in a longer cooling time of the weld metal. The cooling time $t_{8/5}$ during the transition from 800° to 500°C correlates with the weld metal hardness and is determined by thermocouples. The thermocouple is formed by a NiCr-wire and the steel plate to be welded. The wires are pressed against the plate and the junction is formed in the moment the laser passes the wire. Figure 1 depicts schematically the experimental set-up with four NiCr-wires for statistics reasons.

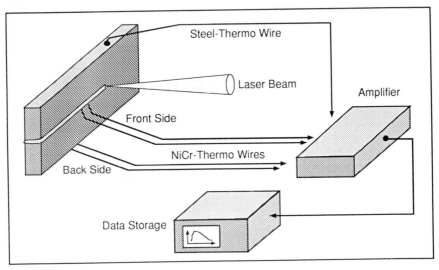

Figure 1: Set-up for Thermal Measurements

There are few publications dealing with the determination of the cooling with respect to the weld metal hardness /3/.

Moreover, a complete understanding of the dependence between the line energy and the cooling time requires the quantification of the absorbed energy by the workpiece. One characteristic feature of laser beam welding is the formation of a vapour capillary penetrating the workpiece. A certain amount of laser radiation is transmitted at the bottom of the capillary depending on the process parameters. An enhanced line energy, either by an increased laser power or by slower travel speeds influences the amount of absorbed and transmitted energy. The efficiency of the process depends on the absorbed share of the applied energy so it is of interest to measure this portion of energy. The complete set of weld runs is carried out again and the absorbed energy is quantified by a calorimeter instantaneously after welding. The calorimeter consists of a thermal isolated can filled with oil of a defined

temperature. The increase in oil temperature after immersing the welded specimen is a measure for the absorbed energy. The heat radiation losses prior to the calorimetry are assessed by thermographic pictures of the plates and turned out to be neglectible.

Furthermore, the transmitted portion of radiation is quantified. It is done for two reasons: first, it is the aim to gain an insight into the distribution between the energy absorption, transmission and others, and secondly, the signal of the transmitted laser power indicates a welding process, in which formation of pores might occur. Therefore, the transmission measurements are carried out time resolved. A mirror at the back of the weld focusses the transmitted radiation into an Ulbricht-sphere, which is equipped with a calibrated pyro- and thermo-detector.

Welding Results

The cooling time is an indicator for the weld metal hardness obtained. It is determined from the course of temperature during welding as it is exemplary shown in figure 2.

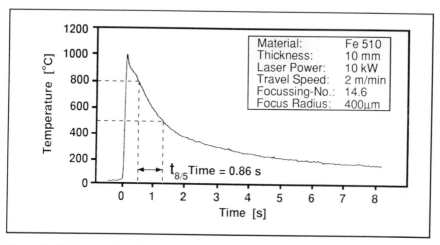

Figure 2: Determination of the $t_{8/5}$-Time from the Thermal Cycle

A 10 mm plate is welded at a laser power of 10 kW and a travel speed of 2 m/min resulting in a cooling time $t_{8/5}$ of 0.86 s. In distinction to the cooling times of 20 s to 30 s common to the submerged arc welding, the laser process has $t_{8/5}$-times of 0.5 to 10 s, depending on the welding parameters and the material thickness. The cooling time $t_{8/5}$ correlates the line energy to the weld metal hardness as shown in the figures 3 to 5.

The cooling time $t_{8/5}$ increases progressively with higher line energies. The heat flow into the base material depends on the ratio of the travel speed to the thermal

146

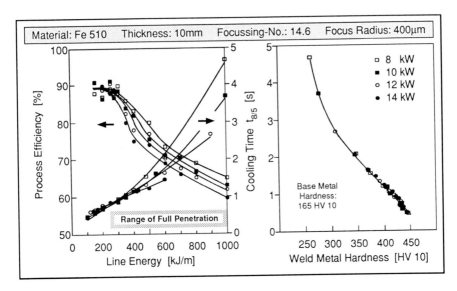

Figure 3: Influence of the Line Energy on the Process Efficiency
and the Weld Metal Hardness; 10 mm Material Thickness

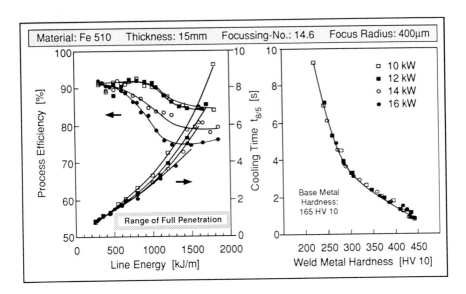

Figure 4: Influence of the Line Energy on the Process Efficiency
and the Weld Metal Hardness; 15 mm Material Thickness

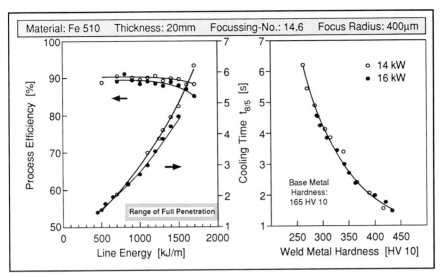

| Material: Fe 510 | Thickness: 20mm | Focussing-No.: 14.6 | Focus Radius: 400μm |

Figure 5: Influence of the Line Energy on the Process Efficiency
and the Weld Metal Hardness; 20 mm Material Thickness

conductivity of the material. Therefore the cooling rate for a given line energy decreases with laser power. The higher the line energy, the bigger is the difference in cooling time between the power levels.

A cross over to the right side of the figures 3 to 5 establishes the correlation between the welding parameters and the weld metal hardness. The Vickers hardness is measured at one third of the material thickness below the surface. The values do not depend on the beam power and range between about 250 HV 10 and 450 HV 10. For some applications, the welding codes recommend hardness values below 350 HV 10 /8/. In order to fulfill these requirements, the diagrams 3 to 5 can be interpreted "backwards": The maximum hardness is related to a minimum line energy for a given laser power.

Table 2 lists representative welding parameters applied to the three steels of table 1 to obtain the hardness levels 300/350/400 HV 10. For each hardness level, the travel speed is almost independent from the material thickness on condition of a constant ratio of beam power to material thickness (8kW/10mm, 12kW/15mm, 16kW/20mm).

The results show that a weld metal hardness of 350 HV 10 is obtainable by an increased line energy (travel speed of about 0.8 m/min). In the individual case, it has to be considered from the economical point of view, whether a reduced travel speed without further heat treatment of the material or a high travel speed in conjunction with a suitable treatment is to be prefered.

148

Laser Power P [kW]	Plate Thickness s [mm]	Ratio P/s	Line Energy E [kJ/m]	Travel Speed v [m/min]	Weld Metal Hardness [HV 10]
8	10	0.8	730	0.658	300
12	15	0.8	1070	0.673	300
16	20	0.8	1400	0.686	300
8	10	0.8	570	0.842	350
12	15	0.8	800	0.900	350
16	20	0.8	1050	0.914	350
8	10	0.8	390	1.231	400
12	15	0.8	540	1.333	400
16	20	0.8	650	1.477	400

Figure 6: Welding Parameters for Defined Hardness Levels

Figures 3 to 5 also give the efficiency of the process which is defined by the absorption by the workpiece of the applied energy. The absorption amounts to about 90 % in the incomplete-penetration regime at low line energy. The efficiency starts to decline when full penetration occurs. This point shifts towards lower line energy with increasing laser power as a consequence of the higher ratio of travel speed to thermal conductivity. A comparison of efficiency at very high energies reveals a significant decline for the 10 mm plate (≈ 65 %), whereas the 20 mm thick material is welded at about 85 % efficiency.

Porosity

The pores are one important quality mark of the joint and occur more often in welding thick plates than in welding plates less than 5 mm thick. The porosity at the above mentioned sets of parameters is determined by X-ray examination transverse to the seam. Here, the porosity is defined as the ratio of overall pore area to the longitudinal cross section of the weld. Figure 7 shows the porosity versus the line energy for welds of 10 mm and 15 mm thick material.

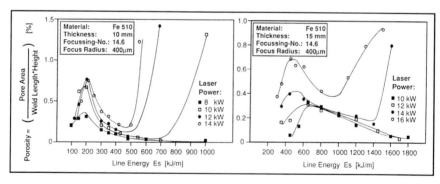

Figure 7: Formation of Pores

The local maximum of porosity occurs, when the penetration depth is equal to the material thickness and the capillary just opens at the bottom of the weld. A higher line energy increases the opening of the capillary, thus improving the outgasing of the capillary. Consequently, the formation of pores is reduced.

Figure 9 depicts exemplary time resolved signals of the transmitted laser power. All welding parameters except the material thickness are kept constant. At 15 mm thickness, the penetration is about the material thickness and only some transmission with large fluctuations is registrated. The peaks correspond to the states of the capillary when it briefly opens. These unstabilities of the capillary affect the melt pool dynamics causing an increased amount of porosity. In case of 6 mm thickness, a quite stable and high level of transmitted radiation is obtained leading to less porosity. At very high line energies, the porosity and the average pore size raise significantly. It is the welding regime, in which the intensity of the transmitted laser power is high enough to initiate the formation of a plasma at the bottom of the weld. The current hypothesis of the sharp increase of porosity is, that the fluctuations of the plasma generate pressure waves, which spoil the degasification of the capillary /9/.

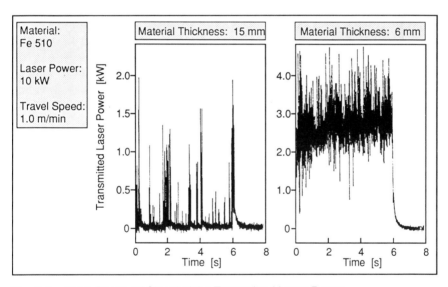

Figure 9: _Time Resolved Signal of the Transmitted Laser Power_

Comparing the porosity for a given line energy, higher power levels (= enhanced travel speed) increase the porosity because the time for degasification becomes shorter. Therefore, it is advantageous to weld at a low laser power in order to minimize the formation of pores.

Conclusions

One essential feature of laser beam welding is the short cooling time $t_{8/5}$ resulting in high weld metal hardness. For plates of 10 to 20 mm thickness, it can be influenced in the time regime between about 0.5 s and 10 s by an increasing line energy at the expense of travel speed. One limit is an abrupt increase of porosity at high line energies and there is a limited process regime where a sufficiently low hardness and a tolerable amount of pores are achievable. The results indicate a decrease in efficiency of laser power utilization in the low hardness regime, which has to be balanced against the expense of additional weld heat treatment.

References

/1/ E. Beyer and G. Herziger
Erfahrungen mit dem 22 kW CO_2-Hochleistungslaser in der Schweißtechnik
Koll. Schweißtechnik, 1989, Aachen

/2/ UP-Handbuch
Publisher: Messer Griesheim GmbH

/3/ E.A. Metzbower
Laser Beam Welding: Thermal Profiles and HAZ Hardness
U.S. Naval Research Laboratory Code 6324

/4/ D.J. Abson and R.E. Dolby
A scheme for the quantitative description of ferritic weld metal microstructure
Weld. Inst. Res. Bull., Vol. 21, No. 4, April 1980

/5/ M.N. Watson and I.M. Norris
Properties of laser welds in structural steels
Int. Conf. Power Beam Technology, September 1986, Brighton

/6/ E.A. Metzbower and D.W. Moon
Mechanical properties, fracture toughness and microstructures
of laser welds in high strength alloys
Proc. Conf. Applications of lasers in materials processing,
April 1979, Washington

/7/ P.E. Denney and E.A. Metzbower
HSLA Steel Laser Beam Weldments
Proc. ICALEO '83, Vol. 38, November 1983, Los Angeles

/8/ AD-Merkblatt HP 2/1, Ausg. 7.89
Verfahrensprüfung für Schweißverbindungen

/9/ G.U. Kalla and E. Beyer
Influence of the Welding Gas Parameters on the Weld Quality
Using High Power Lasers
Proc. 10th Int. Congr. LASER '91, June 1991, München

The Laser - A New Tool for Welding Aluminium Materials

Der Laser - Ein neues Werkzeug zum Schweißen von Aluminiumwerkstoffen

J. Berkmanns, K. Behler, E. Beyer

Fraunhofer - Institut für Lasertechnik Aachen

1. Introduction

During the last years an increasing degree of laser welding applications in industry can be noticed. Due to small focus diameters combined with high laser power deep and narrow weld seams can be realized at high welding speeds. Apart from these advantages the use of laser beams as welding tools leads to specific problems. For example joining of butt joints geometries requires an expensive clamping device, to minimize the gap between both parts. Often a gap within the scale of the focus diameter leads to not acceptable seam geometries with weld sink and root suck-up.

The use of filler material can eliminate such geometrical failures. Furthermore a burn-out loss of alloying elements can be avoided and the sensitivity against hot cracks can be put on a lower level. Many aluminium alloys contain alloying

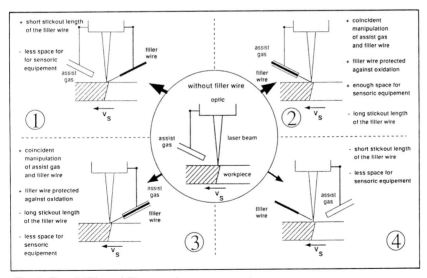

Fig. 1: Possibilities of filler supply arrangements

elements with low boiling temperature like magnesium and zinc (Al-Mg, Al-Mg-Mn, Al-Zn-Mg) or are susceptible to hot-cracking (Al-Mg-Si). Due to these facts the paper will discuss the laser beam welding process of aluminium, in particular using filler material. It will deal with different arrangements of filler supply, the process parameters and the quality of the welding results.

2. Experimental Investigations

Several arrangements of filler supply have been choosen and tested. Coming from the normal laser beam welding process without filler material, which is shown in the middle of fig. 1, four possibilities of filler supply arrangement are possible, which differ in the positon of wire- and gas nozzle, and offer different advantages concerning stick-out length and handling. In this experiments the alternatives 1- 3 have been investigated.

The first significant influence of the welding result is given by the direction of the wire feed in relation of the welding direction. As it can be seen in fig. 2 arrangements like 1 and 3, wich use a dragging wire supply, achieve good welding results with smooth seam surfaces and homogeneous weld metal, caused by a good dilution of base- and filler material. The welding result is nearly independent of the choosen gas supply, so that in principle an extra stigging gas nozzle can be used as well as a gas flow coaxial to the additional wire. In practice one may prefer the stigging nozzle because a shorter stick-out length of filler wire can be realized.

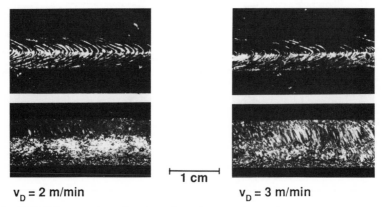

$v_D = 2$ m/min $\qquad$ 1 cm $\qquad$ $v_D = 3$ m/min

Fig. 2: Welding results using dragging wire supplies (v_D= wire feeding speed)

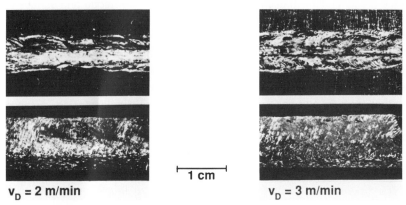

$v_D = 2$ m/min $v_D = 3$ m/min

Fig. 3: Welding results using stigging wire supplies (v_D= wire feeding speed)

In case of a stigging filler supply, when the laser beam is led in front of the wire, a good dilution cannot be reached. As the longitudinal section on the right hand in fig. 3 shows there is partly no connection between filler and base material. In both cases (dragging / stigging supply) the filler wire has been fed to the point, where the laser beam impinges the surface of the workpiece.

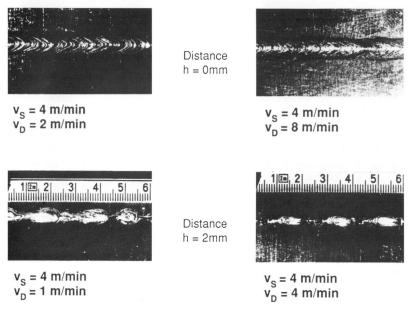

Distance
h = 0mm

$v_S = 4$ m/min
$v_D = 2$ m/min

$v_S = 4$ m/min
$v_D = 8$ m/min

Distance
h = 2mm

$v_S = 4$ m/min
$v_D = 1$ m/min

$v_S = 4$ m/min
$v_D = 4$ m/min

Fig. 4: Influence of the distance between wire and plate on the welding result

154

In the dragging configurations the distance between wire edge and surface of the workpiece is the second important process parameter. Due to the heat of the plasma the wire begins to melt before reaching the laser beam. Caused by the oxid film on the surface of the wire a drop of molten metal is formed in front of the laser beam. Contacting the surface of the plate the drop is ripped of the wire and mixed with the base metal by the following laser beam. The dilution of the drop material and base material becomes worse with increasing distance between wire and plate surface. Distances between wire and plate surface up to 0.4 mm are tolerable (fig. 4). Above this value the metal transfer takes place discontinously so that parts of the welding seam, which contain no filler material, are followed by parts of pure filler material /1/.

As it was mentioned in the introduction seam geometry and structural constitution can be influenced in a positiv way by the use of filler material. This is shown in fig. 5. Due to a lower temperature gradient and the improved chemical composition a finer grain structure is formed in the middle of the seam and no hot cracks occure. It is a matter of fact, that the welding speed decreases about 40%, if filler material is used. If the process parameters are choosen

without filler material   with filler material

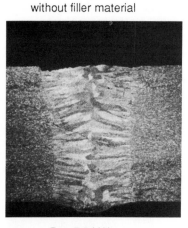 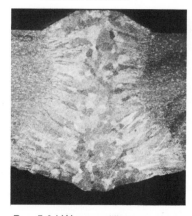

P_L = 5.0 kW P_L = 5.0 kW   filler:
v_s = 5 m/min v_s = 3 m/min   S-AlSi5
f = 150 mm   f = 150 mm   d = 1.6 mm
s = 3 mm   s = 3 mm v_D = 3m/min

Fig. 5: Seam geometry and structural constitution

gap width: 0.5 mm

v_S = 3 m/min

v_D = 5 m/min

P_L = 5.0 kW

gap width: 0.7 mm

v_S = 2 m/min

v_D = 6 m/min

P_L = 4.5 kW

base material: AlMgSi1

s = 2.5 mm

filler material: S-AlSi 5

d = 1.6 mm

Fig. 6: Laser beam weldings of aluminium butt joints with different gap width

adapted to the surrounding conditions the weldability of configurations with 1.2mm gap width are possible (fig. 6).

Fig. 7 compares the tensile strength and the proof strength of the laser butt joints to the values of the base material. The tensile strength of the weld seams achieve 75% of the base material, the proof strength 57%. In contrast to earlier investigations /2/, wich were carried out without filler material, the elongation reaches higher values up to 6.8% and achieves the minimum values, which are demanded for constructions with TIG and MIG welding seams /3/.

3. Conclusion

The results of the investigations can be summarized as follows:

- Using a dragging filler supply a continous metal transfer is possible

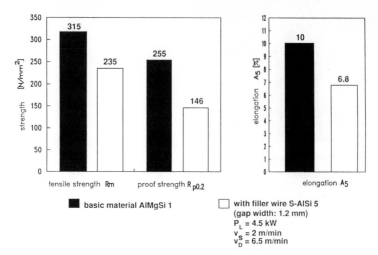

<u>Fig 7</u>: Mechanical properties of laser welded aluminium butt joints

- The wire has to be added to the impingement point of laser and workpiece to avoid discontinous metal transfer and drop appearance

- The weld metal is characterized by fine grained micro structure.

- At weldings speeds of 2 m/min 1.2 mm gap width can be filled up without a lack of fusion at the sides of the groove

- No hot cracks occure

- The mechanical properties achieve the demanded values for constructions with TIG and MIG welding seams

4. References

/1/ H.-D. Steffens, F. Hartung, R. Beermann, "Elektronenstrahlschweißen von Aluminium mit drahtförmigem Zusatzwerkstoff", DVS-Berichte Band 135

/2/ J. Berkmanns, K. Behler, E. Beyer, "Mechanical Properties of Laser Welded Aluminium-Joints", Proc. 25th ISATA Conf. 1992, Florence

/3/ W. Hufnagel et all., "Aluminium Taschenbuch", 1988

High power laser welding of microstructure sensitive and dissimilar materials

H.W. Bergmann, T. Endres, D. Müller
Lehrstuhl Werkstoffwissenschaften 2, (Metalle), Erlangen FRG

Abstract

The service requirements such as wear and corrosion resistance of components often vary locally due to different load and environmental situations. Modern design and manufacturing techniques can achieve optimum performance through localised application of appropriate material properties. Laser welding permits these concepts to be achieved through the welding of dissimilar metals with range of material properties. This technique has the advantage of minimal heat input and precise positioning of the weld joint to suit the materials being joined.

This paper discusses the welding of microstructure sensitive materials and stainless steel to case hardening steel and copper, and of case hardening steel to medium carbon and tool steel. Examination of the welds includes hardness, tensile strength testing and the evaluation of microstructures.

1. Introduction and motivation

In material science, especially in the field of automotive industry and energy technology, it is of great interest to develop materials with increased performance, for example better strength to weight ratio or increased service temperatures. The improved capabilities of such materials, however, have to justify their higher price. Possible solutions are the use of highly specialised materials adapted to the required service conditions, the use of compounds or combination of materials which show qualities impossible for one single material. High material properties are often not required for the whole volume of the component. In this case surface techniques like thermal, thermochemical or chemical layers can be applied [Ref. 1]. In many cases special material properties are not required all over the surface of the component. In this case local treatments or combination of the materials is very useful. Joining is often carried out mechanically (eg. bolting, riveting) while in other cases a metallurgical connection may be required eg. hard soldering or welding. Welding of metallic components can be carried out by CO_2 laser and its main advantage in modern industry is that of cost-effective production. In addition the low heat input and the specific resolidification behaviour of the welded material allows joining of materials and material combinations, which can not be welded by conventional techniques. In earlier works Bimberg [Ref. 2] showed that more material types can be welded with a laser then with conventional techniques. In recent years high power CO_2-lasers with an output power up to 25 kW have become available. The high power densities achievable allow welding depths > 20 mm. The question is, however, whether the increased number of materials and material combinations can still be joined or the allowed materials is reduced to the number of the conventional welding techniques. The presented paper tries to answer this question with typical examples of iron based materials.

158

2. Welding of microstructure sensitive materials.

For demonstrating the capability of laser welding at power levels over 5 kW the directionally recrystallised ODS-alloy is chosen. The task is to preserve the orientation of the grains of the not heat affected zone in the resolidified weld material [Ref. 3]. One result arising from the use of laser welding is the well known epitaxial crystallisation of the melted material on the surface of the unmelted surrounding substrate [Ref. 4]. It is possible to establish a directed solidification over the melted area of the welded path. In this case it is necessary to establish sufficient high

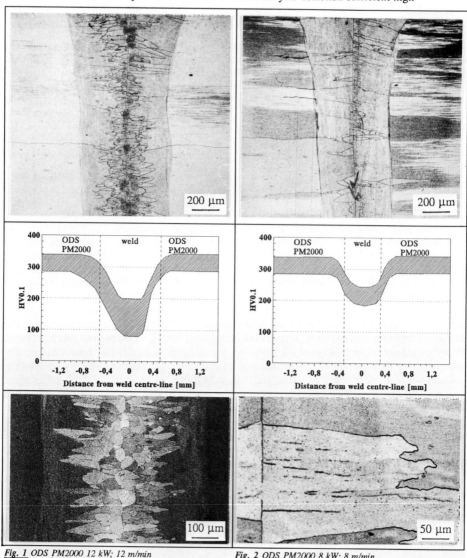

Fig. 1 ODS PM2000 12 kW; 12 m/min   *Fig. 2* ODS PM2000 8 kW; 8 m/min

high resolidification rates to achieve extensive supercooling without the inoculation and growth of new grains. The shielding of the weld with inert gas is critical to avoid the generation of detrimental inoculation oxides. It is also necessary to use high feed rates to avoid effects of segregation and phase separation. Figures 1 and 2 compare the typical microstructures of a welded ODS PM2000. Increasing the feed rate at constant heat input increases the width of the weld and the ratio of the weld depth to weld width changes. In both cases the crystallisation of the melted material starts epitaxially at the solid interface of the surrounding material. At high feed rates new crystals are formed in the core of the weld. At lower feed rates, however, the original grains are reestablished. This is possible because of the short solidification path resulting from the narrow geometry of the weld zone achievable by high power laser welding. High welding speed in combination with high laser output powers cause high supercooling, and therefore the creation of small sized grains is favourable. These two types of microstructures also cause a significant difference in hardness profiles (see Figures. 1, 2).

3. Dissimilar materials

3.1. Combination of iron and non iron-based materials

The typical process of welding dissimilar materials is the joining of iron with non-iron based materials on the one hand and welding of different iron based alloys on the other hand.

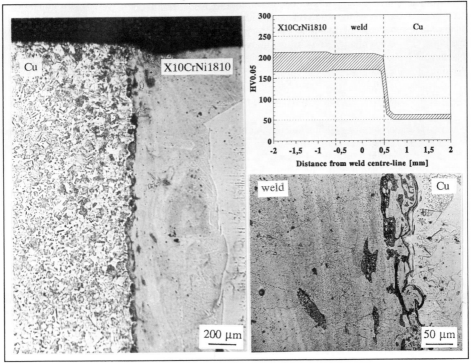

Fig. 3 Overview (upper left); Hardness (upper right); Detail (lower right): stainless steel - copper

The welding of such combinations is limited by thermodynamic effects like differences in melting point, solid solubility and the creation of brittle intermetallic phases. The transformation mechanisms of different iron alloys is a function of their content of alloying elements.

The welding of stainless steel to copper represents the first type of material combinations with the additional advantage of similar crystal structures of both materials and one alloying element (Ni) being soluble in both partners. Figure 3 gives an overview of this combination.

Mechanical properties of a Cu - stainless steel weld are significant. Bending angles up to over 60 degrees can be tolerated. In the middle area of the melted material the microstructure is mainly supersaturated iron γ-solid solution with a significant amount of precipitations. The stainless steel side of the weld shows a very smooth transition zone in microstructure. The copper-rich side contains additionally a pearlitic grain type. Therefore the hardness profile of the weld zone is very constant at the level of the hardness of the γ-iron. The hardness of the pearlitic microstructure is shown in Figure 3.

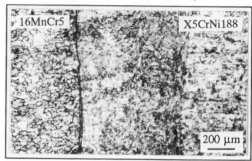

Fig. 4 X5CrNi188 - 16MnCr5

Alloy combinations which have the capability of creating intermetallic phases can be welded by applying high laser output powers and high feed rates. Creation of these intermetallic phases can be reduced or suppressed by high resolidification rates because of the time necessary to establish the crystallisation of complex intermetallics. Eg. conventional iron-chrome alloy with body centred cubic lattice structure builds an intermetallic σ-phase only at low cooling rates. The welding of these materials is therefore even possible in combinations.

4.   Iron based materials with different transformation behaviour

The most frequently used combination of materials with different transformation behaviour is found in the joining of ferritic to austenitic steels. Figure 4 demonstrates that this is possible without problems, and the positioning of the weld is not critical in this case. The mechanical properties are good and limited by the weaker partner, Figure 5. More critical is the joining of low carbon transformable steels (eg. 100Cr6 or St14) to wear resistant high carbon steels (eg. X210Cr12 or 21MnCr5). For these combinations the positioning of the weld is very important if a satisfactory weld is to be achieved, Figure 6.

If the weld is biassed towards the transformable steel, excessive weld hardness results and cracking is likely. If the weld is biassed towards the carbon-rich steel, segregations can occur although cracking is generally

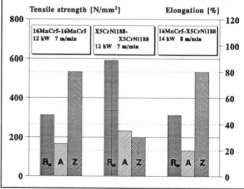

Fig. 5 Mechanical properties

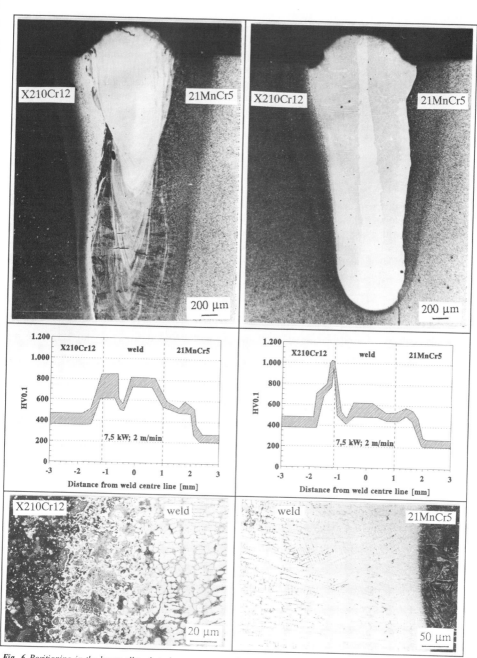

Fig. 6 *Positioning in the lower alloyed steel*

Fig. 7 *Positioning in the high alloyed steel*

biassed towards the carbon-rich steel, segregations can occur although cracking is generally avoided. Because of the high solidification rates involved and the high carbon contents a microstructure with γ-dendrites and a very fine interdendritic eutectic is formed.

This eutectic structure has a very fine grain size and has good deformability. Figure 6 shows details of the zone between the ferritic and the ledeburitic steel (Figure 7). Hardness profiles across welds made at different positions relative to the gap are shown in Figures 6 and 7. In the transition zones of both welds the hardnesses are high compared to the unaffected material. In the first case however, there are very high hardnesses due to martensitic transformations. In the second case the hardness is lower resulting from a mixture of the γ-solid solution and the very fine ledeburitic eutectic [4].

5. Conclusions

The welding of different material combinations is more easily accomplished with lasers than with conventional welding techniques, however weld joint positioning is very important. The small weld zones achieved with a laser allow precise positioning of the weld joint to minimise weld hardnesses and the high powers available permit reproducible resolidification conditions and therefore controlled creation of microstructures. The exact positioning of the weld however needs to be adapted to suit the final weld geometry of the component.

6. Acknowledgement

The authors thank A. Endemann and A. S. Bransden (GEAT-Industrielaser) for help with the experimental work.

7. References

[1] H.W. Bergmann et. al., Laser hardfacing, Steel & Metals Magazine, Vol. 28, 1990, p 275
[2] D. Bimberg, Materialbearbeitung mit Lasern, Expert Verlag, Ehningen, 1991, p 100
[3] R.F. Singer, G.H. Gessinger; Powder Metallurgy of Superalloys, Butterworth, London, 1984 p 213
[4] H.W. Bergmann, Habilitationsschrift, Untersuchungen an schnell abgeschreckten metallischen Systemen, Clausthal, 1983

Mechanical Properties of Laser Welded Al-Alloys

A. Lang, H.W. Bergmann, Lehrstuhl Werkstoffwissenschaften 2, (Metalle), Erlangen FRG

Abstract

The present paper outlines the mechanical properties of CO_2-laser welded aluminum alloys like AA 2090, AA 8090, AA 2091, AA 6082 and AA 5083. Radiographs, microstructure, hardness, tensile test, fatigue behaviour and maximum bending angle are reported. Good results were obtained for the alloys AA 8090, AA 6082 and AA 5083. The alloys AA 2090 and AA 2091 show a reduced weldability. A heat treatment after the welding process leads to an improvement of the mechanical properties.

Introduction

Riveting and glueing are the most commonly applied joining techniques for aluminum aerospace components. The disadvantages of these methods are the poor degree of automation and the long processing times. The use of aluminum alloys in the car industry would be significant for weight reduction. To execute this step a suitable welding process is necessary. Aluminum alloys are difficult to weld due to their high heat conductivity and high gas absorption. In Table 1 welding characteristics using TIG, laser or electron beam are given. Electron beam welding avoids the gas absorption, however, a disadvantage arises from an expensive vacuum technique. Therefore applications are limited to small components. This favours the laser for car welding applications.

Parameter	TIG-welding	Electron beam welding	Laser beam welding
Power density (W/cm^2)	10^5	10^9	10^7
Welding speed (m/min)	< 1	1 - 10	1 - 10
Atmosphere	Shielding gas	Vacuum	Shielding gas

Table 1: *Typical parameters of various welding techniques*

Poor absorption of the CO_2-radiation and a high reduction of solubility of hydrogen at the cristallization point of aluminum are the main problems in laser welding of aluminum alloys. To increase the absorption of the radiation a high power density is necessary which generate a surface plasma causing a nearly complete absorption of the CO_2-radiation. The high decrease of solubility for hydrogen at the cristallization point leads to recombination of hydrogen during solidification which results in a high porosity in aluminum welds (1). To minimize this problem expensive shielding gas arrangements and edge preparations including milling of the welding edges followed by a brushing immediately before the welding operation are necessary (2).

Experimental setup

Different aluminum alloys were used for the welding experiments, in which both conventional alloys like AA 6082 and AA 5083 and aluminum-lithium-alloys are investigated.

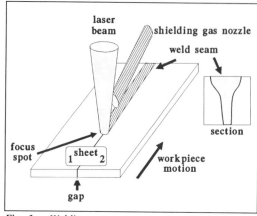

Fig. 1: *Welding arrangement*

The welding experiments were done with a 4 kW CO_2-laser and a handling system consisting a translation stage and a clamp in which sheet samples in butt-joint were fixed. The shielding gas arrangement contains a gas nozzle at the top fed by a mixture of Ar/He and a helium root shielding at the bottom side to prevent atmospheric effects. Fig. 1 shows schematically the welding arrangement. The welding edges were milled and brushed immediately before the welding process. The welds are characterized by metallographic investigations, hardness measurements, tensile tests, scanning electron microscopy of the fracture areas, fatigue tests and evaluations of the maximum bending angles.

Results and discussion

The first investigation of the welded samples is the macroscopic view as can be seen in Fig. 2. All welds have a uniform appearence with nearly oxide-free surfaces. The radiographs show especially at the alloy AA 2091 some pores.

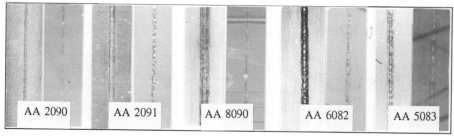

Fig. 2: *Macrographs and radiographs of laser welded sheets*

In the micrographs of Fig. 3 pores are visible. Especially the alloy AA 2091 exhibits a higher porosity effecting a distinct bulge. The microstructures show a fine dendritic cristallization in the weld seam. Only the alloy AA 8090 has a small heat affected zone (HAZ) which is etched a little darker than the parent material. In the hardness measurements, however, a much higher HAZ than in the microstructures is visible. Especially in T3 heat treated alloys (AA 2090, AA 2091, AA 6082) a very large HAZ is noticed which is characterized by a decrease of the hardness values compared to the parent material due to coarsening or solution of the precipition particles. The age hardening alloy AA 8090 which was heat treated for superplastic deformation show small variations in the hardness values. The naturally hard alloy AA 5083 has a similar behaviour.

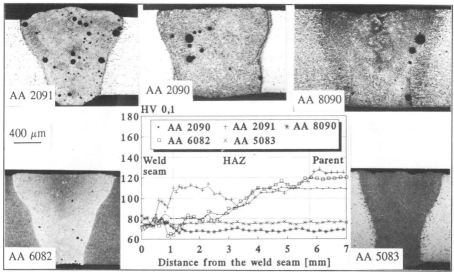

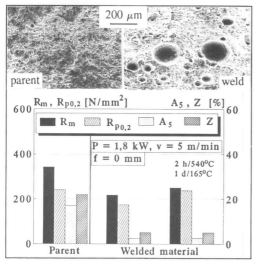

Fig. 3: _Microstructure and hardness_

Fig. 4: _Tensile test of the alloy AA 2090_

In a tensile test the welded specimens of the alloy AA 2090 reach distinctly lower mechanical properties than the parent material. The tensile strength value achieves 210 N/mm² which is 70 % of the parent material's value. The elongation to fracture decreases from 18 % of the parent to 2 % of the weld. After a heat treatment the welded samples reach a higher tensile strength as the non-heat treated samples and the same yield strength as the non-welded material. The reason of the deterioration in mechanical properties is the high porosity of the weld seam, see the surface of fracture. The pores decrease the supporting cross section causing a reduced tensile strength. The notch effect lessens elongation to fracture and necking.

The alloy AA 2091 shows similar results, see Fig. 5. The difference between parent and welded material, however, is higher than by the alloy AA 2090. The tensile strength of the welded sample reachs only 25 % of the parent material's value. Elongation to fracture and necking are minimized. A heat treatment after the welding process leads to a slightly improvement of the mechanical properties. The reason for this reduced properties of the welded samples is a very high porosity of the weld seam which quantity influences the mechanical properties essentially more than by the alloy AA 2090, see the fracture area.

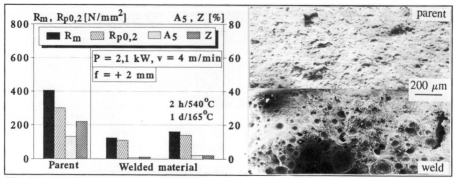

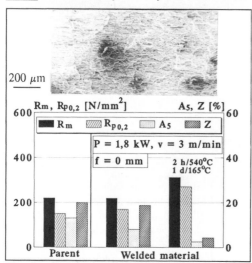

Fig. 5: Tensile test of the alloy AA 2091

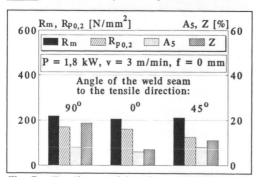

Fig. 6: Tensile test of the alloy AA 8090

The mechanical properties of welded specimens of the alloy AA 8090 are nearly as good as the parent material. There is only a slight decrease in elongation to fracture, see Fig 6. The reason for this behaviour is the previous heat treatment which has generated a microstructure for superplastic deformation. Its good ductility combined with a relative low tensile strength effects the plastic deformation and the fractures occur in the parent material. After a heat treatment an increase of the tensile strength appears where as the elongation to fracture show a reduction due to the changing of the superplastic deformation microstructure to an age hardened one.

Fig. 7 shows tensile tests of the alloy AA 8090 in which the weld seam has different directions relative to the tensile direction. The maximum differences occur by the necking. Especially the sample having the weld seam parallel to the tensile direction the necking is low. The reason is in fact that compared to the samples with rectangular direction the deformation cannot take place in the parent material.

Fig. 7: Tensile test of the alloy AA 8090

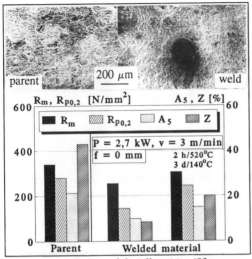

Fig. 8: Tensile test of the alloy AA 82

Welded samples of the alloy AA 6082 attained good results in a tensile test, see Fig. 8. The tensile strength values reach 80 % of the parent material's value. Elongation to fracture and necking values of the welded samples which achieve nearly 10 % demonstrate a sufficient plasticity. The slight decrease arises from the minimized ductility of the weld seam. After a heat treatment all mechanical properties can can be forced up to the range of the parent material. There is one pore in the fracture surface of the welded sample visible which is too less to affect the mechanical properties.

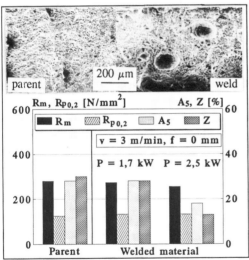

Fig. 9: Tensile test of the alloy AA 5083

The alloy AA 5083 shows the best results in the tensile test, see Fig. 9. The welded samples with the lower laser beam power reaches the parent material's values. The tensile strength achieves a value of 270 N/mm^2 in which elongation to fracture and necking reach nearly 30 %. The fracture appears outside of the weld seam in the parent material at the sample which was welded with the lower laser power. The 2.5 kW welded sheet shows a higher porosity in the surface of fracture. This leads to a deterioration of elongation to fracture. Now the fracture appears in the weld seam causing a decreased necking.

Alternating bending tests were carried out for the alloys AA 8090 and AA 6082. Fig. 10 exhibits Wöhler curves for the alloy AA 8090. The values of the welded samples are only about 10 - 20 % lower than the parent material's values. The fatigue strength reaches 50 N/mm^2 for the welded material. Heat treated samples achieve the same fatigue strength as the parent material.

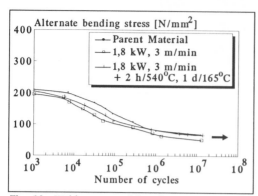

Fig. 10: *Wöhler curve of the alloy AA 8090*

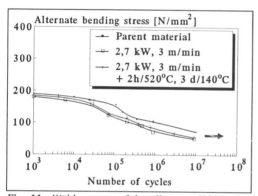

Fig. 11: *Wöhler curve of the alloy AA 6082*

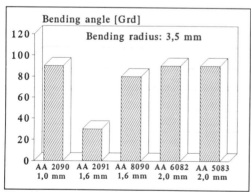

Fig. 12: *Maximum bending angle of different alloys*

The alloy AA 6082 show a similar fatigue behaviour as the alloy AA 8090 (Fig. 11). Welded samples achieve nearly 70 % of the parent material's fatigue strength. A heat treatment leads to a slight improvement of the fatigue behaviour.

Fig. 12 shows the maximum bending angle of the investigated alloys. Welded samples were bent parallel to the weld seam until cracks in the surface are visible. Only the alloy AA 2091 has a low bending angle where as the alloy AA 8090 endure a bending angle of 80°. It could be proved that for the other alloys using a maximum bending angle of 90° there are no cracks detected.

Conclusions

A good weldability was found for the alloys AA 8090, AA 6082 and AA 5083. For the alloys AA 2090 and AA 2091 a high porosity leads to a deterioration of the mechanical properties.

Acknowledgements

These investigations were financially supported by the BMFT (Förderkenn-zahl: 13N5654-2) and the companies Messer Griesheim GmbH, Puchheim, FRG and Dornier Luftfahrt GmbH, Friedrichshafen, FRG.

References

(1) Aluminium-Zentrale, Düsseldorf [Hrsg.], Aluminiumtaschenbuch. Düsseldorf (1988), 378

(2) B. Biermann, R. Dierken, H.W. Bergmann, Metall **45** 4 (1990), 328-336

III.
Surface Melting, Alloying and Hardening

Metallurgical Aspects of Laser Surface Treatment

B. L. Mordike, Institut für Werkstoffkunde und Werkstofftechnik, Technische Universität Clausthal, FRG

Introduction

Laser surface treatment involves the application of high energy density electromagnetic radiation to the surface of a material. This can induce changes in the metallurgical structure of the surface layers and thus change the properties. It is important to understand the interaction between the beam and the material surface, the processes which are initiated and the correlation between properties and microstructure if a controlled surface treatment with a specific property profile is to be at all possible. There are now three types of laser in common use for laser surface treatment - excimer laser, neodymium-YAG laser and CO_2 laser. They differ in wavelength and mode of use (pulsed or continuous) and hence the applications differ in detail but not necessarily principally.

The structural changes which occur can take many forms eg change in dislocation density (work hardening), by generation of a shock-wave, production of internal stresses, modification of the Bloch wall structure, initiation of solid state transformations, selective ablation of a constituent phase as well as the whole range of metastable microstructures obtained by rapidly quenching. Equally important is a treatment to produce a particular surface roughness or treated pattern. It is thus not surprising that lasers are being applied to a very wide range of problems, nor is it surprising that the assessment of the applicability of a specific laser application is a long and tedious process.

The aim of this paper is to discuss the nature of some of these structural changes and how depending on the desired application they can be encouraged and exploited or minimised to avoid deleterious effects.

Hardening by Shock Waves

An interesting application of lasers is shock wave treatment [1]. If a laser beam is applied for a short time to a surface it is possible to generate a shock wave. The energy densities required are about 10^7-10^{11} W/cm^2. The length of the pulse is between some pico seconds to a few hundred nano seconds. Nd:YAG lasers are generally used but CO^2 or KrF (0,25 μm) lasers can also be used. Frequency doubling with Nd:YAG also increases the range available 1.06μ, 0.51μ, 0.35μ and 0.26μ.

It is known, Fig. 1, that dynamic deformation produces various types of lattice defects eg dislocation cells or twins [2]. In Fig. 1 the critical pressure for twinning is plotted as a function of the stacking fault energy. The shock produced by a laser is much shorter than those produced by conventional methods.

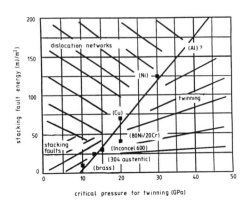

Fig. 1: Various forms of dynamic deformation.

Plastic deformation takes place in two stages: plastification in
compression, then plastification in traction. This can result in a
work hardening of the irradiated area. It is also possible to
produce compressive residual stresses (Fig. 2) in the surface.

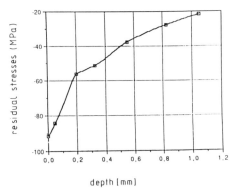

Fig. 2: Compressive residual stresses produced in aluminium Copper by simultaneous
pulses of 15 ns at an intensity of 6 GW/cm^2.

For this it is necessary to separate the thermal and shock wave
effects. Fig 3 shows the tensile stresses obtained on irradiating
a surface with and without an absorbant.

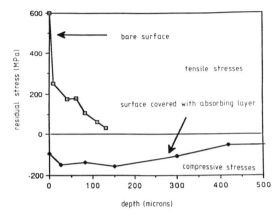

Fig. 3: Application of absorbant coating to
ensure formation of compressive residual
stresses.

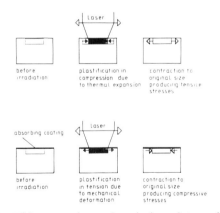

Fig. 4: Illustration of origin of tensile
and compressive stresses with and without
absorbant layer.

Fig. 4 shows the effect of the absorbant. Only the shock wave penetrates the material. The area attempts to deform in tension but
is restrained by the surrounding material. The restrained area is

thus a region of compressive residual stresses. If heat enters the irradiated area expansion results in deformation. On cooling tensile residual stresses build up (1).

The production of compressive stresses is one possible application. Compared with shot peening similar stress levels can be achieved but with less damage to the surface. Greater penetration depths are achieved.

Transformation Hardening in Iron Base Alloys

Laser transformation hardening of iron base alloys involves austenitisation and self quenching of a surface layer. Laser treatment implies rapid heating with a short austenitisation time. The degree of austenitisation determines the hardening response on subsequent rapid cooling. The hardness achieved can thus be significantly different from that obtained using a conventional process. There are several aspects to the problem. Rapid heating displaces the transformation temperatures to higher temperatures. This can be by as much as 100-150°C. The time spent above A_3 is thus short and this limits the processes of dissolution of graphite or carbides and homogenisation. The initial microstructure assumes an importance not necessary in conventional hardening processes. Microstructural heterogenities, coarse carbides or pearlite exacerbate the problem of homogenisation.

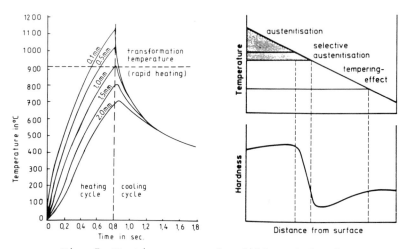

Fig. 5: Heating curves for different depths below the surface and the corresponding microstructural changes at different depths for a given heating rate.

Ideally, the microstructure should be specified, in addition to the composition, for alloys which are to be laser transformation hardened. Otherwise, the hardening is likely to be inhomogeneous with all attendant problems, eg irregular internal stresses. The heating rate is different at different distances from the surface (Fig. 5). The transformation temperature, the temperature attained, the degree of homogenisation and the subsequent cooling rate are all dependent on the depth from the surface (12). Fig. 5 (a) and 5 (b) shows the heating curves for different depths below the surface and the corresponding microstructural changes at different depths for a given heating rate. If these problems are taken into account it is possible to employ laser transformation hardening. It has been successfully applied to carbon and alloy steels, tool steels, case carburised steels and cast irons eg (3, 4)

Control of Magnetic Domain Size

The Institute for Laser Technology in Aachen has developed an industrial process for determining the domain size in transformer sheets (Fe3%Si) (5). The process is exploited by Thyssen steel company. Such steel sheet is coated with an insulating coating. A highly focussed laser beam is scanned over the sheet at right angles to the rolling direction. The insulating coating is unaffected; the sheet shows a significant decrease in magnetic lasers by reducing the eddy currents.

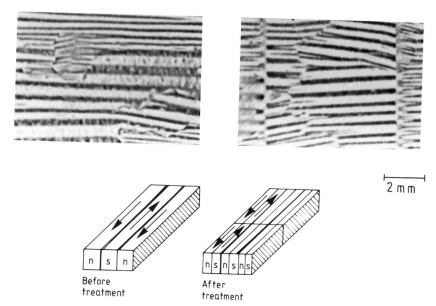

2 mm

Before treatment

After treatment

Fig. 6: Domain pattern in a Fe3,0%Si steel sheet before and after laser scribing.

The laser beam induces thermal stresses which are considered to initiate localised dislocation movement, thereby subdividing the domains. This is turn limits the domain width. By varying the scan separation the optimum domain size can be determined. Fig. 6 shows the a steel before and after laser scribing (6).

Surface Melting

Applications involving surface melting rely for improvement of the surface properties on one or more factors. In the case of cast materials, remelting the surface eliminates casting defects such as porosity, inclusions cold shuts etc. The main reason for laser surface melting is to obtain better properties as a result of the rapid solidification. The quenching rates can be as high as those achieved in other rapid solidification techniques eg melt spinning or powder atomisation.

There are the following techniques for treatment of surfaces involving rapid solidification

    simple surface melting (7, 8)
    surface alloying (9, 10)
    cladding (11)
    amorphisation

Important for a discussion of the constitutional effects is the composition of the melt. Rapid quenching influences the nature and stability range of phases, also their size, morphology and distribution. The final microstructure depends, in addition to the composition, on the process parameters. The phases which form can be non crystalline (amorphous) or crystalline (stable or metastable). Other effects such as extension of the range of solid solubility or suppression of phases may be observed. Equilibrium diagrams can only be employed an aid when considering the possible outcome of a laser surface treatment. The microstructure which is produced in laser surface melting is the result of many competing processes. There is nucleation which can occur either heterogeneously on the substrate or within the melt, or homogeneously. Thereafter there are the various forms of growth.

The solidification behaviour is influenced mainly by three parameters - quenching rate, ε, solidification rate, R, and temperature gradient, G. These three parameters are interrelated in the expression $\varepsilon = R \cdot G$. The temperature gradient in laser surface melting depends on the energy density and the solidification rate, apart from stationary pulsing, on the displacement rate.

Solidification proceeds either as a stable planar front or as an unstable front, which leads to dendrites or cells. Which process prevails depends on the nature of the system and whether constitutional supercooling is possible. Constitutional supercooling arises when the actual temperature gradient is less than the liquidus temperature gradient due to the change in composition as a result of partition at the interface.

The mode of solidification depends on kinetic considerations. Retention of solute requires a plane solid solidification front. A dendritic front leads to segregation in form of coring often resulting in the formation of a second phase from the residual liquid.

There are two conditions for a stable front movement.

(i) low growth velocities 2 DG/ΔT. D is the solute diffusion coefficient in the melt and ΔT is the temperature interval between liquids and solidus. The temperature gradient must be large enough to prevent constitutional supercooling (12). This growth condition is termed conditional stability.

(ii) hogh growth rate velocities such as are possible in laser treatment also result in planar growth. This condition of absolute stability is fulfilled if the velocity > D ΔT/ Γk_0 where Γ is the Gibbs-Thomson coefficient (ratio $\gamma/\Delta s$ of interfacial energy to entropy of fusion) and k_0 is the distribution coefficient.

This is demonstrated in Fig. 7 for Al-Cu alloys.

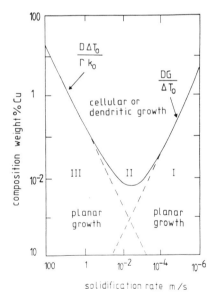

Fig. 7: Interface stability as a function of the interface velocity V and Cu consideration, could under a temperature gradient in the liquid of 2.0×10^4k/m.

Fig. 8 shows the different conditions for planar or cellular and dendritic growth as determined by G and V in Al-Cu alloys. The lines separating two regimes correspond to the locus of conditional stability (12).

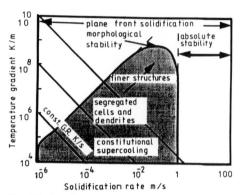

Fig. 8: Plot of temperature gradient versus solidification rate and solidification morphology.

The microstructures produced is thus dependent of many factors. Examples of the type and site of nucleation can be found in the literature. Similarly examples of the influence of the temperature gradient and solidification front rate also numerous. The conditions in the laser melted pool vary from point to point so that often many types of growth are observed depending on the local temperature gradient.

These considerations apply to all forms of surface melting irrespective of whether the composition has been changed (alloying or cladding) or not.

If simple remelting cannot achieve the desired improvement in properties - by structural refinement of modifying the solidification process eg cast iron aluminium silicon alloying can change the relative stability of phases leading to the formation of new phases eg Al-Si-Ni of Ti-N or even enabling amorphisation. The conditions for amorphisation are shown in Fig. 9. The various forms of growth are shown as a function of solidification rate and composition for a simple binary system. It is necessary for amorphisation that the T_0 curves are not continuous or do not cross. Partionless solidification is not possible over a wide range of composition at any growth velocity.

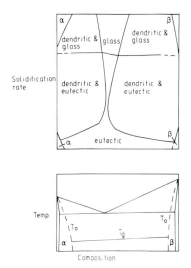

Fig. 9: Solidification rate versus composition with associated equilibrium diagramm
for amorphisation.

Surface Texturing

The surface roughness is an important parameter for many applications. Lasers can be used to create a pattern on the surface by
selective ablation. An interesting application is texturing rolls
for sheet rolling. Rolled sheets will assume this dimple pattern.

Paint adhesion is improved as is lubrication in the pressing operation. The roll is rotated and traversed relative to the laser
beam. A chopper is used to obtain the desired pulsing effect.

There are other surface treatment involving a change in microstructure and or surface structure which have not been discussed
in depth, but which are interesting applications eg: reversible
change in reflectivity on amorphisation (can be used as necessary
storage plates), selective removal of cast iron to lay bare graphite lamella, patterned treatment of surfaces instead of complete
treatment (skelton treatment) as used by MAN on cylinder liners.

References

(1)   J. Fournier, PhD Thesis, 1989, Ecole Polytechnique Paris

(2)   L. E. Murr, Shock waves in condensed matter, Ed. S. Schmidt,
      N. Holmes, Elsevier BV 1988, 315-320

(3)   W. Amende, Der Laser in der industriellen Fertigungstechnik,
      Ed. H. Treiber, 1990, Darmstadt: Hoppenstedt Verlag 193-233

(4) B. L. Mordike, 1987, Laser M2P Conference Proceedings Journal de Physique, C7, 37-43

(5) A. Gilner, K. Wissenbach, E. Beyer, G. Vitr, Proceedings 5th Int. Conf. Lasers in Manufacturing (LIM5), Edit. H. Hügel, Stuttgart, Sept. 1988, Publ. IFS Ltd (U.K.), 137-144

(6) B. Weidenfeller, TU Clausthal, Private Communication

(7) H. W. Bergmann, NATO School on Laser Applications, San Miniato, Italy, Applied Sciences NATO ASI Series E, Ed. C. Draper and P. Mazzoldi, Publ. Martinus Nijhoff, Dordrecht, 1986, 351-368

(8) R. Sivakumar, B. L. Mordike, 1988, J. Surface Engineering 4, 127-140

(9) B. L. Mordike, D. Burchards, Industrie-Anzeiger, 1988, Nr. 73, 10, 11, 17

(10) S. Mordike, SURTEC Berlin '89, Carl Hanser Verlag München

(11) D. Burchards, A. Hinse, B. L. Mordike, Mat-wiss und Werkstofftechnik, 1989, 20, 405-409

(12) S. R. Coriell, R. F. Sekerka, Rapid Solidification Processing: Principles and Technologies II, Ed. R. Mehrabian et al Claitors Baton Rouge La, 1980, 35-49

Characterisation and Wear Resistance of Cobalt Base Coatings Deposited by CO2 Laser on Steam Turbine Components

P.A. Coulon, GECAlsthom, Steam Turbine Research, 90018 Belfort, France

J. Com-Nougue and E. Kerrand, Alcatel Alsthom Recherche, Centre de Recherche, 91460 Marcoussis, France

G. Thauvin, GECAlsthom, CERM, 90018 Belfort, France

Abstract

A laser coating process has been set up and its performances evaluated in the case of a cobalt base alloy to be deposited on steam turbine components. The experiments were carried out using a CW CO_2 laser of 5 kw maximum power connected to a 5 axis workstation with CND. The laser beam was partially focussed on the workpiece using a ZnSe lens and the prealloyed powder was delivered by a vibrating feeding system and injected outside the nozzle at the level of the beam impact on the workpiece. The coatings generally exhibited an excellent compactness and were free of porosity. Single layer coatings up to 1.8mm thick were achieved, characterized by a very fine dendritic structure and a highly regular interface with the steel substrate showing a very low dilution. Microanalysis performed on both sides of the interface indicated that the inter-diffusion zone did not exceed 20 microns. The coatings generally exhibited microhardness values above 600 Hv but the progressive heating of the specimen during a multitrack coating was shown to induce a slight increase in the hardness from the first track towards the last one. The residual stresses induced by the coating process were also determined. Typically, the coating itself exhibited tensile stresses which reached a peak of 600 MPa while compressive stresses (max. 110 MPa) were produced in the substrate hardened zone followed by a tensile stress peak and again by compressive stress.

In the present paper, CO2 laser coating of a cobalt base alloy on a 12% chromium steel with a view to applying it to the protection of steam turbine blades against wear was investigated in order to achieving coatings thicker than 1.5mm and 20mm wide and applying the process to test-blades for evaluating their abrasion/erosion resistance.

Introduction

The leading edge of "12 % Cr" steel LP blades is protected against erosion by the soldering of a grade 6 stellite plate (1). The welding alloy used (Ag-Cu-Zn-Cd) has a liquidus temperature of around 650°C, close to the tempering temperatures of the "12 % Cr" steels. Incidents showed that it was difficult to keep to the soldering temperature. Whenever the temperature is exceeded, considerable structural modifications bring about the formation of extremely hard, corrosion-sensitive martensite.

This may affect the performance of the blades during operation.

GECAlsthom, in collaboration with the Alcatel-Alsthom Recherche, has developed a procedure for depositing cobalt base alloy powders on to "12 % Cr" steel with, a CO² laser. These depositing techniques and the related technology are dealt with in numerous Marcoussis Laboratories publications (2 and 3). In this document, GECAlsthom is interested in the erosion resistance of the deposits made, and their characterisation after the erosion tests.

This laser depositing technique may in fact be an excellent replacement for the technique of the soldered plating, the tests showing in particular that the juxtaposition of the high degree of hardness of the deposited metal, together with the very fine structure obtained by the laser would make for outstanding resistance to wear by erosion.

Cobalt base alloy

The filler to be used is in the form of powdered grade 6 stellite, grain size : 25 to 125 μm.

Composition as follows : 63 % Co - 29 % Cr - 4.5 % W - 1.2 % C - 1.2 % Si.

Depositing conditions

The following parameters were used :

- Diameter of laser spot : 3 mm

- Speed : 40 cm/min⁻¹ to 60 cm/min⁻¹

- Pitch between passes : 2 mm

- Laser power : 3.5 kW

- Linear energy :   4.3 kJ.cm⁻¹

Various depositing procedures have been studied in order to perfect the process of application onto the blades. These procedures are described in figure 1 :

- Procedure 1 : consists in carrying out two lateral passes followed by a central pass with a power of 2.7 kW on a spot 3 mm in diameter, at a speed of 40 cm/min. The coating thus obtained, is surfaced by two re- fusion passes performed at 2.5 kW (spotdiameter : 4 mm) and at 65 cm/min.

- Procedure 2 : a variant of Procedure 1 with 2 extra passes.

- Procedure 3 : the first pass is made in the axis of the part and is followed by 2 lateral passes and a fourth central pass made at a higher speed (50 cm/min instead of 40 cm/min).

In all the cases, before each laser pass the part was positioned under the laser beam, so that the beam could be aimed at the part at an angle perpendicular to the surface, and so that the height of the surface to be treated remained constant in relation to the focal points of the optic system.

The deposits obtained are illustrated in figure 2.

Erosion tests (by impact)

The impact erosion tests were carried out by DORNIER (4-7) on samples 17 mm in diameter and 6.5mm thick, coated in stellite on one side.

Each test was made on two samples under the following conditions :

- V = 600 m.s⁻¹

- angle of incidence = 90°C

- diameter of water droplets = 1.2 mm

- test pressure lower than atmospheric pressure in order to avoid heating

- initial surface state of ground Ra = 2.6 μm

- water concentration = 4.5.10⁻⁶ (volume of water in relation to volume of air)

with measurement of weight loss against time.

The results obtained are shown in figure 3. The laser-welded resurfacing proved to have a high resistance to erosion, a long incubation period (> 30 min) and a slow rate of actual erosion. It had better resistance characteristics, at comparable parameters, than solid forged stellites, or casted stellites of the same grade.

The erosion resistance of the laser deposits, already demonstrated by the results of the Dornier tests, is thus shown to be excellent.

Metallurgical tests

Examination of substrates

Stellite resurfacing was carried out on two types of "12 % Cr" steel.

For steel type Afnor Z20 CNDV 12.1 (Din X19CrNiMoV12.1), the maximum hardness in the heat affected zone is very high (670 HV), much higher than the maximum hardness of the grade (600 HV) determined during a metallurgical study on another type of steel with an identical carbon content.

The difference may stem from the temperature reached during resurfacing, allowing a higher level of solubility of the carbides and carbonitrides, and/or carbon diffusion from the stellite.

For steel type Afnor Z12 CNDV 12.2 (Din X12CrNiMoV2.2) maximum hardness in the heat affected zone is identical to that measured on a martensitic structure (500 HV).

Tests on "stellite" deposits

Laser stellite resurfacing is characterised by :

- a fine-grained dendritic structure

- improved mean hardness.

Figure 4 enables comparison of the laser resurfacing structure with the structure obtained with a grade 6 stellite deposited using an oxyacetylene torch and a grade 6 casted stellite.

The usual-hardness value of stellite 6 is 400/500 HV.

Laser welding makes it possible for an improved degree of hardness (570 HV) to be obtained, principally due to a finer structure. The ductility of the resurfacing, tested by bending, is close to zero. This resurfacing prevents deformation. Some faults in consistency were observed in the new surface coating (blistering). It should be noted that the depositing conditions allowed resurfacing without cracks to be carried out.

Hardness

Hardness was measured on the original deposit and after heat treatment. Table 1 gives the average hardness of stellite and the maximum hardness measured in the heat-affected zone. In the heat -affected zone, steel type Afnor Z20 CDV 12.1 has a maximum hardness of 670 HV, type Afnor Z12 CNDV 12.2 of 500 HV. After tempering at 640°C, hardness is around 350 HV for both types of steel. Stellite hardness is 580 HV as deposited and 550 HV after treatment. Hardness charts are plotted in figs. 5 and 6.

	TYPE Z20 CDV 12.1 STEEL	TYPE Z12 CNDV 12.2 STEEL
Max. hardness in heat affected zone	670	500
Max. hardness in heat affected zone 4h - 640°C	360	345
Average stellite hardness	577	577
+ 4 h - 640°C	546	546

Table 1   HV hardnesses

Characterisation after erosion

After erosion the surface showed some parts intact and some with large cracks   .Examination under SEM shows that the erosion originates on both sides of the ghost lines.   One may observe numerous cracks on the walls.   No clear relationship can be seen between the pattern of erosion and the structure; a very fine erosion micro relief can, however, be seen, probably related to the size of the stellite grain structure.

Bending tests

The test was carried out using a roller 100 mm in diameter (10e). The coating was ruptured on samples with sideways welded resurfacing with no permanent deformation, and on a sample with lengthwise resurfacing with a permanent deformation corresponding to a deflection of 1 mm.

Conclusions

The resurfacing samples of "12 % Cr" steel with grade 6 stellite by laser welding created a good quality assembly free from cracking (in heat affected zone and on new surface coating). The high degree of hardness in the heat affected zone may make a tempering treatment necessary after welding.

The resurfacing showed a few very minor defects in consistency, a higher degree of hardness and a fire-grained structure. It proved to be highly erosion-resistant, which was the aim of the testing (Ref 8-9) The required thickness of 1.5 mm was reached without difficulty with a single application, or better with two applications, the second pass causing (moderate) heat treatment of the initial pass.

This process of laser welding should eventually make it possible to obtain coatings showing sufficient erosion resistance with various grade stellites of better ductility (grade 21 for example).

Bibliography

(1)   P.A. Coulon, "Erosion-Corrosion in Steam Turbines : A Problem Largy Resolved", Lubrication Engineering, June 1986 Vol 42 N° 6

(2)   J. Com-Nougue, E. Kerrand, "CO2 Laser Deposition of a Cobalt Base Alloy on a 12 % Chromium Steel", Proceedings of the 3rd International Conference on Lasers, 3-5 June 1985 Paris, pp 191-195

(3)   J. Com-Nougue, E. Kerrand, J. Hernandes, "Chromium Steel Laser Coating with Cobalt Base Alloy", Proceedings of LAMP 87, Osaka May 198 pp 389-394

(4)   Dr. Schroeder, "Handbuch Regen-Erosion", Dornier System GmbH., Koblenz BWB.AT (Dec 1984)

(5)   DORNIER System, "Liquid Impact Investigations, with the Mach 3 Rotating Arm Facility", Angewandte Forschung (1980)

(6)   G.F. Schmitt Jr., "Advanced Rain Erosion Resistant Coating Materials", SAMPE 18, pp 57-75, (1973)

(7)   G.F. Schmitt Jr., "Liquid Impact Erosion - A Unique Form of Wear for Aerospace Material", SAMPE, pp 16-22, (July-Aug. 1977)

(8)   M. Cantarel, G. Coquerelle, T. Puig, "Traitements de Surface par Laser", Matériaux et Techniques, Juillet/Août 1988, pp 17-24

(9)   P.A. Coulon, "The Resistance to Erosion of Multiphase Alloys with Titanium Carbide Base", Lubrication Engineering, April 1988, Vol 44 N°4

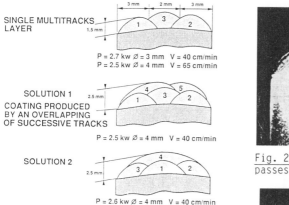

SINGLE MULTITRACKS LAYER

3 mm 2 mm 3 mm

1.5 mm

P = 2.7 kw Ø = 3 mm V = 40 cm/min
P = 2.5 kw Ø = 4 mm V = 65 cm/min

SOLUTION 1

2.5 mm

COATING PRODUCED
BY AN OVERLAPPING
OF SUCCESSIVE TRACKS

P = 2.5 kw Ø = 4 mm V = 40 cm/min

SOLUTION 2

2.5 mm

P = 2.6 kw Ø = 4 mm V = 40 cm/min
P = 2.6 kw Ø = 4 mm V = 50 cm/min

Fig. 1: Cross-section of 3 types successive passes

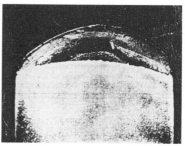

Fig. 2a: Cross-section of successives passes

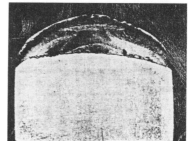

Fig. 2b

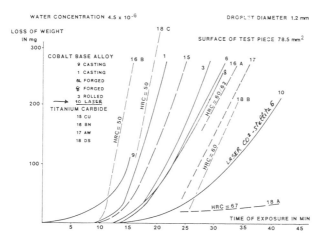

WATER CONCENTRATION 4.5 x 10^{-6}

DROPLET DIAMETER 1.2 mm

LOSS OF WEIGHT
IN mg

SURFACE OF TEST PIECE 78.5 mm^2

300

COBALT BASE ALLOY
9 CASTING
1 CASTING
6L FORGED
8 FORGED
3 ROLLED
10 LASER
TITANIUM CARBIDE
15 CU
16 BN
17 AW
18 DS

200

100

18 C
16 B 1 15 6 16 A 17
3 8
HRC = 50
HRC = 50
HRC = 60/62
9
18 B 10
LASER CO2-stellite 6
HRC = 60
HRC = 67 18 A

TIME OF EXPOSURE IN MIN

5 10 15 20 25 30 35 40 45

Fig. 3: Impact erosion of cobalt alloys

186

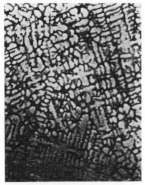

Casted stellite

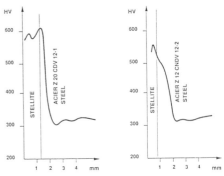

Fig. 5a: Microhardness HV 50 N
Without tempering

Laser deposition
Fig. 4: Comparison between
resurfacing structure with
stellite Gr 6

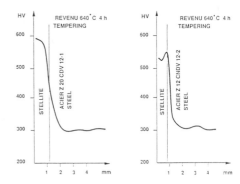

Fig. 5b: Microhardness HV 50 N
after tempering 640°C/4h

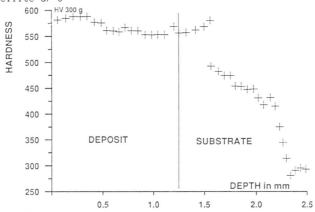

Fig. 6: Evolution of hardness HV as a function
of the depth

Laser Alloyed High Speed Steels

R. Ebner, B. Kriszt, Christian Doppler Laboratorium für "Lasereinsatz in der Werkstofforschung" at the Institut für Metallkunde und Werkstoffprüfung, Montanuniversität Leoben, A - 8700 Leoben, Austria

1. Introduction

Most of the conventional high speed steels (HSS) have alloying element contents which approach the metallurgical limits (1-5). Carbon and strong carbide forming elements like vanadium, tungsten, molybdenum etc. form a coarse interdendritic carbide eutectic under common solidification conditions which are responsible for a deterioration of the workability and toughness.

The outstanding process characteristics of laser treatment offer new possibilities in the development of new high alloyed HSS (6-7). Most important are the high melt pool temperature at the interaction zone with the laser beam, which enables the dissolution of thermodynamic stable chemical compounds like monocarbides, and the high cooling rate, which is responsible for the fine solidification structure in laser alloyed HSS. The typical cooling rate during solidification in laser surface treated HSS lies between 10^3 and 10^4 K/s (6,8). In comparison cooling rates lie for conventional HSS between 10^{-1} and 10^1 K/s (9,10) and for powder metallurgical (PM) processed HSS between 10^3 and 10^7 K/s (9-11).

Structure and properties of laser alloyed HSS strongly depend on the solidification process and solid state phase transformations after the solidification. As discussed in previous papers (6,7), alloying element additions significantly change the constitution of the HSS used as substrate which is itself a multicomponent system. Therefore the solidification sequences are also altered and the kind and the volume fraction of the various phases are changed.

Most of the as solidified structures contain plate-like carbides resulting from a eutectic solidification even for the high cooling rates in laser treated HSS. In addition usually relatively high amounts of retained austenite are present combined with a reduced hardness (12). Heat treatment subsequent to laser alloying can therefore be performed to modify the structure and the properties. In principle tempering as well as rehardening followed by tempering are possible (13,14).

In the following, typical laser alloyed structures in the as solidified condition are discussed and the effects of various heat treatment procedures are reported.

2. Experimental

Laser alloying was performed by using a 2,5 kW CO_2-Laser with a TEM_{20} beam mode in combination with a 6-axes CNC machine and a powder feeder in a so called "one step" process. Alloyed melt tracks are formed by blowing carbide powder into the melt pool which is created by scanning the laser beam across the HSS surface. Details of the alloying process are summarized in prior papers (8,13).

An annealed AISI M2 HSS was taken as substrate. Niobium carbide (NbC), vanadium carbide (VC) and molybdenum carbide (Mo_2C) were chosen as additions. Therefore, the adding of carbide forming elements simultaneously causes an increase of the carbon content. The process parameters were chosen so that nearly all injected particles were dissolved in the superheated melt. This means that a dilution of the elements of the substrate material occurs in the melt tracks. Typical chemical compositions of alloyed tracks are reported in (7,8).

The as solidified and the heat treated microstructures were examined by means of scanning electron microscopy and quantitative electron beam microanalysis. Further X-ray diffraction was performed to identify the crystallographic structure of the microstructural constituents.

The properties of the alloyed tracks were characterised by means of vickers indentation hardness testing with a load of 29.43 N (3 kp).

3.Results

The laser alloyed tracks exhibit a nearly semicircular cross section, the width is about 2 mm and the depth about 1 mm measured from the surface of the melt track. This leads to a cooling rate of about $6 \cdot 10^3$ K/s for the parameters chosen (6-8).

Solidification structure:

Fig. 1 shows the microstructure of the surface melted M2 HSS without additions. The high cooling rate causes a significant structural refining. For instance, in the surface melted AISI M2 a typical dendrite arm spacing of about 3μm can be found. A slight supersaturation of the dendrites occurs as a result of the high cooling rate. The alloying element concentration in the dendrites is about 4 to 20 % higher than in the conventional casted HSS ingots. The supersaturation is different for the various alloying elements (8). Further it has to be noted, that the amount of the carbide eutectic is reduced due to the high cooling rates from about 13 vol. % to about 5 vol. %.

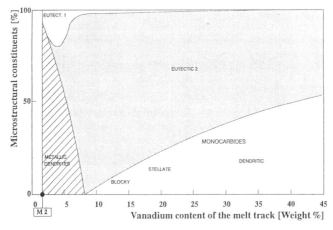

Fig. 1: Microstructure of the laser surface melted M2 HSS, backscattered electron image

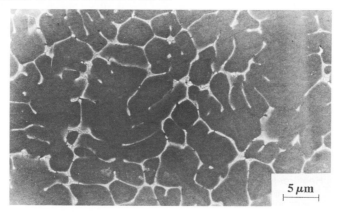

Fig. 2: Structure diagram for a M2 HSS laser alloyed with vanadium carbide (cooling rate $6 \cdot 10^3$ K/s)

Microstructural investigations of the laser alloyed melt tracks reveal that the effect of carbide additions depends extremely on the kind of the added carbides. As shown in prior papers the solidification structures can be interpreted well on the basis of equilibrium phase diagrams (7,8). But the addition of the various carbides changes the constitution of the new formed alloys in a rather different way. Consequently the kind of phases and their volume fraction vary with increasing carbide additions. So called "structure diagrams" were developed to describe the microstructural changes. Structure diagrams for NbC, VC and Mo$_2$C additions were reported in prior papers (6-8). Fig. 2 for instance shows the changes in the solidification structure as a function of the mean vanadium content of the alloyed tracks. The structure for 2 wt. % vanadium equals that of the laser surface melted M2 HSS. As shown in fig. 1 the surface melted M2 exhibits a dendritic microstructure with about 5 vol. % of an interdendritic eutectic. Increasing vanadium and corresponding carbon content lead to a reduction of the volume fraction of dendrites and to an increase of the amount of eutectic. But it has to be noted, that above about 4 wt. % vanadium a second kind of eutectic is created which forms eutectic cells additionally to the interdendritic eutectic. Fig. 3 shows an example of a laser alloyed melt track with about 10 wt.% vanadium. The dark phases are vanadium rich MC$_{1-x}$ monocarbides which are precipitated directly from the melt. These monocarbides are surrounded by eutectic cells which also contain MC$_{1-x}$ carbides. Among these cells a small amount of a molybdenum rich eutectic is still present. The morphology of the primary monocarbides changes with increasing vanadium content from blocky (fig. 3), to stellate and dendritic shape at the highest vanadium contents (8). Microanalyses revealed that the monocarbides dissolve appreciable amounts of the other alloying elements. Especially tungsten enters the monocarbides which causes a tungsten depletion in the metallic components.

Fig. 3: Microstructure of a M2 HSS laser alloyed with VC, 10 wt. % vanadium, backscattered electron image

Fig. 4: Microstructure of a M2 HSS laser alloyed with NbC, 6.6 wt. % niobium, backscattered electron image

The structures of niobium carbide alloyed tracks are very similar to that of vanadium carbide alloyed tracks (6). Primary niobium rich monocarbides precipitated from the melt can be observed for niobium contents above about 4 wt. %. An example of the solidification microstructure of a melt track with about 6.6 wt. % Nb shows fig. 4. The MC_{1-x} monocarbides appear bright in the backscattered electron image. It has to be noted, that the niobium rich monocarbides dissolve significantly smaller amounts of the other alloying elements than the vanadium rich monocarbides.

The effect of alloying with Mo_2C on the structure is rather different to that of alloying with VC or NbC. Mo_2C additions lead to an increase of the eutectic, but there are no indications for a change in the kind of the eutectic. In the case of tracks with higher molybdenum contents X-ray diffraction reveals that the eutectic is a M_6C-γ-eutectic. At about 30 wt. % molybdenum the structure is fully eutectic, a further increase of the molybdenum content leads to a precipitation of ternary M_6C carbides from the melt (6,7). Fig. 5 for instance shows the microstructure with about 17 wt. % molybdenum consisting of dendrites embedded in the M_6C-γ-eutectic. Some of the dendrite centers in fig. 5 appear darker in the back scattered electron image than the outer regions. Metallographic investigations revealed that the dendrite centers consist of retained deltaferrite which has not transformed to austenite.

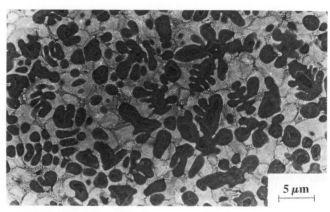

Fig. 5: Microstructure of a M2 HSS laser alloyed with Mo_2C, 17 wt. % molybdenum, backscattered electron image

Following the phase diagrams, the metallic dendrites are deltaferrite in the first step of solidification. In a second step, the deltaferrite transforms into a peritectic reaction to austenite. In most cases this transformation into austenite completely occurs, but as shown in fig. 5 some exceptions are possible. During further cooling, austenite partially transforms into martensite.

Heat treatment of laser alloyed HSS:

Various heat treatment procedures can be performed to modify the structure and properties of the laser alloyed tracks. In principle single or multistep tempering as well as rehardening followed by tempering are possible (13,14).

Fig. 6 shows the effect of a two step tempering on the hardness of a VC alloyed M2-HSS. The tempering curves reveal a pronounced secondary hardening above 500°C tempering temperature which is combined with a significant decrease of the amount of retained austenite. Fig. 6 also indicates the low hardness of the laser melted M2 in the as solidified condition. The hardness is about 100 to 200 HV lower than that of the solid state quenched M2. The low hardness in the laser melted HSS is a result of the high amount of retained austenite. But whereas the hardness can be significantly raised by tempering, the shape of primary or eutectic carbides remains unchanged. These brittle plate-like carbide phases in combination with the secondary hardened martensitic matrix reduce the toughness.

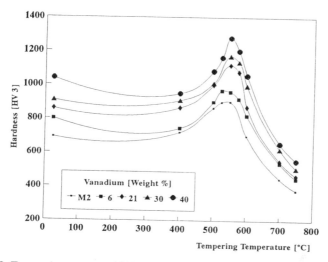

Fig. 6: Tempering curves of M2 HSS laser alloyed with vanadium carbide

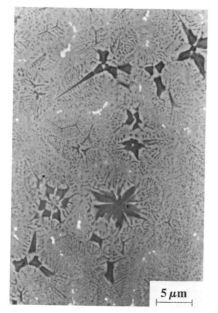

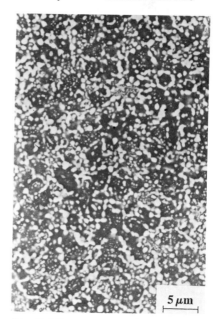

Fig. 7: Microstructure of a M2 HSS laser alloyed with VC and rehardened at 1100°C for 3 min, 10 wt. % vanadium

Fig. 8: Microstructure of a M2 HSS laser alloyed with Mo₂C and rehardened at 1180°C for 3 min, 17 wt. % molybdenum

A significant change of the carbide shape requires higher temperatures. To spheroidize the eutectic carbide plates usually temperatures above 1000°C are necessary. Fig. 7 shows

192

the structure after rehardening of a vanadium carbide laser alloyed M2 at 1100°C for 3 minutes. The solidification structure of the heat treated tracks in fig. 7 is compareable to the microstructure shown in fig. 3. As indicated in fig. 7 the carbide plates transform into small spherical carbides at this relatively low hardening temperature whereas at the primary carbides no significant morphological changes can be observed. A further example for the effect of rehardening on the microstructure is shown in fig. 8 for a Mo_2C alloyed M2. The solidification structure of the track in fig. 8 equals the structure shown in fig. 5. As indicated by fig. 8 not only a spheroidization of the eutectic carbide plates occurs but also a precipitation of carbides from the supersaturated dendrites.
Tempering experiments revealed that rehardened laser alloyed HSS exhibit a pronounced secondary hardening similar to conventional HSS.

Summary

Laser alloying of an AISI M2 high speed steel with NbC, VC and Mo_2C is discussed. The carbide additions to the M2 HSS significantly change the constitution. Therefore the kind and the volume fraction of phases are altered in the solidification structure. The rapid solidification of laser alloyed tracks causes fine solidification structures, supersaturated crystals and reduced amounts of eutectics.
The contents of the alloying elements can be increased significantly over that in conventional or even PM HSS. For instance melt tracks with vanadium contents up to 40 wt. % could be realised, but increased cracking probability above about 20 % vanadium has to be taken into account.
As a result of the supersaturation the as solidified laser melted and alloyed structures exhibit relatively large amounts of retained austenite which reduces the hardness. Heat treatment can be performed to modify the structure and the properties. The as solidified structures exhibit a pronounced secondary hardening on tempering. To spheroidize the brittle carbide plates heat treatments at temperatures above 1000°C are necessary.

Acknowledgements:

Financial support by the Austrian Industries, the Ministry of Science and Research (BMWF) and the Austrian National Bank (Österreichische Nationalbank) is gratefully acknowledged.

Literature:

(1) K. Bungardt, H. Weigand, E. Haberling: Stahl und Eisen 89 (1969) 420
(2) A. Randak, W. Verderber: BHM 121 (1975) 379
(3) H. Brandis, E. Haberling, H.W. Weigand: Proc. Processing and Properties of High Speed Tool Steels, The Metallurgical Society of AIME (1980) 1
(4) R. Riedl, S. Karagöz, H. Fischmeister, F. Jeglitsch: BHM 129 (1984) 71
(5) H.J. Becker, H. Brandis, P. Gümpel, E. Haberling: Thyssen Edelst. Techn. Ber. 11 (1985) 83
(6) G. Hackl, R. Ebner, F. Jeglitsch: Z. Metallkde. 83 (1992) in press
(7) R. Ebner, B. Kriszt: Proc. Int. Conf. Tooling Materials, Interlaken 1992, in press
(8) R. Ebner: Proc. 2nd European Conf. on Advanced Materials and Processes, Cambridge (1991) 115
(9) M.C. Flemings: Proc. Metallurgical Treatises, Beijing, The Metallurgical Society of AIME (1981) 291
(10) R. Riedl: Dissertation Montanuniversität Leoben (1984)
(11) H. Gahm, K.D. Löcker, H. Fischmeister, S. Karagöz, A. Gruber, F. Jeglitsch: Sonderbände der Praktischen Metallographie 18 (1987) 479
(12) R. Ebner, E. Pfleger, F. Jeglitsch, K. Leban, G. Goldschmied, A. Schuler: Pract. Met. 25 (1988) 465
(13) R. Ebner, G. Hackl, E. Brandstätter, F. Jeglitsch: Proc. 1st Int. High Speed Steel Conf. (1990) 81
(14) B. Kriszt, R. Ebner: VDI Bericht Nr. 917 (1992) 363

Load Bearing Capability of Laser Surface Coatings for Aluminium-Silicon Alloys.

E. Blank, O. Hunziker and S. Mariaux, Ecole Polytechnique Fédérale de Lausanne, Département des matériaux, CH-1015 Lausanne (Switzlerland)

Introduction

In mechanical engineering, there are numerous problems where frictional wear due to the relative motion of sliding interfaces is accompagnied by high local pressure in the contact area. Situations of this kind are encountered in fretting and machining. Also, the moving components of combustion engines and pumps undergo the same type of loading. The role of surface treatment is to protect the components either by applying a thin high-strength coating to an already strong substrate material, or by applying a relatively thick load-bearing coating if the substrate does not exhibit the necessary strength.

Laser surface remelting and alloying with Fe, Ni and Co of AlSi cast alloys has been shown to significantly improve the wear resistance under the conditions of unlubricated friction against steel, and in scratch tests [1-5]. Both surface remelting and surface alloying result in reduction of wear rates up to a factor of 2-3 [1, 4, 5]. Most prominently however, the critical contact pressure characterizing the transition from mild to severe wear increases by a factor >5, i.e. from <2 MPa to >10 MPa [5]. At these high contact pressures coatings in the order of 100 mm thickness as obtained by laser remelting and alloying do not protect any more the substrate from indentation by wear debris and from plastic deformation. Accordingly, investigation of the load bearing capability of coatings up to 2 mm thickness will be the main purpose of this paper. Eutectic AlSi12 cast alloys are taken as substrate materials which will be coated with AlSi60 layers.

Previously, the problem of the load bearing capability of protective layers has been dealt with by studying hardness indentations as a function of the applied load. Bückle [6] first has proposed a model describing the hardness of a coated material by the weighted sum of the hardnesses of a stack of 12 layers situated below the indenter, all having the same thickness equal to the penetration depth. This model reproduces the famous 1/10 rule indicating that the substrate will not influence hardness as long as the penetration depth does not exceed 1/10 of the coating thickness. In order to predict the critical depth of indentation not influenced by the substrate Lebouvier et al. [7] and Edlinger et al. [8] use a kinematic model in which the substrate and layer are assumed to be rigid-perfectly plastic materials. The apparent hardness is predicted by minimization of the plastic work rate using a rigid block velocity field. The results apply to wedge and pyramid indentation and depend strongly on the hardness ratio between layer and substrate. They compare satisfactorily with experimental measurements.

Experimental

The powder injection technique was used to deposit layers with a nominal composition of AlSi60 on flat plates (dimensions 6*4*1 cm) machined from cast AlSi12 alloy. The operating conditions of the CO_2-laser (Rofin-Sinar; 1.6 kW cw-power) were: Power: 1.5 kW; beam diameter: 1mm; scan velocity: 2m/min . Single-layers of 0.7mm thickness as well as double- and triple-layers of 1.4mm and 2-2.3 mm thickness were produced in this way. Layer thickness here is defined as the distance between the surfaces of the coated and non-coated specimen. Prior to processing, the substrates were preheated to 300°C in order to avoid crack formation.

Hardness was measured on polished surfaces as a function of load, using pyramid (Vickers) and ball indenters (2.5mm ball diameter). A stylus-type profilometer (Perthometer) was used in order to determine the surface profile around the indentation. A malfunctioning of the hardness tester was detected after the tests had been performed, i.e. all graphs in this paper show a jump in hardness at loads of 50 and 60 kg. The effect of the contact load on the elastic deformation of the coating-substrate system was calculated by FEM using the software program "Flux-Expert".

Results

The coatings were produced without almost any dilution (<3%). Between substrate and coating, there was a remolten layer of about 60 μm thickness next to the substrate, followed by a second layer of similar thickness in which the silicon content reached its nominal value. The primary silicon particles had a platelike morphology (Fig. 1).

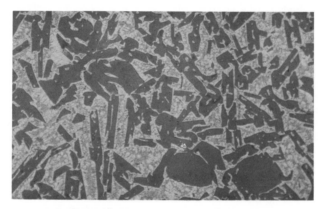

Fig.1 Microstructure of AlSi60 coating

Pyramid hardness (Vickers) strongly depends on the applied load and on the direction of indentation. While microhardness (30 grams) in transverse sections of the coatings is in the order of 250 HV, indentations perpendicular to the treated surface exhibit outstandingly high hardness values of >600 HV at 1 kg load. Beyond, hardness drops sharply to attain a plateau value of ~250 HV for loads >5kg. A logarithmic plot of the hardness evolution shows that the plateau represents a ramp of slowly decreasing hardness values, see Fig. 2. No plateau is observed in the specimen with 0.3mm layer thickness. Instead, the hardness of this sample gradually decreases with increasing load, approaching the substrate hardness, but still higher by a factor of 2 compared to the substrate at a load of 100 kg. Regarding the thicker coatings, the plateau regions extend to some critical load indicating the beginning of a second drop of hardness (Fig. 2). The critical load depends on coating thickness. For a coating with 0.6mm thickness, it is found somewhere between 60 and 100 kg. In the case of a 1.5mm coating, the corresponding value amounts to ~150 kg. The 2 mm coating resists loads even higher than 250 kg. It can also be seen from Fig. 2 that the thicker the coating the more shallow is the slope of decreasing hardness beyond the plateau region.

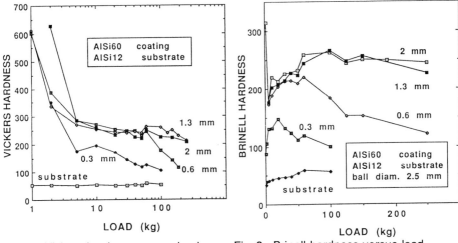

Fig. 2 Vickers hardness versus load   Fig. 3 Brinell hardness versus load

In ball indentation, carried out in normal direction only, the hardness values of all specimens including the substrate increase with increasing load. Hardness drops beyond some critical load (Fig. 3). The critical loads in ball and pyramid indentation are identical. As before, the thinnest coating (0.3mm) does not reach the same hardness as the thicker coatings.

Surface profiles obtained by traversing a stylus across the indentations provide information about the deformation fields. While little information is obtained from traces across the diagonal of pyramid indentations, traces perpendicular to the rims of the indentations reveal lips created by upward flow of material displaced by the indenter (Fig. 4 a). The height of the lips diminishes as load increases. Concomitantly, a second crater is formed with a radius significantly larger than the lip radius. At high loads, i.e.

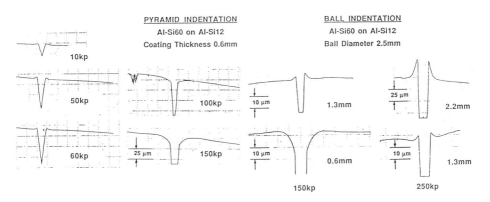

Fig. 4 Surface profiles around hardness indentations; a) Vickers; effect of load on coating of 0.6mm thickness; b) Brinell; effect of load and coating thickness

100 kg and 150 kg in Fig. 4 a, the lips eventually disappear. Only an outer and inner crater are left, the inner one being the imprint of the indenter. Similar observations are made in ball indentation, as resumed in Fig. 4 b. There it is shown that formation of the outer crater is sensitive to load and coating thickness. Concerning the specimen of 1.3 mm thickness, for example, beginning of crater formation is perceived at 150 kg, but the lips related to the inner crater still exist at 250 kg where the second crater clearly is distinguished. The lip has disappeared completely in the case of the specimen with 0.6 mm coating thickness, indented with a load of 150 kg. The metallographic section given by Fig. 5 shows that the coating section next to the indenter has been pushed right into the substrate. The corresponding surface profile reveals that the material removed from the inner and outer crater has been displaced into a broad dam built around the periphery of the wider crater (Fig. 4 b: 0.6 mm coating loaded with 150 kg). At the other extreme, if the coating is very thick, the coating withstands high contact pressures without formation of the second crater, see 2.2 mm coating loaded with 250 kg in Fig.4 b.

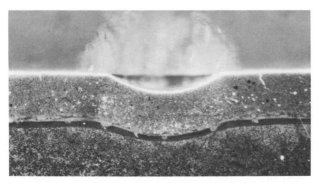

Fig. 5   Coating of 0.6mm thickness indented with a load of 150kg

Discussion

The evolution of hardness as a function of load clearly is different from what would be expected for a homogeneous material. In pyramid indentation, these differences refer to the drop of hardness in the load range up to 5 kg and to the lack of a load-independant plateau. Ball indentation leads to a relatively strong increase of hardness with load. Finally, a critical load for plunging of the indenter into the substrate has been found, the load values being identical for ball and pyramid indentation.

No detailed analysis concerning the drop of Vickers-hardness at low loads has been undertaken. As microhardness (30 grams) measured on metallographic transverse sections of similar specimens lies in the range of the plateau value (~250) one might argue that the high hardness found in normal indentation is related to internal stresses. This hypothetical argument clearly needs further confirmation.

As far as the influence of the substrate on hardness is concerned, satisfactory agreement with the theoretical calculations given by Lebouvier et al. [7] is obtained. For a hardness ratio of 5 between coating and substrate the calculations predict that hardness should decrease, i.e. influenced by the matrix, if the ratio of indentation depth to coating thickness exceeds a value of 0.13 . The model by Edlinger et al. [8] allows to calculate critical loads which should be 50 kg for a layer of 0.6mm and 200 kg for a layer of 1.3mm thickness. Again, theory and experiment compare well even if the critical load for the thicker layer is somewhat overestimated.

There remain nevertheless serious problems in modelling hardness indentations of coated materials. First, the thinnest layer of0.3 mm thickness does not fit the predictions of the model which applies to bilayers and therefore does not take into account the interface as a region of finite thickness and varying composition. Second, the experimental observations suggest two competing modes of deformation, lip formation around the indenter and formation of a large shallow crater. Lip formation is the onliest mode of deformation considered by the models but the experiments show that lips completely disappear beyond the critical load. The observation that softening occurs at the same critical loads for pyramid and ball indentation suggests that crater formation is independant of the deformation field associated with the indenters which seems a reasonable assumption for deformation occuring far from the indenter. Indeed, surface profiling supported by microscopic observation of the indentations under oblique illumination prove that crater formation begins at loads significantly smaller than the critical loads determined from hardness measurements. The steady decrease of pyramid hardness in the "plateau" region possibly reflects this effect. Similarly, the relatively strong increase of ball hardness in the same load range might be related to a contact area which grows owing to upward flow of material (similar to the lips in Vickers hardness) *and* formation of the shallow crater of larger diameter.

Stimulated by these observations we have simulated by finite element analysis the elastic deformation of AlSi60/AlSi12 bilayers in order to determine the critical loads for which the flow stress of the substrate is attained. Simplifying assumptions have been made. The coating is supposed to remain in the elastic regime. Load is applied through a punch, assuming a rectangular stress distribution over the contact area. The size of the contact area is given by the as-measured projected area of the hardness indentation. Concerning the elastic modulus of AlSi60, a rule of mixtures between AlSi12 and pure Si has been applied. Van Mises stresses at the interface between coating and substrate have been calculated for samples of axisymmetric geometry (Fig. 6a). According to this simple model, supposing a substrate flow stress of 210 MPa, the load at the onset of substrate deformation corresponds to approximately half the experimentally determined critical load (Fig. 6b). This result rationalizes the experimental observation that substrate deformation associated with the formation of the large crater occurs before the critical load for penetration of the local deformation field of the indenter into the substrate is reached [7,8].

Conclusions

The load bearing capability of hypereutectic AlSi60 coatings deposited by laser processing on soft AlSi12 substrates has been investigated by studying pyramid and ball indentations up to high loads. The critical load for breakthrough of the indenter increases almost linearly with coating thickness and compares well with theoretical calculations by Lebouvier et al. [7] and Edlinger et al. [8]. Plastic deformation of the substrate however occurs at smaller loads, leading to the formation of a large shallow crater of several times the diameter given by the hardness indentation itself. Laser cladding using the powder injection technique is shown to be an adequate process for depositing thick layers guaranteeing the protection of aluminium based components.

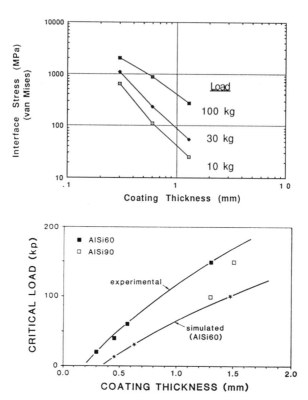

<u>Fig. 6</u> Effect of coating thickness on substrate deformation; a) calculated stress at the
coating-substrate interface for various loads; b) critical load (experimental) and
critical load (simulated) corresponding to substrate flow stress at interface.

<u>References</u>

(1)   G. Coquerelle and J. L. Fachinetti: Proc. 1st Europ. Conf.on Laser Treatment of
       Materials (ECLAT '86), DGM, Oberursel (Germany), (1986), p. 171-178
(2)   H. Haddenhorst, E. Hornbogen and H.W. Vollmer: Proc. 2nd Europ. Conf.on
       Laser Treatment of Materials (ECLAT '88), DGM, Oberursel (Germany), (1988),
       p. 97-99
(3)   M. Pierantoni and E. Blank: Proc. 2nd Int. Federation on Heat Treatment
       Seminar "Surface Engineering with High Energy Beams", Lisbon (Portugal),
       (1989), p. 415
(4)   E. Blank, M. Pierantoni and M. Carrard: Proc. 3rd Int. Conf. Surface Modification
       Technologies III, Neuchâtel (Switzerland), (1989), TMS, Warrendale (PA), p 493
(5)   E. Blank and M. Pierantoni: Techn. Rundschau 15 (1991) 86
(6)   H. Bückle: "L'essai de dureté et ses applications", Publications scientifiques et
       techniques du ministère de l'air, N.T.90, Paris, 1960
(7)   D. Lebouvier, P. Gilormini and E. Felder: Thin Solid Films 172 (1989) 227
(8)   M.L. Edlinger and E. Felder: "A theoretical approach of hardness distribution in
       rigid perfectly plastic materials", Leeds Symp. on Materials, Elsevier, (1990)

Laser Hardening - An effective Method for Life Time Prolongation of Erosion Loaded Turbine Blades on an Industrial Scale

B. Brenner, G. Wiedemann, B. Winderlich, S. Schädlich, A. Luft, W. Reitzenstein, Fraunhofer-Einrichtung für Werkstoffphysik und Schichttechnologie Dresden, FRG
W. Storch, ABB-Bergmann-Borsig GmbH Berlin, FRG
D. Stephan, Institut für Festkörper- und Werkstofforschung Dresden e. V., FRG
H.-T. Reiter, Pädagogische Hochschule Dresden, FRG

Introduction

Erosion by impact of water droplets is a serious problem which limits life time of blades in low pressure stages of steam turbines (e.g.(1,2)). As a commonly used method for reducing the erosion, the peripheric zones of the leading edge are shielded by a hard alloy, normally consisting of stellite. This process however is connected with some disadvantages as e.g. rising the production costs in an undesirable degree, insufficient increase of wear resistance for severe erosion conditions, lowering the fatique limit of the blade, limiting the possibilities of renewal the protection.

Some years ago it was proved that laser surface hardening is applicable to turbine blade hardening (3,4) and offers the possibility to reach a remarkable reduction of erosion without these disadvantages (5,6). Therefore with the aim of process industrialization a complex approach had been untertaken to investigate the metallurgical response of steel X20Cr13, the best geometry and hardness distribution of the hardened layer, the wear and fatique properties both of the hardened layer and the hardened turbine blade as well as the demands for both an optimized blade hardening technology and a laser work station design (7,8). According to the aim of the conference, this paper focusses its attention on material aspects and mechanical properties resulting from laser hardening of turbine blade leading edges.

Results

i.) **Wear resistance:** Water droplet impact tests proved that optimal laser hardened specimens show a superior wear behaviour (Fig. 1). It's evident that their mass loss remarkably falls below those of stellite W3K (russian equivalent to stellite 6). Preconditions for this high resistance against local structural damage are: high hardness of 600-700 HV0.05, sufficient depth of this high hardness level and minimum hardness fluctuations along the surface.

ii.) **Microstructure:** TEM-investigations showed that the microstructure of the base material originally consisting of elongated and globular (Fe,Cr)-carbides embedded in an highly tempered ferritic structure (see Fig. 2a) is replaced by a carbide-free, high dislocated, fine dispersed martensitic structure in the zones of high hardness (see Fig. 2b). The origins of the high hardness and wear resistance are therefore: The high interstitial

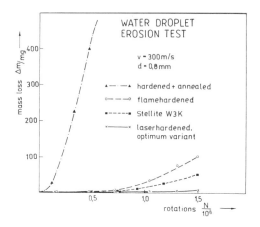

Fig. 1: Comparison of wear resistance of several turbine blade protection methods tested in an accelerated water droplet impact test (v = droplet velocity, d = droplet diameter)

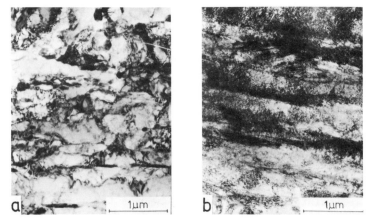

Fig. 2: Transmission electron micrograph of the annealed and tempered martensitic state (a) of the base material X20Cr13 and the laser hardened state (b)

carbon concentration in the matrix, the fine dispersed martensite and the high dislocation density resulting from high transient thermal stresses occuring during self-quenching.

iii.) **Hardness, hardness distribution and track geometry:** The correlation between carbide dissolution, surface hardness level and wear resistance, respectively, demands a constant maximum temperature of short time hardening cycle across the track surface. Therefore an high-grade protection of all erosion endangered parts of the turbine blade leading edge requires a local adaption

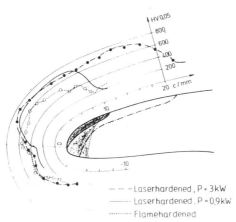

— — — Laserhardened, P = 3kW

·········· Laserhardened, P = 0,9kW

·········· Flamehardened

Fig. 3: Hardness distribution of surface hardened leading edges, measured on a cross section in a distance of 0.05 mm from the surface

of the heat input to the complex geometry of the leading edge depending furthermore on the distance from the blade tip. We solved this problem by
- a processing time, carefully adapted to the blade thickness,
- a variation of the processing time, of the impingement angle and of the leading edge distance of the laser beam along the leading edge,
- a computer controlled simultaneous movement of the blade in 4 axis
- and a beam shaping attachment, which enables to change the power density distribution across the feet direction.

Fig. 3 and 4 demonstrate the realizing of the nessecary high hardness level across the laser track and of the track geometry optimally adapted to the variation of the wear intensity in dependence from the distance to the leading edge. In contrast to flame or induction hardening the more flexible laser hardening is able to produce such a hardness profile and geometry of hardening zone over the whole erosion endangered length of the leading edge.

iv.) **Fatigue:** The industrial use of laser hardened turbine blades presumes the guaranty of cyclic load bearing capacity. Therefore in a many-stage program the fatigue limit of the laser hardened structure itself and the mean stress sensitivity of the fatigue limit, the endurance limit of laser hardened turbine blades, the overload capacity and the influence of technological parameters (e.g. maximum austenitization temperature, influence of location and angle of track tail) were estimated.
The results can be summarized as follows:
- The fatigue limit of the laser hardened zone exceeds those of the base material for about 60 % (in detail see (9)).
- The fatigue resistance of a laser hardened turbine blade loaded in its first harmonic is remarkably higher than those of a conventional protected one (Fig. 5).

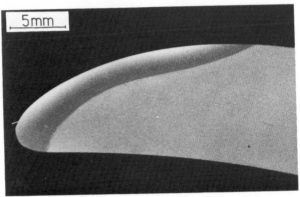

Fig. 4: Etched cross section of a laser hardened turbine blade
leading edge

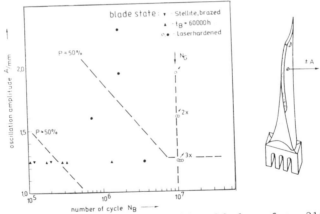

Fig. 5: Fatique resistance of turbine blades of a 210MW-turbine
in several stages; (▼) - new blade, leading edge brazed with
stellite plates, (▲) - as (▼), after an operation time of 60000h,
(O,●) - laser hardened

- Laser hardened specimens, in geometry, laser treatment parameters and loading stress gradient similar to a turbine blade
  sustain a higher stress amplitude at the tensional service
  mean stresses (Fig. 6).
- If the maximum peak temperature of the short time hardening
  cycle exceeds the melting temperature locally the sustainable
  stress amplitude and/or mean stress diminish slightly (Fig.
  7).However it's remarkable that the cyclic load bearing capacity of this not desired state is higher than those of the conventional hardened and tempered state.

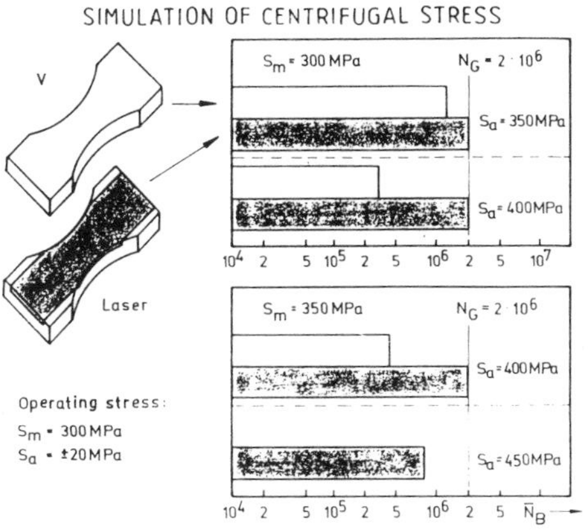

SIMULATION OF CENTRIFUGAL STRESS

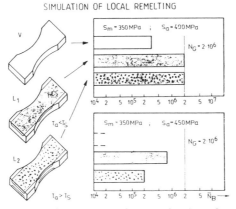

Fig. 6: Influence of laser hardening (L) on the sustainable stress amplitude S_a at high tensional mean stresses S_m (V = conventional hardened and annealed); in contrast to (9) fatigue tests were carried out under quasi-deformation control

SIMULATION OF LOCAL REMELTING

Fig. 7: Influence of a local remelting during laser hardening on the sustainable stress amplitude S_a at high tensional mean stresses S_m (T_s = melting temperature; T_a = austenitization temperature); quasi-deformation controlled experiment

v.) **Residual stresses:** It is well known, that residual tensile stresses within a hardened component increase the susceptibility against stress-corrosion cracking. Therefore within surface hardened zones residual tensile stresses should be avoided. Fig. 8 proves that within the laser hardened zone residual tensile stresses don't exist.

204

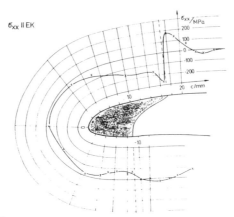

Fig. 8: Distribution of the residual stress component σ_{xx} in direction of the centrifugal stress in dependence of the distance c to the leading edge

vi.) **Process industrialization:** A 25000 h test in an extremely erosion endangered type of steam turbines proved the meet of all requirements of industrial application (7). Up to now more than 1300 laser hardened turbine blades have been delivered up for industrial use. They are employed in four steam power plants, one waste combustion plant and one chemical plant in Germany, The Netherlands and South Africa.

References

(1)   B. Stanisa: Strojarstvo 26(1984) 11
(2)   H. Jesper, W. Hennecke, H. Remmert: VGB Kraftwerkstechnik 69(1989) 911
(3)   M. Roth, M. Cantello: 2nd Internat. Conf. "Lasers in Manu-facturing" (ed. by M.F.Kimmit), Birmingham 26.3.-28.3.1985, Conf. Proc., p.119
(4)   V. Bedogni, M. Cantello, W. Cerri, D. Cruciani, R.Festa, G. Mor, F. Nenci, F.P. Vivoli: Conf. Proc. "Laser Advanced Materials Processing'87" Osaka, May, 1987, p.567
(5)   B. Brenner, W. Reitzenstein, G. Wiedemann, W. Storch, H. Junge: 4. Tagung "Lasertechnologie" Jena, 2.-4.11.1987
(6)   W. Storch, G. Blum, B. Brenner, G. Wiedemann: Mitteilungen aus dem Kraftwerksanlagenbau 28(1988) 16
(7)   B. Brenner, W. Storch: VGB-Konferenz "Werkstoffe und Schweißtechnik im Kraftwerk" Essen, 9.-10.1.1991, Conf. Proc., p. 11.1
(8)   F. Duscha, W. Göttert, W. Storch, B. Brenner: ZIS-Report 1(1990) 536
(9)   B. Winderlich, B. Brenner, C. Holste, H.-T. Reiter: in this vol.

Microstructures and Properties of Surface Layers Produced on Steels by Different Laser Melting Techniques

A.Luft, W. Löschau, K.Juch, R. Gassmann, V. Fux, W. Reitzenstein

Fraunhofer-Einrichtung für Werkstoffphysik und Schichttechnologie, Dresden, Germany

1. Introduction

There is a widespread interest in using high power lasers for alloying and hardfacing of steels in order to enhance hardness and improve wear and corrosion resistance. These advanced technologies require not only knowledge of the processing techniques themselves but also a detailed unterstanding of the factors which control the microstructures and the mechanical properties of the modified surface region.

The paper reports on technological and microstructural investigations performed to produce wear resistant surface layers on steels using the following laser surface modification techniques:

(i) incorporation of hard particles into steel surfaces by either melting-in a pre-coated layer or injecting particles into the melt pool

(ii) hardfacing of steels with Ni- and Co- base alloys by dynamic feed of either powder or amorphous band in the laser spot.

Special attention is devoted to the behaviour of hard particles in the melt and to the role and nature of the interface region between the melted surface layer and the substrate.

2. Incorporation of TiC - base hard particles in steel surfaces

It is well recognized that the wear resistance of metals can be substantially improved by adding hard particles into surface layers exposed to wear. Therefore, the incorporation of hard particles using high power lasers to form wear resistant composite layers has been the subject of many investigations. The result of the laser treatment depends sensitively on the chosen process and material parameters, such as laser power, spot size, scan speed, intensity distribution, type and size of particles, powder feed rate and thickness of preplaced layer, respectively. From the metallurgical point of view the most difficult problem is, on the one hand, to avoid the dissolution of the hard particles within the steel melt and, on the other hand, to produce composite layers which are sufficiently thick for technical application and contain the hard particles in a homogeneous distribution and with high volume concentration.

We performed experiments to incorporate hard particles on the base of TiC and WC into various carbon steels (C45, 90MnV8, 80CrV3) using different CO_2 lasers in the range from 1 to 6kW. The particles were either deposited on the steel surface by a screen printing methode prior to the laser melt treatment or injected by an Ar or N_2 gas stream in the laser generated melt pool (1,2). The most homogeneous microstructures containing a fine dispersion of small re-solidified TiC crystals could be produced by the melt treatment of screen-printed layers consisting of a mixture of fine grained TiC particles (grain

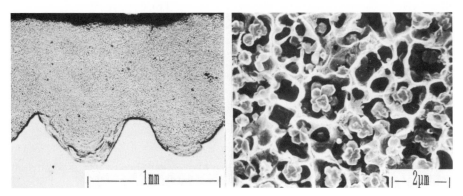

Fig.1: Alloying of steel 90MnV8 with TiC/Fe-4%B by melting-in a screen-printed layer using a 1kW laser
a. Cross section of overlapped tracks   b. Solidification microstructure

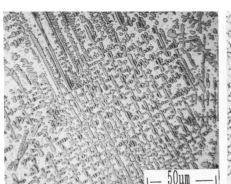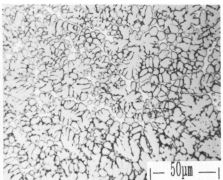

Fig.2: Typical Microstructures obtained by TiC particle injection into steel C45 using 5kW laser power
a. v = 10 mm/s   m = 0.05 g/s   b. v = 13 mm/s   m = 0.09 g/s
   Dendritic growth of TiC from the   High volume fraction of primary
   melt   and re-grown TiC particles

size < 20µm) and the eutectic Fe-4% B alloy. As shown in Fig. 1, nearly all primary TiC particles dissolved in the melt and new TiC crystals formed on cooling from the melt. The solidification of the melt proceeded by cellular or celldendritic growth of γ-crystals (austenite) associated with the formation of an interdendritic eutectic. The austenite within the cells has been mainly transformed to high carbon martensite. The microhardness was higher than 1200 $HV_{0.05}$ exceeding substantially that achievable by only transformation hardening.

The use of coarse grained particles (60 to 100 µm) and the deposition of thick layers (500 µm or more) yielded microstructures densely packed with partially dissolved primary TiC particles and re-grown new crystals. The concomitant suppression of melt convection promotes the formation of sintered

conglomerates and the occurrence of inhomogeneities in the particle distribution (2).
The injection of TiC particles having grain sizes between 45 and 90 µm into C45 steel using 5 kW laser power leads also to a substantial dissolution of the particles. Figure 2 shows typical microstructures achieved with different process parameters. The dendritic growth of large TiC crystals occurs in a rather homogeneous manner throughout the whole melt zone (about 1 mm deep and 3 mm wide) if the powder feed rate and the transverse speed are low. On the contrary, the increase of powder feed rate and / or scan speed makes melt convection and particle dissolution more difficult resulting in a distinct enrichment of the upper part of the melt zone with primary and secondary hard phase particles (2).
It is worth noting that the tendency for cracking rises with the increase of volume fraction of hard phase being much more pronounced for WC than for TiC additions.

3. Laser powder cladding of valves

An investigation has been made of the application of the laser powder cladding technique for hardfacing of valves used in combustion engines. Laser cladding promises advantages in comparison with other surfacing techniques e.g. TIG or PTA deposition welding. The easy control of location and delivery of energy in connection with high power density leads to improved layer properties such as good fusion bonding with low dilution, enhanced solid solubility and refined microstructure due to high cooling rates. From the economical point of view the processing time should be as short as possible what requires the use of high power lasers and the careful optimization of process parameters with the aim at increasing the scan speed as high as possible. Furthermore, it has been established from the experiments that the control of all process parameters in dependence on the rotational angle is a prerequisite for hardfacing of small valves (fast integral heating) and for that faultless start/end overlap regions be achieved (3,4).
Experiments have been conducted to optimize the process parameters for the laser powder cladding of stellite 6 and stellite F Co-base hard alloys onto the austenitic valve steel X45CrNiW 18.9. Within the parameter range studied a dendritic solidification structure has been obtained consisting of dendritic Co solid solution and interdendritic eutectic enriched with the hard carbide phases. The volume fraction of eutectic is higher in stellite F than in stellite 6. It was found that the temperature-time regime exerts a particular influence on the structural processes which proceed near the fusion boundary between clad layer and substrate. To summarize, three typical states can be characterized:
(i) Strong temperature-time regimes (high temperature, long life time of melt) lead to wide transition regions and a high dilution of clad alloy by substrate material. The eutectic network of the solidified stellite changes gradually into the grain boundary network of the steel.
(ii)Accelerated welding provides conditions under which a good fusion bonding without remarkable mixing between clad alloy and substrate can be achievied. In this case a narrow zone about 10µm thick consisting only of Co solid solution without any carbides is formed by a planar solidification mode. For more details see (5).
(iii)Insufficient temperature-time cycles result in the formation of bonding faults in form of thin cavities along the clad layer/substrate interface.
All these phenomena can be encountered across a single track deposited on valves having a deep circular groove as commonly made for TIG deposition

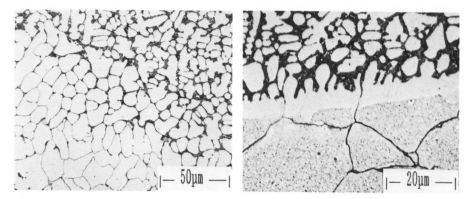

Fig.3: Transition zones between laser clad layer (stellite F) and steel substrate (austenitic steel X45CrNiW 18.9) for different local temperature-time
cycles

a. gradual transition with high   b. sharp transition with narrow
   dilution of base material   carbide-free zone

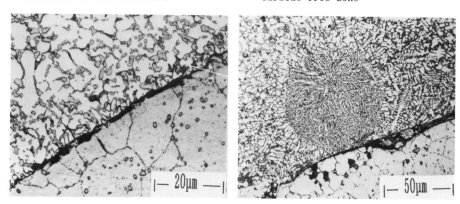

Fig.4: Examples of insufficient fusion bonding at the clad/substrate interface.
Note the cellular solidification microstructure of the stellite F layer and the
existence of undissolved carbides in the steel due to low local temperature.

welding. The reduced heat conduction at the most outer part of the valve
favours local overheating associated with the wide transition region and the
high amount of mixing (Fig.3a). Going in direction to the centre of the groove, optimal local temperature-time cycles are achieved providing the desired
properties of the layer/substrate interface (Fig.3b). However, in the central
deepest part of the deposited layer bonding faults can appear if the local
temperature does not exceed the melting temperature of the steel (Fig.4).
Careful microscopic examinations at high magnification are nesessary to
detect the thin cavities along the melt boundary.
It can be concluded from theses observations that the geometry of the groove
has to be adjusted to the laser generated temperature field and that an on-
line temperature control of the cladding process is necessary in order to
guarantee faultless fusion bonding also at high welding speeds.

4. Laser hardfacing by dynamic feed of amorphous band

An interesting variant of laser hardfacing consists in the dynamic feed of alloy material in form of a flexible band into the laser spot. The thickness of the deposited layer can be simply controlled by changing the proportion of band feed rate to transverse speed of the substrate. Compared to laser powder cladding the loss of alloy material is negligible. Another advantage of band cladding is the easy adjustment of band width to the requirements of technological application under consideration. However, the main problem for the realization of such a technology is the production of flexible band because the conventional hardfacing alloys are inherently brittle due to their high content of hard phases. One of the possible ways to overcome the brittleness is the use of rapid quenching in order to change the alloy into the amorphous state. For this purpose the conventional hardfacing Co-base alloys have to be modified by additions of glass formers, e.g. C, Si and B. A careful and systematic optimization of alloy composition is necessary with the aim at minimizing the content of nonmetals in order to reduce their embrittling effect in the solidified crystalline clad layer.

Starting with conventional hardfacing alloys like stellite 6 and stellite F the composition has been changed in order to facilitate the production of thin amorphous bands. Figure 5 shows the solidification structure of a clad layer obtained from such a CoCrNiWSiCB alloy by laser deposition welding. The structure consists of the dendritic Co solid solution (bright areas), the coarse eutectic with the needle - like chromium carbides known from the conventional stellites and additionally a second very fine eutectic (grey areas in the optical micrograph distinctly resolved in the SEM micrograph).

By optimization of the alloy composition it was possible, on the one hand, to realize just the amorphous band and, on the other hand, to obtain a sufficient large volume fraction of solid solution crystal in the solidification

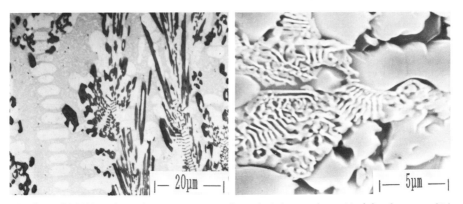

Fig.5: Solidification microstructure of a clad layer deposited by laser melting of an amorphous band made of a modified stellite F alloy (CoCrNiWCSiB)
a. Optical micrograph   b. SEM micrograph

210

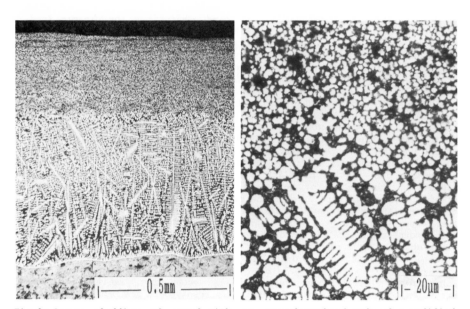

Fig.6: Laser cladding of steel with an amorphous band made of a modified
Co-base hard alloy with an optimized composition
a. Cross section of clad layer with   b. Transition region between dendritic
   two different solidification modes   and fine cellular microstructure

structure of the clad layer. Figure 6 shows an example of such a clad layer
which exhibits sufficient ductility without cracking. Two different solidification modes can be observed: In the lower part of clad layer the solidification of the melt occurs by dendritic growth of solid solution crystals followed by the eutectic solidification of the interdendritic rest melt. Mor-
pholgy and proportion of phases are very similar to those found in conventional stellites accounting for the observed ductility. The upper part of the
clad layer is characterized by a cellular fine-crystalline structure which
has been formed probably by spontaneous crystalization from the undercooled
melt due to the lack of crystallization nuclei. The properties of clad layers
having such a small scale crystalline structure are now under investigation.

References

(1) W.Löschau, A.Luft, H.Heinze: Tagungsband 9.Int. Pulvermetallurgie-Tagung,
    Dresden (1989) 301
(2) W.Löschau, K.Juch, W.Reitzenstein: this volume
(3) R.Gassmann, A.Uelze, St.Nowotny, W.Pompe: DVS-Bericht 135 (1991) 240
(4) R.Gassmann: unpublished report (1991)
(5) R.Gassmann, A.Luft, St.Schädlich, A.Techel: this volume
(6) V.Fux, W.Schwarz, St.Nowotny, D.Pollack: this volume

Improvement of mechanical properties of steels by addition of tungsten carbides : laser cladding and laser welding.

J.M. Pelletier*, D. Dezert**, F. Fouquet*, M. Robin*** and A.B. Vannes****

*GEMPPM-CALFETMAT, INSA-Lyon, France,
**Technogenia, St Jorioz, France
***IUT-CALFETMAT, Villeurbanne, France.
****MMP-ECL, CALFETMAT, Ecully, France.

I- Introduction

Improvement of the performances of steels has often to be achieved, since many components are used in severe environments : corrosion or wear are often involved and, even, at elevated temperature. Tools used for machining are typical examples. Two solutions may be retained : either cladding with hard coatings (1), or addition of a very hard tip, either mechanically clamped, brazed or welded (2). Two problems appear :

- what kind of material has to be chosen? For insatnce, in order to enhance the wear resistance, tungsten carbides have several attractive features : high hardness, good chemical stability and good thermal conductivity, limiting therefore the surface temperature increase.
- what is the best suited process? Among many techniques, laser cladding and laser welding have different advantages : flexibility, limited heat affected zone (HAZ) and, therefore, only small deformations induced...

The present study will then be devoted to both laser cladding with a cast tungsten carbide base powder, and laser welding of cemented carbide tips. Substrates are medium carbon steels. A binding element, nickel- or cobalt-base material, is added to the carbides, in order to facilitate the mixing.

II - Experimental procedure :

For laser cladding, an unalloyed steel : Fe - 0.28%C - 0.075%P - 0.062%S was used as substrate. Parallelepipedic samples ($100*30*10$ mm^3) were rectified and sand blasted before laser processing. No absorbing coating was used, in order to avoid any contamination of the sample. Two powders were injected:
- powder A: contains grains of different tungsten carbides (WC, W_2C, WC_{1-x}), coated with electroless nickel;
- powder B (binder) : prealloyed grains of composition : Ni - 14% Cr, 4.2% Fe - 4.2% Si - 3% B - 2.5% Cu - 2.5% Mo - 0.65%C.
For laser welding, core is made of a medium carbon steel with the following composition (wt %): Fe - 0.35 %C - 1% Cr - 0.2% Mo. Parallelepipeds ($50x25x2$ mm^3) were cut from oil quenched and tempered plates. Tips contain cobalt (20%) and tungsten carbide (80%). They are processed by a powder-metallurgy technique and cut in parallelepipeds ($40x8x3.2$ mm^3). Only WC and Co phases are detected by X ray

diffraction prior to welding. Core and tip sides are polished and cleaned prior to welding. Parts are then abutted without gap and mechanically clamped together.

Specimen were mounted on a numerically controlled X-Y table and irradiated by a continuous CO_2 laser, of power up to 3.6 kW. The laser beam was focused by a spherical Zn-Se lens (focusing distance : 254 mm for laser cladding and 127 mm for laser welding). Scanning rate V_T of the sample under the laser beam ranged from 0.2 to 3 m/mn. A variation of the interaction time, defined by $\tau = d/V_T$, where d is the equivalent beam diameter on the sample, was thus possible. During scanning, argon was blown through a nozzle, in order to prevent oxidation.

For laser cladding, the coating material was sprayed with a 10C Twin Plasma Technik system. The angle between the injection nozzle and the beam axis was fixed to 15° while the feeding rate F was a variable parameter.

For observations using optical or scanning electron microscopy and for Vickers hardness measurements (O.2 kg load), specimen were cutted and polished (down to 1 μm). For optical micrography, etching was finally performed. Different reagents were used : Murakami and Struers V2A for coatings, nital for mild steel.

A 840A JEOL microscope was used for scanning electron microscopy; either backscattered or secondary electrons were taken to form the image (BEI or SEI mode, respectively). The same apparatus was equipped with an EDS microanalysis system (Tracor).

X ray diffraction patterns were obtained with a Cu (Kα) radiation and a graphite monochromator.

III - Experimental results for laser cladding :

Different situations can be observed, depending on the processing conditions, especially on the laser power density on the sample (I) and on the powder composition:
- without binder and with a low value of I, a "pure" cladding is achieved: the incoming powder interacts with the laser beam, melts and falls on the specimen, the temperature of which has not reached the melting point. In this case, there is no dilution of the powder into the substrate, but the adhesion of the cladding is weak and some porosities are present, exactly like with a plasma spraying. X-ray diffraction experiments reveal nearly the same phases in the initial powder and in the cladding : W_2C, WC, WC1-x, amorphous Ni-P phase and also W, following the eutectoïdic transformation :
$$W_2C \rightarrow WC + W \qquad \text{near } 1300°C$$
- without binder, but with a high value of I, part of the substrate is heated up beyond its melting temperature. Then the powder penetrates into the melted pool and dissolve partially into it. A metallurgical bonding is achieved and, consequently, adhesion is very good. We observe the existence of different phases: Fe_3W_3C, Fe_2W_4C, Ni_2W_4C, FeW_3C, in addition to the initial phases. Therefore it can be easily concluded that a partial melting of the tungsten carbides occurred and that new complex carbides (M_6C type) formed with the melted part of the substrate. Hardness of these metastable particles is lower than that of the initial particles, and therefore this solution is not optimum.
- with the binder (typicaly 50% of volume fraction) and a low value of I (typically $4.10^4 W/cm^2$), laser-material interaction produces total melting of the binder together with a partial melting of the substrate and avoids the dissolution of the hard tungsten carbide particles. Fig.1 shows an example of a good cladding; it is clearly observed that

carbides are still present and homogeneously distributed. X-ray diffraction experiments confirm that tungsten and carbon elements are present only in tungsten carbides and that no complex $Fe_xW_yC_z$ or $Ni_xW_yC_z$ phases are formed during cladding. Indeed, phases revealed by diffraction are the followings : W_2C, WC, Ni_3Fe, $Ni_{2.0}Cr_{0.7}Fe_{0.36}$....

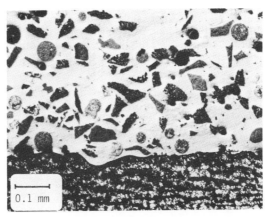

Fig. 1 : micrography of a cladding obtained with a mixture of powder (tungsten carbides + binder); laser power $P = 900$ W; scanning rate : $v_T = 40$ cm/mn; beam diameter on the sample : $d = 2.5$ mm; powder feeding rate : 10 g/mn.

Consequently, the addition of a binder to the initial powder has a very beneficial effect to the cladded layer, since adhesion is strong and no dilution of the carbides has occurred. Hardness values are as follows : about 500-600 HV for the matrix, and higher than 2500 for the carbide particles.

III - Experimental results for laser welding :

A good weld should have at least the following quality features : full penetration, structural homogeneity and no apparent defect (holes, cracks) either in the weld bead or in the heat affected zone (HAZ).

Defects are always present during preliminary attempts. Big holes are often observed in the central part of the weld and cracks appear in the weld bead as well as in the HAZ on the steel side.

After various attempts, we have found more convenient to slightly defocuse the beam in order to enlarge the interaction zone and to shift the beam towards the steel side, in order to first melt the steel and make the melt pool progress more easily into the tip.

The weld bead is homogeneous and a full penetration is achieved. Four different zones are always detected in cross-sections, as illustrated in figure 2 ; from the top to the bottom, we observe :

a - the base steel (HV $\approx$ 350-400) : microstructure is typical of an oil quenched and tempered medium carbon steel, i.e. a fine tempered martensite revealed by etching.

b - the steel HAZ (HV$\approx$ 400-450): as in other laser treated steels (3), classical microstructures can be recognized :

- a martensitic microstructure, near the weld bead; here, a high peak temperature has induced full austenitization upon heating and rapid quenching is achieved by efficient heat transfer to cold steel.

- a bainitic microstructure in various conditions :

. either far from the weld bead. As shown by Ion et al (4), the diffusion of carbon atoms concerns only very limited distances and a complete dissolution of carbides is therefore impossible. Austenitizing is only partial and volume fraction of martensite able to form during cooling is small.

. or because cooling rate is too low, as reported by Pergue et al (5), after laser melting of a medium carbon steel (0.48%).

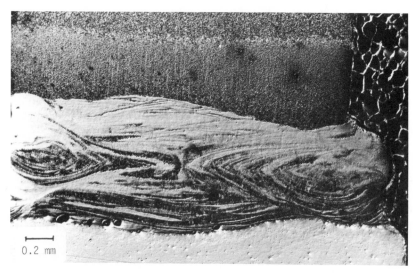

0.2 mm

Fig. 2 : typical micrography of a sound weld ; P = 2040 W; v_T = 2 m/mn; d = 0.4 mm; shift of the beam towards the steel side : δ = 0.3 mm.

c - microstructure of the weld bead (HV ≈ 1000)

This microstructure is mainly dendritic, in spite of some irregular aspect. These irregularities can be attributed to different factors :

- owing to a cross-section bias effect of a three-dimensional growth network, dendrites appear to be randomly oriented ;

- convection flow is produced in the melted pool due to Marangoni effect (6-8). Therefore, temperature gradients exist at the beginning of the solidification phenomena, with different orientations. So solidification is not initiated from a plane interface with an uniform temperature gradient, but on the opposite, boundaries are irregular, directions are nearly random and rates are high.

d - the tip (HV ≈ 900-100)

The spatial repartition of carbides is fairly regular.

Distribution of the different metallic elements in the various zones has been determined by EDS. Due to its low atomic mass, carbon was not detected. Contents of the different elements in the weld bead are as follows (Fig.3): 70% Fe, 25% W, 3% Co (wt %). So, iron is the main element present in the melt.

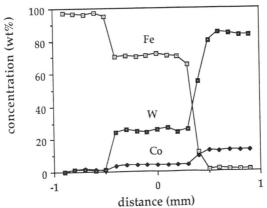

Fig. 3 : concentration profiles, determined by EDS, of the different metallic elements in the various areas of the sample shown in Fig. 2.

Optical microscopy has revealed the dendritic microstructure of the weld bead , but it is best shown by scanning electron microscopy (Fig. 4).

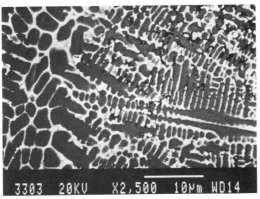

Fig. 4 : dendritic microstructure of the melted zone shown in Fig. 2; scanning electron micrography (BEI mode).

In addition, a chemical contrast appears : tungsten, a heavy element, is bright, while lighter elements (Fe, Co, Cr, Mo, ...) are grey or dark. Therefore, the inner part of the dendrites contains mainly Fe and Co, while W is mainly rejected in the interdendritic zones. However, it cannot be concluded that Fe or Co are not present in these zones ; it is only sure that their concentration is smaller here than within dendrites.

Starting from experimental evidence, the bonding mechanism may be described as follows : the laser beam interacts first with the steel part and induces its melting ; the liquid steel wets the cobalt present in the tip and yields rapid melting of this element (15); convective flow originating in the melt carries the WC particles far away from the cermet boundary and enhances their dissolution rate. An homogeneous liquid phase containing Fe, Co, W and C is thus obtained. After laser-material interaction, cooling begins and solidification starts with the formation of Fe-Co rich dendrites, while the segregating element (W) is rejected in the interdendritic zones. Fe-Co equilibrium phases diagram (9) shows that solubility of Co in Fe is fairly large and even complete at high temperature; it may thus be assumed that dendrites are constituted by an austenitic Fe-Co solid solution, at least at high temperature. Solidification of the interdendritic liquid occurs in a second step and yields probably complex carbides $(Fe,Co,W)_xC_y$.

V - Conclusion :

After various attempts and optimization of the processing parameters - laser power, scanning speed, defocusing distance... - good laser welding or cladding has been achieved, in spite of the very different properties (especially their melting point) of the materials : a medium carbon steel core and a tungsten carbide-rich material.

In laser cladding, a surface dispersoïd alloy is formed : hard tungsten carbide particles (HV > 2500) are embedded in a ductile matrix (HV ≈ 500). Adhesion is good, since a metallurgical bonding is achieved between the substrate and the surface layer.

In laser welding, the weld bead (HV≈ 1000) has a dendritic microstructure, which is likely to be obtained in two main steps. First, formation of an homogeneous Fe - Co - W - C liquid solution after cobalt wetting and melting by liquid and dissolution of WC particles stirred into the melt by convection. Second, solidification leading to a Fe-Co rich solid solution and W rich mixed carbides in interdendritic areas.

References :

1 - T.R. Tucker, A.H. Clauer, I.G. Wright and J.J. Stopski, Thin Solid Films, 118, 73 (1984)..
2 - M.B.C. Quigley, in "Physics of Welding", J.F. Lancaster ed., Pergamon Press (1986), p. 306
3 - M.F. Ashby and K.E. Easterling, Acta Metall., 32, 1935 (1984).
4 - J.C. Ion, M.F. Ashby and K.E. Easterling, Acta Metall., 32, 1949 (1984).
5 - D. Pergue, J.M. Pelletier, F. Fouquet and H. Mazille, J. Mat. Sci., 24, 4343 (1987).
6 - C. Marangoni, Ann. Phys. Chem., 143, 337 (1871).
7 - C. Chan, J. Mazumder and M. Chen, Met. Trans., A15, 2175 (1984).
8 - A. Galerie, M. Pons and M. Caillet, Mat. Sci. Eng., 88, 127 (1987).
9 - T.B. Massalski, "Binary Phase Diagrams", ASM, Metals Park, Ohio (1986).

Acknowledgments :

The authors would like to thank C. Vialle, B. Jacoty, F. Jouanna, M. Aouine, M. Pilloz and M. Benchérif for experimental assistance.

Laser Beam Surface Treatment - Is Wear No Longer The Bug Bear Of Old?

Prof. Dr.-Ing. Dr.h.c. W. König, Dipl.-Ing. L. Rozsnoki, Dipl.-Ing. P. Kirner
Fraunhofer-Institute for Production Technology - IPT, Aachen, FRG

Introduction

Wear is today still the single largest contributory factor resulting in the premature demise of production tools - in stubborn defiance of the deployment of materials of the highest quality to make them. Conventional surface treatment techniques applied with the purpose of improving wear characteristics are successful only in a few isolated areas of application. Subjected to the harsh operating conditions under which hot working tools are required to labour, even nitrate, PVD or CVD coatings which have proved their worth in cold forming operations simply cannot cope.

In such cases, laser beam surface treatment offers new ways of producing wear-resistant surfaces. The causes of the damage can be countered locally to some considerable effect by applying thermal and thermo-chemical methods of modifying the material. The prolongation of tool life thus achieved, which can reach 500 %, represents a decisive increase in the economic efficiency of forging and diecasting operations.

Thermochemical laser beam surface treatment processes

Laser beam surface treatment in the form of alloying, dispersing and coating refining techniques, as well as the variety of thermal processes, affords the opportunity to alter surface layers of materials not merely by modifying the structure or grain of the surface layer, but by also influencing its chemical composition by adding alloying elements.

Due to the kinematic conditions when the laser beam is guided selectively, which are similar to those prevailing during thermal processes, these refining processes also permit the material surface to be machined partially in an area limited to a few tenths of a millimeter. The main factor which distinguishes laser beam surface treatment from laser beam hardening processes and the majority of conventional surface treatment techniques currently in use, lies however, in the operating principle behind these processes whereby modification is always effected via the phases solid ⇨ liquid ⇨ solid. The position of laser beam alloying, dispersing and cladding in the matrix of conventional thermo-chemical surface technologies is shown in **Figure 1**.

Description and definition of various processes

The main difference between the refinement techniques lies in the process design. In alloying processes, the additional material is melted with the base material. Steep temperature gradients and the associated convection flows ensure that the two components are well mixed. After solidification, the base material content predominates in the melting zone. As well as chemical modification of the surface area, the rapid solidification process results in the formation of an extremely finely-grained structure which is both highly tensile and tough.

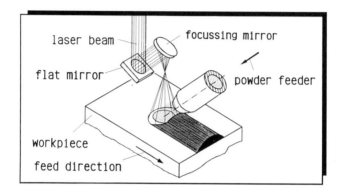

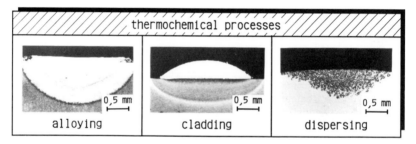

Figure 1: Thermo-chemical surface treatment technologies

Dispersion is a special type of alloying. By incorporating materials which have a high melting point into a matrix with a lower melting point, it is possible to ensure that the particles added become firmly anchored in the matrix, without disolving. The integral percentage of additional material in the surface area is thereby significantly higher than is the case after alloying.

The aim of cladding, however, is to apply as tightly adhesive and as pure a layer as possible. In contrast to the alloying process, that of cladding is designed in such a way as to ensure that slight intermixture occurs only in the transitional zone, in order to produce a metallurgical bond between base and additional materials. The degree of intermixture and, therefore, the percentage of base material in the layer is minimal; the layer hardness depends solely on the additional material used. The cladding technique allows layers ranging between 0.1 mm and several millimeters in thickness to be applied in either single or multi-layer technology. Depending on the material combination and component geometry, the reaction depth in the alloying and dispersing processes is limited to approximately 1 mm.

Additional materials and coating methods

The opportunity to alter parts of the surface chemically and structurally in a variety of ways using laser technology provides the user with a wide range of new approaches to the task of designing requirement-oriented funtional zones. The multitude of base and additional materials, in conjunction with the three refinement techniques allows a matrix to be drawn up from which the appropriate combination for the application in hand can be selected.

All tool steels are suitable for use as base materials in laser surface refinement operations. Additional materials relevant to steel are primarily the following elements: W, Cr, Mo, Ni, Co, Ti and their carbides. Additional materials for laser beam cladding are, in general, characterised by a matrix with a low melting point. This is essential, as is attention to the way in which the process is applied in order to avoid integral intermixing of the base and additional materials. Co, Ni and Fe base alloys as used in resurfacing by welding and coating techniques are particularly suitable.

An additional degree of freedom within the area of refinement techniques, concerns the way in which the additional material is introduced into the process. There are essentially two ways.

In the case of the two stage process, the first stage consists of coating the surface of the base material with the additional material by plasma-spraying. In the second stage, this outer layer is retreated using a laser beam. It is characteristic of the single-stage process that the additional material is introduced selectively into the melt via a nozzle simultaneously with the movement of the laser beam above the surface. Plasma spray powders, whose grain sizes should be as small as possible for alloying processes and as uniform as possible for coating processes are used for this application.

Industrial applications

The multitude of processes which, as has been demonstrated can be performed by combining refinement techniques and additional materials is matched by just as comprehensive an array of obtainable surface properties. This opens up very impressive avenues of industrial application for this process.

Nowadays, it is predominantly CO_2-lasers which are used for laser surface refining in the power range between 2 and 10 kw. In addition to the laser source itself, the handling system exerts considerable influence on the degree of difficulty of the application. While level surfaces can be machined using 3-axis techniques, laser beam cladding of complex component geometries, such as those of sculptured surfaces for example, necessitates the use of a 5-axis workstation.

At present, alloying and cladding account for the majority of laser beam surface treatment processes in industrial applications. The importance of these processes increases particularly in areas in which the tools are subjected during the machining process to temperatures above the tempering temperature of the tool materials in use. An important criterion with regard to the machining of hot forging and pressure diecasting tools, is the high temperature stability and resistance to temperature required of the material in the functional area. The filter bearing areas of a conventionally heat treated pressure diecasting tool and that of a laser beam alloyed one are compared in **Figure 2**. The purpose of the laser treatment was to minimise the severe erosion around the gate mark by partially alloying on tungsten carbide. As a result, the tool life quantity was increased by 45,000 to 125,000.

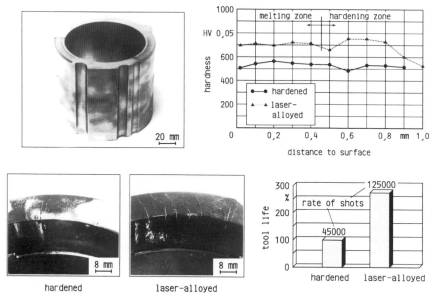

Figure 2: Tool life comparison of differently treated diecasting tools

One example of the laser alloying of hot forging tools is the treatment of die inserts from a multi-stage press. The thermal crack formation is a typical failure of such tools. In the case in question, this is a result of the high thermal shock stress to which the water-cooled tool is subjected. The tool material used, hot working steel, cannot withstand to a sufficient degree the resultant compression / tension loads. The laser treatment of this tool was restricted to the damaged area on the face. Alloying with a WC/Co bond resulted, in association with a laser treatment oriented sequence of thermal treatment, in an increase of over 500 % in tool life quality.

Further applications in the field of forging tools which are subjected to high thermal and mechanical strain are taken from maxi-press tool science. The long pressure dwell with high forming forces, characteristic of this forging process, results in the formation of grooves and of thermal cracking, generally causing premature tool failure. By alloying the wear zones of the die shown in **Figure 3**, which is used for making parts required by the automotive industry, with WC/Co laser beam in conjunction with subsequent nitride treatment, the tool life quantity was increased by more than 100 %. A comparison of the two differently treated tools shows that when the tool which has been only nitrogen case-hardened reaches its total tool life quantity, the laser-treated and nitrogen case-hardened tool demonstrates considerably less wear progression around the mandrel.

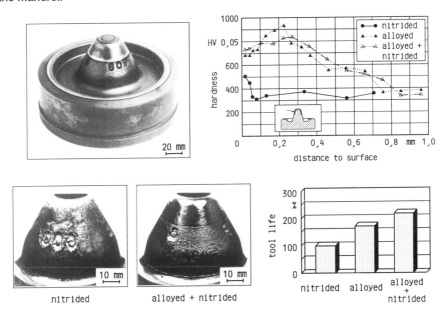

nitrided   alloyed + nitrided

Figure 3: Tool life comparison of differently treated forging tools

The use of laser power in the range > 5 kw permits even tools with large dimensions to be treated. In the case of the roller shown in **Figure 4** (left), the weight of the tool (1 t) made it impossible to handle during the machining operation. The degrees of freedom required, were, in this case, provided by a CNC-controlled gantry robot capable of guiding the laser beam 5-axially. Here too, wear characteristics were greatly improved by alloying an additional material containing carbide onto the roller.

The CNC programs required for the movement of the laser beam or component, are written mainly in teach-in mode. If tool geometries are made up of complex sculptured surfaces as is the roll previously mentioned, the amount of programming work necessary increases out of all proportion. Ratios of programming to laser machining times of 50 : 1 are not unusual. The development of suitable CAD/CAM modules with which to ensure the profitability of laser surface refinement even for such applications, is highly interesting.

In the area of laser beam treatment of complex level tools, a module for CAD/CAM-aided machining of such tools is already in existance. The forging die shown in Fig. 4 on the right, was first designed using a 2D - CAD system. In a second operation, the traverse paths were generated by a CAD program and the control system-specific machining program was subsequently calculated via an appropriate postprocessor.

222

This meant that the ratio of programming work time and laser machining time could be reduced in initial trials by factors of between 5 and 10.

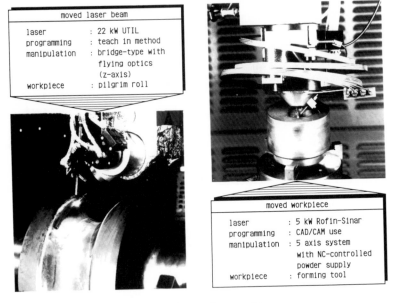

```
        moved laser beam

  laser         : 22 kW UTIL
  programming   : teach in method
  manipulation  : bridge-type with
                  flying optics
                  (z-axis)
  workpiece     : pilgrim roll
```

```
        moved workpiece

  laser         : 5 kW Rofin-Sinar
  programming   : CAD/CAM use
  manipulation  : 5 axis system
                  with NC-controlled
                  powder supply
  workpiece     : forming tool
```

Figure 4: Laser beam alloying of complex tool geometries

Summary

It can be said, in conclusion, that wear remains a threat to tool life. The examples presented, however, are undisputable evidence that laser beam surface treatment does indeed permit the advance of tool wear to be slowed down substantially. As well as increasing tool life in the field of forging and diecasting tools, laser beam surface refinement can make a major contribution to increased economic efficiency especially when tool manufacture costs are high or when tool changes result in protracted machine down-time.

References

/1/ König, W.; Rozsnoki, L.: Laserstrahloberflächenbehandlung auf dem Weg zum Fertigungsverfahren, Stahl und Eisen 109 (1989) 21, S. 15-17

/2/ König, W.; Rozsnoki, L.; Treppe F.: Standmengen von Schmiedegesenken durch Laser-Behandlung erhöhen, Werkstatt und Betrieb 124 (1991) 10, S. 831-834

/3/ König, W.; Rozsnoki, L.; Treppe F.: Standzeiterhöhung von Werkzeugen zur Warmumformung durch Laserbehandlung, Vortragsband zum 4. Umformtechnischen Kolloquium, 1991, Darmstadt

LASER CLADDING WITH PREHEATED WIRES

A. Hinse-Stern, D. Burchards, B.L. Mordike

Institut für Werkstoffkunde und Werkstofftechnik, Technische Universität Clausthal, Germany

Abstract: The present article concerns the laser hot wire cladding process, using alloy filler wire. This is a highly efficent cladding and coating technique, using a CO_2- laser as a thermal energy source. Preheating of the alloy wire to temperatures exceeding 1000°C was found to improve the cladding process. A chromium–hard–alloy, a strain hardening capable chromium–manganese steel and a ferritic cladding material containing tungsten carbides were used as alloy filler wires. These commercially available filler wires are used to improve the wear properties of components exposed to impact conditions, pressure loads and abrasive wear respectively.

1. Introduction

The laser hot wire cladding process is an interesting and powerful method of surface improvement (1),(2). Laser Hot Wire Cladding (LHWC) technique is an alternative to laser powder spraying and to the two step laser processes for the production of thin films of about 0,2 to 2 mm. The continuous nature of the LHWC gives it an economic advantage over the two step processes. Unlike laser powder spraying, where powder losses are incurred, hot wire cladding is almost 100 % efficient in the use of additional material. These are both one step processes and lead to low distortion of the workpiece. Heat input is reduced by a factor of 2 to 4 times compared with standard cladding techniques, e.g. arc welding or TIG.

Fig. 1 shows schematiclly the LHWC–process. The preheated wire is fed into the interaction zone, where it is melted and forms a melt–metallurgical bond with the substrate. The numerous variables such as focus size, laser intensity, laser mode, substrate and wire feed velocity, preheating temperature of the wire, degree of overlapping and shielding gas require an optimisation. The duration of the laser generated melt pool plays an important role in the optimisation of the process parameters. Suficient time must be allowed for the removal of gases by "bubbling" out, however the hard paticles must not be dissolved too much. The dilution of the clad by the substrate should be minimised, while still achieving a good bond.

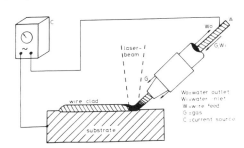

Fig. 1: Laser Hot Wire Cladding Process (schematic)

Up to 50 % of the process energy required can be provided by preheating the wire (3). The preheating is carried out by resistance heating, and depends on the nature of the wire. The laser is then only required to melt the additional material to form the bond. This preheating allows faster velocities and also improves the cladding homogenity by forming a smoother interface than is otherwise possible. Preheating the wire removes residual stresses giving more control for positioning. Furthermore the absorption of CO_2-laser radiation on the annealing wire surface is improved. Residual oils contaminating the wire are removed during preheating.

2. Experimental Detail

2.1 Characterisation of used filler wires

Three different filler wires were investigated for wear application, their composition is shown in table 1. The filler wires are tubes, formed from metal strips filled with alloying element powder and closed with an overlap. During production some porosity remains in the powder of the filler wire.

Filler wire	Diameter	Composition
Gridur F-C24	2,4 and 2,8 mm	Chromium hard alloy with 1,5 wt.% C and 27 wt.% Cr
Gridur F44	2,8 mm	Ferritic matrix with 70% enclosed tungsten carbides
Gridur F48	2,4 and 2,8 mm	strain hardening capable alloy austenitic chromium-manganese steel 0,5 wt.% C; 15 wt.% Mn; 15 wt.% Cr; 0,8 wt.% Mo

Table 1: Spezification of used filler wires

Gridur F-C24 is a filler wire based on a chromium-hard alloy and is used for welding wear resistant hardclads on steel machine parts. The cladding provides good protection against conditions of abrasive wear with low impact. On cooling from the liquidus hard chromium carbides compounds can be formed, due to the high chromium and carbon content. These carbides embedded in the alloyed steel matrix form the abrasive wear resistance.
Gridur F44 has a ferritic coat and contains of about 70 % tungsten carbide particles. Alt-hough it is good for abrasive conditions it not suitable for impact conditions or metal to metal wear. The claddings can only be machined by grinding.
Gridur F48 is a strain hardening capable chromium-manganese filler wire. The clads show very high impact and pressure toughness, and so can be used on crusher rolls or hammering mills. It can also be used as a protection against metal-metal wear and abrasive media. The high wear resistance of manganese hard steels come from the stress induced martensitic transformation on cold working. This produces a very hard surface while the untransformed cladding below remains ductile.

2.2 Laser Cladding

The cladding experiments were carried out using a 6 kW HERAEUS-C76 Laser. This transverse flowed CO_2-Laser produces a rectangular multimode beam. The wire transport was achieved using a commercially available four-roll-feeder. The feed velocity can be continuously varied between 50 and 3000 mm/min. Wire diameters between 1,2 and 3,2 mm can be transported. The preheating of the wire is realised with a one-phase-transformer using the ohmic resistance. The current source produces a maximum current of 200 Amperes at a maximum voltage of 30 Volts. The preheating temperature of the fed wire can be regulated by adjusting the applied voltage. Figure 2 shows the cladding process. The holder of the cladding gun allows exact positioning of the wire in the interaction zone.

Fig. 2: Laser hot wire cladding of St52 with Gridur F-C24

2.3 Cladding behaviour of the filler wires
The cladding behaviour of the tested filler wires is very different. All wires were melted with a laserpower of about 4 kW and a focus size of 5x7mm². The preheating temperature was about 900°C to 1000°C.

The worst cladding behaviour was exhibited by Gridur F44. Without substrate preheating cracks formed and pores remained. Preheating the substrate to 400°C produced a better clad with fewer pores and minimal cracking.

A better result was achived with Gridur F-C24, which showed a uniform wire melting by the laser beam. It produced more homogeneous clads with a relatively smooth surface, requiring little machining. The use of CO_2 as shield gas leads to further improvement.

By far the best results were achieved with the chromium-manganese hard steel Gridur F48.

3. Discussion
3.1 Metallographical examination and hardness
The microstructure of the chromium hardalloy is composed of martensite with finely distribu-ted special alloy carbides. On subsequent process previous tracks are thermally affected thus tempering the martensitic structure – this results in lower hardness. The heat affected zone

226

(HAZ) is about 700 μm deep. Fig. 3 shows the martensitic structure of Gridur F-C24. The hardness profile as a function of the distance from surface for Gridur F-C24 and Gridur F48 is shown in Fig. 4. It was measured in a track where annealing took place. The hardnes of Gridur F-C24 is about 890 HV$_{0,3}$, is constant through the thickness of the clad and then falls to the level of the substrate.

The microstructure of Gridur F48 is strongly dependend on the post-cladding treatment. The austenitic structure, attained by cladding can be transformed to martenstite by mechanical working, e.g. sand blasting. These two different structures are shown in Fig. 5 a) and b). Hardness testing was carried out on the as-clad samples. These samples were then sandblasted (30 sec), ground on fine SiC-paper and polished with 3 μm diamond paste. They were then retested. The hardness profiles are shown in Fig. 4. The hardness was increased by about 450 HV0,3.

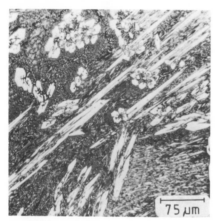

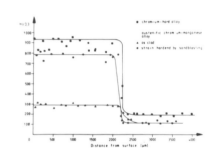

Fig. 3: Gridur F-C24   Fig. 4: Microhardness of Gridur F-C24 and Gridur F48

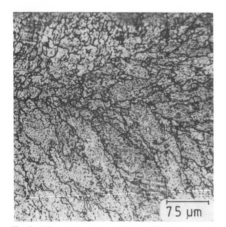

Fig.5: Microstructure of Gridur F48 a) as-clad,   b) strain hardened

The micrstructure of Gridur F44 consists of a ferrite matrix surrounding non fused tungsten carbides. Finer resolidified tungsten carbides are also present. Diffractometrical analysis revealed, that the carbides present are WC and W_2C. The mixed carbide $(Fe,W)_6C$ was also present. Fig. 6 shows a SEM micrograph of Gridur F44 clad. The WC content is minimised by the formation of a tungsten rich ferritic matrix, which exhibits a high hardness. This "matrix" hardness is due to the finely distributed resolidified carbides. The clad has

Fig. 6: SEM–micrograph of Gridur F44 clad

an average hardness of about $1100HV_{0,3}$. Variability in the measured hardness may be attributed to large carbide particles lying just below the surface at the mesurement points. The hardness of the large carbides has been determined as about 2600 $HV_{0,3}$.

4. Summary and conclusions

The LHW cladding using a multimode CO_2–laser is a workable method for the cladding of high alloy filler wires. Wire use efficiencies of up to 100% can be achieved using this "clean" one step process. Melt rates of up to 3 kg/h have been reached – this is low in comparison with other cladding processes, but the heat input is correspondingly small.

The wire preheating is a major factor in the lowering of the laser energy input, allowing higher feed rates of wire and substrate material. With preheating good results can be achieved using the 2,8 mm thick wire. Such good results are difficult using this wire cold. There is little noticable difference in the preheating behaviour of the wire. With clean contact between the wire and the substrate immediate heating occurs. If, however, the wire is displaced or melting of the wire occurs above the substrate, causing droplets of melt, the preheating current is not constant and the cladding formed is nonuniform. Due to the need for good contact between substrate and wire this technique is most suitable for rotationaly symmetrical workpieces, e.g. tubes, valves, disks. This technique can be used for multitrack cladding of a planar surface, but this requires the wire to be stopped at the end of each track to achieve a smooth surface. Bidirectional scanning is also possible, allthough surface quality is still poor.

The austenitic chromium–manganese hard alloy Gridur F48 and the chromium hard alloy Gridur F–C24 are good additional materials for this technique. The hard and wear resistant clads formed were free from pores and cracks. Preheating of the substrate is necessary for Gridur F–C24, if thick clads (> 1 mm) are desired. Acceptable clads were not obtained with the tungsten carbide containing Gridur F44 without preheating of the substrate.
It may be possible to overcome the pore- and crack problem by using another matrix, e.g. cobalt or nickel. Initial experiments using the substrate to dilute the additional material by up to 50 % show no cracks and no pores. This cladding is however not as hard. It is possible to use two hot wires simultanously in a one step process (4).
By using the tungsten carbide filler wire together with, for example, a cobalt full wire it may be possible to overcome the problem of cracks, by diluting the WC in a ductile cobalt matrix. This method is also more efficient in its use of the rectangular laser focus. The substitution of the wires by alloy ribbons (preheatable) would be advantageous but the availability of these ribbons is limited.

5. Future aspects
Work is currently being undertaken on applying this two wire technique to the cladding of valve seats. We reached cladding times of abaout 10 seconds laser working time. It should be possible to reduce this time by optimizing the process.

6. References
(1)   D. Burchards, A. Hinse, B.L. Mordike: Laserdrahtbeschichten; Mat.-wiss. u. Werkstofftech. 20, 405–409 (1989)

(2)   Ch. Binroth, R. Becker, G. Sepold: Zusatzdraht erweitert die Einsatzgebiete der Lasermaterialbearbeitung, ECLAT 90, Erlangen, Vol.2, S.717–719

(3)   A. Hinse-Stern, D. Burchards, B.L. Mordike: Laserdrahtbeschichten mit vorge-wärmtem Zusatzwerkstoff; Mat.-wiss. u. Werkstofftech. 22, 408–412 (1991)

(4)   A. Hinse-Stern, D. Burchards: Leistungsfähiges Laserdrahtbeschichtungsverfahren mit Tandemheißdrahtzufuhr; to be publ. in Proc. of. LASER 91, Springer Verlag

7. Acknowledgements
This work was carried out with wires kindly donated by Dipl.-Ing. G. Juretzko from Messer-Griesheim, Frankfurt and Dipl.-Ing. V. Groß from Thyssen Schweißtechnik GmbH, Bochum. Thanks to Mr. Kley from OSU Schweißtechnik GmbH, Castrop-Rauxel for providing the two wire feed gun for laser cladding experiments.

Anschrift: B.L. Mordike, D. Burchards, A. Hinse-Stern, IWW – Institut für Werkstoffkunde und Werkstofftechnik, TU Clausthal, Agricolastraße 6, 3392 Clausthal, Tel.: 05323/72-2120

Laser Surface Alloying of Titanium with Nickel and Cobalt

A. Weisheit, B. L. Mordike, Institut für Werkstoffkunde und Werkstofftechnik, Technische Universität Clausthal, Germany

Introduction

Gas nitriding of Titanium to improve wear properties is nowadays an established laser surface treatment /1/. The formation of hard TiN dendrites embedded in a ductile matrix of an interstitial α solid solution leads to an increase in resistance especially against abrasion. In this respects also those substitutional alloying elements are of interest, which form hard phases with Titanium. A promising alloy should fulfill the following requirements:

– the formation of a hard phase rich in Titanium to minimize the alloy concentration

– the primary solidification of the hard phase, because solid state transformation is aspected to be incomplete considering the rapid cooling rates during laser melting

– a sufficient solubility of the alloying element in the α-or β-phase to allow solid solution hardening and to retain ductility simultaneously

The transition metals Nickel and Cobalt fulfill these requirements. The Titanium rich parts of the binary systems were closely investigated in the fifties and sixties /e.g. 2,3/. Because of the brittleness of the intermetallic compounds Ti_2Ni and Ti_2Co, these alloys were of no technological interest for bulk volume components. But considering a small laser alloyed layer on a ductile base material Cobalt and Nickel gain attention again.

Both binary systems have a similar structure. An intermetallic compound of the type Ti_2X melts incongruently at 33.3 % (all compositions are given in at.-%, if not marked otherwise). The stoichiometrical range is very narrow (below 1 %). The compound is hard but also brittle, due to its complex lattice structure (96 atomes per cell). With increasing alloy concentration the compound solidifies in a eutectic (Ni:10–24 %, Co:14.5–23.2 %), primary from the melt (Ni:24–32 %, Co:23.2–27.1 %) and finally in a peritectic reaction with the equiatomic phase TiX (Ni:32–49.5 %, Co:27.1–49 %). With more than 5 % Cobalt and 6 % Nickel respectively the β–phase can be retained at room temperature after rapid cooling /4/.

This presentation introduces results of the one–step and two–step laser alloying with Nickel and Cobalt. For the basic research work (microstructure, hardness, brittleness) c. p. Titanium was used as substrate material. Wear tests under abrasive and erosive conditions were performed on alloyed TiAl6V4.

Experimental procedures

The experiments were carried out on a 5.5 kW CO_2–laser with a rectangular beam shape and a multimode energy distribution. In the one–step process the laser spraying technique - described elsewhere /5/ - was used. In the two–step process plasma sprayed layers were remelted. So far the two–step process proved to give better reproducible results. The thickness of the plasma sprayed layers was easier to control than the feed rate of the powder feed unit. Alloying requires very low feed rates (< 0.1 g/s) which lie near or even beyond the lower limit of cenventional powder feed units designed for plasma spraying. However, this is only a technological problem. The laser power density was varied between $5 \cdot 10^4 - 1 \cdot 10^5$ W/cm^2 and the scanning rates between 0.25–2 m/min. Melted depths from 0.3 to 0.8 mm could be achieved. Alloy concentrations were measured by EDX analysis. Because no standard were available the ZAF–correction had to be used, which gives results of less accuracy. However, these measurements proved to be useful for the identification of the phases and the understanding of the solidification process.

Mechanical properties were investigated measuring micro hardness (HV0.3) and fracture toughness and by performing a scratch test with a sintered WC/Co–tip. Wear tests were performed on a grinding machine using SiC emery paper as abrasive medium and in a blasting machine under different jet angles using corundum as blast compound. This wear test involves surface fatigue as well as abrasion.

Results and discussion

The microstructure of Nickel alloyed Titanium is shown in Figure 2. The β–phase solidifies in free dendritic growth (1 a). The eutectoid transformation is suppressed. The eutectic solidifies in a discontinuous way (1 b). This structure was described by /6/ as "chinese script" and is typical for a not completely coupled solidification. The spacing between the two phases is very small, even at low scanning rates (0.5 m/min) less than 1 μm were measured. The intermetallic compound forms in free dendrites or sometimes cellular dendritic growth. The dendrite tips tend to form facettes, (1 c). The equiatomic phase TiNi eventually solidifies as dendrites without facetting (1 d). Due to the small dendrite arm spacing the diffusion pathes are short and the peritectic transformation can perform extensively despite the rapid cooling rates. Alloying with Cobalt leads to identical microstructure.

Cooling rates for hypoeutectic alloys were calculated on the basis of results of /7/, who found for a Ti–Ni4.5 and a Ti–Co7.4 alloy the following relation between the dendrite arm spacing λ of the β–phase and the cooling rate ε:

$$\lambda = 50 \, \varepsilon^{-0.35} \tag{1}$$

The cooling rates were calculated using a 3–dimensional heat flux model. This involves the use of thermo–physical data, which are different for the present concentrations (18–22 %). The deviations had to be neglected, but this will not change the magnitude of the results. For the scanning rate range of 0.25 to 2 m/min cooling rates between 10^3 and 10^5 K/s were calculated for both alloy systems.

The hardness of near–eutectic composition (18–27 % Ni, Co) lies between 550 and 620 HV0.3 (c. p. Titanium: 180 HV0.3). In stochiometrical compositions maximum hardnesses of 675 HV0.3 (Ti$_2$Ni) and 720 HV0.3 (Ti$_2$Co) were measured. A further increase of the alloy concentration leads to a drop in hardness due to the formation of the ductile

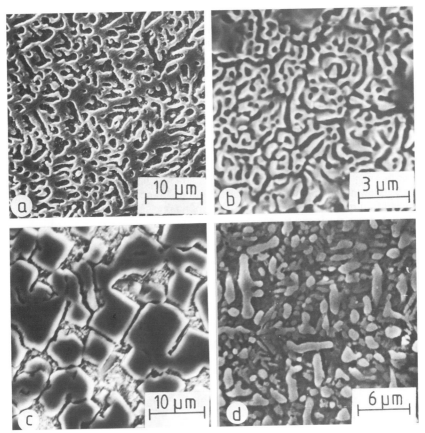

Fig. 1: Solidification structures of Ni–alloyed Ti; a) hypoeutectic, β–Ti dendrites + eutectic; b) eutectic; c) Ti$_2$Ni dendrites + eutectic; d) peritectic, primary TiNi dendrites

TiX phase (CsCl–type). Especially in the case of TiNi, known as memory alloy, the decrease is very significant, e. g. for a concentration of 46 % Ni the hardness drops down to 360 HV0.3.

Alloying of TiAl6V4 leads to identical microstructures and only slightly increased hardnesses. In hypoeutectic alloys the hardness lies well above 600 HV0.3, in hypereutectic alloys the maximum hardness reaches 720 HV0.3 (Ti–Ni) and 750 HV0.3 (Ti-Co) respectively.

Cracking during cooling was found to be dependent on the amount of the intermetallic compound within the microstructure. Hypoeutectic alloys showed only a few cracks perpendicular to the scanning direction, whereas in hypereutectic alloys a network of cracks formed. Preheating up to 400°C can reduce the crack sensivity due to the reduction of tensile thermal stresses. Crackfree layers could be obtained for all concentration, except for Ti–Ni–alloys near the stoichiometrical composition of Ti_2Ni.

The scratch tests were carried out under a load of 20 N and with a velocity of 0.5 m/min. Investigation of the scratches revealed some more information about the mechanical properties. In hypoeutectic alloys microcutting and -ploughing were observed, which are typical for ductile materials. In hypereutectic alloys the depth of the scratches is lower because of the higher hardness, but traces of microcracking could be found. This behaviour can be attributed to the brittle intermetallic compounds Ti_2Ni und Ti_2Co.

K_{IC} values were determined using the indention crack method. /8/ was the first who showed that the length of indention cracks could be used as a measure for the crack sensivity. Growing understanding of crack origin and propagation under a sharp indentor has led to a quantitative analysis. Based on the principles of the fracture mechanics it is possible to calculate the fracture toughness /9/:

$$K_{IC}{}^* = (\pi^{3/2} \alpha^* \cdot \tan \beta)^{-1} \qquad (2)$$

$$\alpha^* = \Delta P/\Delta C^{3/2} \; ; c = (d_1 + d_2 + \Sigma l) / 4$$

β is the half tip angle of the Vickers pyramid, P is the load, d is the length of the indention diagonal and l is the crack length. /9/ showed that this $K_{IC}{}^*$ value is comparable to that measured conventionally in a bending test, provided that no significant residual stresses are present. $K_{IC}{}^*$ values were determined for layers containing more than 90 Vol.% of the compound Ti_2X. In Ti_2Ni the first cracks appeared at a load of 3 N, in Ti_2Co at a load of 5 N. The maximum applied load was 20 N. For Ti_2Ni a value of 2.1 $MN/m^{3/2}$ and for Ti_2Co a value of 2.8 $MN/m^{3/2}$ were calculated. These values must be regarded probably to low because the laser induced tensile stresses make crack origin and propagation easier. However, the magnitude is reliable considering the complex lattice structure, which makes plastic flow very difficult. Compared to other materials the $K_{IC}{}^*$ values are similar to those of ceramics like SiC (3 $MN/m^{3/2}$) or Al_2O_3 (3–5 $MN/m^{3/2}$).

The wear performance of Ti–Ni and Ti–Co alloys was compared to that of the base material and laser gas nitrided layers. Figure 2 shows the results of the sliding tests against SiC emery paper. Only alloys near maximum hardness were tested. The wear rate of all layers is reduced approximately by a factor of 1/3. The differences lie within the standard deviations. In this test pure abrasion is responsible for material removal, thus the wear resistance is mainly dependent on the indention depth of the abrasive particles, which itself is a function of hardness. All worn surfaces showed only signs of microploughing and -cutting, no evidence of microcracking could be observed. The brittleness is therefore of less importance. The impact test involves two wear mechanisms /10/. At jet angles above 75° wear by surface fatigue dominates. At lower

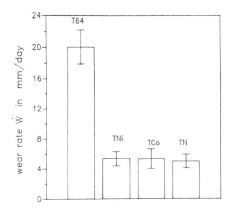

Fig. 2: Wear rate of laser alloyed
samples against SiC (P150);
P = 8KPa
T64 : 330 ± 10 HV0.3
TNi : 710 ± 15 HV0.3
TCo: 720 ± 26 HV0.3
TN : 675 ± 15 HV0.3

angles the sliding contact between the abrasive particles and the sample surface becomes more important, material removal due to abrasion will than be also involved. At an angle of 0° the wear rate (theoretical) must be zero, because the momentum of the particles vertical to the surface is also zero. Resistance against surface fatigue is mainly influenced by the (elastic or plastic) deformability of a material, whereas abrasion is mainly dependent on hardness. Therefore the wear rate as a function of the jet angle allows a characterisation regarding ductility and brittleness. The results of the impact tests are shown in Figure 3.

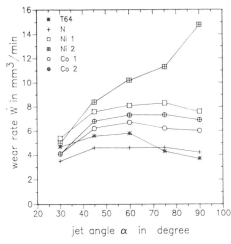

Fig. 3:
Impact wear rate of laser alloyed
samples as a function of the jet
angle
abrasive medium : Al_2O_3
velocity : 24 m/s
T64 (age hard.) : 415 ± 10 HV0.3
TCo1 (hypereut.) : 644 ± 18 HV0.3
TCo2 (hypoeut.) : 719 ± 23 HV0.3
TNi1 (hypoeut.) : 613 ± 18 HV0.3
TNi2 (hypereut.) : 733 ± 34 HV0.3
TN : 675 ± 28 HV0.3

The performance of the hypereutectic Ti–Ni–alloy (> 80 Vol.% Ti_2Ni) is typical for a brittle material with low elasticity. The wear rate increases continuously with increasing jet angle. The material is obviously very sensitive to surface fatigue. The energy of the abrasive particles is mainly converted into crack formation energy leading to massive delamination. This behaviour corresponds with the results of the scratch test and the fracture toughness. The hypoeutectic Ti–Ni performs much better at angles above 30°, but still no improvement compared to the base material is achieved. This means that still surface fatigue is the dominating wear mechanism. Both Ti–Co alloys show lower wear rates than the Ti–Ni alloys, although the hardnesses are nearly the same (and so are the volume fractions of the phases). Considering the crack sensivity during melting and the fracture toughness, this behaviour confirms that Ti_2Co is less brittle than Ti_2Ni. However, an improvement of wear resistance compared to the base material could only be achieved at an jet angle of 30°. The best performance is observed for the Ti–N alloy. Only at angles above 75°, when surface fatigue dominates, the wear rate exceeds that of the base material. Obviously this alloy represents the best compromise (among the investigated alloys) between ductility and brittleness. The volume fraction of the hard phase in the microstructure is less than that in the hypereutectic alloys Ti–Ni and Ti–Co (< 50 Vol.%). However, similar hardnesses can be achieved due to the high absolute hardness of TiN.

Conclusions

Laser alloying of Titanium with Nickel and Cobalt was investigated as a potential alternative to laser gas nitriding. Comparable hardnesses can be achieved. The hardening is mainly due to the formation of the intermetallic compounds Ti_2Ni and Ti_2Co. Additionally the β–phase is retained at room temperature, whereas nitrogen stabilizes the α-phase. Crack formation during solidification and scratching and the determination of the fracture toughness revealed a high brittleness of both intermetallic compounds. Under pure abrasive wear conditions this is of less significance, but when surface fatigue is involved the wear rate is higher than that of nitrided layers with similar hardnesses. Beyond this results the determination of the fracture toughness and the impact wear test have proved to be useful methods to characterize the mechanical properties of laser treated surfaces.

References

/1/ Final Report of the BRITE/EURAM Project RI. IB.015 C(H) "Surface Engineering of Titanium Components", 1991

/2/ F. L. Orrell jr., M. G. Fontana: ASM, 47, (1955), 554–564

/3/ D. M. Pool, W. Hume–Rothery: J. Inst. Metals, 83, (1954–55), 473–480

/4/ N. V. Ageev, L. A. Petrova: Dokl. Akad. Nauk. USSR, 138, (1961), 359–360

/5/ A. Weisheit, B. L. Mordike: Proc. Conf. ECLAT, (1990), 605–611

/6/ B. Chalmers: Principles of Solidification, J. Wiley & Sons, 1964, 211–213

/7/ W. A. Baeslack, e. a: Proc. Conf. Rapid Solidification Technologies, (1986), 97–110

/8/ S. Palmquist: Arch. Eisenhüttenw., 33, (1962), 629–633

/9/ K. H. Zum Gahr: Z. Metallkunde, 69, (1978), 534–539

/10/ H. Uetz: VGB, 49, (1969), 50–57

Pulse Laser Alloying Energy Spatial Distribution vs. Concentration Fields of Alloyed Elements

Smurov I., ENISE, Saint-Etienne, France
Covelli L.,IMU-CNR,Milano,Italy
Tagirov K.,Aksenov L., A.A.Baikov Institute of Metallurgy,
Academy of Sciences, Moscow, Russia

Mathematical Model

In the absence of extensive evaporation and concentration capillary phenomena, the thermocapillary convection plays the main role in melt hydrodynamics in laser produced molten pool. It is induced by surface temperature gradient and corresponding surface tension gradient (i.e., $\sigma(T,C) \sim \sigma(T)$) [1,2].

In surface treatment, the parameters of pulse laser action are optimized to obtain shallow pools (which diameter D is much larger than depth S). For these typical melting regimes, the thermal problem can be considered separately from the hydrodynamic one. This opportunity occures when heat conduction in the shallow pool predominates over convective heat transfer , i.e., when the melt velocities are fairly low (or the pool is shallow enough). The thermal problem determines the hydrodynamic problem that, in its turn, as a first approximation, does not react to the previous one.

Let us consider the two-dimensional (because of the cylindrical symmetry) melting of the massive metallic body by laser irradiation (surface heat source). We assume here that the thermal conductivity λ_i and thermal diffusivity a_i of the liquid phase (i=1) and solid phase (i=2), the dynamic viscosity η and kinematic viscosity ν of the melt, and the density ρ are constants; that the surface tension depends linearly on the temperature: $d\sigma/dT = -\alpha$ = const, and that the free surface of the liquid is not curved.

In the cylindrical coordinate system r,z,ϕ, where the z axis is directed into the target material and the laser beam is absorbed on the surface z=0, the phase transition front is determined by the equation z=S(r,t), where t is time. The maximum depth S_{max} is small if compared to the radius of the pool R(t), so it will be possible to assume, in the heat - conduction equation, the terms corresponding to the conductive heat transfer in the radial direction, are small enough if compared to heat conduction values along the z axis.

The hydrodynamic problem presents two characteristic linear dimensions, one of them exceeding the other. So the condition that the inertia terms appear small when compared to the viscosity terms, has the form $Re^* \ll 1$, where $Re^* = (S/r_0)^2 (V_r r_0 /\nu)$ is the reduced Reynolds number, r_0 is the radius of laser beam (for example, for Gaussian spatial distribution: $q(r)=q_0 *exp(-kr^2)$, $r_0 \sim (k)^{-1/2}$), V_r and V_z are the melt velocity components along the corresponding axes, respectively.

For liquid metals the Prandtl number $Pr=\nu/a_1 < 1$; so, in this particular case, we obtain the product $Re^* Pr \ll 1$. It means that in the melt the convective heat transfer is small if compared to conduction (Pe number is much less than unity). Because of this, the heat - conduction equation for the liquid phase does not contain the velocity components, i.e., the thermal and hydrodynamic problems are not coupled. The solution of the thermal problem determines the coordinates of the phase - transition surface and the temperature distribution along the free surface of the melt, which are used in the boundary conditions for the Stoke's equations [3,4].

As a consequence of the above-mentioned reasons, for the heat task, the appropriate governing equations are:

$$\frac{\partial T_1}{\partial t} = a_1 \Delta T_1 \quad , \quad \frac{\partial T_2}{\partial t} = a_2 \Delta T_2$$

$$-\lambda_1 \left.\frac{\partial T_1}{\partial z}\right|_{z = S(r,t)} = -\lambda_2 \left.\frac{\partial T_2}{\partial z}\right|_{z = S(r,t)} + \rho_2 L_m \frac{\partial s / \partial t}{1 + (\partial S / \partial r)^2}$$

$$-\lambda_1 \left.\frac{\partial T_1}{\partial z}\right|_{z = 0} = q_0 f(r)$$

$$T_1 (r, z = S, t) = T_2 (r, z = S, t) = T_m$$

$$T_2 (r = \infty, z, t) = T_2 (r, z = \infty, t) = T_2 (r, z, t = 0) = T_0$$

Here: Δ - Laplace operator; T_i - temperature of the liquid and solid phase, respectively; T_m - melting temperature; T_0 - initial temperature; L_m - latent heat of melting; $f(r)$ - spatial distribution of absorbed energy flux q_0. We assumed the isothermal boundary condition on the melting front (Stefan type boundary condition). Another consequence of the shallow molten pool assumption is the following: $Re^* \ll 1$, $V_z \ll V_r$, $\partial V_j / \partial r \ll \partial V_j / \partial z$, $j = r, z$; so that the hydrodynamic task can be presented in the following form :

$$\frac{\partial v_r}{\partial t} = -\frac{1}{\rho} \frac{\partial p}{\partial r} + V \frac{\partial^2 v_r}{\partial z^2}$$

$$\frac{\partial v_z}{\partial t} = -\frac{1}{\rho} \frac{\partial p}{\partial r} + V \frac{\partial^2 v_z}{\partial z^2}$$

$$\frac{1}{r} \frac{\partial}{\partial r} (r v_r) + \frac{\partial v_z}{\partial z} = 0$$

$$\left.\frac{\partial v_r}{\partial z}\right|_{z=0} = \frac{\alpha}{\eta} \frac{\partial T_1^{(0)} (r, 0, t)}{\partial r} + \frac{\partial \delta T_1^{(0)} (r, 0, t)}{\partial r}$$

$$V_r [z = S_1 (r, t)] = V_z [z = S_1 (r, t)] = V_z [z = 0] = 0$$

Here p - is a pressure. The solution of the system is based on the method discussed in details in [4,6,7]. The problem of convective mass transfer of admixture was solved by the method of particles (markers, for example [8]).

Results and Discussion

In the processes of pulse laser alloying of metallic materials, the spatial distribution of laser beam intensity plays an important role (together with the pulse duration and energy density flow value). Namely spatial distribution of laser energy determines the structure of melt convection. In the cylindrically symmetric case the number of toroidal vortexes in the molten pool equals to the number of local extremes of energy flux (fig. 1). In each vortex the melt moves near the surface from the region of local maximum of temperature towards the neighbouring local minimum, then deeply into the melt; in the regions of surface temperature maximum the melt raises to the surface. During the initial stage of melting, when, as usual, the zone of action consists of individual separated rings of the melt (corresponding to the regions of maximum values of energy flow), the sizes of the vortexes are determined (limited) by the melted zones. As the irradiation continues, the rings transform into the circle and the vortex sizes are determined by the distance between the neighbouring extremes of surface temperature.

The structure of spatial distribution of energy flow determines the structure of alloying element distribution in the molten pool. During the initial stage of alloying, when the melted zone consists of individual separated rings, alloying elements are introduced into the bulk of material only from their border parts (near the melting front boundary). Furthermore, the alloying elements are carried inside the melt in all the regions of local minimal values of energy density flow (even if far away from the edges of molten pool), and the most intensive remixing takes place in the zones with the largest values of energy and corresponding temperature gradients.

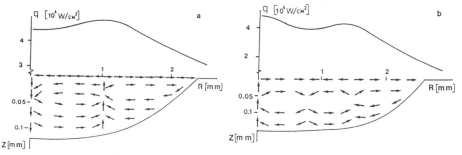

Fig. 1 Field of directions for velocity vector of surface tension melt flow of titanium at time moment t=3.5 ms (a), t= 4.3 ms (b) for the absorbed energy flux radial distribution.

By varying the positions of local extremes of energy flux (for all the same average values of treatment parameters) it is possible to change sufficiently the distribution of alloying elements in the zone of action. Depending on the conditions of laser alloying, it is possible to obtain the alloying zones of the different shapes: layers, spirals, drops, etc. To identify the alloyed regions and to verify the mathematical model, experiments on Titanium have been carried out. The material being irradiated was previously coated with a 10 μm thick layer of graphite. A solid state Nd:YAG laser source (400 W average), operated in pulsed mode was used. The beam was delivered either by an optical fiber (10 meters long) or directly from the source head to the moving tables. Different energy spatial distributions were obtained and focused through a 100 mm focal length. A wide range of pulse duration (0.2 ms -10 ms) as well as different values for energy density flux ($0.8-5 \times 10^5$ W cm^{-2}) were explored. The photos 5 are used mainly to illustrate the above discussion concerning the influence of spatial distribution of laser beam (and the condition of focussing) on the alloyed zones structures (black zones correspond to grafite penetration). Generally all the experimental results can be

238

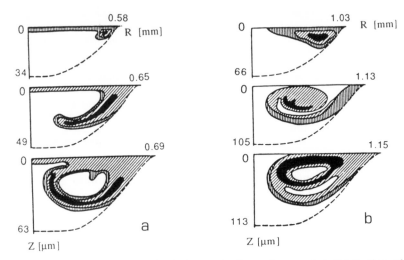

Fig. 2. Concentration field of admixture for Gaussian spatial distribution of laser beam; base material - Ti.
(a) - alloying from predeposited coating; $q_0 = 5*10^4$ W/cm^2; k= 100 cm^{-2}; corresponding moments of time; 2.1, 3.4, 3.7 ms.
(b) - alloying from gas phase; $q_0 = 2.5*10^4$ W/cm^2; k= 150 cm^{-2}; corresponding moments of time; 2.8, 3.6, 4.4 ms.

separated on two large families: (1) - the penetration of admixture into the melt mainly in the centre of the zone of laser action (photos a,b fig. 6); (2) - mainly in (and from) the peripheral part of the molten pool (photos c,d). Photo 6 (a) corresponds to the initial stage of remixing; photo 6 (b) presents the further development of the central zone remixing. On the photo 6 (b), also it is possible to see the second (smaller) zone of graphite penetration into the melt. It is closer to the melt boundary and is situated near the surface. It correspondes to the melt remixing in a form of two toroidal vortexes similar to fig. 1 (a). Probably the similar type of flow pattern is presented on fig. 5 (a). It is necessary to underline that generally the alloyed zones are not exactly axial simmetric, oftenly the penetration of admixture occurs in a form of separated "tongues" starting from the surface. By using optical fibres it is possible to obtain large temperature gradients in the peripheral part of the melting pool: the vortex structure of the convection (three revolutions) in the region of the beam edge easily can be seen (photo 6 d). The diameter of this vortex is much smaller than the pool's diameter. In all the other zones of melting the vortex structure of remixing is less evident, the nitrogen is penetrated into a thin layer near the free surface (more deeper in the central part of the melt). Three revolutions of the melt, detected by the admixture remixing (this is our record value from a large number of experiments), can be seen only in the region of large temperature and energy flux gradients, corresponding to the peripheral part of the beam delivered through an optical fibre. It is possible to note that even in the case of melt remixing corresponding to one toroidal vortex flow pattern, an admixture can enter the bulk of the pool from different points of the surface (photo 6c): close to the centre, and close to the boundary of the pool. In the left part of this photo it is possible to see the other (symmetric about the centre) vortex.

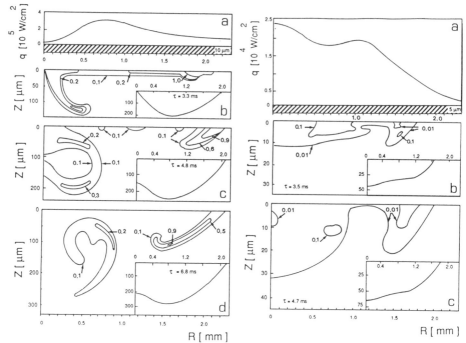

Fig. 3 Concentration field (b,c,d) of the alloying element at corresponding time moment t= 3.3 (b), 4.8 (c), 6.8 (d) ms. Base material Fe; coating thickness 10 μm; continuous coating; a - spatial distribution of energy density flow and of the coating initial position. The corresponding shapes of molten pool are shown in the frames in the reduced scale.

Fig. 4 Concentration field of the alloying element at corresponding time moment t= 3.5 (b), 4.7 (c) ms. Base material Ti; coating thickness 5 μm; continuous coating; a - spatial distribution of energy density flow and of the coating initial position. The corresponding shapes of molten pool are shown in the frames in the reduced scale.

References

(1) C.Chan, J.Mazumder, M.Chen: Metall. Trans. (1984) 15A, 2175.
(2) C.Chan, J.Mazumder, M.Chen: Materials Science and Technology (1987) 3, 305.
(3) A.A.Uglov, I.Yu.Smurov, A.G.Guskov: Sov. Physs.Dokl. (1988) 33, 755.
(4) I.Yu.Smurov, A.G.Guskov: in Physics-Chemical Processes of Materials Machining by Concentrated Energy Flows (1989), "Nauka", Moscow, 25.
(5) A.A.Uglov, I.Yu.Smurov, A.M.Lashin and A.G.Guskov: Modelling of Thermal Processes under Laser Action on Metals (1991) Moscow, "Nauka".
(6) A.A.Uglov, I.Yu.Smurov, K.I.Tagirov and A.G.Guskov. Int. J. Heat Mass Tranfer (1992)
(7) I.Yu.Smurov, L.Covelli, K.I.Tagirov and L.Aksenov: J. Appl. Phys. (1992).
(8) D.Potter. Computational Physics (1973) Imperial College. London. A Wiley-Interscience Publications.

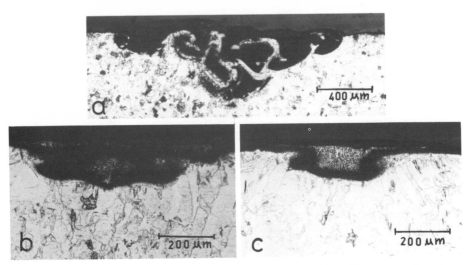

Fig.5 Alloying of Ti from graphite coatings (10 μm thick). All cross-sectional photos correspond to the same parameters of laser action: E= 32 J (35 J through the optical fibre, 12% losses), t = 10 ms, in the flow of nitrogen (50 lt/min); (a) free delivery beam; (b) and (c) through an optical fibre; (a) and (b) 100 mm focal length, (c) 50 mm focal length.

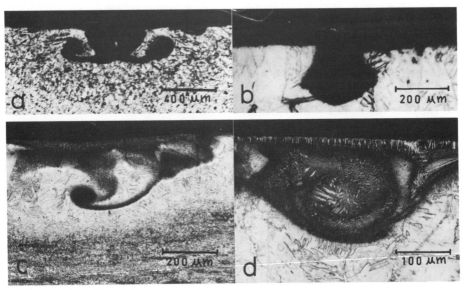

Fig. 6 Alloying of Ti: (a),(b) from graphite coatings (10 μm thick); (c),(d), from the nitrogen atmosphere; (b),(c), free delivery beam; (a),(d) through an optical fibre; (a),(c) all the zone of action;(b) central part; (d) edge of molten pool.

The Application of Laser Technology for Improvement High Load Friction Joints Tribotechnical Parameters

M. Ignatiev, E. Kovalev, V. Titov, A. Uglov, Institute of Metallurgy, Moscow, Russia
I. Smurov, Ecole Nationale d'Ingenieurs, Sent-Etienne, France
S. Sturlese, Centro Sviluppo Materiali, Rome, Italy

Introduction

Various coatings and conversion laser-assisted techniques are being used to modify the friction and wear properties of surfaces. These coatings are of two general types: hard coatings; diffusion barrier coatings (1). The wear tests results were used to examine the wear resistance of the titanium nitride hard coating produced by laser irradiation. The developed laser surface alloying technique may be also used to provide low shear coatings based on the extended solid solution of metals which have a high ductility: Sn, Pb, In. The technique allows to vary, for example, the Sn content up to 30% . The wear tests results of these coatings are also presented.

Experimental details

Materials used and wear tests.
Ball-bearing steel with chemical composition 0.55% C, 0.8% Si, 0,55% Mo, 0,1% V was selected as the test material. All the wear tests were carried out without lubrication in a dry air at room temperature. The specimens (outside diameter,40 mm; width 10 mm) was installed on a wear tester of type SM-2 designed to produce rolling contact. Each test was performed at 1000 rev/min, under a load of 300 MPa. The weight loss was measured with a precision balance.
Processing method.
2.5 kW continuous wave carbon dioxide gas laser (10.6 μ wavelength) was used in the present work. The Gaussian beam configuration is the typical for power range 1.0-1.7 kW. The ball-bearing specimens without absorption coating were mounted on a numerically controlled (CNC) x-y table and were irradiated with the laser using appropriate optical tools. The scanning beam rate was varied from 100 mm/min to 400 mm/min. During laser treatment , a helium jet was introduced coaxially with laser beam to protect a metal surface from oxidation. The gas expenditure was varied from 10 l/min to 35 l/min.
The structure and phase composition investigations of modified zone have been done by metallography and X-ray micro probe analysis. A comparative study of hardness of the steel and the alloyed layer was made by obtaining microhardness data from a Leitz miniload testing machine.

Results and discussions

It has been found at NASA that low shear coatings for some applications involving friction and wear can often be best applied using a plasma-related process (2). However the technique can't be used for tribotechnical joints with large permissible

wear value (several hundreds microns).
There is a futility of attempting to laser surface alloy Sn
(low melting point; high vapor pressure) into ball-bearing
steel (high melting point). This suggests that such low melting
point materials might prove unsatisfactory as a metal choice
for laser surface alloying (LSA) (3).
Compositional analysis of the alloyed layer. Microhardness.
The developed laser-assisted technique allows to vary Sn content up to 30 wt% at various parameters of LSA. The thickness
of the modified layer is up to 1 mm.
Figure 1. shows the Sn distribution over extended solid solution region produced by LSA. The compositional analysis was
carried out along the transverse section of the alloyed surface
layer at the 0.5 mm depth. One can see that rather full homogenization of the layer has occurred.

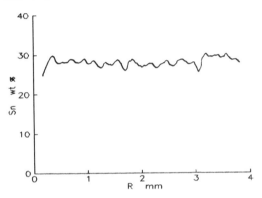

Fig. 1: Sn distribution over the
transverse section of the alloyed
zone. Laser power 1.5 kW.
Traverse speed 100 mm/min.

In Fig 2. the hardnesses of the treated zone are plotted vs the
thickness of the modified layer. Figure 2 shows that the hardness of the alloyed with Sn zone has a maximum value of 600 Hv.
The minimum value of 400 corresponds to matrix hardness. The
heat- treated zone has an intermediate hardness value.
The experiments proved the advantages of the posterior laser
modification of titanium plasma-sprayed coatings in a nitrogen
jet. The technique provides the synthesis of nitride layer of
anomalously high thickness (up to 200 microns). The hardnesses
values of 2000 Hv were obtained.
Wear resistance.
The samples produced by LSA and by titanium coating nitridi-
zation were subjected to wear tests. Figure 3 shows wear rate
curves in terms of reduction of weight vs time for: 1-heat
treated ball-bearing steel; 2-composite with TiN coating; 3-
alloyed steel with Sn.

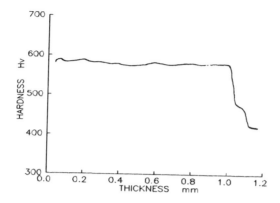

<u>Fig. 2</u>: Effect of LSA on the micro-
hardnesses of the matrix regions.

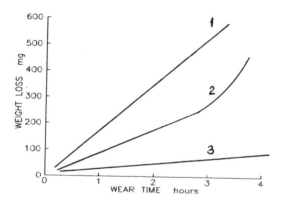

<u>Fig. 3</u>: Relationship between wear
time and weight loss.

Some industrial application.
The developed techniques were successfully used to enhance the
wear resistance of bearings operated as a shaft of turbodrill.
This type of turbodrill usually work in water medium at high
loads (the axial load 15 kN, the first contact stress 2500 MPa)
and have rather high allowed rate of wear (up to 3 mm). The
internal and external surfaces of ball races that were operating in contact with the balls had been alloyed with Sn by
means of laser irradiation. The thickness of the alloyed layer
having extended Sn solid solution was also near 1 mm (Fig. 1,
Fig. 2).
During exploitation the intensive tin transfer between friction
elements was observed. The corrosion-fatigue and adhesion pro-

cesses were reduced to minimum. The wear of alloyed surface did not influence on the wear rate as the balls took the tin on both sides from contact point and transfered it to worn-out parts of friction details. The sudden wear velocity increase was observed only after total wear of alloyed layer. The industrial wear tests showed that the increase of the longevity of turbodrill axial bearings working in water medium at high loads was more than 10 times

Conclusions

The techniques of laser alloying of bearing steels by low shear solid lubricants and laser nitriding of predeposited titanium coatings were developed. These methods allow to increase the wear resistance of friction joints working at severe conditions (high load, chemical active and low viscous medium etc.) up to 4-10 times in comparison with conventional heat treatment. The laser alloying with Sn of bearings was sucssesfully used in industry to increase the longevity of turbodrills

References

(1) B. Mordike, H. Bergman: Mat. Res. Soc. Symp. Proc. 28 (1984) 45.
(2) D. Mattox, J. Greene, D. Buckley: Mat. Sci. and Eng. 70 (1985) 79.
(3) C. Draper, J. Poate: Int. Metals Rev. 30, N 2, (1985) 85.

Laser Fusion of Ceramic Particles on Aluminium Cast Alloys

H. Haddenhorst, E. Hornbogen, Ruhr-Universität Bochum, Institut für Werkstoffe, Bochum, FRG

Introduction

Strength and hardness of Al-alloys with comparable microstructure is only one third of that of steels (1). This is due to the lower energy of bonding, elastic moduli and consequently the lower value for theoretical limits for plastic shear and fracture of crystals of aluminium. It excludes Al-base alloys from applications for which a very high absolute strength or hardness is required. This is the case for example in tribological systems in which abrasion plays a role, in ball bearings or gear wheels.

Laserremelting is one possible way to improve surface properties of aluminium alloys. Thereby different types of resolidificated microstructures are possible, these may become crystallin, quasicrystallin (2) or amorphous structures (3). Experiments proved, that all these structures can occur after laserremelting of different aluminium-alloys.

The current investigation is dealing with laserfusion of hard ceramic particles into surface layers of aluminium cast alloys with the aim to improve surface properties (i.e. hardness, friction and wearresistance) as much, that steels can be replaced in tribological systems.

Material and Experimental Procedure

Most aluminium cast alloys are based on binary aluminium silicon alloys. These elements form a eutectic system with its eutectic point at 11,7 wt% Si and therefore three binary alloys containing 8 wt% (hypoeutectic), 12 wt% (eutectic) and 17 wt% (hypereutectic) Silicon are choosen as substrates. Former investigation on simple laserremelting of these alloys showed, that an extended range of solid solution of silicon in aluminium associated with a displacement of the eutectic microstructure to higher silicon content leads to an increase of hardness of about 100%. Nevertheless these are too small absolute values (up to 200 HV) insufficient for new applications to these alloys (4).

To increase hardness further ceramic particles (SiC, TiC, B_4C, ZrO_2) are incorporated in-situ into the surface layer during laserremelting. As good mixing is aimed for, the specimens are prepared in a special way. Firstly saw cuts of 3 mm depth and different widths are put into the surfaces. Secondly these are filled with ceramic particles. The as-prepared specimens are remelted along the saw cuts whereby the liquid is heated up to the boiling point. Laserremelting is done with a Rofin-Sinar 5000 W CO_2-Laser, the parameters are listed in table 1.

The microstructures of the lasertracks are firstly analysed using light- and electron-microscopy, supported by EDX. Secondly hardness and microhardness are determined. To probe the micromechanisms of abrasive wear a metallographic scratching method is used (5). The specimen is moved parallel and straight under a diamond pyramid, which is loaded with a force normally to the specimens surface. The plane becomes scratched showing three possible mechanisms of damage, namely microploughing, microcutting or microcracking. Ideal ploughing leads to no disspation of matter, it is pressed aside by plastic deformation. If microcutting occurs, the material is separated by combined shearing and cracking. Brittle fracture is the dominating mechanismen for microcracking.

Table 1: Parameters of laser treatment

Laserpower	4,8 KW
Thrust	60 mm/s
d (Focus)	0,41 mm
Position of focus	-1 mm
Focal length	200 mm
d (mirror)	34 mm

The normal load acting on the diamond in this investigation is 1 N, the sliding velocity 1,8 μm/s. A frictional force results, which depends on the groove path. An abrasive coefficent of friction is defined as ratio of measured frictional force and applied load. The environment of the grooves is investigated by electron-microscopy.

Experimental Results
The resulting lasertracks are more than 3 mm deep and more than 1 mm wide, and can be subdiveded into two areas, the central zone containing a mixture of ceramic and substrate and an outer zone containing only rapidly quenched aluminium silicon (Fig. 1). Depending on width of the saw cut, different microstructures can be fabricated in the central zone. Fig. 2 shows the resulting structure using the largest saw cut. Coarse particles are luted by rapidly quenched aluminium-silicon. This structure is well known from cermets. These areas show high hardness of about 900 HV_{10} but combined with brittle behavior in wear tests.
A smaler saw cut leads to a better mixing of the different materials. The resulting microstructures (e.g. Fig. 3) consist of small partly remelted ceramic particles, rapidly

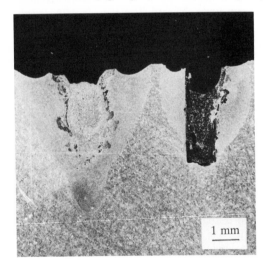

Figure 1:
Lasertracks consisting of two areas: a central zone containing a mixture of ceramic and substrate and an outer zone without ceramics molten

quenched aluminium-silicon and a needle like phase which is presumably of ternary composition. This structure occurs, if SiC, TiC or B_4C is alloyed. Alloying with ZrO_2 does not lead to complete impregnation by the Al-based melt. Evaporation causes rapid grooved surfaces. Table 2 shows the measured microhardness obtained in the central zones.

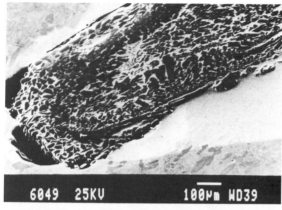

Figure 2:
See right track (Fig. 1) Cermet-like structure, the substrate alloy is melted at the right hand side and has partially impregnated

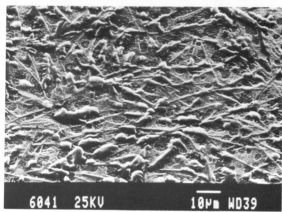

Figure 3:
See left track (Fig. 1) Structure of the central zone: small partly remelted ceramic particles, rapidly quenched aluminim-silicon and a needle like phase.

Table 2: Microhardness $HV_{0,025}$ measured in the centered zones

	AlSi 8	AlSi 12	AlSi 17
Substratum	68	80	--
add. SiC	--	406,6	211,2
add. B_4C	482,2	500	515,3
add. TiC	--	248,7	--

Scratchtests show, that good abrasive wear properties can be expected after alloying of SiC, TiC or B_4C. All these ceramics lead to similar results. Microcutting is the predominating wearmechanism with additional small amounts of microploughing and microcracking. In table 3 the abrasive coefficients of friction and the measured scratch widths

Table 3: Abrasive coefficient of friction/scratch width obtained in scratchtests [μm]

	AlSi 8	AlSi 12	AlSi 17
Substratum	0,35 / 54	0,36 / 71	-- / --
add. SiC	0,13 / 10	0,10 / 11	0,12 / 12
add. B_4C	0,14 / 12	0,16 / 12	0,12 / 11
add. TiC	0,22 / 20	0,15 / 21	0,15 / 19

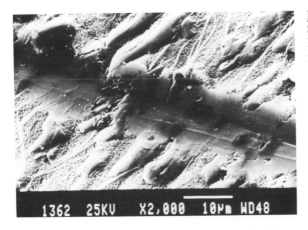

Figure 4:
Scratchgroove in AlSi 12 + B_4C, microcutting predominates

are listed. Obviously B_4C alloyed specimens have the best wear properties. Fig. 4 shows a typical scratchgroove in AlSi 12 + B_4C. Rapidly quenched eutectic microstructure shows microcutting with small amounts of microploughing, ternary phase shows pure microcutting while B_4C-particles are mostly pressed aside. Only a few particles crack during scratching.
Additional experiments using a steelball instead of the pyramid lead to the conclusion, that adhesive wearresistance is improved correspondingly. The best properties are provided in this case by the cermet-like structure where no weardamage can be observed.

Discussion and Conclusions
A three step process has to be considered for an interpretation of the observed microstructures: firstly rapid heating, secondly rapid mixing and thirdly rapid quenching.

Rapid heating is limited by the boiling temperature of aluminium (2467°C), which is lower than melting or decomposition temperatures of most ceramics. This in combination with the fact, that only a few particles closed to the surface are highly heated by the laser-beam lead to only partially melting of ceramic particles.
As the duration of liquid mixing is limited by the process, there are zones of high alloy content in the adjacent liquid. It is insufficient to mix the melt completely by convection.

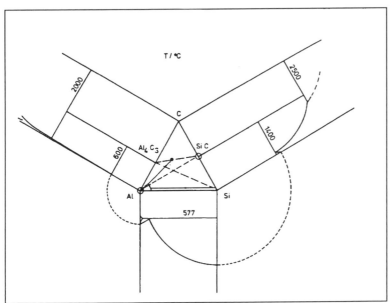

Figure 5: Tendative Al-Si-C diagram

As soon as the laser is moved away, the process of rapid solidification starts favored by exellent thermal conductivity of aluminium. As the melt has variable composition and temperature, the conditions are very complex. Heterogenous nucleation starts at undissolved particles and at the coastline of the unmelted substrate. Also homogenous nucleation may occur because of large undercooling.
Fig. 5 shows the ternary system Al-Si-C to explain the solidification process. The influence of Boron or Titanium in quaternary systems is neglected. The first phase which forms could be a metastable ternary phase, which forms around the quasibinary Al_4C_3-SiC eutectic. It should decompose into these two phases (6), but rapid cooling prevents this. In hypereutectic alloys, this phase supports the nucleation of silicon. All reforming silicon crystals grow on its surfaces. Furthermore an important observation is the change in morphology of eutectic microstructures, which solidify last. The new extremly fine netstructure should provide the desired mechanical surface properties. This structure results evidently from homogenous nucleation of supersaturated aluminium, which is immediately followed by crystallisation of one or two compounds from the residual liquid.

Investigations of hardness and wear properties of these resolidificated structures show, that the surface properties of hardened steels are achieved. This considerable raise in hardness is due to dispersion hardening by the small unremelted ceramic particles in addition to the ternary phase. The melting temperature of TiC is about 650°C higher than those of SiC and B_4C. Due to this fact, only a small amount of TiC is melted with the consequence that particles are larger and ternary phase is hardly detectable. Therefore hardness and tribological properties are not as good as they are, if the other ceramics are used.

As wear resistence is not only depending on hardness but also on local plasticity and fracture toughness, the suppresion of microcracking leads to a remarkable improvement. Brittle ceramic particles are so small, that necessary deformation can occur in the surrounding net-structure, which bears a small amount of plasticity.

Additions of nickel led to the formation of small Al_3Ni and Al_3Ni_2 -particles in the resulting microstructures with a positiv effect on hardness and wear properties (7). Therefore a further increase in surface properties might be reached by a combination of ceramics and nickelbased intermetallics. Such investigations are under way.

Acknowledgement
Laser treatments were carried out by Dörries & Scharmann GmbH, Mechernich

References
(1) E. Hornbogen, A.G. Crooks: "Metallurgical aspects of laser alloying of Aluminium", Proc. ECLAT 90 (1990) 613
(2) E. Hornbogen, H.W. Vollmer: "Microstructural effects in laser heated Al-Si and Al-Si-Mn alloys", Proc. ECLAT 88 (1988) 137
(3) N. Jost, H. Haddenhorst: "Laserglazing of Aluminium-based alloys", J. Mat. Sci. Let. 10 (1991) 913
(4) H. Haddenhorst, E. Hornbogen, H.W. Vollmer: "Investigations of laser trated surfaces of aluminium silicon alloys by scratch-, friction- and wear-tests", Proc. ECLAT 88 (1988) 97
(5) H. Nöcker, E. Hornbogen: "Reibungs- und Verschleißmessungen mit Hilfe einer metallographischen Kratzermethode", Sonderband 19 der praktischen Metallographie, 103
(6) L.L. Oden, R.A. McCune: "Phase Equilibra in the Al-Si-C System", Met. Trans. A 18 (1987) 2005
(7) B. Velten: unpublished results, Ruhr-University Bochum

Synthesis of nitride and carbide compounds on titanium by means of a solid state laser source

Smurov I.
ENISE, 58 rue Jean Parot, 42023 Saint-Etienne, France

Covelli L.
IMU, C.N.R., 56 via Ampere, 20131 Milano, Italy

The physics-chemical processes during laser alloying of titanium are considered. Pulse (Nd:YAG, energy per pulse 5-45 J, pulse duration t up to 10 ms) and pulse-periodic laser action (scanning velocity 5-30 mm/s, frequency 5-20 Hz) are used. The laser beam is delivered either directly or through an optical fiber (10 m long, 0.6 mm core,12% power losses). The solid state laser beam acts both independently and together with a continuous wave CO_2 laser (max. power 2.5 kW).
The alloying elements are in a gas phase (flow of N_2, CO_2), and as predeposited coating (graphite layers, about 10 μm thick).
The relationships between the parameters of laser treatment and the structures of the alloyed zones are determined. It is shown that the redistribution of the alloying elements (both from gas phase and from the predeposited coatings) in the laser affected zone is determined mainly by the convective mass transfer under the action of surface tension force [1,2]. The dynamic of pulse laser alloying is modelled by varying the pulse duration and energy input, but by keeping the constant energy density flow. The influence of the spatial distribution of laser beam intensity, pulse duration and energy density flow value on convective mass transfer is analyzed. The existence of threshold of admixture remixing versus pulse duration and energy density flow (keeping the other parameters constant) is shown. By varying the parameters of the process, it is possible: (a) - to obtain comparatively large melting zone practically without remixing of admixture; (b) - to stimulate the propagation of admixture in the center of the molten pool; or (c) - in its peripheral part; or (d) - from both of them simultaneously. The flow pattern can be presented by one, two, or three toroidal vortexes.
The 3-D shape of alloyed zone is reconstructed on the base of different cross sections. During one experiment (keeping the chosen parameters) a series of laser pulses (without any overlapping) was performed on a Titanium plate. Then the plate was cut on different parts close to the individual zones of action, and later each zone was polished in one of the orthogonal directions. After polishing on the depth of about 50-70 μm the process was interrupted to make a photo. By this way, a series of photos of different parallel cross-sections of one alloyed zone separated by the above mentioned distances was obtained. Because of the identical conditions of experiment, it is possible to assume, that we obtained series of cross-section's photos practically of the same zone, along three orthogonal axes. The results corresponding to Ti alloying from graphite layer are reported on Fig. 1. Because of the limited space only a few photos are presented. On the base of the photos 3-D structure of the alloyed zone can be reconstructed. Neglecting the details, it can be represented as a kind of a "mushroom" growing into the bulk of material. The "foot" is situated approximately in the center (photos 1a, e); visible distance between the "foot" and the "hat" can be recognized (photos 1a, e). This structure of the similar type also is presented on photo 3a, but it corresponds to the initial period of remixing ("mushroom's" growth). The "hat" is convex down surface, that is why when it is crossed by plane (by cross-section) in the middle appears a kind of a window (see photos 1b, c). Usually the peripheral part of the "hat"

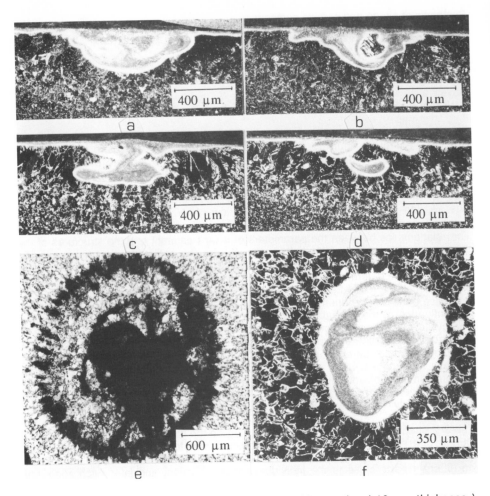

Fig. 1. Pulse Nd:YAG laser alloying of Ti from graphite coating (10 μm thickness), pulse duration τ = 10 ms, pulse energy E = 35 J, free delivery beam, air. Cross-sectional photos correspond to the following planes (axes x, y are on the surface, z directed inside the material): (a) - x = 0, (b) - x = 300 μm, (c) - x = 600 μm, (d) - y = 300 μm, (e) - z = 0 μm, (f) - z = 300 μm

reaches the free surface of the melt. It is necessary to underline that the above discussed structure of the alloyed zone is typical for a wide range of parameters, but it is not universal. This one and the similar types of structures can be explained on the base of flow pattern that consists of: (a) - one toroidal vortex transporting admixture from the surface deep into the pool towards its center; (b) - the same vortex in the central-part

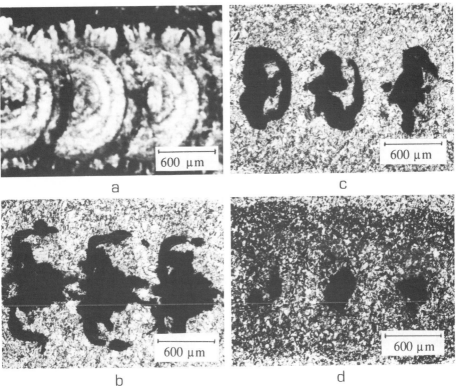

Fig. 2. Pulse Nd:YAG laser alloying of Ti from graphite coating (10 μm thickness), pulse duration τ = 6 ms, pulse energy E = 25 J, frequency 10 Hz, scanning velocity 0.5 m/min., beam delivered through an optical fiber. Cross-sectional photos correspond to the z axis, directed inside the material): (a) - z = 0, (b) - z = 30 μm, (c) - z = 60 μm, (d) - y = 120 μm.

(but of the smaller size, that remixes the melt near the center) plus another one (much larger), situated in the outer part of the pool) rotating in the opposite direction [2]. The later vortex moves the admixture from the peripheral part of the pool down and towards. The influence of pulse overlapping of the structure of the alloyed zones was evaluated on the example of pulse-periodic treatment of Ti slab covered by graphite layer. The corresponding cross sections (parallel to the surface) were prepared in the same way as - on photo 1. The influence of overlapping is easy to see up to the depth of about 40 μm, deeper in disappears because on this level the zones of intensive remixing do not come into contact. The absence of axial symmetry on photo 2b is the result of "cleaning" of irradiated surface from graphite layer by the preceding pulse, that remixes the coating with the base material. So, the successive pulse acts on the zone that is only partially covered by graphite (coating exists on the left part of the laser irradiated zone and destroyed on its right part, see photo 1a).The penetration of graphite from the rest part of the coating in a shape of semicircle can be see on photo 1b. This semicircle

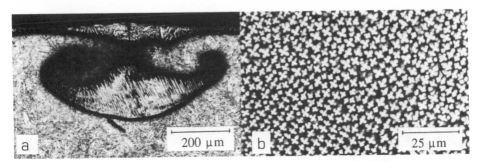

Fig. 3. Pulse Nd:YAG laser alloying of Ti from graphite coating (10 μm thickness), pulse duration τ = 10 ms, pulse energy: (a) - E = 40 J, (b) - E = 35 J; (a) - optical fiber, (b) - free delivery beam, air. Cross-sectional photos correspond to: (a) - the center of melted zone, (b) - microstructure in the plane z = 300 μm

corresponds to the peripheral vortex of melt convection. Photo 2c presents the central part of the alloyed zone where overlapping does not influence on. Its ellipsoidal shape corresponds to the inner vortex of melt convection. It is possible to make a conclusion that in the present case melt convection is determined by two embedded vortexes of different sizes and intensities of remixing. The alloyed zones are essentially not uniform, a wide range of different structures can be seen. They are determined by the convective heat and mass transfer in the laser produced melt. In dependence of the alloying elements (N_2,C) concentration, and cooling solidification conditions, the observed zones (structures) can be roughly separated on: (a) - clear dendritic morphology (central part of photo 3a); (b) - solid solution with low concentration of N_2, C in Ti with low differences from the pure remelted Ti (black areas in the peripheral parts of the alloyed zone, photo 3a); (c) - intermediate zones usually with fine structure probably determined by diffusion phenomena (photo 3b). The zones (a) with high concentration of admixture can be found deep inside the melt (close to the bottom of the pool) and, as a rule, correspond to the path of convective flow (of alloying element) from the surface down the pool. The thin layer (5-30 μm) on the irradiated surface often consists of dendrites oriented normally to the surface (along the temperature gradient). The same phenomenon determines the orientation of dendrites (perpendicular to the stream of hot metal from the heating surface) in the central part of photo 3a. The microhardness inside the zones of melting, that generally do no coincide with the alloyed (remixed) zones, is essentially not uniform. This is the result of the variation of the structures and chemical composition along the melt pool. The microhardness is very high in the dendritic zones: the maximum values in pulse alloying of Ti from graphite coatings are about 2800 HV, for Ti alloying in the flow of N_2: 2200 HV. In the central part of the pool, close to the above mentioned dendritic zones, it is possible to measure the microhardness close to the base material values. See, for example, local minima on the curves: microhardness versus depth (Fig. 5,6). Often these points are close to the center of the toroidal vortex of the melt that transport admixture from the surface inside the melt. That is why it is more correct to measure the average values of microhardness, for examples on the depth of 100 μm below the surface: for example, it increases from 200 till 450 HV (the system Ti+graphite coating in the of N_2) with the increase of pulse energy from 20 till 40 J. On the contrary, in the same conditions the average microhardness decreases from 650 till 300 HV for the system Ti+graphite

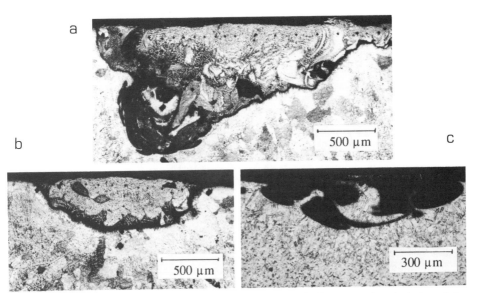

Fig. 4. Alloying of Ti from the flow of nitrogen (50 l/min). Cross-sectional photos: (a) - combined CO_2 (power 1400 W, scanning velocity 0.2 m/min.) + Nd:YAG pulse-periodic laser action (pulse energy E = 25 J, pulse duration τ = 10 ms, , frequency 4 Hz, beam delivered through an optical fiber); (b) - single CO_2 laser action with the same parameters; (c) - single Nd:YAG pulse-periodic laser action with the same parameters.

coating in the flow of Ar; and keeps about 500 HV for Ti in the flow of CO_2. It means that gas atmosphere is the one more important parameter to determine the results of laser alloying.

Laser synthesis of titanium nitrades and carbides was provided also in the conditions of combined laser action: both 2.5 kW CO_2 and 400 W Nd:YAG lasers sources were focused on one zone. The irradiation of solid state laser was delivered through an optical fiber (10 m long). CO_2 laser beam was oriented normally to the surface, Nd:YAG to an angle of 45°. The cross-sectional photos are presented on Fig. 4. Easily can be seen a qualitative difference in the structure of alloyed zones (even in shape and dimensions), that corresponds to the different way of melt remixing. Pulse-periodic action with high energy density flow, produces large temperature gradients and intensifies thermocapillary convection. This is a good example of the strong influence of spatial distribution of energy density flow on melt convection [2]. It is necessary to underline that in the present experiment the average power of Nd:YAG laser composed 8% from CO_2 laser source. The large increasing of the melt pool dimensions also can be explained by the heat released as a result of TiN formation. The corresponding values of microhardness are presented on Fig. 5. The lower microhardness in the zone of combined action (relative to single CO_2) can be explained by its sufficiently larger sizes and more homogeneous distribution of the alloying element (as a result of intensive remixing).

256

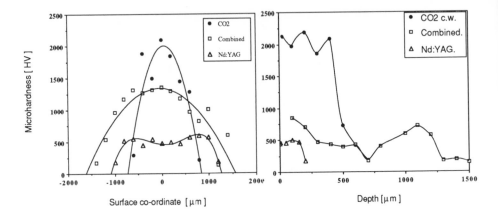

Surface co-ordinate [μm]

Depth [μm]

Fig.5. Alloying of Ti from the flow of nitrogen (50 l/min). The dependence of microhardness versus: (a) - radius, (b) - depth; curve (1) - combined CO_2 + Nd:YAG pulse-periodic laser action (conditions as reported on Fig. 4); (2) - single CO_2 laser action with the same parameters; (3) - single Nd:YAG pulse-periodic laser action with the same parameters.

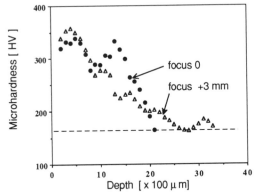

Depth [x 100 μm]

Fig. 6. Alloying of Ti from the flow of nitrogen (50 l/min), pulse energy 21 J, pulse duration τ = 7 ms. The dependence of microhardness versus depth; curve (1) - free delivery beam, focussed on the surface; curve (2) - beam focussed 3 mm over the surface; - - - - - microhardness of base Titanium.

References

(1) J.Mazumder: Optical Engineering. (1991) 30, 1208
(2) I.Yu.Smurov, L.Covelli, K.I.Tagirov and L.Aksenov: J. Appl. Phys, 71 (7), (1992).

Thermochemical nitridation of TA6V Titanium Alloys by CO2 Laser

P. LAURENS, D. KECHEMAIR, L. SABATIER, T PUIG*, L. BEYLAT*,
P. JOLYS**, P. LAGAIN**
  LALP, Unité mixte ETCA-CNRS
* CREA/Physique des Surface
ETCA,16 av. Prieur de la Côte d'Or, 94114 Arcueil Cedex France
** Aérospatiale, Centre commun de recherche Louis Blériot, 12 rue
Pasteur, BP 76, 92152 Suresnes Cedex France

Introduction

Due to its numerous properties (low density and good wear
resistance),titanium nitride found applications in the aerospace
industry. It can be elaborated by deposition technics or by
thermodiffusionnal process which can be plasma or laser assisted.
Laser nitriding was mostly achieved in the liquid state which lead
to the formation of a few hundred micrometers thick nitride layers
[Ref.1,2]; however, due to changes in the surface roughness and
substrate dimensions, a post machining step often has to be
performed on the treated samples. To avoid this, a <u>laser assisted
solid state TA6V nitridation process</u> could be an interesting
alternative. The present investigation aimed to prospect the
potentialities of such a solid state treatment. The nitriding
process will be characterized in terms of gaseous protection
efficiency, substrate thermal cyles and laser-surface coupling.
The performances of the nitride substrates were tested through
nitride thickness, microhardness, wear resistance and fatigue
lifetime measurements.

Experimental Set Up

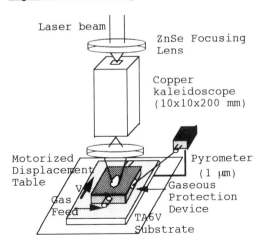

Laser beam

ZnSe Focusing Lens

Copper kaleidoscope (10x10x200 mm)

Motorized Displacement Table

Pyrometer (1 μm)

Gaseous Protection Device

Gas Feed

TA6V Substrate

Fig. 1 : Experimental Set-up

The experimental set up
is presented on Fig. 1. The
CO_2 laser beam was
homogenized through a
kaleidoscope device enabling
to obtain large irradiated
areas (typically 6 x 6 mm
square). Power densities
were ranging between 1.4 and
11.1 kW/cm^2. TA6V substrates
were moved under the laser
beam at a speed ranging
between 0.01 and 0.5 m/min.
Treatments were achieved in
pure nitrogen atmosphere (D
= 100 to 200 l/min). The
TA6V basic material had an
equiaxed($\alpha+\beta$) structure.Its
mechanical characteristics
after the preparation

thermal treatment were : Rm = 1150 MPa, A% = 12 %, Re = 1050 MPa
and β transus temperature = 995°C.

Laser Process Characterization

* Gaseous protection
Measurements of O_2 contamination in the N_2 treatment atmosphere
revealed that a nozzle type protection device ensured only a
moderate protection from ambient contamination (0.1% O_2), whatever
were the N_2 flowrate and the nozzle inclinaison. A "bell shape"
device covering the TA6V substrate had thus been designed ; it
allowed to decrease the O_2 contaminaison to 3.10^{-3} % and therefore
to avoid substrate partial oxidation.

* Thermal cycles investigations
Depending on the laser process paramaters, maximum surface
temperatures in the irradiated area were ranging between 1100 and
1720°C.

Température (°C)

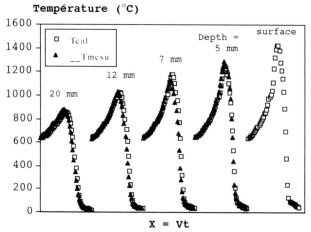

Fig.2 :Measured and
simulated thermal
cycles for the TA6V
nitridation.

Treatmentconditions:
V = 4 mm/s
E = 4.3 kW/cm2
Fx = 6 x 6 mm2
Milled TA6V surface

(0.5 mm or 0.125 s between measurements)

Satisfying thermal cycles simulations had been achieved in a
non linear stationary bidimensionnal configuration taking into
account radiation and convection losses, variations of the TA6V
physical properties with temperature and energy delivered by the
CO2 beam (Fig. 2) [Ref.3]. Moreover, it has been pointed out that,
due to multiple reflexions onto the metallic "bell shape"
protection walls, approximatly 1% of the incoming energy density is
redistributed out of the surface irradiation zone ; this resulted
in a typically 100°C substrate preheating.

* CO2 laser - TA6V surface coupling
From thermal cycles measurements and simulations, numerical
identification of absorption coefficients with spacial resolution
had been performed [Ref.3,4]. A typical example of the variations

of the absorption coefficient, A, along the displacement axe during nitridation is presented on Fig.3.

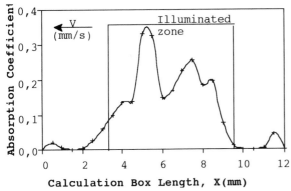

Calculation Box Length, X(mm)

Fig. 3 : Identification of the variations of the absorption coefficients during the solid state nitridation of TA6V on milled surfaces.(Nitriding conditions:V = 4 mm/s, E = 4.3 kW/cm2, Fx = 6 x 6 mm2).

Variations of A in the irradiated zone should probably be attributed to inhomogeneities in the incident distribution energy, whereas low, but non zero, absorption out of the irradiation zone resulted from multiple reflections on the protection device. Average calculated absorption coefficients were in good agreement with measurements under an integrating sphere.

They both pointed out that the growth of a few microns nitride layer did not modify the laser - surface coupling : average absorption coefficients were respectively 0.2 ± 0.1 and 0.4 ± 0.1 for milled and sandblasted surfaces.

Nitrided TA6V Alloys Characterization

SEM analysis of the nitrided TA6V alloys revealed the successive presence (from the outer surface) of : a few micrometers thick chemically affected zone (δ-TiN, nitrogen solid solution α-Ti(N) and an acicular α'-Ti+N), an α' martensitic layer and the initial $\alpha + \beta$ structure.

Nitrogen diffusion profiles obtained from microprobe nuclear analysis pointed out that the total chemically affected zone was in the order of 40 μm thick (Fig.4). The thickness of the δ-TiN + α-Ti(N) layers was found to increase with the energetic density, the substrate displacement speed and the number of successive passes. Typical microhardness in the superficial nitrided layers ranged between 750 and 900 $HV_{0.025}$ (Fig.4). Moreover, a correlation between the substrate hardness increase and the substrate nitrogen percentage had been clearly evidenced. Taking into account that for the aerospace applications, a minimum of 600 $HV_{0.025}$ had to be reached in the nitrogen affected zone, we can therefore consider that a minimum 4 % nitrogen concentration had to be achieved. Under this standart, the affected zone was thus ranging between 10 and 20 μm thick.

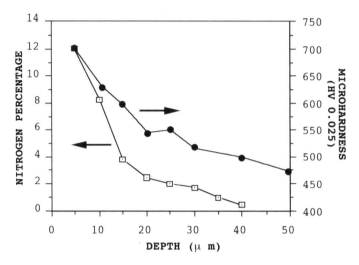

Fig. 4 :
Evolution of N2 percentage and Vickers microhardness with analysis depth.
Treatment :
E =6.6 kW/cm2,
V = 4 mm/s,
Fx = 6x3 mm2,
3 passes,
sandblasted
TA6V.

* Wear resistance of the laser nitrided substrates

The wear resistance tests were achieved on a pin - disc tribometer without lubrification (gliding speed : 0.65 m/s, contact pressure : 0.56 MPa, test time : 20 hours). Pins were either annealed TA6V, furnace or laser nitrided TA6V or 1800 MPa 35NCD16 steel. Wear tests were achieved for different laser solid state nitridation treatment conditions on grinded, milled or sandblasted surface substrates ; the results were compared to furnace nitrided surfaces (furnace nitriding conditions : T = 775 °C, P = 500 mbar, Nitrogen feed rate = 100 l/h, t = 30 hours, δ-TiN layer thickness = 1 μm, total nitrogen diffusion depth = 40 μm).

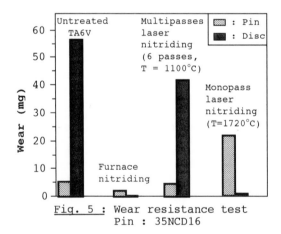

Fig. 5 : Wear resistance test
Pin : 35NCD16

For all the tested laser and furnace nitrided discs, high friction coefficients(between 0.5 and 1) and improvement of the wear resistance from 20 to 100 % (compared to TA6V annealed discs) were obtained. Despite some lack of reproductibility of the laser nitrided tested discs, some first tendencies could be evidenced concerning the influences of surface roughness and laser nitriding conditions on the substrate wear resistance (Fig.5):

- for the multipasses laser nitridation of sandblasted surfaces, the better wear resistance of the total pin + disc system was obtained in the case of a laser nitrided pin (compared to TA6V or 35NCD16 pins).
- a better wear resistance was always noticed for a high temperature (T > 1500°C) monopass laser nitriding on finely milled discs compared to that obtained for a lower temperature (T = 1100°C) multipasses laser nitriding on sandblasted discs. Unfortunatly, it is yet not possible to infer this better resistance to one of these two parameters (surface roughness or laser nitriding conditions).
- in the case of a high temperature (T > 1500°C) monopass laser nitridation on finely milled discs, it appeared that the higher was the treatment temperature, the better was the wear resistance (at, 1530°C, the nitride layer is completly removed after the wear test, whereas at 1720°C, the wear test did not induce a complete destruction of the nitrogen diffusion layer).This probably resulted from a thicker nitrogen diffusion depth at higher treatment temperature.

Laser solid state nitridation therefore permitted to significantly increase the TA6V substrates wear resistance ; in the case of high temperature (1720°C) monopass treatment, comparable wear resistances were obtained for laser or furnace nitrided substrates. The limitation of the increase of the wear resistance should probably be attributed to the moderate thickness of the nitrogen diffusion layer.

* Fatigue lifetime of the laser nitrided substrates

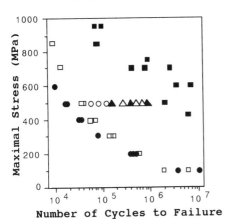

Number of Cycles to Failure

■ Annealed TA6V □ Furnace TiN
● 6 laser overlapped passes(1100°C)
○ △ Laser monopass (1500,1720°C)
▲   He Laser Quench (1720°C)

Fig.6 : Wöhler curves

The fatigue lifetime test was achieved with a three points bending fatigue test machine, both for untreated TA6V, laser quenched and, furnace or laser nitrided TA6V substrates (Fig.6). Wölher curves pointed out that any laser or furnace nitriding treatment lead to a decrease of the fatigue lifetime compared to that of untreated substrates. This can be attributed, either to the relaxation of the - 470 MPa initial superficial compressive stress and to the appearance of a superficial traction stress due to the thermal cycle, or to the nitride layer formation.

A comparison of the fatigue lifetime of laser treated substrates in helium or nitrogen atmospheres pointed out that <u>the lifetime decrease should be mainly attributed to the effect of the thermal cycle rather than to the superficial nitride formation</u>, since laser nitrided or only laser quenched TA6V substrates have shown the same fatigue lifetime (500 MPa for 5.10^5 cycles).In the case of laser nitrided substrates, SEM analysis of the tested substrates pointed out that cracks were initiated in the acidular α' + Ti(N) zone under the saturated TiN layer. The lower the maximum temperature during the laser nitriding treatment, the better the resulting fatigue limit (for 500 Mpa : T = 1530°C, number of cycles : 5.10^5 ; for T = 1720°C, number of cycles : 9.10^4). Moreover, a large monopass laser nitriding treatment permitted to improve the fatigue lifetime compared to a laser treatment with overlapping of small nitrided tracks.

Conclusion

The potentialities of the TA6V solid state nitridation by CO_2 laser for aerospace applications (requiring a good wear resistance without a catastrophic decrease of the fatigue lifetime) had been studied.

The characterization of both the laser process and the nitrided substrates had pointed out the possibilities to growth few microns thick nitride layers ($HV_{0.025} > 600$) on 6 mm large tracks with an efficient gaseous protection device which also ensured a substrate preheating step. These nitrided layers could be either achieved at low substrate temperature (T = 1100°C) in a multipasses mode or a high temperature (T = 1500 - 1750°C) in a monopass mode. The better mechanical properties, in terms of wear resistance and fatigue lifetime, were obtained for high temperature monopass treatment. Sample performances were thus comparable or better than those of furnace nitride samples. Increase of the fatigue lifetime of the laser nitrided samples could be achieved by a laser shock post treatment to induce compressive stresses in the thermally affected zone.

References

(1) B.L. Mordike : "High Power lasers", A. Niku-Lari & B.L. Mordike ed., Pergamon Press (1989) 3.
(2) S.Z. Lee, H.W. Bergmann : "Sixth World Conf. on Titanium", Cannes (France) (1988) 1811.
(3) D. Kechemair, P. Laurens, L. Sabatier, J.P. Ricaud : to be published in : Proceedings of LAMP'92, Nagaoka (Japan) (1992).
(4) D. Kechemair : Thèse de l'Université Paris XI (1989).

Hardness Improvement in Laser Surface Melted Tool Steel by Cryogenic Quenching

Costa, A.R.* ; Anjos, M.A.*; Vilar R. **

* Universidade Federal de Ouro Preto - Escola de Minas
  35400 - Ouro Preto - MG - Brasil
** Instituto Superior Técnico - Departamento de Engenharia
  de Materiais
  1096 - Lisboa - Portugal

Abstract

Samples of a tool steel containing 0.53%C, 3.80%Ni and 1.65%Cr were surface melted using continuous wave CO_2 radiation at a power level of 600W. The investigated structure of the laser modified layer was characterized by optical microscopy and vickers microhardness indentation. In previous works [4,6,8] it had been characterized by scanning electron microscopy (SEM) and X ray diffraction (CuKα radiation).

In the laser-melted zone the microstructure contains δ-ferrite, residual austenite and martensite.

Cryogenic quenching in liquid N_2 caused an increase in hardness in both melted and heat affected zone (HAZ).

Introduction

Laser melting surface technique by scanning with a continuous-wave laser beam is an excellent method for producing rapidly solidified surface layers. The process is controlled and reproducible. Very interesting non equilibrium structures are produced: a columnar structure grows epitaxially on the matrix, which may contain martensite and substantial amounts of retained δ-ferrite and austenite. It depends, obviously, on the chemical composition of the steel as well as on the treatment parameters. Improvement in wear resistance, corrosion and fatigue behavior are noticed as a consequence of this new technical application in manufacturing.

The aim of this work was to compare the microstructural features of laser surface melted tool steel before and after its quenching in liquid N_2.

Experimental Method

In the present investigation we used a self-hardening tool steel with the following chemical composition in weight percentage: 0.53C, 0.5Mn, 0.33Si, 3.80Ni, 1.65Cr, 0.27Mo, 0.003W, 0.17Cu, 0.01Al, 0.01P, 0.01S. The specimens used for surface melting were in the form of 60mm x 30mm x 15mm plates. They were submitted to surface metallografic preparation before irradiation with a continuous wave CO_2 laser beam of 10.6μm wavelength and 600W power to provide melting. The focal length of the lens was 127mm. The specimens were mounted on a numerically controlled x-y table and were submitted to multiple-pass scanning (42cm/min) under argon protection.

After surface melting, the specimens were studied by optical microscopy, vickers microhardness, scanning electron microscopy and X ray diffraction. Conventional metallographic preparation techniques were used for the optical and microhardness studies.

After cryogenic treatment in N_2, the specimens were studied by optical microscopy and vickers microhardness. Specimens for optical microscopy were mechanically polished and etched in a special reagent for revealing martensite.

Results and Discussion

On the transverse cross section of a multiple pass melted layer different zones can be distinguished after etching with sodium bisulfite reagent: basically, a heat affected zone, an intermediary layer (with columnar crystals) and a central zone with a cellular dendritic structure. X ray analysis revealed a bcc and a fcc phase. The high temperature phase δ-ferrite is retained in the quenched state. The occurrence of retained austenite is attributed to higher quenching rates than obtained in conventional heat treatments.

A close observation of the cross section shows that near the intermediary zone, a region of cellular dendrites develops. In the central region of the melted zone an apparently equiaxed structure appears which consists of cross-sections of the dendrites grown parallel to the pass of the laser beam.

The microhardness survey revealed a significant increasing hardness profile in the melted layer. The hardest region is in fact the heat-affected zone where the hardness may reach ~700VHN. In contrast, the prodominantly δ-phase-rich region, near the surface is softer, with a hardness of ~600VHN. Nevertheless there was a significant increase in hardness, since the base metal hardness was of ~250VHN.

In the samples quenched in liquid N_2 significant changes were observed in the laser melted region. The microhardness profile revealed an increase in both melted and heat affected zone (HAZ); and the decrease in microhardness at the interface MZ-HAZ noticed before quenching became more evident after it. Hardness reached a maximum of almost 1000 VHN in the HAZ. As to the MZ microhardness it reached ~800VHNS, that means, higher than the maximum values registered before quenching. The two microhardness profiles are shown in Fig. 1.

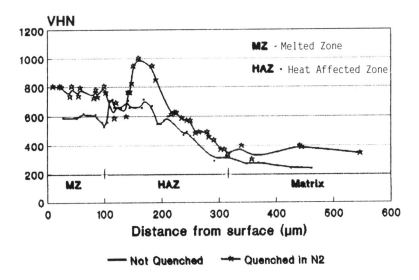

Fig 1: Hardness profiles of specimens before and after quenching in liquid N_2

Optical microscopy showed martensite in the MZ which had not been observed before quenching (Figs. 2A e 2B).

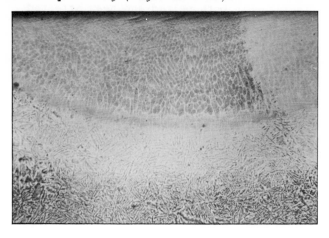

50 μm

Fig. 2A - Transverse cross section of the laser treated surface.

50 μm

Fig. 2B - Transverse cross section after quenching in N_2, showing martensite in the melted zone

We attribute the changes to:

A) Presence of retained austenite in the MZ and HAZ which transformed into martensite during quenching.

B) Little amount or absence of retained austenite at the interface MZ-HAZ.

The residual austenite decompositon was induced by the high cooling rate. Thus, this additional hardening is related to lattice distortion. The thin low-hardness layer at the interface MZ-HAZ became more evident because of the relative increase in hardness of the two neighbor regions.

Conclusions

The laser surface melting of a self-hardening tool steel produces a thin layer of rapidly solidified alloy with interesting mechanical properties related to a new microstructural arrangement.

The melted zone has a cellular dendritic structure.

In the laser surface melted layer the main constituents are δ-ferrite, austenite and martensite.

Cryogenic quenching produces an increase in the hardness of the laser melted layer leading to values well above those of the matrix.

References

(1) R.W. Cahn, P. Haasen (editors): Physical Metallurgy II, Elsevier Science Publishers - Amsterdam, 1779-1851 (1983).

(2) P.A.Molina; Scripta Metall. 15, 1101-1104, (1981).

(3) P.R. Strutt, H. Nowotny; Mat. Sc. Eng., 36,217-222 (1978).

(4) A.R.Costa, R.P. Domingues, R.A.P.Ibañez, R.M.Vilar; Proc. Jornadas Metalúrgicas - SAM - B. Aires - Argentina, 70-80, mai (1989).

(5) N. Rykalin, A. Uglov, A.Kokora; Laser Machining and Welding, MIR Publishers - Moscow (1978).

(6) R.Vilar, R.M. Miranda, A.S. Oliveira; Proc. Conf. Laser Technologies in Industry - SPIE - (1988).

(7) P.A. Molian; Scripta Metall. 17, 1311-1314, (1983).

(8) A.R. Costa, R.P. Domingues, R.A.P. Ibañez, R.M.Vilar; Proc. Jornadas Metalúrgicas - SAM - Córdoba - Argentina, 301-303 mai (1990).

Solidification in Laser Surface Alloying of Al and AlSi10Mg with Ni and Cr

E. W. Kreutz, N. Pirch, M. Rozsnoki, Lehrstuhl für Lasertechnik, Rheinisch-Westfälische Technische Hochschule Aachen, Aachen, FRG

Introduction

The more complex a component's function is and, consequently, the more demanding the structural and functional requirements on the material properties are, the closer the correlation between those properties, the technologies needed in machining that component and the methods of production of components materials. The potential function and reliability of highly developed technical systems, in many cases, only can be realized if the properties of the components employed meet the highest requirements. In the case of aluminium and aluminium alloys their properties limit the improvements in performance properties. Ceramics, polymers, laminates, and composites are emerging as strong, new competitors of aluminium and aluminium alloys at an unprecedented rate. Thus, there exists a critical need for structural materials to be not only cost effective, but also provide superior high temperature strength, reduced density, and improved stiffness properties. These requirements, coupled with an increased emphasis on efficiency and reliability, have engendered considerable interest in the development and use of improved and novel techniques for the production of aluminium alloys having properties superior to those of existing commercial aluminium alloys.

The critical need for high strength, high performance, and cost effective materials has resulted in the use of novel processing techniques for the development of new aluminium alloys for structural applications. While continous casting has long been the dominant production method, it is limited by the slow rates of solidification which generate the large degrees of segregation and the coarse microstructures normally associated with casting processes. Finer microstructures and reduced microsegregation, beyond the levels possible with continous direct-chill casting, can lead to reduced homogenisation times during solution heat treatment and to superior fracture related properties. These refinements can best be achieved by rapid solidification processing by means of processing with laser radiation taking advantage of rapid heating, cooling and quenching of thin layers and surface layers, respectively. The highly attractive combinations of microstruture and mechanical properties achievable through rapid solidification technology

have prompted the study and application of rapid solidification
(1) as a means of improving the behaviour of existing aluminium
alloys. The rapid solidification simultaneously renders possible the developing of novel alloying compositions.

Processing with laser radiation may include the solid, the melt
and the vapour state of matter. In the case of melting the
macroscopic properties temperature distribution in the melt,
pool geometry and flow field in the molten pool govern locally
the rapid solidification with large deviations from
equilibrium, as evidenced by the extension in solid solubility
limits, sharp reductions in grain size, and a reduction in the
size and number of segregated phases with concomitant development of non-equilibrium phases.

The present paper reports on alloying of aluminium (Al99.99)
and Al-based alloy (AlSi10Mg) with CO_2 laser radiation as a
function of processing variables considering the gainfull utilization of the generated aluminium alloys by rapid solidification. Heat flow calculations (2,3) including the Marangoni convection in the melt were carried out in order to relate phase
apprearence and morphology to solute distribution, local solidification and quenching rate. The alloyed zone was investigated by optical microscopy, electron microscopy and measurements
of mechanical-technical properties. The microstructures are
discussed as a function of the local cooling and solidification
rate ruled by melt dynamics and the geometry of the solid
liquid phase front as the driving boundary condition of solidification.

Rapid Solidification Processing

The rapid solidification involves the rapid extraction of
thermal energy during the transition from the liquid state at
high temperature to the solid state at ambient temperature.
This energy transfer permits non-equilibrium states with large
deviations from equilibrium state of matter, which consequently
offer the advantageous possibilities of the extension of solid
solubility, the reduciton in grain size, the reduction in the
number of segregated phases, the reduction in the size of
segregated phases and the production of new phases.
However, these advantageous improvements do not necessarily all
occur simultaneously.

In practice, there are at least three different approaches in
order to achieve rapid solidification such as imposing high
undercooling before solidification imposing a high velocity of
advance during continuous solidification and imposing a high
cooling rage during solidification. The last approach is

perhaps the most widely used because of the numerous techniques to be used and the wide range of alloy compositions to be processed. The main advantage is that the cooling is rapid before, during and after solidification increasing thereby the probability of retaining the microstructural and constitutional characteristics of the rapid solidification stage. The essential requirement of imposing a high cooling rate during solidification is achieved in practice by making for example, one dimension of the solidifying volume very small. Since the nucleation and growth of solid phases from the melt, as for processing with laser radiation, are closely linked with evolution and dissipation of thermal energy, unusual structures are formed under these extreme non-equilibrium conditions.

Characteristic for processing with laser radiation are the short times for heating and quenching. The short duration time of a volume element in the liquid phase due to the different melting of the structural constituents of an heterogeneous alloy can result in incomplete melting of the structural constituents as well as incomplete remixing. Therefore, the original structure has to be taken into account for the choice of processing parameters. The resulting microstructure is determined by the melt pool geometry, the transient evolution of temperature profile and convective transport mechanisms within the melt pool driven by surface tension and buoyancy forces. Since the liquid melt is adjacent at the solidification front to the resolified structure of the same material epitaxial growth without nucleation is possible. Which of the preferred cristallographic directions succeed in growing depends on the local orientation of the cristallographic direction to the local heat flow direction. The resulting microstructure (eutectic interphase spacing, primary and secondary spacing of dendrites, average grain size) depends on the local solute distribution. Along the solidification front a thermal, solutal and momentum boundary layer evolve which mutual influence of each other. In addition, a so-called mushy zone forms direct adjacent to the solidification front with non-Newtonian flow properties. Assuming the local growth parallel to the local heat flow direction (4) the local solidification rate local solidification rate

$$v_s = v \cdot \frac{\delta T}{\delta x} / |\nabla T| \qquad \text{Equ. 1}$$

follows as projection of the processing velocity v onto the normal vector $\vec{n}$, and the local quenching rate

$$T' = v \cdot \frac{\delta T}{\delta x} \qquad \text{Equ. 2}$$

272

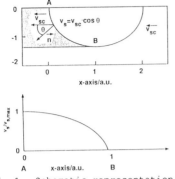

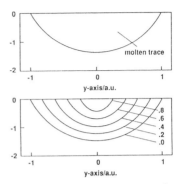

Fig.1: Schematic representation of geometry and local solidification rate parallel to the processing direction.

Fig.2: Schematic representation of geometry and local solidification rate perpendicular to the processing direction.

as temperature decrement in direction of the normal vector. Fig. 1 shows the local quenching and solidification rate along the solidification front in the longitudinal section of the processed geometry. The solidification rate and the quenching rate become zero at the deepest point of the melt pool and reach their maximum values at the melt pool top surface. Fig. 2 shows the local solidification rate as projection from the solidification front onto a cross sectional plane.

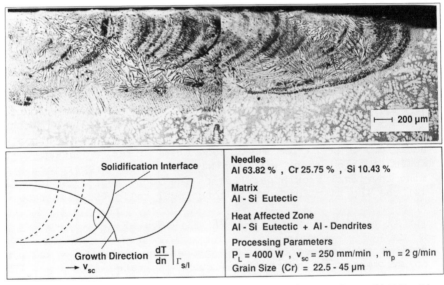

Fig.3: Micrograph and schematic representation of solidification parallel to the processing direction for surface alloying of AlSi10Mg with chromium by CO_2 laser radiation.

Since the quenching as well as the solidification rate increase from the boundary to the middle of the molten region (Fig. 2) the microstructure exhibits an according refinement coupled with a tendency of microstructure parallel to the local heat flow direction as shown in Fig. 3.

The diffusive and convective energy transport in the molten pool mainly govern the mixing of the alloying solute throughout the molten pool (5). Depending on the distribution of the alloying solute within the melt pool surface alloying with laser radiation originates in various phases and solidification structures (4), which show the strong interaction of the convective flow in the molten pool and the rapid solidification depending on the processing variables used (Fig. 3). The dendritic structure exhibits a refining from the solid liquid boundary to the middle of the alloyed track and a preferential orientation of the dendrites normal to the solid liquid interface indicative for a textural growth of crystallites and dendrites dominated by the solid liquid boundary (3). Close to the boundary the epitaxial growth originates in large crystallites and dendrites with the orientation dominated by the direction of the existing surface in the solid liquid boundary (Fig. 3), far away from the boundary the resolidifying growth results in small crystallites and dendrites with the orientation dominated by the crystallographic direction of the grown surfaces of the crystallites and the convective flow in the melt pool. The competition between convective flow and solidification consequently results in various areas within the processed geometry, where the structures and phases are dominated either by convective flow or by solidification (Fig. 3). Low order areas are indicative for the former one, high order areas for the latter one. The dendrites/needles, growing throughout the melt pool (Fig. 3) mainly antiprallel to the actual heat flow and perpendicular to the solid liquid boundary, exhibit, for example, the dominating rapid solidification prevailing the convective flow. In addition, the vigorous agitation of coexisting liquid and solid phases possibly results in breakage of the dendrite crystals to from a slurry of larger particles, which are following the convective flow (6).

Solidifified Phases

Depending on melt dynamics and solidification various phases (Figs. 4 and 5) are observed in the alloyed region strongly dependent on the concentration of the alloying solute. The overall temperature distribution and the concentration of alloying solute result in the formation of various phases and

274

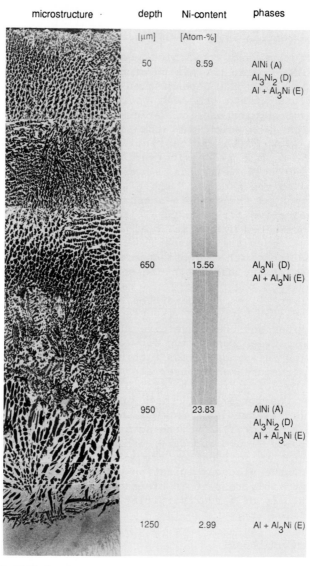

microstructure	depth [μm]	Ni-content [Atom-%]	phases
	50	8.59	AlNi (A) Al_3Ni_2 (D) $Al + Al_3Ni$ (E)
	650	15.56	Al_3Ni (D) $Al + Al_3Ni$ (E)
	950	23.83	AlNi (A) Al_3Ni_2 (D) $Al + Al_3Ni$ (E)
	1250	2.99	$Al + Al_3Ni$ (E)

Al with Ni alloyed ⊢——⊣ 20 μm   (A) agglomerate (D) dendrite (E) eutectic

Fig.4: Microstructure, average concentration profile and phases of Ni alloyed Al99.99 throughout the alloyed zone perpendicular to the surface.

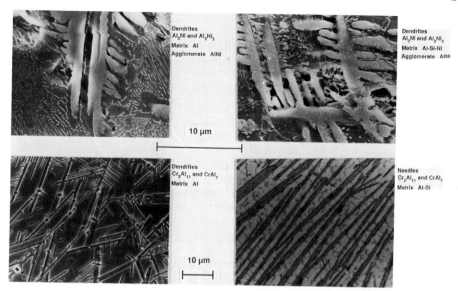

Dendrites Al$_3$Ni and Al$_3$Ni$_2$
Matrix Al
Agglomerate AlNi

Dendrites Al$_3$Ni and Al$_3$Ni$_2$
Matrix Al-Si-Ni
Agglomerate AlNi

10 µm

Dendrites Cr$_2$Al$_{11}$ and CrAl$_7$
Matrix Al

Needles Cr$_2$Al$_{11}$ and CrAl$_7$
Matrix Al-Si

10 µm

Fig.5: Microstructure of Ni and Cr alloyed Al99.99 and AlSil0Mg.

structures (Fig. 6) taking into account the local solidification. In the case of Ni and Cr as alloying elements intermetallic compounds and metal eutecticum are observed with the generated structures of dendrites and needles depending on the type of Al-alloy under investigation (Fig. 6).

The cooling rates and the resulting microstructures strongly depend on the processing velocity (7,8). High processing velocities with small interaction times result in high heating and cooling rates yielding a metallurgical microstructure of fine graininess, low processing velocities with large interaction times result in low heating and cooling rates yielding a metallurgical microstructure of coase graininess, all these effects in agreement with modeling calculations (9).

The diffusive and convective energy transport in the molten pool govern the mixing of the alloying solute throughout the molten pool. Diffuse transport originates in a weak mixing on a long-term time scale, convective one in a strong mixing on a short-term time scale depending both on the overall temperature distribution and the melt pool geometry. The diffusive and convective energy transport in the molten pool results in a mixing of the alloying elements yielding in the case of aluminium alloying an average Ni concentration profile throughout the laser alloyed zone exhibiting a maximum near the solid liquid boundary (4) Surface alloying with laser radiation originates in various phases and solidification structures, which show the strong interaction of the convective flow in the molten pool and the rapid solidification depending on the overall temperature distribution generated and the melt pool geometry in combination with the processing variables used.

The convection also influences the geometry at the solid liquid boundary as the resolidifying interface changing the local growth rate over the pool depth. At the solidification front the convection causes an isotherm compaction near the surface and increases by the higher temperature gradients the cooling rate for teh subsequent solidification. The temperature profile, the composition (Fig. 7) and the rate of solidification govern the possible formation of metastable phases, their composition and the scale of metallurgical microstructures (Figs. 4 to 6) according to the corresponding phase diagrams (Fig. 7). The cooling rates, which are derived along the solid liquid boundary, show no significant influence of the convection in the molten pool since the melt pool geometries are very similar (9) either for diffusive or for diffusive and convective energy transport.

 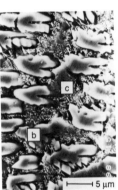

Fig.:6 Microstructure of the Ni alloyed Al99.99 zone showing AlNi agglomerates (a), intermetallic dendrites (b) and interdendritic eutecticum (c).

In spite of the low average Ni concentration the phase AlNi is precipitated as agglomerate (Fig. 6) with subsequent dendritic growth of Al_3Ni_2 due to the peritectic reaction

$$L + AlNi \rightleftharpoons Al_3Ni_2 \qquad \text{Equ. 3}$$

with the formation of AlNi due to the supressed mixing of dissolved particles. The dendritic structure exhibits a refining from the solid liquid boundary to the middle of the alloyed track and a preferential orientation of the dendrites normal to

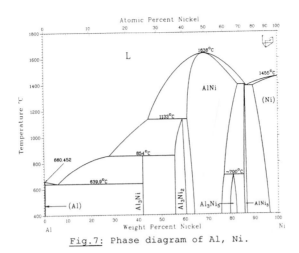

Fig.7: Phase diagram of Al, Ni.

the solid liquid boundary (3). Close to the boundary the epitaxial growth originates in large crystallites and dendrites with
the orientation dominated by the direction of the existing surface in the solid liquid boundary (Fig. 4), far away from the
boundary the resolidifying growth results in small crystallites
and dendrites with the orientation dominated by the crystallographic direction of the grown surfaces of the crystallites and
the convective flow in the melt pool. In addition, the vigorous
agitation of coexisting liquid and solid phases possibly results in breakage of the dendrite crystals to form a slurry of
larger particles, which are following the turbulent flow (6).
This semi-solidification processing has attracted increased attention as a new process to realize the seemingly impossible
ability to freely process metals under weak forces just like
clay modeling (10). The interdendritic eutectic structure (Fig.
6) is in the range of 250 nm. For AlSi10Mg the determination of
the microstructure within the interdendritic spacings was not
rendered possible by REM and TEM analysis. The alloying with Cr
leads to the formation of the intermetallics Cr_2Al_{11} and $CrAl_7$,
in the form of fine dendrites for Al99.99 and of coarse needles for AlSi10Mg (Fig. 5).

The rapid solidification enables the formation of a wide variety of chemical and strutural phases such as different concentration profiles of the alloying element throughout the molten
region as well as extended solid solutions. The extension of
the range of existence of phases and structures as obtained by
time-temperature-transformation processing in the liquidus range under equilibrium conditions for processing velocity $v = 0$

is clearly seen from the Al-Si-phase diagram (9).For processing velocities v = 0 the solubility of silicon in the aluminium matrix becomes enhanced for time-temperature-transformations under non-equilibrium surface processing with laser radiation including the liquidus range (9).

The refinement of the microstructure commonly causes the increase in hardness. The most important mechanism is the hardening by changes of grain boundaries (11). According to the high cooling rates with subsequent rapid solidification a strong undercooling of the melt occures with a large number of nucleus, which result in the growth of a large number of small clusters or crystallites. Since these clusters or crystallites render impossible the movement of grain boundaries, the surface melting increases the hardness from approximately 60 HV of the substrate up to 110 HV (9). Higher hardness values are obtained by alloying of the close surface region with the microhardness increased from 55-65 V (0.1) to 350 HV (0.1) for Ni (20%) and to 250 HV (0.1) for Cr (15%). As seen from disc on disc testing the alloyed surfaces show a significant improvement of the wear resistance compared to the untreated ones (9).

<u>Conclusion</u>

Surface alloying with Cr or Ni using CO_2 laser radiation is an appropriate means for surface modifications of aluminium and aluminium-based alloys. The alloyed surface layers, which show a low surface roughness, are crack and pore free either in the case of single or overlapping tracks. The convective energy transport in the melt pool mainly governs the mixing of the alloying elements. The temperature field, the processed geometry and the flow field represent the macroscopic starting and boundary conditions for the microscopic solidification with the resulting formation of strutures and phases. Following the phase diagrams the solidification yields fine coarse microstructures composed of intermetallics in form of dendrites and needles and eutecticum in between. The formation of intermetallics leads to a significant improvement of the microhardness and wear resistance depending on the processing variables, expecially powder feeding rate and degree of overlap in multitrack processing.

<u>References</u>

(1) E.J. Lavernia, J.D. Ayers, T.S. Srivatsan: Internatinal Materials Reviews 37 (1992) 1.

(2) E.W. Kreutz, N. Pirch: SPIE Proceedings Series 1276 (1990) 343.

(3) E.W. Kreutz, N. Pirch: SPIE Proceedings Series 1502 (1991) 160.

(4) E.W. Kreutz, B. Ollier, N. Pirch: Proceedings International Conference Laser Advanced Materials Processing 1 (1992) 353.

(5) C.W. Draper, J.M. Poate: International Metal Reviews 30 (1985) 85.

(6) E.W. Kreutz: to be published.

(7) G. Backes, A. Gasser, E.W. Kreutz, B. Ollier, N. Pirch, M. Rozsnoki, K. Wissenbach: SPIE Proceedings Series 1276 (1990) 402.

(8) G. Backes, E.W. Kreutz, B. Ollier, N. Pirch, M. Rozsnoki, A. Gasser, K. Wissenbach: Aluminium 67 (1991) 1008.

(9) E.W. Kreutz, M. Rozsnoki, N. Pirch: Proceedings International Conference Laser Advanced Materials Processing 2 (1992) 787.

(10) K. Kitora: Aluminium 66 (1990) 758.

(11) P. Furrer, H. Warlimont: Zeitschrift für Metallkunde 62 (1971) 12.

Alloying and Dispersion of TiC-Base Hard Particles into Tool Steel Surfaces
Using a High Power Laser

W. Löschau, K. Juch, W. Reitzenstein
Fraunhofer-Einrichtung für Werkstoffphysik und Schichttechnologie, Dresden (FRG)

1. Introduction

It's the aim to increase the resistance of tool steels against abrasive wear by means of hard material incorporation into surface layers. There is a close correlation between the abrasive wear of metals and their hardness and it also depends on the grain size of the abrasive. For ferrous materials a hardness increase beyond the Martensite hardening is possible by the incorporation of hard carbides into the melting zone produced by a laser beam. The wear properties depend, in a complicated way, not only on the size, distribution and concentration of the carbides but also on the hardness and toughness of the matrix surrounding them.

By a complete dissolution of fine-grained primary carbides (< 20 µm) in the steel melt (alloying) fine dispersed homogenous structures of TiC crystals (< 1 µm) were produced which crystallize during cooling (1,2). In the fine grain abrasive wear test according to the abrasive paper method these solidification microstructures had two to fivefold lower wear rates depending on the laser parameters than the same conventionally heat-treated steels without hard material (1). For coarse-grained abrasives and impact load these fine dispersed structures didn't have, however, much better wear properties. That's why the tests were orientated at the production of coarse-grained dispersion structures with a tough matrix.

2. Experiments, materials

The hard material powder was either directly injected into the substrate melt in an inert gas stream by means of a disc powder feeder or coated before laser treatment. The layer was produced by means of screen printing of thixotropic pastes. The method was developed in a way that makes possible to process powder particles within a grain size range of 1 to 75 µm and to print thicknesses of layers up to 750 µm in one or several stages as they are needed for some kW of laser power. For direct injection the grain size range is limited by the disc feeder to about 40 to 90 µm. As substrate material carbon steels or tool steels C 45, 90MnV8 and 80CrV3 in soft state are used.

TiC was mainly used as hard material. TiC has a high hardness of 30 GPa and a broad range of homogeneity. It has a much lower solubility in iron melts than WC. TiC is mainly obtained in a fine-grained form in the synthesis. For the production of coarse-grained material it was reduction annealed without and with Ni as sinter assisting material (0.2 or 0.8 %, resp.). In addition, commercial coarse grain powders were used:
a) Amperite 570.2 of Starck Co. (pure TiC, sintered, 45 - 90 µm),
b) angular TiC powder coated with 30% Ni of VALCO Co. (45 - 90 µm).

For the injection tests the powders a) and b) were used as they were supplied and comparatively a sintered nodular WC-6% Co-powder (90 - 160 µm). For screen printing fractions of

32 - 40 µm, 40 - 63 µm and 63 - 100 µm were screened and layers with an thickness of 270, 500 and 750 µm were printed. However, the screen fractions contained remarkable portions of fine particles.

The laser irradiation was carried out with CO_2 - CW lasers in the defocused circular beam (double-ring type) under N_2- or Ar-protecting gas. The laser power was 2.0 to 5.5 kW.

Beam intensity I: $2.0 ... 3.7 \times 10^4$ W/cm^2 (injection)
                      $2.5 ... 8.0 \times 10^4$ W/cm^2 (screen printed)
Traverse speed : $3.5 ... 13.5$ (20) mm/s (injection)
                      $5 ... 20$ mm/s (screen printed)
Powder stream m: $0.02... 0.09$ (0.14) g/s for TiC injection
Gas stream V   : $30 ... 110$ l/h for TiC injection
For wear tests the whole surface of samples with 50% track overlapping was irradiated.

3. Results

3.1. Melt path geometry

Depending on laser parameters I and v melting paths having a width of 2.0 to 3.7 mm and a depth of 0.8 to 2.4 mm were produced with a laser power of 5 kW for particle injection (figure 1). With increasing powder stream the coat over the original surface becomes more and more distinct (0.2 to 1.0 mm). For 2 kW the melt depths were 0.4 to 1.0 mm. For screen printed layers the useful melt depth increases progressively with decreasing traverse speed: its minimum value is 0.6 mm for 20 mm/s and its maximum 2.2 mm for 5 mm/s (for 2 kW 0.9 mm as a maximum). The melt depths reached here may be sufficient for many applications (for instance, forming and cutting tools).

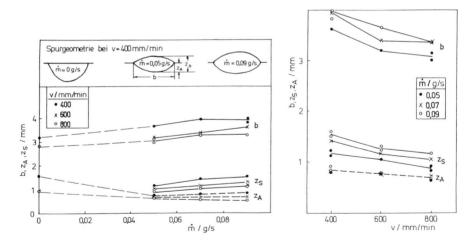

Fig. 1: Melt path geometry. TiC injected in C45 steel, particle size 45 ... 90 µm, P = 4.7 kW, I = 2.6 x 10 ^{4} W/cm^2

3.2. Solidification microstructures

The structures show similar features for melting screen printed layers as well as for direct injection concerning their dependence on the laser parameters and the hard material feed (paste thickness and powder stream, resp.). They contain more or less dissolved primary carbides in different arrangement and concentration. Between them there are secondary carbides with cubic, feather or dendrite like habit the reason for which is the dissolution of partially available fine carbide fractions or the partial dissolution of the big primary carbides in the steel melt (figure 2). Dependig on the interaction time, temperature and convection of the melt the primary TiC particles are angular, rounded or surrounded by secondary carbides as a centre of heterogeneous nucleation. At higher intensities and low traverse speeds the hard material distribution becomes more homogeneous but the primary paticles are mainly dissolved. With increasing hard material feed (high paste thickness/high powder stream) there is a supersaturation of the melt: the number of primary particles increases; the secondary carbides get a dendrite and later feather-like shape; the total hard material concentration exceeds 50 vol% partially.

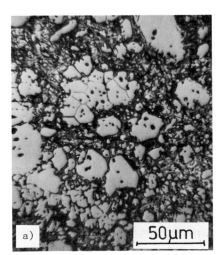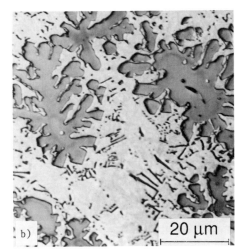

Fig. 2: Solidification microstructures after incorporation of coarse-grained TiC screen printed layers in 90MnV8 steel, P = 5.3 kW, I = 3.3 x 10^4 W/cm². a) Layer thickness 760 µm, v = 10 mm/s. b) 500 µm, 5mm/s.

Especially at low intensities and high paste thicknesses massive sintered compacts from primary carbides were observed. Sintering starts already in the screen printed layer. The sintered agglomerates are not completely dissolved any more in a melt with a low convection. Pores were observed for both feeding methods due to the pore volume of the printed layer or gas carried away by the powder transport. Pore seams appear with low intensities at the lower melting margins by sealing the pore volume of the printed layer as well as with an accumulation of primary particles (increase of the melting viscosity). Cracks are only rarely observed in single paths in TiC and TiC/Ni whereas in WC-Co cracks are created at a high particle concentration due to microstructure-embrittling WC-mixed carbide phases in the matrix (3).

3.3. Hardness, abrasive wear

According to the heterogeneous microstructure in the melt zone the mirohardnesses are exposed to strong local fluctuations. The non-dissolved TiC grains have values of about 3500 $HV_{0.05}$ typical for TiC. The hardnesses in the matrix which contains in addition to Martensite, Austenite also recrystallized TiC correlate with the process depending microstructure features. In the screen printed layers the matrix hardnesses scatter at a traverse speed of 20 mm/s from 700 to 1400 $HV_{0.05}$. The mean values increase with increasing paste thickness. With TiC injection the matrix hardnesses in the upper range of the melting zone are much higher (up to 1800 $HV_{0.05}$) than on the melting bottom and they increase with increasing powder mass stream (figure 3a). In the case of the injection of Ni-coated TiC the matrix becomes austenite-like and its hardnesses are only near the surface due to a high fraction of secondary carbides with about 1300 $HV_{0.05}$ above the hardness of the pure remolten basic material C 45 while they drop to 700 $HV_{0.05}$ at the melt bottom (fig. 3b).

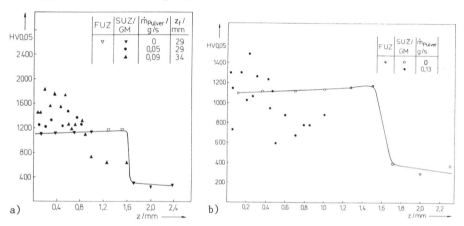

a) b)

Fig. 3: Microhardness in the matrix between primary hard particles injected into C 45 steel. P = 4.7 kW, I = 2.6 x 10^4 W/cm^2, v = 10 mm/s. a) pure TiC, b) TiC/Ni coated (70/30). (SZ: melt zone, FUZ: heat affected zone).

Wear is no material property as such. It can be very different with different loading conditions where hardness and grain size of the abrasive are important. A rotating disc with wet SiC grinding paper against which the sample was pressed was used as wear test device. The wear mechanism is equivalent to a micro cutting, i. e. , a pure abrasive wear. By comparison, conventionally hardened 90MnV8 tool steel sample (wear factor 1) and samples with fine-grained, completely dissolved TiC powder were used (1). Injected, coarse-grained, pure TiC powder had the lowest wear rates (factor 0.10 to 0.20) where a high traverse speed and low beam intensity had a positive effect (fig. 4). The injected TiC/Ni powder resulted in worse values due to the lack of the supporting effect of the soft purely austenite-like matrix. Here a clear decrease of the wear rate can be observed with increasing powder stream, i.e., with increasing hard material concentration. The micoscopic explanation of the wear mechanism depending on the abrasive grain size is the subject of further studies.

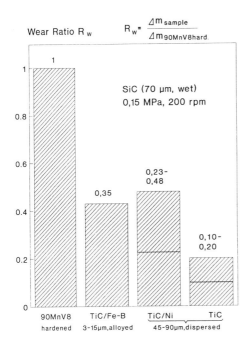

Fig. 4: Wear ratios of different TiC alloyed or injected layers
on steel samples in an abrasive wear test

4. Acknowledgement

The authors express their thanks to the Federal Minister of Research and Technology for promoting this study.

5. References

(1) W. Löschau, A. Luft, H. Heinze,
   Tagungsband 9. Int. Pulvermetallurgie-Tagung Dresden 1989, 301-312
(2) A. Luft, B. Brenner, W. Löschau, W. Reitzenstein
   Microstructure and Mechanical Properties of Laser Processed Steels and Cast Irons, in
   Kohler: Jahrbuch Laser, 2. Ausgabe, Vulkan-Verlag Essen 1990, 295-299
(3) R. P. Cooper, J.D. Ayers, Surface Engineering (GB) 1(1985)4, 263-272.

Microstructure and Properties of Laser Processed Composite Layers

A. Schüßler, Kernforschungszentrum Karlsruhe, Institut für Materialforschung I, Karlsruhe, FRG

Introduction

Metal-ceramic particle composites have a large potential for industrial applications, especially as anti-friction and anti-abrasion materials. Forming of particulate composite layers using the laser melt particle injection process have been reported since the 80´s (see ref. 1-5). Differing from this method composite layers described in this paper were obtained by liquid metal penetration into predeposited coatings of fine hard particle powders. Reduced bath convection due to the present technique results in a different surface structure and less particle dissolution compared to conventional methods of particle incorporation. This may be of advantage in reducing friction and wear intensity of composite layers.

Processing

A tool steel 90MnCrV8, containing 0.9% C, was used as a substrate. A slurry consisting of ceramic particles (table 1) of a size between 2 and 30 μm and alcohol was applied to the substrate surface in thicknesses of about 30 to 60 μm prior to laser processing. Laser treatment was carried out using a 3.5 kW continuous wave CO_2-laser with a beam integrator and an argon gas shield. Power densities (of about 7 to 10 kW/cm^2) and scanning speeds (100 mm/min) were chosen to melt the substrate surface and allow the liquid metal to penetrate into the predeposited coating as result of capillary attraction (fig. 1, see e.g. 6). During processing, the melt pool remains covered by the ceramic coating and surface tension forces and consequently Marangoni effects are suppressed. For this situation, melt pool convection is mainly controlled by buoyancy forces. Relative velocity amplitudes were shown (7) to be 3 to 4 orders of magnitude smaller compared to powder feeding processes (8,9), where strong thermocapillary effects are responsible for an intense flow inside the melt pool.

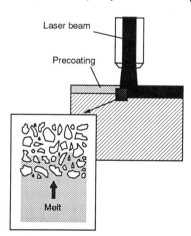

Laser beam

Precoating

Melt

Fig. 1: Liquid metal penetration into ceramic particle coating using laser radiation (schematic).

Wetting, Particle Dissolution and Microstructure

Fundamental requirements for laser processing of composite layers are complete wetting of the hard particles by the melt and prevention of an excessive particle dissolution. Oxides (alumina, zirconia) on the one hand could not be incorporated into steel surfaces due to poor wetting, and covalent hard phases, such as boron carbide and silicon carbide, on the other hand exhibited a high solubility in steel melts. However, particle composites incorporating transition metal carbides were obtained (table 1).

Hard phase/Group		Melting point (°C)	Hardness (HV)	Contact angle* (deg)	Dissolution during processing**
TiC		3160	3200	28 - 41	low
ZrC	IVB	3530	2560	45	none
HfC		3890	2700	45	none
VC		2830	2950	20	low
NbC	VB	3500	2400	25	none
TaC		3780	1790	23	none
Cr_2C_3		1850	2280	0	complete
Mo_2C	VIB	2400	1950	0	complete
WC		2600	2080	0	high
TiN		2950	2450		none

*) With iron melts (see ref. 10), low contact angle=high solubility

**) Results are valid for laser treatment using 10 kW/cm^2 und 100 mm/min

Table 1: Properties of carbides of the transitions metals and TiN used for laser processing.

In accordance with their low contact angles (values <60° apply for good wettable particles) no wetting problems had been observed during laser processing. It is important to note, that properties like hardness, contact angle of wetting and consequently particle dissolution change according the portion of the metallic bond, which increases from the IV. to the VI. group of the periodic system. The increasing tendency of hard material dissolution with increasing metallic bonding behaviour were confirmed in the own experiments. No dissolution (fig. 2a) or low dissolution (fig.2b and 2c) was observed for the incorporation of carbides of the IV. and the V. group, whereas Cr_2O_3, Mo_2C and WC (fig. 2d) exhibited high dissolution and are therefore not suitable for laser processing under the experimental conditions used.

Reduced bath convection results in a different structure of laser processed surfaces compared to conventional methods of particle incorporation. Surface structure obtained is characterized by an distinct demarcation between particulate composite and the underlying melt pool. Basically, three zones can be distinguished (6): the composite layer, melting zone and thermally affected zone. Except for laser treatment of Cr_2O_3 and MoC_2 particle coatings (melting depth corresponded to alloying depths) this structure was observed for all hard materials listed in table 1.

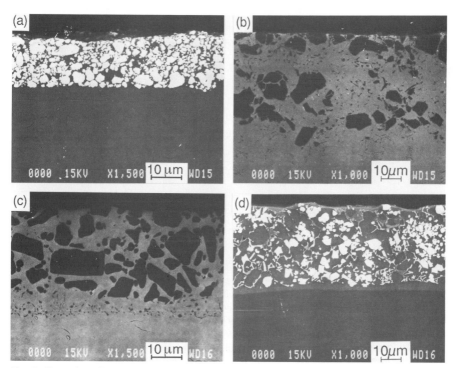

Fig. 2: Scanning electron micrographs (cross section) of laser processed composite layers showing different particle dissolution. (a) TaC - no, (b) TiC and (c) VC - low, (d) WC - strong dissolution in the steel matrix.

Differences in the depths of melting and heat affected zones caused by different absorption behaviour have been revealed depending on particle idendity and particle size: Depths of melt pools varied from 40 μm (TaC, VC) to about 400 μm (TiC, TiN). Depths of thermally affected zones, i.e. hardened zones, varied from 2.0 to 3.1 mm for laser treatment using 10 kW/cm^2 und 100 mm/min. Single composite layers were in the range between 30 and 50 μm with particles of volume fractions from 40 to 60 Vol%. homogeneously distributed in the metallic matrix. Steel matrices of particle composites and melting zones revealed martensitic microstructures with no quantitatively determined retained austenite contents. Microscopical investigation revealed no cracks and an ideal layer bonding to the substrate (fig. 2).

By repeating the two step treatment procedure (coating and laser treatment) multilayer structures up to 5 single layers have been produced up to a total thickness of about 200 μm. By applying multilayers, hard phase incorporated in metal matrices can be varied gradually in type, size and volume fraction from the surface to the substrate (11). Figure 3 shows a composite surface, wherein the size of particles hase been varied in three layers from the substrate (1.7 μm) to the surface (10.3 μm).

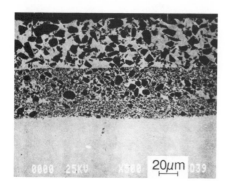

Fig. 3: Composite layer with increasing size of TiN particles from the substrate to the surface.

Wear

Wear behaviour of particulate composite layers obtained by the method described were reported in detail elsewhere (12, 13, 14). Results can be summarized as follows: Abrasive wear of tool steel against hard abrasive SiC grits was improved by embedding TaC and TiC particles. Wear loss of composite layers containing 30 μm and 3 μm TiC reinforcing particles (fig. 4) is significantly reduced compared to hardened tool steel and in the same range in comparison with commercial TiC-steel composites (TICALLOY) of the same TiC volume fraction (fig. 5). A recent investigation (14) revealed the complex effect of size of reinforcing TiC particles, abrasive grit size as well as matrix structure and hardness on wear behaviour. Wear resistance of composite layers could be further improved by heat treatment after laser treatment. Depending on particle size, i.e. mean free path length in the matrix, a martensitic / retained austenite microstructure may be beneficial due to strain-induced γ to α transformation during wear testing.

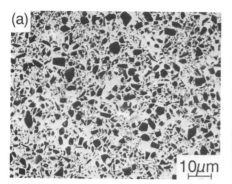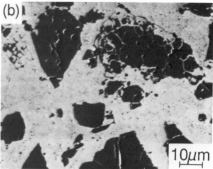

Fig.4: Scanning electron micrographs (top view) of TiC composite layers used for wear testing. (a) 3 μm reinforcing TiC particles, (b) 30 μm TiC particles

Resistance to oscillating sliding wear of tool steel (90MnCrV8) was also improved by the incorporation of TiC reinforcing hard phases. Composite layers reduced friction and wear in dry oscillating sliding contact with of Al_2O_3 and 100Cr6 steel balls compared to the martensitic tool steel under the experimental conditions used (13). Whereas wear intensity of composite layers was not decisively affected by the size of reinforcing particles, wear loss of counterbodies and consequently total wear of the sliding pair was significantly lower using 3 μm TiC phases compared to 30 μm TiC particles (fig. 6).

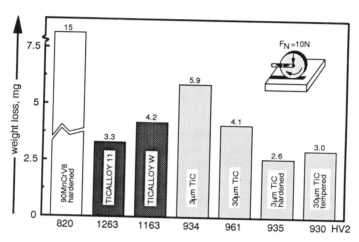

Fig. 5: Weight loss of TiC composite layers worn by 320 mesh SiC abrasive grits in comparison with commercial TiC-steel composites and hardened tool steel.

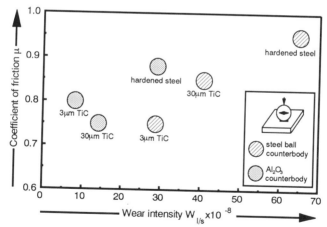

Fig. 6: Total wear intensities (plate plus counterbody) and coefficients of friction of TiC composite layers and hardened steel in dry oscillating sliding contact with steel and Al_2O_3 balls.

References

(1)   J.D. Ayers, T.R. Tucker, Thin Solid Films, 73 (1980), 201-207.
(2)   K.P. Cooper and J.D. Ayers, Surface Engineering, Vol. 1, (1985) 263-272.
(3)   V. M. Weerasinghe and W. M. Steen, Proc. ECLAT, Bad Nauheim 1988, 166-174.
(4)   J.H.Abboud and D.R. West, Materials Sci. and Technology, Vol. 5, 725-728, 1989.
(5)   A. Weisheit and B.L. Mordike, Proc. ECLAT, Erlangen 1990, 605-611.
(6)   A. Schüßler, K.-H. Zum Gahr, Proc. ECLAT, Erlangen 1990, 581-592.
(7)   A. Schüßler, P.H. Steen und P. Ehrhard, J. Appl. Phys. 71 (1992) 1972-1975.
(8)   P.A. Molian, in Surface Modification Technologies, ed. T.S. Sudershan, Marcel
        Dekker, New York 1989, 421-492.
(9)   P.M. Moore, L.S. Weinmann, Proc. Phot. Opt. Instr. Eng. (Spie) 198 (1979) 120-125.
(10)   G. Geirnaert, Bulletin de la Societe Francaise de Ceramique. Paris 41 (1975) 7-50.
(11)   A. Schüßler und K.-H. Zum Gahr, 2nd Int. Conf. Laser M2P, Grenoble, France, July
        1991, Journal De Physique IV, Colloque C7, supplément au Journal de
        Physique III,Vol. 1, (1991) 121-126.
(12)   A. Schüßler and K.-H. Zum Gahr, Mat.-wiss. u. Werkstofft. 22 (1991), 10-14.
(13)   A. Schüßler, J. Bartos und K.-H. Zum Gahr, Prakt. Met. Sonderbd. 23 (1992) 547-
        558.
(14)   N. Axén and K.-H. Zum Gahr, Wear, to be published

Laser Carburizing of a Low Carbon Steel

D. Müller, H.W. Bergmann, T. Endres, T. Heider

Forschungsverbund Lasertechnologie Erlangen, Universität Erlangen-Nürnberg,
Lehrstuhl Werkstoffwissenschaften 2, Metalle, Erlangen, FRG

1.   Introduction

Low carbon steels are cheap construction materials comprising good weldability and deformability. During the last years the interest for this group of materials grows more and more especially for applications in power transmitting heavy duty parts. Hollow and therefore weight reduced shafts can be processed via internal high pressure deformation /1/. In many cases, however, the application of such parts is opposed by the low wear behaviour of low carbon steels. Therefore, it is necessary to improve the tribological properties by a thermochemical surface treatment such as case hardening or nitriding possibly followed by a thermal laser treatment /2,3/. Another way to produce carburized layers is alloying of the surface via the liquid state using a carbon source (eg. gas /4/ or graphite) and an appropriate heat source as induction heating /5/ or high power lasers /6,7,8/. Especially the CO_2 laser is a suitable tool to produce local carburized layers up to ledeburitic carbon concentrations /7/.

The present work shows the possibilities to control the composition of the layers by using several alloying additions and gives the limits of the process.

2.   Experimental Procedure

A low carbon steel (St52-3; Mat.no. 1.0570) with a composition of 0.14% C, 0.15% Si, 1.02% Mn and <0.04% P and S was used for the investigations. The alloying powders were fixed on the ground surface of the samples (200 x 40 x 6 mm^3) via screen printing. The alloying procedure was performed using high power CO_2 lasers (4 and 14 kW maximum output power) focused to a line of 6 and 15 mm length respectively. Shielding gas protected the melt bath against oxidization and burning of the carbon.

3.   Results

Fig. 1 shows the variety of microstructures and the corresponding hardness profiles achievable by alloying graphite into the surface of the St52-3 steel. The microstructures shown exhibit martensite (A), martensite with retained austenite (B), perlite with secondary cementite (C), eutectic (D) and hypereutectic ledeburite (E). The control of the composition is possible by controlling the process paramaters, especially the thickness and graphite content of the printed layer, the volume of the melt bath via output power and feed rate and the interaction time (intensity distribution and feed rate). In Fig. 2 a nearly linear relationship between the resulting carbon concentration at the surface and the feed rate can be seen. By using such relations it is possible to predict the necessary process parameters to achieve a layer of a certain depth and composition.

The microstructure of the layer depends essentially on the composition (carbon content) and the solidification and cooling rate. The influence of the carbon content can be seen in Fig. 1 (A to E increasing carbon concentration). Fig. 3 shows the influence of the feed rate and therefore of the solidification and cooling rate on the hardness and microstructure of a layer containing appr. 1.6% C. A low feed rate (0.1 m/min) results in a microstructure containing fine lamellar perlite and secondary cementite. With increasing feed rate areas of martensite and retained austenite appear, additionally resulting in an increase of hardness.

294

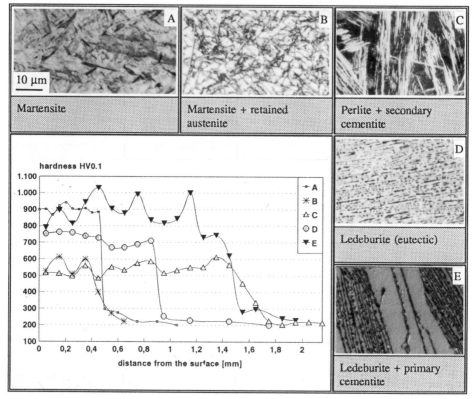

Fig. 1: Microstructures and hardness profiles of laser carburized layers on a St52-3 steel; A-E increasing carbon concentration

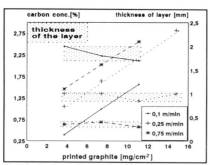

Fig. 2: Carbon Concentration and thickness of layers as function of the amount of printed graphite and feed rate in laser carburizing of St52-3 (output power 10 kW, optic 1 x 15 mm)

Layers with a carbon content up to eutectic concentrations show a homogeneous microsrtructure, whereas in hypereutectic layers a concentration gradient from the surface to the substrate can be observed (see Fig. 4).

The quality of layers regarding surface roughness, pores and cracks and the geometry of the layer depends on different parameters. The roughness depends mainly on the feed rate (see Fig. 5a) and is decreasing from values of R_z = 80 - 90 µm at 0.1 m/min to less than 20 µm at 0.5 m/min. Cracks could only be observed in hypereutectic structures and were minimized by preheating of the samples at temperatures of 400 - 500 °C. The appearance of pores increases using thick printed layers and low feed rates (see Fig. 5b) and could possibly be avoided by a following remelting step. The geometry of the melt bath is essentially influenced by the intensity distribution and therefore by the quality of the laser beam and the line focussing optic.

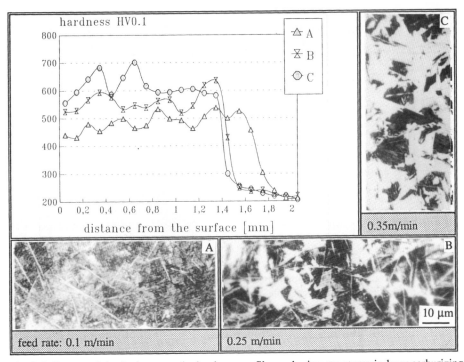

Fig. 3: Influence of the feed rate on the hardness profiles and microstructures in laser carburizing of St52-3; carbon concentration 1.6%, output power 10 kW, optic 15 x 1 mm

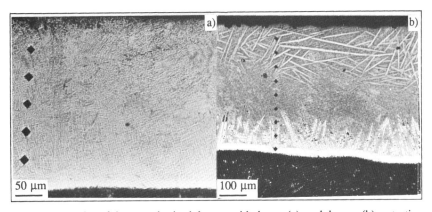

Fig. 4: Macropraphs of laser carburized layers with hypo (a) and hyper (b) eutectic microstrucures

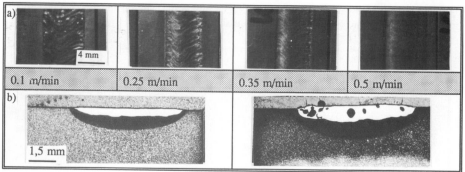

Fig. 5: a) surface quality of laser carburized St52-3

b) cross section of a laser carburized track; left hypo eutectic and right hyper eutectic carbon cocentration

Cast iron powder and cast iron/graphite powder mixtures were also investigated and were fixed on the surface via screen printing as alloying addition with the aim to avoid pores in the layer. These additions show no pores because of the low melting point of the cast iron and the therefore appropriate melting behaviour. Fig. 6 shows an example for a laser carburized layer. The microstructure exhibit primary Fe-dendrites in a ledeburitic matrix. Similar microstructure are obtained from TIG and laser remelting of cast iron /9/ are suitable and nowadays used for applications as e.g. camshafts.

Further investigations were carried out to improve the quality of the layer by using other alloying additions. Good qualities of the layers can be achieved by using powders of high alloyed steels and mixtures of such powders with graphite and/or further additions (e.g. Cr). Using this procedure it is possible to generate nearly any desired steel composition at the surface of a St52-3 steel. An example is shown in Fig. 7 where

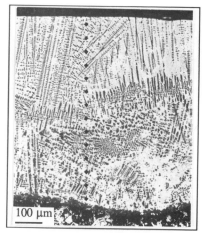

Fig. 6: Macrograph of a laser carburized layer using a powder mixture of cast iron and graphite

a mixture of HSS powder (S-6-5-3) with graphite and Cr was used. As explained above for the the alloying of graphite the composition of the layer depends on the amount of additions and the volume of the melt bath. Increasing alloying concentrations and increasing hardness values are achieved with increasing feed rates (using the same amounts of alloying additions) due to increasing cooling rates and amount of carbides. A further optimization of the quality of the layer may be realized by a post heat treatment.

4.   Summary

It is possible to adjust the carbon level of a surface layer to a desired value by laser carburizing of low carbon steels. To improve the quality of the layer furthermore additional alloying elements can be mixed so that nearly any composition can be generated on the surface. In some applications laser carburizing seems to be an efficient alternative process to case hardening for the production of local, high quality surface layers. Ledeburitic layers enable further possible applications in the field of combustion engines, eg. cam shafts. Before realizing heavy duty applications the residual stresses in the laser carburized layers and their effect on the dynamic behaviour has to be examined.

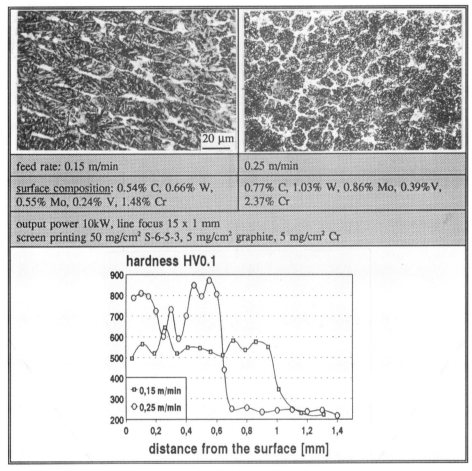

Fig 7: Microstructure and hardness profiles of a laser alloyed St52-3 using a powder mixture of HSS, graphite and chromium

Acknowledgements
The authors like to thank the companies Mauser Werke Oberndorf GmbH, Thyssen HOT Nürnberg and ATZ-EVUS Vilseck for the support of the investigations.

Literature

/1/ F. Dohmann, P. Bieling: Grundlagen und Anwendungen des Innenhochdruckumformens, Blech Rohre Profile, 38(1991)5, S. 379-385

/2/ R. Zeller, D. Müller, H.W. Bergmann: Combination of Thermochemical and Laser Treatments, LASER'91, June 1991, München, in print

/3/ H.W. Bergmann, D. Müller, H.U. Fritsch: Experimental Investigations and Mathematical Simulation of the Carbon Diffusion During Laser Hardening, ECLAT'90, Sprechsaal Verlag, 1990, Coburg, S. 123-136

/4/ B.L. Mordike, H.W. Bergmann, N. Groß: Gaseous Alloying with Laserheating, Z. Werkstofftechnik, 14(1983), S. 253-257

/5/ V. Auerbach, J. Grosch: Möglichkeiten des induktiven Randschichtumschmelzlegierens von Stahl, HTM, 39(1984)

/6/ A. Walker, D.R.F. West, W.M. Steen: Laser Surface Alloying of Ferrous Materials with Carbon, Laser 83 Optoelektronik, ed. W. Waidelich, Springer-Verlag, 1984

/7/ A. Walker, D.R.F. West, W.M. Steen: Laser Surface Alloying of iron and 1C1.4Cr Steel with Carbon, Metal Technology, 11(1984)

/8/ J. Shen, F. Dausinger, B. Grünenwald: Möglichkeiten zur Optimierung der Randschichteigenschaften eines Einsatzstahles mit CO_2-Lasern, Laser u. Optoelektronik, 23(1991)6, S. 41-49

/9/ D. Müller, J. Domes, H.W. Bergmann: Eigenspannungen und Gefügeausbildung nach dem Randschichtumschmelzen von Nockenwellen mit linienfokussierter CO_2-Laserstrahlung, HTM, 47(1992)2, S. 123-130

The Micro Characterization of Laser Gas Nitrided Titanium

Arvind Bharti, Manish Roy and G. Sundararajan, Defence Metallurgical Research Laboratory, Kanchanbagh, Hyderabad-500 258, India

Introduction

Due to concern for stategic raw materials such as Mo, Co, Mn, Cr, Ni, surface alloying (SA) is receiving great attention (1, 4). Laser gas alloying (LGA) which offers many advantages over diffusion controlled SA methods, is slowly emerging as a viable process for many applications. It has been reported to substantially increase surface hardness and change surface chemistry in many alloy systems (5-6). Most of the published studies on LGA have correlated surface properties such as hardness, melt depth (6), wear and corrosion resistance (8,9) with laser parameters. There is a need to have focussed studies to develop a comprehensive understanding of issues like the mechanism of solute transport, origin of surface roughness, compositional inhomogeneity and formation of pores during LGA. In tune with such an overall objective, the present work has the limited objective of characterizing the melt zone formed in CP titanium by single pass laser gas nitriding in terms of its microchemistry, microstructure and micromechanical properties.

Experimental

A 5 KW CW CO_2 laser has been used to surface alloy cold rolled CP-Ti plates of hardness 200 VPN with N. CP-Ti (99.95 % Ti) plates of 100 mm x 100 mm x 8 mm size were melted with the laser under N_2 atmosphere. Laser beam and gas delivery nozzle (GDN) were held stationary while the substrate was moved with the help of a 5-axis CNC work station. Angle between axis of GDN and melting surface and distance of the GDN from melting spot were kept at $0°$ and 20 mm respectively. At low pressure a high flow rate of N_2 was maintained on and around the melting spot. Scanning speed was varied from 20 inches per minute (IPM) to 150 IPM while the laser power was kept constant at 3 KW all through this study. Single tracks were made at each power and scan speed combinations. X-ray diffractograms were taken from alloyed surface and transverse sections (TS) were cut and polished to evaluate the microstructure, microchemistry (EPMA) and microhardness.

Results and Discussion

The variation of the experimentally observed melt depth (defined in the inset of Fig. 1 with the laser beam scan speed in the case of titanium is illustrated by solid circles in Fig. 1. A sevenfold increase in the scan speed, decreases the melt depth approximately by a factor of three. A number of models capable of predicting the melt depth as a function of laser processing parameters have been proposed (9-13). An analytical, one-dimensional heat flow model, due to Easterling and Ashby (14), is one such model. The melt depth predicted by this model, assuming the melt depth to be the maximum depth upto which the melting point isotherm penetrates during the interaction time, is shown as unfilled circles in Fig. 1. For the above prediction, the room temperature values of specific heat (520 J/kgk), density (4510 kg/m^3), thermal conductivity (0.97 J/smk) and melting point (1943 K) relevant to CP titanium were used. In addition, values of 1 mm for the beam diameter and 0.6 for the absorptivity were chosen on the basis of previous measurement (14).

It is clear from Fig. 1 that the model underestimates the melt depth by 20 to 25 % at all the scan speeds. The inclusion of the latent heat of fusion in the model (using the equivalent melting point concept) or the use of the melting point of Ti alloyed with N to its maximum solubility limit (equals ≈ 2300 K) instead of pure titanium does not alter the results significantly. Thus, it is more likely that the melt depth is underestimated by the model because it does not account for the extra heat generated either by the nitrogen going into solution in titanium or by the formation of titanium nitrides (both being exothermic reactions).

Typical microstructural features observed in alloyed layers are shown in Fig. 2 & 3. It essentially shows distinct regions (pockets) of different microstructures (Fig. 2). Careful examination on SEM showed epitaxially grown α and transformed β grains. The transformed β grains contain α structure in some regions (Fig. 2b, 3b top right and 3d centre) and α^m (Fig. 3b bottom right, 3c bottom left) in other regions. The melt zone also shows dendrites and few laths. In general dendrites are growing from surface (Fig. 2, 3a and b) but are also observed in other locations along with small 3 & 4 fold growth patterns (Fig. 2b, 3b right and 3d). Volume fraction of laths is low and they are observed more near melt interface (Fig. 3d).

Due to availability of N$_2$ even after laser beam has crossed the pool (Fig. 4) N concentration and melting point (MP) of the melt continuously increases. The fluid element at surface remains at highest temperature but due to increase in MP at some point of time dendrite starts growing. Individual dendrites (Fig. 2b, 3b, d), dendrite colonies (Fig. 2a, 3c), broken dendrite tips (Fig. 2 & 3) and laths (Fig. 3d) can be swept to different locations by moving fluid elements were they freeze. All this results in non-uniform microstructure and N concentration across the alloyed layer.

The microstructural features in the melt zone were also analysed for microchemistry using EPMA. Fig. 5 shows the characteristic X—ray counts (XRC) of nitrogen at random locations in the melt (or alloyed) zone. These counts do not show any obvious trend and therefore do not support the presence of a nitrogen concentration gradient within the melt zone. This observation is consistent with the fact that the convection in the melt pool is responsible for nitrogen transport.

Formation of structural pockets and variation of N XRC in alloyed layer can be explained by considering the convection in meltpool.

Fluid flow patterns in laser melting (10-12) and alloying (10-12) have been modelled. These models consider gradient in surface tension as driving force. Assuming rectangular (13) or spherical (12) melt cavity with plane surface (10-13) on which gaussian (10-13) or sinusoidal (13) heat source, they solve energy and momentum equations for different (boundary) conditions and generate velocity of fluid elements and isotherms. The most relevant results that all the referred models predict is presence of reverse vortices (RV) during melting and solidification. The shape, size and location of RVs depends on thermal diffusivity of substrate, power, energy distribution and interaction time of the beam. Fig. 4 schematically shows RVs obtained by gaussian beam (12). The arrows on the same figure show streamlines. The transport of N to the RVs is predominantly diffusion controlled due to the existence of boundary layer between RV and streamlines. Therefore, the liquid contained in RVs is likely to have lower N and is likely to have different microstructure compared to remaining melt. Since different RVs are formed at different locations and at different point of time, therefore, they are likey to have different N concentration and consequently different microstructures. The observed structural pockets (Fig. 2, 3) are possibly manifestations of RVs formed during melting. Because energy distribution of the beam is non—symmetric (ring mode with hot spots) resultant pattern of RV formation is expected to be very complicated.

X—ray diffractograms were obtained from the laser alloyed surface and also as a function of depth by successively removing the material from the alloyed surface. The results of this work, to be detailed elsewhere (16), can be briefly summarized as follows. The CP titanium laser alloyed with nitrogen at 50 IPM, showed the presence of titanium oxide (TiO_2) and titanium nitride (TiN) on the surface. Though TiO_2 formation was restricted only to the near surface regions ($\approx 20 - 30$ µm), the presence of TiN can be observed well below the surface upto a depth of 225 (±75) µm. The volume fraction of TiN phase continuously decreases with increasing depth from initial maximum value at the surface and becomes zero at an intermediate depth (225 $\pm$ 57) µm. On the contrary, Ti(α) phase does not exist on the surface and can be detected only beyond a depth of 150 µm and at larger depth (>300 µm) only Ti (α) phase is observed.

Microhardness measurements were also carried out on the sectioned and polished surfaces of the melt zone to evaluate the variation of microhardness as a function of depth. Figure 6 presents the microhardness—depth profiles obtained at the scan speeds of 20, 50, 100 and 150 IPM. Compared to the base microhardness of around 200 KHN, the near—surface hardness lies in the range 600 to 1600 KHN. More importantly, there is a general tendency for the near surface hardness to decrease with increasing scan speed (except for the 150 IPM data). The very high near—surface hardness obtained at the scan speeds of 20 and 50 IPM is most probably caused by the TiO_2 layer. The smooth downward gradient in the hardness profile beyond a depth of about 50 μm, is most likely due to the decreasing volume fraction of TiN phase with increasing depth as suggested by the XRD work. The general upward shift of the microhardness-depth profile with decreasing scan speed must also be related to the increased formation of TiN phase at lower scan speeds.

Conclusion

1. Gas alloying of N increases surface hardness of CP-titanium from 200 KHN to 600-1600 KHN.

2. The variation in volume fraction of TiN with increasing depth from melt surface results in smooth downward gradient in microhardness.

3. Patterns of convection (RV and streamlines) in liquid pool result in non-uniform distribution of N wich manifests itself as highly non-uniform microstructural pockets.

Acknowledgement

Authors express their gratitude to Shri S. L. N. Acharyulu, the Director, DMRL for granting permission to present and publish this work.

References

1. Mordike, B. L., Z Werkstofftech, V 14 (1983) pp. 221-228.

2. Draper W, and Poate J. M., Int. Met. Rev. V 30 (2) (1985) p. 85-107.

3. Majumder J., Singh J. and Raja, Laser Cladding (1987) Annual Report AN ADA 1796069 XSP, Air Force Office of Scientific Research, Bolling APB, DC.

4. Bharti A., Goel D. B. and Sivakumar R., Jr. of laser Appl. V. 1 (2) (1989) p. 41.

5. Katayama S., Matsunova A., Morimoto A., Ishimoto S. and Arata Y., in Laser Processing of Material, eds. Mukherjee ad Majumder J., Pub. TMS of AIME (1985) pp. 159-168.

6. Mirdha S. and Baker T. N., Mat. Sci. Engg. A 142 (1991) p. 115-124.

7. Parvathavarthini N., Dayal R. K., Bharti A., Sivakumar R. and Kamachi Mudai U., Accepted in Mat. Sci. & Tech.

8. Yerramareddy S. and Bahadur S. in Wear of Materials, Ed. Ludema, K. C., ASME (1991), pp. 531-540.

9. Cline H. E. and Anthony T. R., J. Appl. Phys. V.48 (1977) p. 3895.

10. Chan C., Majumder J. and Chen, M. M., in Lasers in Material Processing, ed. Metzhower E. A., ASM (1983) pp. 150-157.

11. ibid, Mett. Trans. V 15a (1984) pp. 2175-2184.

12. ibid, (1986) Proceedings III Engineering Foundation Conference in Modelling and Control of Casting and Welding Process, Pub. AIME, to be published.

13. Srinivasan J. and Basu B., Int. J. Heat & Mass Transfer V 29(4) (1986) pp. 563-572.

14. Ashby M. F. and Easterling K. E., Acta Metall. V. 31 (11) 1984, pp. 1935-1948.

15. Rambabu D., Bharti A. and Sivakumar R., unpublished work on absorptivity measurement.

16. Manish Roy, Bharti A., Sundararajan G., unpublished work.

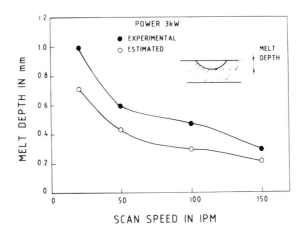

Fig. 1: Depth of alloying as a function of scan speed at 3 kW power.

304

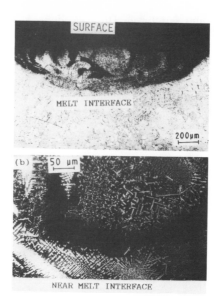

Fig. 2: Optical micrograph of the transverse section of the alloyed layer at 3KW 50 IPM scan speed.

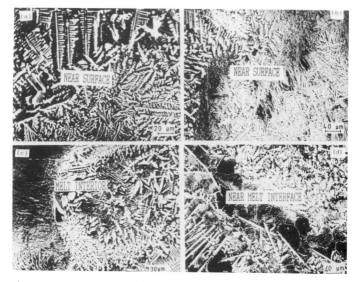

Fig. 3: High resolution SEM micrographs showing typical microstructural features observed in alloyed layer (a) dendrites from surface; (b) α and α^m structure in pure β grains; (c) broken dendrite tips and other growth forms in a freezed RV; and (d) laths.

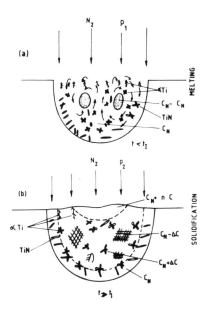

Fig. 4: A schematic of the convection patterns in melt pool, created by a moving laser beam of gaussian energy distribution, estimated by a numerical model (12). Arrows show streamlines, broken line shows the location of solid liquid interface at two different times during solidification. C_N represents average N content of the melt pool.

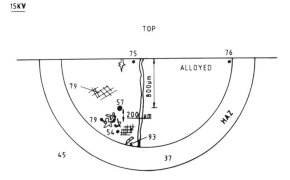

Fig. 5: A schematic represantation of the N X-ray counts at random locations on the transverse section of the alloyed zone.

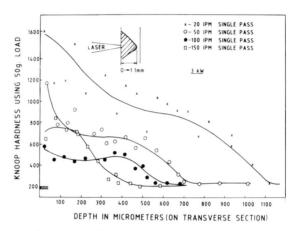

Fig. 6: The variation of the microhardness as function of melt depth on transverse section of single laser tracks with different scan velocities at 3kW power.

Laser Heating and Structure of Steel

V.D. Sadovsky, V.M. Schastlivtsev, T.I. Tabatchikova,
I.L. Yakovleva, Institute of Metal Physics Ural Division
of Russian Academy of Sciences

Introduction

Laser action on metal is characterized by extremely high rates
of heating and cooling, which are not observed in any other
type of heat treatment. Hence, phase and structural transformations due to laser treatment exhibit a number of specific
features. In this monograph consideration is given to the peculiarities of phase transformations in steels due to the
superhigh rate of laser heating.

Materials

Commercial and experimental steels, including constructional
carbon and alloy steels were used in the investigations:
(Fe-0.2C; Fe-0.4C; Fe-1.0C-0.4Cr; Fe-0.36C-1.3Cr-3.2Ni;
Fe-0.37C-1.3Cr-3.2Ni-0.4Mo-0.1V, for example).
The specimens were preliminarily either annealed or quenched
in oil from 1200°C.
The continuous-wave CO_2-laser had an output of 1.0-1.5kW. The
velocity of the irradiated specimen was 100 m/h. As a result,
a strip of rapidly melting and rapidly solidifying metal
emerged on the specimen surface. On both sides of it there was
a thermally affected zone - the main object of our investigation.

Experimental results

Laser heating of steels which have been annealed preliminarily
and wich have had an initial coarse ferrite-pearlite structure
leads to the conservation of coarse inhomogeneities of composition and hardness. In this zone, which was heated up to temperatures a little exceeding A_{c1}, austenite was formed in place
of pearlite, while the initial ferrite remained untransformed
(Fig.1).

Fig.1 : Structure of laser-
treated zone in Fe-0,2C
steel after preliminery
annealing at $12oo^\circ$C.

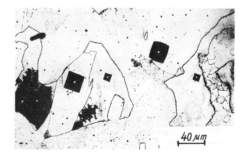

At heating temperatures corresponding to the middle of the intercritical interval Ac_1-Ac_3 one observes a structure made up of patches of high-carbon martensite and ferrite which have undergone recrystallization with grain refinement. Correspondingly, the microhardness of patches of former pearlite becomes very high, while the microhardness of patches of ferrite remains practically unchanged (Fig.2). With rising temperature ferrite transforms into austenite, while on quenching a low-carbon martensite we have patches of high-carbon martensite which emerge in place of former pearlite colonies. Such an homhgeneous structure remains right up to the melting temperature.

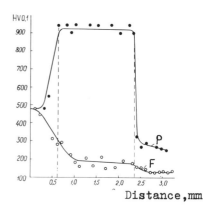

Fig. 2: Distribution of micro-hardness values in laser-treated zone in Fe-0,2C steel after preliminary annealing at $1250^{\circ}C$ (F - microhardness in ferrite sectors, P - micro-hardness in pearlite sectors).

In the molten metal the carbon concentration equalizes rapidly and the hardness of the martensite formed on cooling in more uniform.

Evidently, during laser heating there is not enough time for the equalization of the carbon concentration. This leads to the conservation of structural inhomogeneity in laser-quenched steels which had an initial ferrite-pearlite structure. Hence, to ensure a uniform incomposition martensitic structure it is necessary to carry out prior to laser quenching a preliminary heat treatment including quenching.

In case of laser heating of steels which have been preliminarily quenched the hardness of the laser-affected zone is more uniform (Fig.3). However, special features related to the manifestation of structural inheritance have been revealed when steels with an initial crystallographically ordered structure are heated. Experiments with laser heating of preliminarily quenched steels have shown that at the border of the affected zone

Fig.3 : Distribution of
laser-treated zone in
Fe-0,4C steel after pre-
liminary quenching in
water from 950°C.

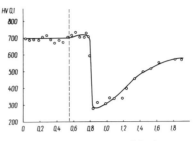

Distance, mm

austenitic grain pecular to the initial condition is reprodused.
Austenite restored during ultra-rapid laser heating is found to
be structurally unstable.
With a small rise of temperature, achieved in the laser-affected
zone, it undergoes spontaneous recrystallization coused by
internal cold-working (Fig.4).

Fig.4 : Structure of laser-
quenched zone in Fe-0,37C-
-1,3Cr-3,2Ni-0,4Mo-0,1V
steel after preliminary
quenching in oil from 1200°C.

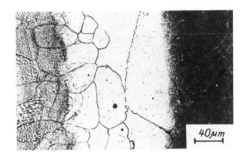

40μm

Because of this, the patch which was heated to higher temperatures, close to fusion point, asquires a fine-grained structure.
Most often the melt crystallized on a substrate, which in this
case is the fine-grained recrystallized austenite. Sometimes,
when the heating rate of quenched steel is high enough, the
recrystallization of austenite is suppressed, and then the
restored austenite serves as substrate for the crystallization
of the melt (Fig.5). The microhardness in the laser-affected
zone drops at first as a result of tempering, but laser starts
rising because of the formation of newly quenched austenite.
The microhardness is found to be practically the same in the
zones with restored coarse grain and fine recrystallization

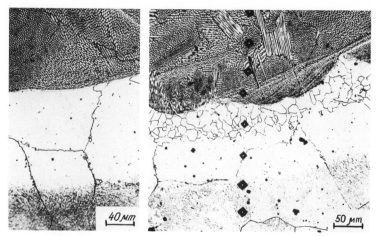

Fig.5 : Structure of laser-quenched zone in Fe-0,36C-1,3Cr-3,2Ni steel after preliminary quenching in oil from 1200°C.

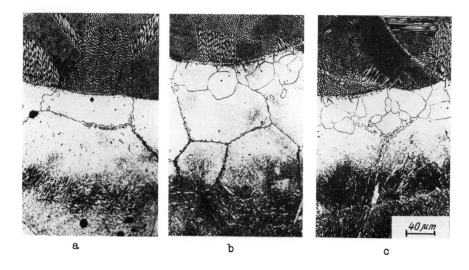

a   b   c

Fig.6 : Structure of laser-quenched zone in Fe-0,37C-1,3Cr-3,2Ni-0,4Mo-0,1V steel after preliminary quenching in oil from 1200°C and ageing at 300°C(a), 400°C(b), 500°C(c).

grain and is considerably higher than the initial values. Nevertheless, one should try to obtain a fine-grained structure when using laser treatment.

Grain refinement in the zone of laser quenching can be effected by either pre-tempering, or plastic deformation of quenched steel (Fig.6,7). Increasing the temperature of tempering brings about the widening of the fine-grained strip of recrystallized metal. In a highly-tempered steel fine grain is formed in the entire laser-affected zone.

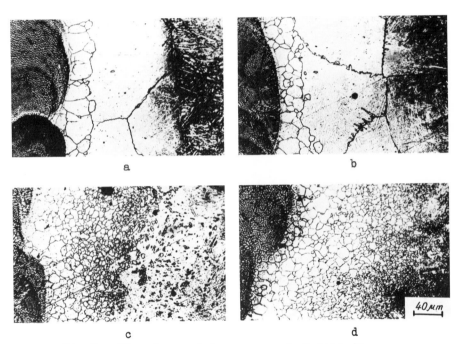

Fig.7 : Structure of laser-quenched zone in Fe-0,37C-1,3Cr-3,2Ni-0,4Mo-0,1V steel after preliminary quenching in oil from 1200°C and tension straining at 0%(a), 8%(b), 15%(c), 20%(d).

Similarly, increasing the degree of plastic deformation leads to a drop in the starting temperature of recrystallization down to the alpha-gamma transformation interval. Hereat, the entire affected zone acquires a superfine-grained structure. The emergence of a fine-grained structure in the laser-affected zone during treatment of pre-quenched and deformed steel is the result of recrystallization of austenite, which inherited in part the defects add up with those due to phase cold-working.

Effects of transformation-hardening without absorptive coatings using CO_2- and Nd:Yag-Lasers

Th. Rudlaff, Zentrum Fertigungstechnik Stuttgart, Stuttgart, FRG
F. Dausinger, Universität Stuttgart, Institut für Strahlwerkzeuge, Stuttgart, FRG

Introduction

Laser Transformation Hardening as a new processing technique for production is known and has been in use since the early 1970's. The breakthrough of this technique, however, has not yet happened. This is due to the very high price of the laser and because some advantages of the process are reduced when using the often required absorptive coatings. For example, the possibility of integrating the laser hardening into automated production lines and the process reproducibility are affected by this requirement.

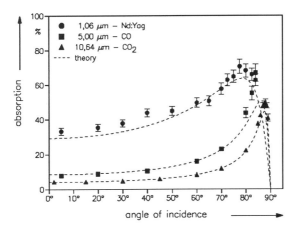

Fig. 1 Absorption of blank steel surface (C45) at different angles of incidence with parallel polarized laser beam.

The possibility to utilize the enhanced absorption by oblique incidence of polarized laser beam during CO_2-laser-hardening (wavelength: 10.6 µm) was shown a couple of years ago [1]. Absorption measurements on blank steel surfaces show, that the absorptivity of the surface can be increased from 5-10% to 20-35% when using a polarized laser beam [2]. For this effect the angle of incidence should be at least 75° and the vector of polarisation in the plane of incidence (Fig. 1).

From the measurements it also can be seen, that a Nd:Yag-laser with a wavelength of 1.064 µm increases the absorption at normal incidence to values of about 30%.

Additional to the increase of absorption by oblique incidence or short wavelength another enhancement due to the formation of an oxide layer and due to the higher temperature during hardening can be observed. The oxide layer formation and the absorption enhancement is dependend on the interaction time between the laser beam and the surface, the temperature distribution during hardening and a shielding gas, which could be used.

Distortion of intensity distribution due to oblique incidence

If a laser beam hits the surface at a certain angle of incidence, the distortion of the intensity distribution is dependend on the relative position of the surface to the focal point of the optic. If the surface is positioned directly in the focal plane, the only distortion is a change of the width to the hight of the laser-beam, whereas outside of the focus an intensity maximum can appear, leading to surface meltings when hardening with this setup (Fig. 2).

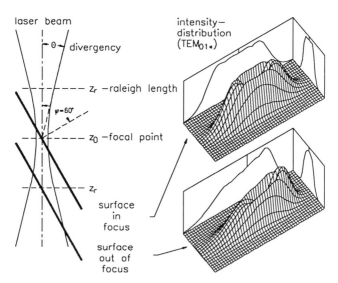

Fig. 2 Distortion of intensity distribution at different positions in the beam caustic due to oblique incidence.

Because of this effect for hardening with inclined incidence a focussing optic with long focal length should be used, forming a big spot size in the focus and a large focus-depth (raleigh-length).

Hardening experiments with CO_2-laser

The linear polarized laser-beam of a CO_2-laser with 5 kW maximum output power and a laser-mode near TEM_{01*} was focused with a mirror optic of 1 m focal length onto the surface of a perlitic steel (C45). The experiments were done with an angle of incidence of $75°$, because this angle already shows an absorption enhancement in comparison with normal incidence and is better to handle due to greater tolerance in positioning. The scanning direction and plane of incidence were chosen perpendicular to each over, with the polarization direction in the plane of incidence. Because of the long focal length of this optic, the whole irradiated zone was in a region with nearly constant beam diameter of 2.8 mm.

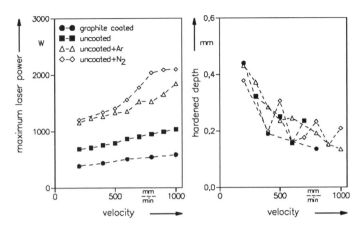

Fig. 3 Maximum Laser power and achieved hardening depth versus hardening velocity with different coatings (CO_2-laser, C45, $75°$ incidence).

The experiments were performed with uncoated, cleaned surfaces and with graphite coated surfaces. At the experiments with the uncoated surfaces a couple of trials were done using a shielding gas, either argon (Ar) or nitrogen (N_2). During the experiments at each scanning velocity the maximum laser power with which the surface could be irradiated without surface melting was determined. These samples were examined.

Fig. 4 Photograph of surface with formation of oxide layer and melting.

The maximum laser power (P_L) versus the travel velocity (v) of the laser beam is shown in Fig. 3 together with the achieved hardening depth (Rht) of the hardened tracks. It can be seen, that the lowest power is needed with graphite coating. Nearly double the power was needed to achieve the same surface temperature without coating and without shielding gas and with shielding gas at least 3 to 4 times of power was necessary. It can be observed, that the factor in power needed between the different surfaces remains rather constant.

At the uncoated surfaces without shielding gas the formation of an oxide layer at the surface was observed, leading to an enhancement of absorption, which could lead to meltings of the surface during the process. An example of this effect is shown in Fig. 4. During hardening on coated surfaces this effect was not observed and with a shielding gas the oxidation could be fully suppressed, such leading to a clean and shiny surface.

Fig. 5 Cross-section through track hardened with CO_2-laser (C45, $75°$ incidence, N_2 shielding gas, 1 m/min velocity, 2000 W laser power, scale: 1 part=0.1 mm).

The achieved hardening depths with same velocity and different surfaces are in the same region. A maximum depth of 0.5 mm was reached, due to the small width of the laser-beam in the direction of travel. The cross-section through the hardened tracks was symmetric. An example is shown in Fig. 5

Hardening experiments with Nd:Yag-laser

With a 1.3 kW cw Nd:Yag-laser hardening experiments on the same material as above were carried out. The laser beam was guided through an optical fiber with 1 mm core diameter and focused with a lens optic with 100 mm focal length (1:1 optic). Because the laser was not polarized, only experiments with normal incidence were performed.

The treatment was done using a defocused laser-beam with different diameters on the surface. Here the experiments with a diameter of 7.5 mm are shown. The surface of the mild steel was used with and without a graphite coating. At the uncoated surface a nitrogen shielding gas was used at a couple of experiments. Like in the trials described bevore at every velocity the maximum laser power was determined, with which the surface could be irradiated without the appearance of surface meltings. A cross section of a hardened track is shown in Fig. 6.

During the experiments without coating a formation of oxide on the surface was observed, similar to the experiments with CO_2-laser. The difference was in the time of the oxide formation. At this experiments the formation happened nearly instantaneous, such avoiding an effect of suddenly surface meltings during hardening. In opposite to the trials with CO_2-laser the oxide formation also appeared during hardening with shielding gas, but the oxide layer was thinner and harder to remove.

Fig. 6 Cross section through track hardened with Nd:Yag laser (C45, normal incidence, 250 mm/min velocity, 730 W laser power, 7.5 mm beam diameter on surface)

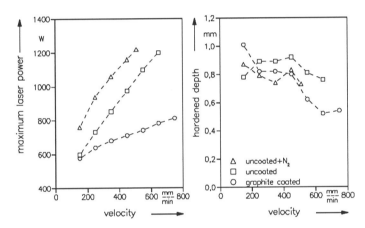

Fig. 7 Maximum laser power and hardening depth versus velocity (C45, Nd:Yag laser, normal incidence, 7.5 mm beam diameter).

The maximum laser power also showed a somewhat different behavior than during CO_2-laser hardening (Fig. 7). The uncoated surface with and without shielding gas showed a constant difference in laser power, whereas the factor between coated and uncaoted surface varies between 1 and 1.6. This means that at the slow velocity, which means that at long interaction times with formation of thick oxide layers, the absorption of this two different surfaces is nearly equal.

Discussion of results

Laser hardening without absorptive coating is possible with CO_2-laser with linear polarized laser beam and inclined incidence. With application of this "brewster-effect-hardening" the coupling efficiency is lower than with graphite coated surfaces.

A hardening without coating and without shielding gas is rather unreproducable with CO_2-lasers due to the formation of an oxide layer and melting of the surface. Better results are achieved by application of a shielding gas to prevent the surface from oxidation and should also be achieved by simultaneous temperature control with pyrometric measurement of surface temperature.

The results of hardening with shielding gas are oxide-free surfaces with reproducable results in hardening depth and cross-section of the hardened track. The lack of absorption in comparison with coated surfaces can be equalized through the advantages of the hardening without coating: no coating has to be applied and removed and the surface can be hardened in the finished state.

To achieve hardened tracks with symmetric cross-section under inclined incidence great attention has to be paid to the beam shaping. One possibility is the use of focussing optic with long focal length and positioning the workpiece in the focal point of this optic. In this case the beam interaction time is rather short, leading to slow hardening velocity and low hardening depths. Better results would be expected form a cylindrical optic or a scanner optic [3].

With solid state lasers a hardening without absorptive coating is possible at normal incidence. Because the difference in absorptivity between blank and oxidized surface is smaller than with CO_2-lasers the possibility of unexpected melting of the surface due to absorption anhancement is reduced. The handling of the process is much better.

The difference between hardening with and without coating is much smaller and dependend on the formation of the surface layer. If deep hardenings are wanted there is nearly no difference in absorptivity between graphite coated and uncoated surface. Together with some other advantages of the Nd:Yag laser, like the fiber beam delivery, this type of laser would be the best choice in hardening on blank surfaces and of small areas where low average power is needed.

References

1. F. Dausinger, Th. Rudlaff: "Novel transformation hardening exploiting Brewster absorption", Proc. of LAMP´87, Osaka, 1987, S.323.

2. F. Dausinger, M. Beck, J.H. Lee, E. Meiners, T. Rudlaff, J. Shen: "Energy Coupling in Surface Treatment Processes", Journal of Laser Applications, 2 (1990) Nr.3-4, S.17.

3. Th. Rudlaff, W. Bloehs, F. Dausinger, H. Hügel: "Hardening with CO_2-Lasers and Flexible Beam Shaping Optic", Proc. of LAMP´92, Nagaoka, 1992 (to be published).

High Intensity Arc Lamps for Transformation Hardening and Remelting of Metals

H.K. Tönshoff, M. Rund, Laser Zentrum Hannover e.V., Hollerithallee 8, 3000 Hannover 21, Germany

Introduction

To increase the performance of components, surface treatments such as transformation hardening, remelting, cladding, alloying etc. are often carried out. Apart from conventional methods, beam technologies, namely laser and electron beam processes, have acquired an increasing significance for thermal surface treatment. The advantages of these new methods using the laser beam as a flexible tool are opposed to the disadvantages of the high costs and the relatively small output available. Lately, CO_2 lasers with an output of up to 25 kW have appeared on the market. However, the necessary investment costs for such an installation are considerable. A new radiation method using a high power arc lamp which promises to make thermal surface treatments of large areas economically viable is currently being developed at Laser Zentrum Hannover.

The High Power Arc Lamp

Fig. 1 shows the principle on which surface treatments with long arc lamps are based.

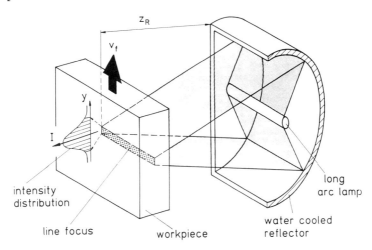

Fig. 1: Scheme of the surface treatment with high power arc lamps

A plasma arc powered by direct current is the radiation source. The arc operates between two tungsten electrodes fixed at a distance of 110 mm (7). Radiation is concentrated within a reflector and focussed onto the working area, resulting in a line focus of 100 mm length and a width of 40 mm. In the direction of the small axis, the working area has a Gaussian intensity profile.

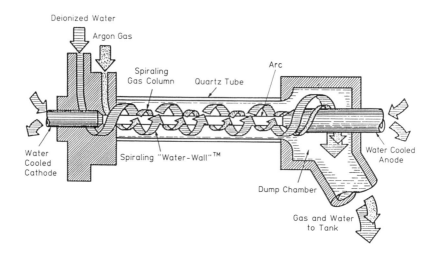

Fig. 2: Principle of the high power long arc lamp

Investigations at the Laser Zentrum Hannover showed that for hardening purposes, a laser beam with a minimum power density of 1400 W/cm^2 (2,6) is needed. The high radiation intensity on the specimen surface places extremely high demands on the lamp light arc. The operation of the lamp is explained with the help of a cross section of the lamp head. On the left hand side is the cathode, and the anode can be seen on the right. The arc is confined in a quartz tube 21 mm in diameter and runs in argon. The lamp current and gas pressure determine the radiation output of the plasma arc. To achieve a high light yield, the lamp is operated with an argon pressure of 0,7 MPa.

Lamp power cannot be increased at will because it is limited by the ability of the electrodes to withstand thermal loading. For the construction shown, hollow copper electrodes with tungsten tips were used which were water cooled; the nominal current is 1200 A. Despite high arc power and the high thermal load of the lamp electrodes, life expectancy at a intensity of 2200 W/cm^2 reaches 10-20 hours.

Using this construction, an extremely high arc performance of 300 kW with an arc diameter of 11 mm is achieved. Because of the high power concentration of the plasma, the quartz tube has to be cooled. On the cathode end of the lamp head, deionised water is injected with high angular velocity, covering the inner wall of the tube with a thin protective film of water which spirals along the length to leave at the anode side. Argon is injected similarly to the water and flows to the anode end inside the water film. This construction tightens the arc and produces a constant emission along its length. At the anode end, water and argon drain into a chamber. They are conveyed into a processing container and recycled. The lamp head is constructed in such a way that it can be run in any direction.

Of the 300 kW input power, 30 kW are absorbed by the electrodes, and 150 kW by the water film. With an emitted output of 120 kW, the lamp possesses an electric efficiency of 40%. 40-60 kW are absorbed by the reflector (dependent on the type of reflector which is used); 10-40 kW are lost in diffused radiation which leaves the reflector at the front. 40-50 kW remain available as useable power in the working area.

Depending on the reflector type, the intensity distribution in the line focus perpendicular to the beam axis is Gaussian or even homogenous. Fig. 3 shows intensity distributions for three types of reflectors.

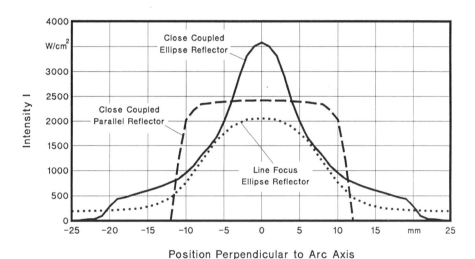

Position Perpendicular to Arc Axis

Fig. 3: Intensity distribution of the focussed radiation for different types of reflectors

A material factor concerning the efficiency of beam techniques is the degree to which energy can be absorbed by the surface of the specimen. Regarding metals, the shorter the wave length of the radiation, the higher the degree of absorption. The lamp emits a continuous spectrum ranging between 200 and 1400 nm. In comparison, the CO_2 laser has a longer wave length of 10600 nm. The continuum is overlaid with intensive spectral lines covering a range from 700 to 900 nm. An estimation shows that approx. 15 % of the lamp output is emitted from the UV-range. This produces, as far as the nominal value is concerned, an output of 7.5 kW in the UV-range. If desired, the UV emission can be avoided by using a different glass material for the quartz tube.

Applications of High Power Lamp for Surface Layer Treatment

Based on the system data, the high power arc lamp shows a high performance capacity, which is substantiated by the experiments on thermal surface treatments (5). Areas of application are transformation hardening, hardening by remelting and alloying by remelting thermally-sprayed layers.

The surface hardening experiments were done using a tool steel 90MnV5, which was available as a sheet material in a thickness of 9.5 mm. On account of the short wave length of the radiation, it was possible to dispense with absorption-improving coatings.

The hardening teached values of up to 900 $HV_{0.2}$ and was virtually constant over the entire hardening depth. The depth of hardening is shown as a function of the transverse velocity at three different arc powers in Fig. 4.

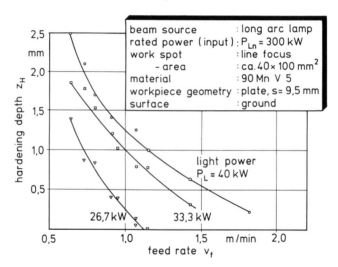

Fig. 4: Hardening depth as a function of transverse velocity for different arc powers

The diagram does not differentiate between solid phase and liquid phase transformation processes. The maximum hardening depth achieved was z_H = 2.5 mm using a radiation power of 40 kW and a transverse velocity of 0.66 m/min. With a decrease of power and an increase of transverse velocity, the depth of hardening drops. However, using a velocity of v_f = 1 m/min still achieves a depth of hardening of 1.25 mm. Taking a beam width, i.e. a working width of 100 mm, this corresponds to an area rate of 1 m^2 in 10 minutes. Using quenching media, hardening depths on metal rods of 3.5 mm were obtained.

Applications for this new process for transformation hardening are on tools for agriculture, transmission shafts and tools for forging.

Cast iron was remelted in order to produce a ledeburitic structure in the surface layer, which has a fine grain size, a high hardness and high wear resistance. At a lamp power of P_L = 36,7 kW and a feed rate of v_f = 0.6 m/min, remelting depths of 1.8 mm were obtained. Near the surface, hardness values over 800 $HV_{0.2}$ were occasionally found.

Moving components along the long arc axis, small components such as rocker arms have been treated to produce tracks 5-20 mm wide (see Fig. 5).

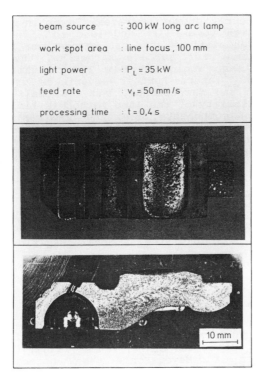

beam source	: 300 kW long arc lamp
work spot area	: line focus, 100 mm
light power	: P_L = 35 kW
feed rate	: v_f = 50 mm/s
processing time	: t = 0,4 s

Fig. 5: Hardening and remelting of rocker arms

Other applications of the arc lamp for remelting are on rolls for the paper industry, components for earth moving machines and cast iron guideways.

For alloying, the high power arc can be used for post-treating thermally-sprayed coatings by remelting. The rapid solidification of the surface alloy result in microstructural refinement with formation of a submicron grained, highly homogenous phase with very little porosity (8).

Economical aspects

Because of the short wave length of the lamp radiation, there is no need to use coatings for metal surfaces in order to increase absorption. However, this is generally necessary when using CO_2 lasers (3,4). Coatings for metal surfaces represent a considerable time and cost factor because of the additional operations. The area rate of the lamp with a radiation power of 40 kW is larger than that of a 20 kW laser by a factor of 3, in spite of only twice the radiation output (1,5). The reason for that is the larger trace width, higher radiation output and better beam absorption.

When comparing hourly operating rates, lower investments are in favour of the high power lamp and lead to 66% lower costs. A comparison of economical aspects has to consider the higher area rates of the lamp. For surface hardening, the manufacturing costs with the lamp would be only 10% of the costs occured employing the CO_2 laser. Further advantages of the lamp are simple operation and maintenance, and the small area of space needed. The flexibility of the lamp method is limited as regards the geometries of components, i.e. the treatment of the inside of corners and edges. The radiation emitted by the light arc can only be focussed into a relatively small area when working with short focal distances.

References

(1) H.-J. Herfurth: "Treatment of metal surfaces with laser outputs of up to 20 kW," Industrie-Anzeiger, 77 (1990), (in German).
(2) C. Meyer-Kobbe: "Laser hardening in comparison to conventional methods", Steel & Metals Magazine, 28-9 (1990), (in German).
(3) C. Meyer-Kobbe: "Surface hardening with Nd:YAG and CO_2 lasers", Ph. D. thesis, University of Hannover, (1990), (in German).
(4) C. Schmitz-Justen: "Integration of laser hardening into practical production technology", Ph. D. thesis, Technical University of Aachen, (1986), (in German).
(5) H.K. Tönshoff, C. Meyer-Kobbe: "Surface treatment with high power lamps", Stahl und Eisen, 110-4 (1990), (in German).
(6) H.K. Tönshoff, C. Meyer-Kobbe: "New potentials of solid state lasers for flexible manufacturing", Proc. of Manufacturing International '90, March (1990), Atlanta, Georgia, USA.
(7) VORTEK, Product information, Vortek Ltd., Vancouver, Canada.
(8) Z.S. Wronski, K.C. Wang, D.M. Camm, D.A. Parfeniuk, L. Boyd: "Rapid Solidification of Thermal Spray Coatings Using Powerful White Light Sources", Intern. Thermal Spray Conference (1992), Orlando, USA.

Possibilities for the Combinations of Thermochemical Diffusion Treatments with Thermal Laser Treatments

D. Müller, M. Amon, T. Endres, H.W. Bergmann

Forschungsverbund Lasertechnologie Erlangen, Universität Erlangen-Nürnberg, Lehrstuhl Werkstoffwissenschaften 2, Metalle, Erlangen, FRG

1. Introduction

The technologically and commercially most important thermochemical surface treatments are carburizing and nitriding. These processes are often used to improve the wear and/or corrosion resistance on the one hand. Another effect is the generation of compressive stresses in the surface region which causes enhanced properties of dynamically loaded components /1,2/. As shown in /3-6/ high quality layers can be produced by a thermochemical treatment followed by laser beam hardening for both processes - carburizing /3,4/ and nitriding /5,6/.

Based on investigation on carburizing and laser hardening /3/ the authors present now the advantages obtainable by a combination of plasma nitriding and laser hardening. Inspired from results of Zenker /5/ and Bell and Bloye /7/ both sequences - nitriding followed by laser hardening and laser hardening prior to nitriding were performed.

2. Experimental Procedure

steel	%C	%Si	%Mn	%P	%S	%Cr	%Ni	%Mo	%V	%Al
21CrMoV57	0.21	0.23	0.40	0.008	0.003	1.27	0.19	0.67	0.26	0.017
25CrMo4	0.24	0.21	0.74	0.015	0.024	0.97	0.03	0.17	0.004	0.03
34CrMo4	0.35	0.27	0.76	0.010	0.034	1.05	0.21	0.21	0.009	0.03
34CrAlNi7	0.36	0.19	0.62	0.009	0.013	1.55	0.96	0.23	0.008	0.91
37MnSi5	0.36	0.54	1.46	0.016	0.063	0.15	0.09	0.04	0.08	0.02
42CrMo4	0.42	0.25	0.74	0.019	0.021	1.11	0.23	0.24	0.005	0.03
50CrMo4	0.50	0.21	0.70	0.012	0.024	1.00	0.15	0.15	0.004	0.028
50CrV4	0.49	0.26	1.06	0.021	0.026	1.07	0.04	0.01	0.13	0.026
60MnSiCr4	0.60	0.78	1.12	0.020	0.026	0.31	0.03	0.01	0.003	0.007

Table 1:   Measured chemical composition of the investigated steels

For the investigations technical relevant steels were chosen with the compositions shown in Table 1. All steels had an annealed microstructure except the 37MnSi5 with ferritic pearlitic structure. The grinded samples were nitrided in a commercial plasma nitriding unit (Klöckner Ionon). Before laser hardening the surface was coated with graphite to increase the absorption of the 10.6 μm laser radiation. The laser hardening procedure was carried out with a high power CO_2 laser (4kW Messer Griesheim) in combination with a pyrometric temperature control device /8/. The process parameters are listed in Table 2.

laser hardening -> plasma nitriding		plasma nitriding -> laser hardening	
control temperature 1250°C, 1300°C respectively	nitriding temperature 470°C and 520°C	nitriding temperature 570°C and 670°C	control temperature 1000°C to 1300°C
feed rate 0.3 m/min	nitriding time 4h	nitriding time 4h	in steps of 50°C
appr. 1mm resulting hardening depth	gas composition 10% N_2, 90%H_2	gas composition 10% N_2, 90% H_2	feed rate 0.2, 0.3, 0.5, 0.7 m/min

Table 2: Process parameters used for the investigations

3. Results
3.1. Nitriding Followed by Laser Hardening

Fig. 1 shows hardness profiles of the 34CrMo4 obtained by a combination of plasma nitriding followed by laser hardening in comparison to the single processes. It is obvious that the Duplex processing leads to a higher hardness level at the surface than the both single processes. The increase of the surface hardness in Duplex processing is related to an increased nitrogen concentration in

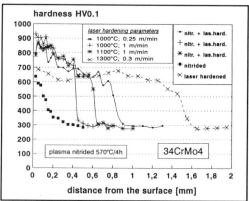

Fig. 1: Hardness profiles of the steel 34CrMo4 after plasma nitriding, laser hardening and a combination of plasma nitriding followed by laser hardening

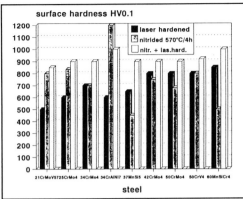

Fig. 2: Surface hardness of the investigated steels after plasma nitriding, laser hardening and a combination of plasma nitriding followed by laser hardening

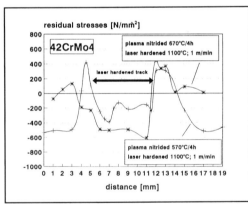

Fig. 3: Residual stresses of the steel 42CrMo4 after the combination plasma nitriding followed by laser hardening across the irradiated track

the formed martensite. The depth of the influenced layer corresponds to the thickness and nitrogen concentration profile of the diffusion zone and is therefore a function of the nitriding parameters. The technical hardening depth of the Duplex treatment is controlled only by the laser treatment (see Fig. 1). The results of all investigated steels are summarized in Fig. 2. In most of the cases the nitriding parameters 570°C/4h and subsequent laser hardening at different control temperatures causes higher surface hardness values similar to the results of the 34CrMo4. Only the 34CrAlNi7 which is an especially designed nitriding steel shows superior hardness after nitriding. Low carbon steels like 25CrMo4 or materials without nitride forming elements show the highest relative increase in hardness.

Nitriding at high temperatures (670°C/4h) prior to laser hardening leads to relative lower hardness values. The increased Nitrogen content causes higher amounts of retained austenite in a surface layer. During laser hardening a destroying of the nitride layer at the surface can be observed caused by diffusion and recombination of nitrogen. This effect increases with increasing hardening temperature and interaction time.

One important aspect except hardness levels is the formation of residual stresses in the produced layers. Fig. 3 shows the resulting residual stresses across a nitrided and laser hardened track. The illustrated curves are similar to those found in laser hardening causing compressive stresses in the track and tensile stresses near it due to the martensitic transformation and annealing processes.

3.2. <u>Laser Hardening followed by Nitriding</u>

Using the process sequence laser hardening prior to nitriding it is possible to create hard nitrided layers on the surface of a steel component supported by a thick layer formed by the laser hardened track annealed during the nitriding procedure. As an example hardness profiles obtained from laser hardening and nitriding (520°C/4h) of the steel 21CrMoV57 are comprised in Fig 4. The effects can be summarized as follows:

- The hardness in the diffusion zone is higher than that of the nitrided and that of the laser hardened sample. This may possibly be caused by smaller nitride precipitations in the laser hardened microstructure.

- Because of annealing processes during nitriding the hardness of the laser hardened case

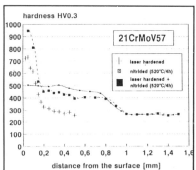

Fig. 4: Hardness profiles of the steel 21CrMoV57 obtained after laser hardening, plasma nitriding and a combination of laser hardening prior to plasma nitriding

decreases. This effect depends on the nitriding temperature and the composition of the steel, especially its retention hardness. The depth of the supporting layer depends on the laser hardening parameters.

The results for the investigations of all steels from Table 1 are shown in Fig. 5a and b. The surface hardness of the combination is most cases (except 60MnSiCr4) higher than that of the nitrided and of the laser hardened samples as well.

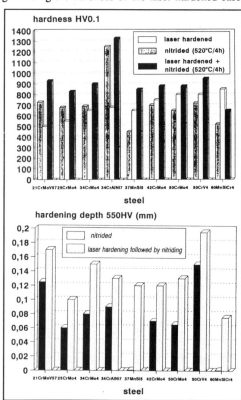

Fig. 5: a) Surface hardness of the investigated steels after laser hardening, plasma nitriding and a combination of plasma nitriding after laser hardening b) Hardening depth 550HV according to a)

Comparing the hardening depth (550HV) after nitriding and after laser hardening and nitriding higher values are always produced by the Duplex treatment. Therefore the process combination can be effectively used to reduce nitriding time because for doubling of the nitriding depth it is necessary to use four times of the nitriding time.

The residual stresses (see Fig. 6) show only compressive stresses across the laser hardened and nitrided tracks with smaller values than those of the purely nitrided surface. Near the laser hardened tracks no tensile stresses are appearing after the

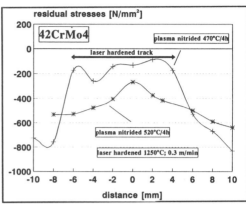

Fig. 6: Residual stresses at the surface of the steel 42CrMo4 after a combination of laser hardening prior to plasma nitriding across the irradiated track

nitriding procedure. The level of the compressive stresses seems to depend on the nitriding parameters. Using low nitriding temperatures (470°C/4h) a difference of 500 N/mm² can be observed between the purely nitrided and prehardened surface whereas at nitriding conditions of 520°C only a difference of about 200 N/mm² remains.

4.   Summary

The possible process combinations of laser hardening and nitriding can be used to generate surface layers with new properties. With the Duplex treatments, especially by laser hardening followed by nitriding it is possible to combine the advantages of the good tribological properties of nitrided surfaces with high hardening depths produced by laser hardening. Therefore the Duplex treatments can also be economical processes that avoid long nitriding times and reduce the necessary nitriding capacities.

Acknowledgements

The authors want to thank the companies Thyssen HOT Nürnberg and BMW AG München for the support of the investigation

Literature

/1/ K.H. Kloos, E. Velten: Einfluß einer Einsatzhärtung auf die Biegewechselfestigkeit bauteilähnlicher Proben, HTM, 39(1984)3, S. 126-132

/2/ E.J. Mittemeijer: Aufbau und Eigenspannungen von Oberflächenschichten nitrierter Stähle, Neue Hütte, 28(1983)10, S. 393-399

/3/ H.W. Bergmann, D. Müller, H.U. Fritsch: Experimental Investigations and Mathematical Simulation of the Influence of Carbon Concentration Profiles during Laser Hardening, ed.: H.W. Bergmann, R. Kupfer, ECLAT'90, Sprechsaal Verlag, Coburg, 1990

/4/ W. Amende: Härten von Werkstoffen und Bauteilen des Maschinenbaus, Technologie Aktuell 3, ed.: V. Bödecker, VDI-Verlag, Düsseldorf, 1985

/5/ R. Zenker: Kombinierte thermochemische/Hochgeschwindigkeitswärmebehandlung - einige Grundlagen und Behandlungsergebnisse, Neue Hütte, 31(1986)1, S. 1-6

/6/ R. Zeller, D. Müller, H.W. Bergmann: Combination of Thermochemical and Laser Treatments, LASER'91, June 1991, Munich, in print

/7/ T. Bell, A. Bloyce: Nitriding Laser Treated Titanium Bearing Low Alloy Steels, Heat Treatment 84, Int. Conf. London (1984), S. 36.1-36.7

/8/ E. Geissler, H.W. Bergmann: Temperature Controlled Laser Transformation Hardening, 2nd Int. Sem. Surface Engineering with High Energy Beams, Lisbon, 1989, S. 121-132

Treatment of overlapping runs by CW CO2 Laser (1)

F. NIZERY **, G. DESHORS**, J. MONGIS*, J.P. PEYRE*.
*   CETIM, 60300 SENLIS, FRANCE
**   LALP, ETCA-CNRS, 94114 ARCUEIL, FRANCE

Abstract

The quenching of large plane surfaces after multipass surface heating by laser beam is limited by a tempering phenomenon which occurs between passes. In this study, we examined the influence on a 34CrMo4 steel of the overlap ratio on the decrease in mechanical properties and that of the precipitation mechanisms which occur during laser treatment of micro-alloyed METASAFE F1200 steel.

Key words

CO2 laser beam treatment, surface hardening, overlapping runs, tempering, AISI 4135, AFNOR 34CrMo4, micro-alloyed METASAFE F1200 steel, micro-hardness, residual stresses.

I. Introduction

The aim of surface hardening by continuous CO_2 laser beam is to create a layer of martensite on the surface of the part in order to improve the fatigue strength (residual compression stresses) and the wear by abrasion (high hardness level).

However, one of the limitations of this process is the treatment of large plane surfaces. To treat this type of surface, several laser passes must be carried out successively. However, the last pass systematically results in the following (figure 1) [1] [2] [3] :
- re-austenitizing and quenching of the overlap zone of the two martensitic widths (zone 3 : $\theta > Ac_3$) ;
- tempering of part of the martensitic width of the first pass (zone 4 : $\theta < Ac_1$).

These two phenomena have the following results :
- a slight increase in the hardness level of the overlap zone (zone 3) ;
- a considerable decrease in the hardness level of the tempering zone (zone 4) ; this results in a decrease in the fatigue strength and wear resistance [4]. In the case of multipass laser treatment, this is to be found again in zones 4 and 6.

To minimize the decrease in mechanical properties, the overlap ratio of the martensitic widths and/or the steel grade (precipitation hardening) can be varied.

To study the influence of overlapping runs, we selected a quenched and tempered chromium/molybdenum steel grade, 34CrMo4, which is widely used in mechanical construction.

To study the precipitation hardening, we used a micro-alloyed steel, F1200, in two different states i.e. as quenched $\theta^\circ = 850\,°C$ and $\theta^\circ = 1250\,°C$.

In this article we will present the results of the influence of overlapping runs on hardness and residual stresses (34CrMo4) and that of precipitation during tempering of the second pass (METASAFE F1200) on these same properties.

---

(1) Study conducted in collaboration with the LALP (ETCA-CNRS mixed unit) and CETIM

II. Choice of experimental conditions

II.1. Choice of steel grades - Heat treatments

We selected two steel grades :

- a 34CrMo4 steel austenitized at 850°C, oil quenched and tempered at 600°C for 1 hour. This grade enabled us to study the tempering and requenching phenomena produced by the multipass laser treatment in zones 3 and 4 ;
- an F1200 micro-alloyed steel. The mechanical properties of this steel grade are partly obtained by the precipitation of vanadium and/or niobium carbonitrides [5].

For this steel grade, we chose two different metallurgical states before laser beam treatment :

a) oil quenched at 850°C. At this temperature, the vanadium and niobium carbonitrides do not go back into solution, but coalesce during austenitizing, and do not precipitate during the laser treatment ;
b) oil quenched at 1250°C. In this case, the vanadium, niobium carbonitrides and copper go back into solution and are therefore likely to be precipitated in the tempered zone (zone 4) by the second laser pass.

The 1250°C treatment enables the increase in hardness linked to the precipitation of carbonitrides in the tempered zone (zone 4) to be studied and compared with the steel quenched at 850°C without precipitation hardening.

The laser treatment parameters enable a width of martensite 8 mm wide by 0.6 mm deep to be obtained.
The overlapping run, defined as the overlapping width of the quenched areas, varies from -0.5 mm to 4 mm.

II.2. Methods for defining the metallurgical characteristics of laser treatment

The characteristics of laser treatment were defined by :
- a series of microhardness measurements ($HV_{0.1}$) carried out 0.1 mm below the surface at right angles to the laser passes ;
- the micrographic structure of the different zones ;
- the residual stress measurements (σ_{R11} et σ_{R22}) by X-ray diffraction across the treatment pass and on the surface of the part.

To define the decrease in the mechanical properties, three criteria were established (figure n°2) :
- the minimum hardness and maximum residual tensile stress ;
- the difference in hardness ($\Delta HV_{2/4} = [HV_{zone\ 2} - HV_{zone\ 4}]$) and residual stresses ($\Delta\sigma_{Rmax} = [\sigma_{tensileR} - \sigma_{compressiveR}]$) between the value measured at the centre of the second laser pass (zone 2) and the minimum recorded (that of the quenched zone i.e. zone 4) ;
- the width in mm of the quenched zone (zone 4) defined as 80% of the difference in hardness defined above (L_{HV4} is defined for $HV = 0.8 \times HV_{zone\ 2}$) and the width corresponding to zero residual stresses ($L_{\sigma R4}$ is defined for $\sigma_R = 0$).

III. Test results

III.1. 34CrMo4 steel

III.1.1. Description of the curves connecting up the series of microhardness measurements

The general trend of these curves can be described according to seven zones (figure 2) :

Zone 1 + 2 : completely martensitic zones with a hardness in the order of 650 $HV_{0.1}$.

Zone 3 : zone with a martensitic structure re-austenitized during the second laser pass and then quenched, with a hardness level slightly greater than that of zones 1 and 2. (hardness = 680 to 700 $HV_{0.1}$ for 34CrMo4 steel). This gain in hardness is mainly due to the refining effect of the structure by the first quenching cycle).

Zone 4 : first pass zone thermally affected by the second. The maximum decrease in hardness is at the interface of zones 4 and 5 and corresponds to the highest quenching parameter (temperature/time) (hardness = 300 $HV_{0.1}$).

Zone 5 : zone taken into the intercritical temperature range (α - γ). The hardness varies from 300 to 550 $HV_{0.1}$).

Zone 6 : overquenched zone of the base metal which is not visible under our test and measuring conditions (very small width of affected zone).

Zone 7 : base metal (hardness = 300 $HV_{0.1}$) not affected by the laser treatment.

III.1.2. Influence of overlapping

The complete results are given in table 1. The main results are as follows :
- the absence of any obvious correlation between the overlap ratio and the difference in hardness $\Delta HV_{2/4}$;
- the independence of the width of the softened zone (L_{HV4}) and the overlap ratio ;
- an increase in the hardness level of zone 3 (+ 20 $HV_{0.1}$) with respect to zone 2.

III.1.3. Results of residual stresses

The results of a series of residual stresses measurements taken in crosswise direction are given in table 1 and [3]. This figure also shows the series of microhardness measurements taken perpendicular to the treated surface. This figure brings out the following measurements :
- zone 6 has tensile peaks (σ_{R11} about 500 MPa) ;
- zone 4, softened by the second pass, has a tensile peak of the same magnitude (σ_{R11} about 500 MPa) ;
- zone 2 has a residual compression stress plateau (σ_{R11} about -300 MPa and σ_{R22} about -400 MPa) ;
- zone 2 has a higher residual compression stress level than zone 1 (+ 100 MPa). This is due to the increase in the average temperature of the part after the first pass, which produces :
  • an increase in the austenitizing temperature of the second pass, resulting in more homogeneous austenite and therefore, after cooling, in more homogeneous martensite ;
  • shrinkage due to cooling at the core of the part after the martensitic transformation.

III.2. METASAFE F1200 steel

III.2.1. Comparison of micro-hardness results obtained for 34CrMo4 and METASAFE F1200 steels

The minimum hardness noted for the series of micro-hardness measurements obtained for the 850°C-F1200 test specimen is the same as that recorded for the 34CrMo4 steel (300 $HV_{0.1}$) ; but the quenched zone (zone 4) is not as wide i.e. 0.96 mm for the F1200 and 1.3 mm for the 34CrMo4 (see table 1).

On the 1250°C-F1200 test specimen, the minimum hardness is greater than that of the 850°C-F1200, while the quenched zones are virtually the same width (1.06 and 0.96 mm respectively).
An examination of the change in hardness of the METASAFE 1250°C-F1200 steel, as a function of the quenching temperature for a treatment time of one hour, shows a minimum of 380 HV for a temperature of 600°C (figure 3). The minimum hardness recorded for the series of micro-hardness measurements carried out on the laser quenched test specimen is 340 HV i.e. a difference of 40 HV. This is due to the much shorter thermal cycle in the case of laser quenching. Precipitation is therefore probably not at its optimum. By optimizing the parameters for subsequent tempering, we can hope to increase the minimum hardness from 340 HV to 380 HV (minimum recorded on figure 3).

III.2.2 Comparison of the residual stress results obtained for 34CrMo4 and METASAFE F1200 steels

A comparison of the results obtained from measuring the residual stresses on the 34CrMo4 and METASAFE F1200-1250°C steels shows that the general trend of the curves is identical. In particular, the levels of the residual compression stresses are identical ($\sigma_{R11} \approx 340$ MPa et $\sigma_{R22} \approx 400$MPa).
However, the tensile peaks in zone 4 are much smaller in the case of METASAFE F1200. In the crosswise direction, the tensile peak is almost nil (≈ 30 MPa). Better results could therefore be expected for METASAFE F1200 in the case of fatigue stress.

IV. Conclusion

A study of the quenching of 34CrMo4 and METASAFE F1200 steel grades after multipass surface heating by laser beam produced the following main results :
- in the case of 34CrMo4 steel, the variation in the overlap ratio within the limits studied (-0.5 to 4 mm) does not allow the decrease in hardness of zone 4 to be limited ;
- in the case of METASAFE 1250°C-F1200 steel, the precipitation of niobium and vanadium carbonitrides in zone 4 enables the decrease in hardness to be raised ($+40$ $HV_{0.1}$) with respect to the 850°C-F1200 steel. However, it is incomplete since it does not lead to the maximum hardness (340 $HV_{0.1}$ instead of 380 $HV_{0.1}$) observed in figure 3.

Because the heat cycles of surface heating by laser beam are too fast, hardening by the precipitation of micro-alloy elements does not have time to take place. To obtain maximum hardening (380 $HV_{0.1}$) in zone 4, we will attempt to favour the total precipitation of carbonitrides, either by tempering before laser treatment, or by local laser tempering in zone 4 after quenching.

V. References

[1] Y. IINO, K. SHIMODA - "Effect of overlap pass tempering on hardness and fatigue behaviour in laser treatment of carbon steel". Journal of Material Science Letters 6 (1987) 1193-1994.

[2] G.P. MOR, R. PEZZONI - "Residual stresses measurements by means of X-ray diffraction on electron beam welded joints and laser hardened surfaces". Construzioni Aeronautiche Giovanni Agusta, June 1989.

[3] F. NIZERY and al - "Traitement de durcissement après chauffage superficiel par faisceau laser : étude du traitement multipasses". Presentation during the MAT-TEC 91 conference held at the CNIT, La Défense, France, 8th-9th October 1991.

[4] ISH II, IWAMOTO, SHIRAIWA, SAKAMOTO - "Contraintes résiduelles superficielles de l'acier trempé par induction" - SAE - Preprint N° 710 280 - 24th-25th September 1968 - Translation CDM-71 - 7629.

[5] A. MULOT, C. LEROUX - "Exemples d'applications des traitements superficiels aux aciers à dispersoïdes" - Les aciers à dispersoïdes appliqués à la mécanique, p. 77-88, Senlis 24th September 1986.

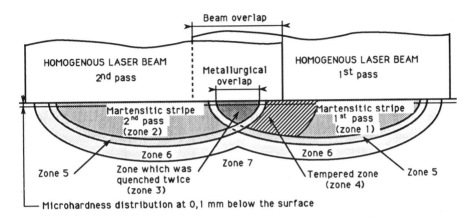

Figure 1 : Sketch of areas quenched and tempered during a multipass treatment.

Steel	Value of the overlap (mm)	Ratio of the overlap	HARDNESS		RESIDUAL STRESSES			
			$\lvert \Delta HV2/4 \rvert$ (HV0.1)	L_{HV4}(mm) for HV = $0.8 \times \lvert \Delta HV2/4 \rvert$	$\lvert \Delta\sigma_{R11max} \rvert$ (MPa)	$L\sigma_{R11}^4$ (mm)	$\lvert \Delta\sigma_{R22max} \rvert$ (MPa)	$L\sigma_{R22}^4$ (mm)
34CrMo4	- 0.5 (spacing)	- 6 %	352	1.5	515	-	-	-	-
	0	0 %	296	1.3	521	-	-	-	-
	1	12,5%	290	1.2	518	-	-	-	-
	1,5	20 %	291	1.2	517	-	-	-	-
	2	25 %	275	1.2	521	750	1.9	500	1
	2,6	33 %	312	1.2	542	-	-	-	-
	3	37 %	341	1.3	517	-	-	-	-
	4	50 %	310	1.4	525	-	-	-	-
F1200 1250°C	1		157	0.96	421	436	1.4	always compressive stresses	
F1200 850°C	1.5			1.06	404	621	1.8	429	0.63

Table 1 : Comparison of micro-hardness results obtained for
34CrMo4 and METASAFE F1200 steels

336

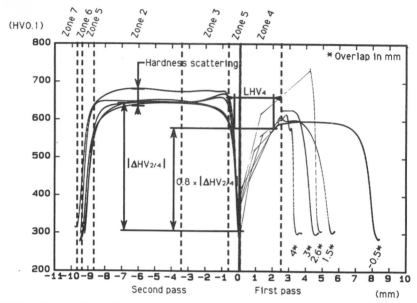

Figure 2 : Series of microhardness measurements (HV0.1) carried out 0.1 mm below the surface.

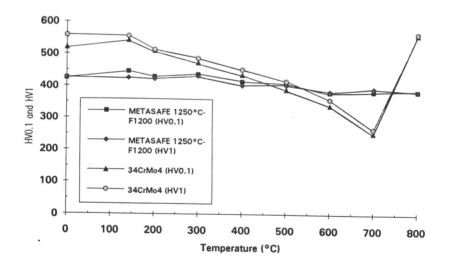

Figure 3 : Hardness of the 34CrMo4 and METASAFE 1250°C-F1200 steels as a function of the quenching temperature for a treatment time of one hour.

Fatigue strength of laser hardened plain carbon and alloyed steels

S. Schädlich, B. Winderlich, B. Brenner, Fraunhofer-Einrichtung für Werkstoffphysik und Schichttechnologie, Dresden, FRG

Introduction

With regard to increase the fatigue strength of structural components by laser transformation hardening an exact prediction of attainable properties may be useful. For smooth specimens from the plain carbon steel C70W2 a simple dependence of the reversed bending fatigue limit R_{bW} on the hardening depth x_L was found (1):

$$R_{bW} = \frac{R_{bWo}(x)}{1 - \dfrac{x_L}{d/2}} \qquad \text{Eq. 1}$$

R_{bWo} - reversed bending fatigue limit of residual stress free core material, d - specimen thickness.

Eq. 1 follows directly from the local fatigue concept (2), which is based on the comparision of the local fatigue strength with the loading stress distribution. A precondition for the validity of Eq. 1 is the fatigue crack nucleation in the region between the laser hardened layer and base material.

The objective of this paper is to examine wether Eq. 1 is valid not only for smooth specimens from plain carbon steels but also for low-alloyed and alloyed ones. Former results from C70W2 are included in this paper too.

Experiments

Beside C70W2 two steels 42CrMo4 and X20Cr13 as typical represantives for low-alloyed steels and alloyed ones were used for experiments. The chemical composition is listed in Table 1.

	C	Si	P	S	Cr	Mn	Ni	Cu	Mo	Al
C70	0,74	0,30	0,013	0,027	0,19	0,33	0,12	0,19		
42CrMo4	0,37	0,28	0,025	0,028	0,99	0,69	0,13	0,12	0,18	0,027
X20Cr13	0,21	0,44	0,025	0,019	12,93	0,34				

Table 1: Chemical composition (wt.-%)

Specimens shown in Fig. 1 were machined from plates and heat treated to normalized state (C70W2) or to quenched and tempered state (42CrMo4, X20Cr13).

The specimens were laser heat treated by a single pass in longitudinal direction on both (one after another) smooth sides (Fig. 1). The aim of this treatment was a martensitic transformation hardening without significant tempering of the first hardened zone during specimens back side treatment. The laser processing parameters used are given in Table 2.

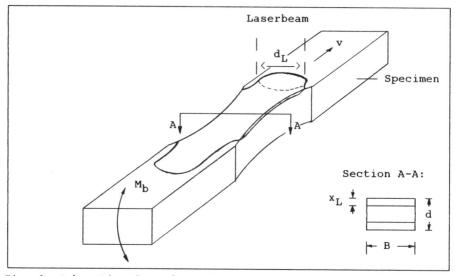

Fig. 1: Schematic view of specimen, laser treatment and specimens cross-section after hardening. Specimens dimensions in the nominal cross-section A-A: $d = B = 4,0$ mm (C70, 42CrMo4), $d = 3,5$ mm, $B = 8,0$ mm (X20Cr13)

	C70	42CrMo4	X20Cr13
Laser	$\longleftarrow$	cw-CO_2	$\longrightarrow$
Power P (W)	1000	2500	2550
Traverse speed v (mm/min)	950	1500-2200	550-850
Focal spot dimension $d_L \| \| v$	5,6	8,0 - 9,5	6,7
" " " $d_L \perp v$	5,6	6,1 - 7,2	20,4*

Table 2: Laser treatment parameters
* - Scanning amplitude 5,75 mm, scanning frequency 210 Hz.

Reversed bending fatigue tests were carried out using an electromechanical testing maschine with an approximately constant loading amplitude and a loading frequency of 25 Hz. The basic number of cycles was choosen $N_G = 2 \times 10^6$.
The fatigue strength was determined in two different ways:
(a) Statistical estimation of the reversed bending fatigue limit using the "Boundary-Method" by Maennig (3) both for all untreated specimens and laser hardened C70W2 and X20Cr13 ones.
(b) Stepwise raising of loading amplitude up to fracture on laser hardened C70W2 and 42CrMo4 specimens. The fatigue limit of an individual specimen is defined to be equal the fracture stress amplitude.
After fatigue failure the fracture surfaces of laser treated specimens were examined by scanning electron microscope.
Vickers hardness depth profiles near the location of crack origin

were measured at the ground and polished fracture cross-section. In accordance with DIN 50190 the hardening depth x_L is defined as the distance from the surface possessing a hardness of 50 HV greater than the core hardness.

Results and Discussion

The experimental results are shown in Fig. 2. The bending fatigue limits R_{bW} referred to the fatigue limit of the untreated material R_{bWo} (fracture probability P = 0,1) are plotted versus the normalized hardening depth at the crack nucleation site $(1 - 2x_L/d)^{-1}$. Taking into consideration the statistical scattering in fatigue results straight lines representing fracture probabilities P = 0,1 (lower line) and P = 0,9 (upper lines) were calculated by Eq. 1 on the basis of the measured untreated materials fatigue limits R_{bWo} for P = 0,1 and P = 0,9 respectively.

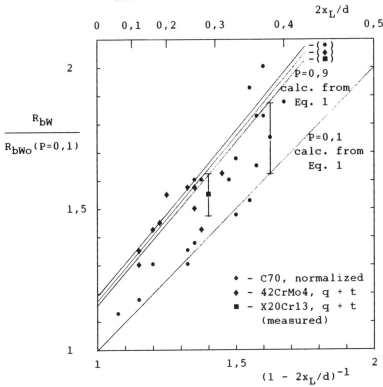

Fig. 2: Normalized increase in bending fatigue limit R_{bW}/R_{bWo} versus normalized hardness depth $(1 - 2x_L/d)^{-1}$

Most of the data points as well as both the scattering bands of the statistically estimated fatigue limits for C70W2 and X20Cr13 fall between the straight lines verifying the linear relationship of Eq. 1. This proves the determining role of hardening depth x_L for the increase in bending fatigue limit of the examined smooth specimens.

In accordance with the C70W2 results (1) fracture surface analysis has shown that in the 42CrMo4 and X20Cr13 specimens the fatigue cracks usually are initiated beneath the hardened layer, thereby sometimes the "fish eye"-phenomenon could be observed.

On the base of the predominate fracture mechanism Eq. 1 can be derived as a simple geometric relation (Fig. 3).

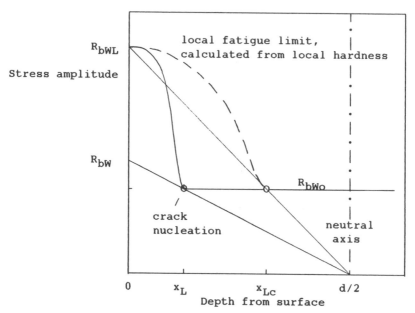

Fig. 3: Schematic diagram of hardening depth effect on fatigue limit, R_{bW} - reversed bending fatigue limit of specimen, R_{bWL} - local reversed bending fatigue limit of the laser hardened case at the surface

A rise of hardening depth x_L results in an increase of the bending fatigue limit R_{bW} correspondig to Eq. 1. If the hardening depth x_L exceeds a critical value x_{Lc} given by

$$x_{Lc} = [1 - \frac{R_{bWo}}{R_{bWL}}] \star d/2 \qquad \text{Eq. 2}$$

the crack will be initiated at the laser hardened surface and the fatigue limit of the specimen approaches the fatigue limit of the hardened layer R_{bWL}.

For hardening depths $x_L < x_{Lc}$, i.e. crack nucleation beneath the hardened zone, the following conclusions can be drawn:

- The fatigue limit of the hardened layer is sufficiently high.
- Surface roughness or surface flaws don't have a determining influence on the specimen fatigue limit.

- Since the sign reversal of residual stresses takes place usually between the hardened layer and the base material, i.e. in
the region of crack nucleation, their effect on the specimen
fatigue limit is expected to be small.

Hence it follows, that in this case the functional optimization
of laser treatment can be reduced to a simple geometrical one and
Eq. 1 can be used for estimation of the fatigue limit of structural components from the specimen fatigue limit (Fig. 4).

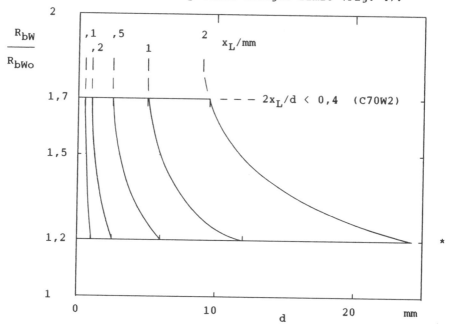

Fig. 4: Possible relative increase in bending fatigue limit
R_{bW}/R_{bWo} of steel components with thickness d due to laser hardening, * - (an increase below 20 % is considered to be technically not useful)

However there are several restricting factors for the increase in
fatigue limit attainable by laser hardening:
- Because of the self-quenching principle the technologically
possible hardening depth is limited to about 2 mm.
For alloyed steels a greater hardening depth may be realized.
- For hardening of thin-walled components it has to be taken into
consideration that the increase of surface layer strength can
be diminished by self-tempering processes, tempering of the
opposite hardening layer or by falling short of the critical
cooling rate for martensitic transformation.
For the carbon steel C70W2 we found a practicable maximum relative hardening depth $2x_L/d < 0,4$ as shown in Fig. 4.
For 42CrMo4 and X20Cr13 especially a greater relative hardening
depth should be attainable.
- From Eq. 2 it can be seen, that the critical hardening depth

x_{LC} depends on the ratio of the base material fatigue limit to the hardened layer fatigue limit. Hence for quenched and tempered core material x_{LC} should be moved to lower values. A decreasing carbon content of the steel also lowers the value of x_{LC}. This results from the fact that the ratio R_{bWL}/R_{bWo} at constant R_{bWo} falls with decreasing carbon content.
Comparising the hardness profiles of the examined steels (Fig. 5) these facts become evident.
Whereas for C70W2 steel inside the marked area in Fig. 4 x_{LC} is not reached for 42CrMo4, X20Cr13 and also for plain carbon steel with a lower carbon content this can't be excluded.
Exceeding x_{LC} for these steels Eq. 1 is no longer valid. In this case surface roughness, surface residual stress state and structural damages at the surface play an important role in the fatigue behaviour of laser hardened unnotched structural components.

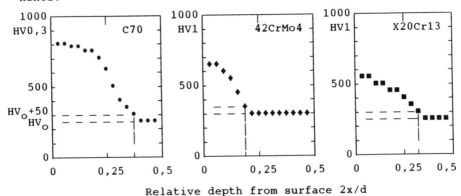

Relative depth from surface 2x/d

Fig. 5: Excamples of hardness profiles HV(x)

Conclusion

For laser hardened smooth specimens from plain carbon steel C70W2, low-alloyed steel 42CrMo4 and alloyed steel X20Cr13 the dependence of the reversed bending fatigue limit on the hardening depth can be described by one and the same relationship.
Considering such limiting factors as attainable absolute and relative hardening depth, carbon content and base material strength the reversed bending fatigue limit of unnotched steel components can be estimated in advance.

References

(1) B. Winderlich, B. Brenner: HTM 44 (1989) 166.
(2) B. Winderlich: Mat.-wiss. u. Werkstofftech. 21 (1990) 378.
(3) W.-W. Maennig: Materialprüfung 12(1970) 124.

Influence of Laser Hardening on the Fatigue Strength of Notched Specimens from X20Cr13

B. Winderlich, B. Brenner, Fraunhofer-Einrichtung für Werkstoff-physik und Schichttechnologie, Dresden, FRG
C. Holste, H.-T. Reiter, Pädagogische Hochschule, Dresden, FRG

Introduction

With regard to improve the fatigue strength of structural components with stress concentrators the local hardening of the highly stressed surface areas by laser beam is of interest.
Whereas the possibility of raising the fatigue limit of smooth and notched specimens by local laser hardening is proved on principle (e.g. 1-6), the knowledge about the behaviour in the fatigue strength region for finite life as well as the effect of mean stresses on the fatigue strength is not sufficient.
Both factors may be important for practical use of laser harde-ning to improve the fatigue strength or fatigue life of structu-ral components.

Experiments

The material examined in this study was common turbine blade steel X20Cr13 in a quenched and tempered condition.
Geometry and dimensions of specimens with a stress concentration factor of 1,7 are shown in Fig. 1.
The specimens were ground along all sides. Semicircular notches were produced by drilling.
Before laser treatment the notch areas to be hardened were cove-red with an absorbent coating from black paint.
For hardening the notch area the laser beam was moved perpendicu-larly to the notch root, whereas an oscillation of the beam parallel to the notch root was realized. In order to attain a symmetrical hardening zone around the notch root the irradiation was started 5 mm in front of the notch edge and the traverse speed was adapted to the local heat removal.
Fatigue tests were carried out using a servohydraulic testing machine. With regard to obtain statistically determined values of fatigue limit the tests were conducted after the "Boundary-Meth-od" by Maennig (7). In the range of finite endurance 7 specimens were tested at each stress level. For statistical evaluation of testing results the arcsin$\sqrt{P}$-transformation (8) was used.
Experimental details concerning chemical composition and mechani-cal properties of base material as well as laser hardening and fatigue tests are summarized in Table 1.

Influence of laser hardening on fatigue strength under symmetri-cal tension-compression

Results of fatigue tests under symmetrical tension-compression loading are shown in Fig. 2. The scattering band includes failure probabilities P from 10 % at the lower level up to 90 % at the

higher one.
For P=50 % an tension-compression fatigue limit increase of about
60 % was achieved.
It should be noted, that laser hardening for symmetric cycling up
to the examined stress level of 610 MPa results also in a considerable improvement of the fatigue strength for finite life. In
relation to the fatigue life this means an increasing factor of
about 15 to 20.
This result may be important for loading conditions with overload
cycles or designing structural components for finite life respectively.

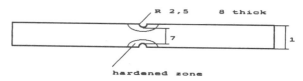

<u>Fig. 1</u>: Specimen

Chemical composition [weight %]					
C	Si	Mn	P	S	Cr
0,21	0,44	0,34	0,025	0,019	12,93

Mechanical properties (base material)		
$R_{p0,2}$ [MPa]	R_m [MPa]	HV10
690	860	260

Laser hardening (cw CO_2 laser)

Power: 900-960 W	Hardening depth: 1,4-1,6 mm
Traverse speed at	Surface hardness: 560 HV0,3
notch root: 80-100 mm/min	
Scanning amplitude: 3,05 mm	
Scanning frequency: 228 Hz	
Spot dimensions: 5,3x11,7 mm²	
Inert gas: N_2	

Fatigue tests

Tension-compression loading: R=-1; -0,2; 0
Sinusoidal load-time-function, load control
Constant mean loading rate: $3,2x10^4$ MPa/s
Basic number of cycles: $2x10^6$
Laboratory environment, room temperature

<u>Table 1</u>: Experimental specification

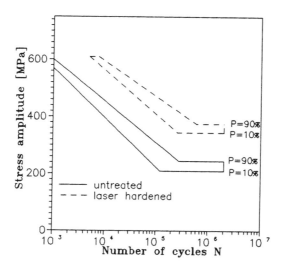

Fig.2: Wöhler diagram for symmetrical tension-compression loading

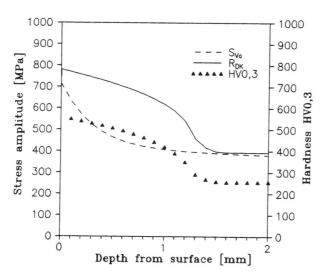

Fig. 3: Local fatigue strength $R_{DK}(x)$ calculated from hardness $HV0,3(x)$ in relation to the loading stress distribution $S_{Va}(x)$ at fatigue limit stress level

Fractographic analysis has proved, that both in the region of finite endurance and infinite one the fatigue cracks started at the specimen surface in the notch root, especially from locally remelted or overheated regions. Hence it follows that the fatigue strength of the specimens is determined by the local fatigue strength of the hardened layer itself.

For the fatigue limit range an assessment of factors limiting the fatigue strength and location of failure origin can be realized by means of the local concept (9).

Fig. 3 shows the local fatigue strength estimated from hardness distribution for residual stress free conditions in relation to the loading stress distribution calculated for the measured fatigue limit stress level. It can be seen, that the value of the measured fatigue limit lies below that one to be expected for residual stress free condition. Since the achieved hardening depth is sufficient it means, that in the notch root tensile residual stresses exist or a structural damage due to overheating happened. On the other hand it appears, that the local reversed fatigue limit in the specimen core is almost exhausted. Therefore in the present case a considerable improvement of fatigue limit can not be expected even after optimization of the irradiation regime with regard to surface hardness and residual stress state.

Influence of mean stresses on the fatigue strength

In order to investigate the influence of tensile mean stresses on the fatigue strength tests with stress ratio R=-0,2 and R=0 were carried out.

The corresponding Wöhler curves for 50 % failure probability are shown in Fig. 4.

In contrast to symmetrical loading pulsating tension results in fatigue limit increase of 19 % only for laser hardened condition compared with the base material. The Wöhler curves for laser hardened specimens and base material ones intersect in the range of finite endurance, whereas with increasing mean stress the point of intersection is shifted to lower maximum stress level.

In Fig. 5 the mean stress effect on fatigue strength is shown by a Haigh diagram. The mean stress dependence both of fatigue limit and fatigue strength for 10^4 cycles is demonstrated.

For both cases influence of mean stress S_m on fatigue strength can be described by a linear equation in form of $R_D=R_W-mS_m$ (R_D -stress amplitude, R_W - reversed fatigue strength, m - mean stress sensitivity). The knowledge of m is necesarry e.g. for assessment of dynamic safety of turbine blades (see also (10)).

It is remarkable, that the laser hardened specimens have a high mean stress sensitivity m=0,61 in relation to the base material with m=0,18.

Consequently tensile mean stresses result in a drastic reduction of the laser hardening effect on fatigue strength.

For tensile mean stresses above 300 MPa the fatigue limit of the laser hardened specimens can be expected to be even below that one of base material.

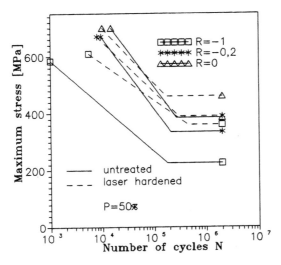

Fig. 4: Wöhler curves (P=50%); R=-1, R=-0,2, R=0

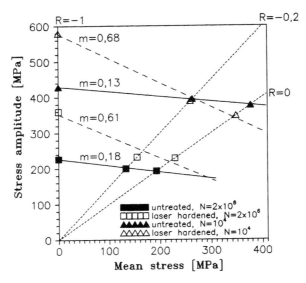

Fig. 5: Haigh diagram for fatigue limit ($N=2\times10^6$) and fatigue strength of finite life ($N=1\times10^4$)

The high mean stress sensitivity for the laser hardened condition remains also in the range of finite endurance, whereas for the base material m decreases slightly. This is also the reason for the intersection of Wöhler curves (Fig. 4).
Calculation of m by the local concept (9) gives with m=0,26 for the base material and m=0,54 for the laser hardened condition values with a right tendency. In contrast to this the use of the well known Goodman relation $R_D=R_W-(R_W/R_m)S_m$ (m=R_W/R_m, R_m - tensile strength) results in dangerous underestimation of the negative influence of tensile mean stresses on fatigue limit in surface hardened specimens (m=0,2).

Conclusions

Laser hardening of notched specimens from X20Cr13 steel provides an 60 % increase of reversed tension-compression fatigue limit.
For symmetrical cycling considerable improvement was obtained also for the finite life fatigue strength. This offers the opportunity to design laser hardened structural components for finite life.
The laser hardened specimens exhibit a high mean stress sensitivity, so that with increasing tensile mean stress the positive effect of laser hardening gets lost.
Therefore the applicability of laser hardening for surface modification of cyclic loaded structural components subjected to high tensile mean stresses is restricted.

References

(1)   H. B. Singh, S. M. Copley, M. Bass: Metallurgical Transactions (1981) 138.
(2)   W. Amende: 2. Europ. Konferenz über Laser-Materialbearbeitung ECLAT'88, Bad Nauheim 1988, 86.
(3)   B. Winderlich, B. Brenner: Härterei-Technische Mitteilungen 44 (1989) 166.
(4)   S. Schädlich, B. Winderlich, B. Brenner: Laser Technologie und Anwendungen, Jahrbuch, 2. Ausgabe, Vulkan-Verlag Essen 1990, 283.
(5)   T. Ericsson, R. Lin: MAT-TEC 91, Ed.: L. Vincent, IITT-International 1991, Gournay sur Marne, France, 255.
(6)   B. Winderlich, S. Schädlich, B. Brenner: DVM - Betriebsfestigkeit "Moderne Fertigungstechnologien" - 1991, Schaffhausen (Schweiz), 197.
(7)   W.-W. Maennig: Materialprüfung 19 (1977) 280.
(8)   D. Dengel: Zeitschrift für Werkstofftechnik 6 (1975) 253.
(9)   B. Winderlich: Materialwissenschaft und Werkstofftechnik 21 (1990) 378.
(10) B. Brenner, et al.: in this volume.

Optimization of the laser hardened track geometry
for different kinds of power density distribution

D. Lepski and W. Reitzenstein
Fraunhofer Institute for Material Physics and Surface Engineering,
Dresden, Germany

1. Introduction

In laser transformation hardening the properties of the laser induced temperature field determine the changes of the metallurgical structure in the heat affected zone and of the residual stresses and thus the resulting mechanical properties of the workpiece such as wear resistance and fatigue strength. The
workpiece surface thereby is usually covered by a pattern of hardened tracks
produced by a moving cw laser beam. The depth and width of these tracks have to
be choosen according to the load the workpiece is to be exposed. The track
geometry e. g. may have to ensure that a large local tensile stress is compensated by compressive stresses due to the martensitic transformation. The achievable track depth increases with increasing surface temperature and laser interaction time. However, melting is to be avoided in pure transformation hardening, and the interaction time is closely related to the cooling rate, which
must be larger than the critical cooling rate for the formation of martensite.
Melting temperature and critical cooling rate, therefore, define the maximum
achievable depth of hardening for each material. The track width, on the other
hand, increases continuously with rising laser beam power and may be influenced
within certain limits also by beam shaping. If the area to be hardened is covered by several more or less overlapping parallel tracks, weakened annealing
zones in the overlap regions are observed. Therefore hardening of the whole
area by one single sufficiently wide track may be necessary in some cases.

Optimization of the laser track geometry means determination of those values of
the processing parameters which yield the required track geometry with minimum
expense of time and energy. For a given interaction time the largest track
depth is obtained if the local surface temperature is as close as possible to
the maximum allowed temperature during most of that time. The surface temperature distribution, of course, depends strongly on the beam intensity distribution on the workpiece surface. Power density distributions yielding such optimized temperature fields for a given speed have been derived by Burger [1]. On
the other hand, the track width is a function of the spot shape and may be enhanced by stretching the effective laser spot across its direction of movement.

Therefore, beam shaping by means of integrators and other devices as well as
beam oscillation in one or two dimensions are suitable to improve the original
intensity distributions of technical laser beams, provided that the user is
able to optimize the processing parameters also for the resulting effective
power density distributions in order to take ful advantage of the beam shaping.

2. Computer aided optimization of laser processing

Optimization of the laser track geometry requires the calculation of the relationship between the processing parameters (including the properties of the effective power density distribution on the workpiece surface) and the thermophysical properties of the material on the one hand and selected quantities of the
laser induced temperature field and track geometry on the other hand. This is a
rather time consuming procedure, which has to be performed for many sets of
processing parameters in order to find the best one. Since in practice there is

not enough time for such lengthy computations, we have carried out them once for all and have stored the results in a data base. The stored information may be evaluated by means of the personal computer program GEOPT, which has been described in detail elsewhere [2]. It enables the calculation of optimized processing parameter sets within about one second using a personal computer.

3. Calculation of the laser induced temperature field

The laser induced temperature field in a solid metal can be calculated sufficiently accurate solving the heat diffusion equation by means of the Green's function method. Such calculations have been succesfully performed for some common intensity distributions such as the TEM_{00} beam or beams with uniform intensity over a square or rectangular spot [3], which occur if a beam integrator is used. For more complex power density distributions, as produced e. g. by two-dimensional beam oscillation [4] or by any technical laser, temperature field calculations have been formerly carried out by numerical methods (FEM), which, however, require much more computing time than the simple Green's function method. Therefore, we derive here from Rosenthal's solution for a point source [5], using the superposition principle, a general integral expression of the temperature field valid for any intensity distribution:

effective power density distribution $p(\vec{r}, t)$:

$$p(\vec{r}, t) = P_{abs.} * \delta(z) * f(\xi, \eta) / (\pi * R^2) ,$$

$$\xi \equiv (x - v_L * t) / R , \qquad \eta \equiv y / R , \qquad \iint_{(\infty)} \frac{d\xi \, d\eta}{\pi} f(\xi, \eta) \equiv 1,$$

laser induced temperature field $\vartheta(\vec{r}, t)$:

$$\vartheta(\vec{r}, t) = \frac{P_{abs.}}{2\pi * \lambda_0 * R} * \iint_{(\infty)} \frac{d\xi' d\eta'}{\pi} f(\xi', \eta') * \exp[-2 * \varkappa * (r + \xi - \xi')] / r, \qquad (1)$$

$$r \equiv \sqrt{[\zeta^2 + (\xi - \xi')^2 + (\eta - \eta')^2]}, \qquad \varkappa \equiv v_L * R / (4 * a), \qquad \zeta \equiv z / R.$$

(R: beam radius, $P_{abs.}$: absorbed laser beam power, v_L: laser traverse speed, a: thermal diffusivity).

This is the solution $\vartheta(\vec{r}, t)$ of the linear heat diffusion equation with a temperature independent heat conductivity λ_0. For most materials, however, the dependence of the heat conductivity on the temperature has to be taken into account. This can be done approximately by KIRCHHOFF's transformation, which connects the true temperature T with the solution ϑ of the linear equation:

$$\vartheta = \int_{T_U}^{T} dT' \, \lambda(T') / \lambda_0 \qquad \text{with} \qquad \int_{T_U}^{T_S} dT' \, \lambda(T') \equiv \lambda_0 * (T_S - T_U). \qquad (2)$$

(T_U: ambient temperature, T_S: maximum surface temperature).

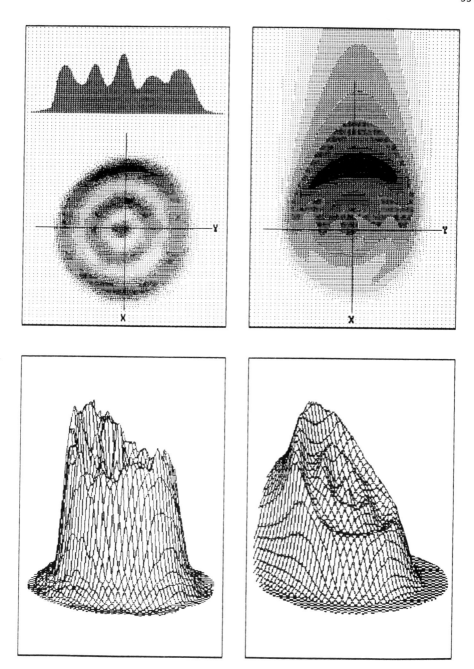

Fig. 1: Plots of the intensity (left) and surface temperature distribution at the Peclet number $\varkappa \equiv v_L * R / 4a = 10$ (right) for a technical laser.

On the basis of the formula (1) a program has been developped which is able to
calculate from the output of a beam diagnostic system the laser induced temperature field for each technical laser beam as a function of the processing parameters and to construct the data base to be evaluated by the program GEOPT.
As an example, Fig. 1 shows contour plots and isometric representations of the
power density distribution of such a technical laser obtained by a beam diagnostic system (left hand side) and the corresponding surface temperature distribution (right hand side) if the beam is moving along the positive X-direction (Peclet number $\varkappa \equiv v_L * R / 4a = 10$). The surface temperature distribution shows the typical ascent during the laser interaction time but also the influence of the double-ring structure of the intensity distribution, which becomes more and more pronounced with increasing speed. Fig. 1 shows the surface
temperature to be close to its maximum value only for a very short time in this
case. Hence a large track depth cannot be expected here. Better results could
be obtained if the maximum intensities were reached at the spot front side.

4. Comparison of different intensity distributions

The transformed volume per unit energy as a measure of beam efficiency in laser
transformation hardening depends strongly on the intensity distributions, most
of which are far away from yielding the optimum surface temperature distribution of a "Burger chair" [1]. The method described here enables a convenient and
rapid comparison of the laser track geometry to be obtained by the different
types of power density distribution. As an example, Fig. 2 shows for the steel
C45 the maximum track depth to be achieved using different laser beams. Thereby
the maximum allowed surface temperature T (solidus temperature) and cooling
time Q (critical cooling time according to the TTT-diagram) have been prescribed. Obviously the TEM_{00}-beam yields the worst result since the slow descenden-
ce of the intensity beside the track causes a strong loss of energy. The technical laser of Fig. 1 and a doughnut beam are lying between the TEM_{00} beam and
the square beam with uniform intensity. As expected, a beam like the "Burger
chair", which has been optimized for a large speed ($\varkappa = 10$), yields the largest
track depth of the compared beams for high values of power and traverse speed.

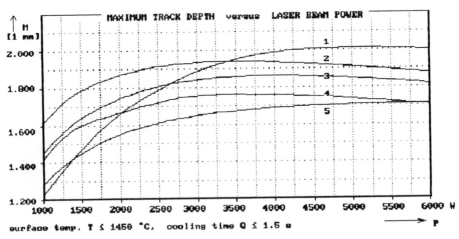

Fig. 2: Maximum achievable hardening depth as a function of laser beam power:
1: "Burger chair" for high speed, 2: uniform intensity (square), 3: doughnut,
4: technical laser of Fig. 1,   5: Gauussian beam (TEM_{00}).

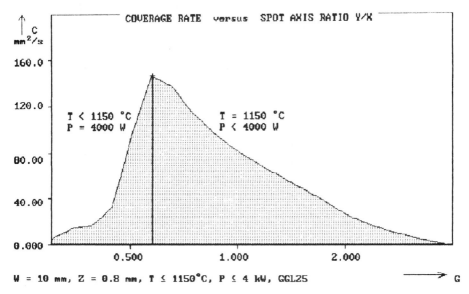

Fig. 3: Coverage rate versus spot axis ratio Y / X for a 10 mm wide and 0.8 mm deep track in transformation hardening of cast iron GGL 25 using a 4 kW laser.

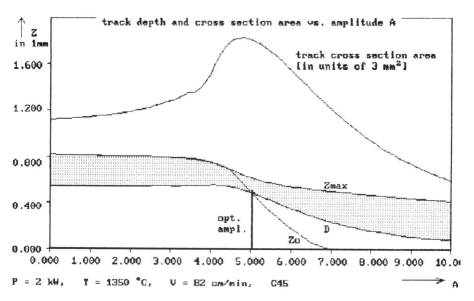

Fig. 4: Maximum (Z_{max}), average (D), and central (Z_0) track depth as well as track cross section area versus scanning amplitude A for prescribed values of power P, speed V, and maximum surface temperature T (steel C45).

5. Beam shaping

From the large number of methods of beam shaping we shall consider here only two examples: (i) beam integrators, which produce an almost uniform intensity over a square or rightangle spot, and (ii) the one-dimensional beam oscillation across the laser track. In both cases the effective spot axis ratio may be varied in order to obtain some prescribed aspect ratio of the track and/or to optimize the transformed volume per unit energy. Besides also the intensity distribution can be improved this way with respect to the original beam.

If, e.g. for reasons of the workpiece geometry, both track width and depth are prescribed, a circular beam will yield suitable results only for a narrow range of aspect ratios, because in most cases either the maximum allowed surface temperature or the available laser power cannot be fully used. This is demonstrated by Fig. 3, where for prescribed values of track depth and width the achievable coverage rate is represented as a function of the spot axis ratio of a rectangular beam. The spot axis ratio yields (beyond power, spot area, and traverse speed) the fourth degree of freedom needed in order to fulfill four conditions (for track depth and width, surface temperature, and power).

One-dimensional beam oscillation distributes the energy concentrated within sharp peaks of the intensity distribution over a larger area and enhances the track width. A harmonic beam oscillation perpendicular to the traverse direction yields an energy density profile across the track with maxima near the turning points of the motion which become larger with increasing oscillation amplitude. For small amplitudes the maximum track depth is achieved in the track center, for larger ones in the vicinity of the turning points. In between there is an optimum value of the amplitude yielding a track of almost uniform depth. About at this amplitude also the track cross section area (and hence the transformed volume per unit energy) reaches its maximum value. As it is visualized for a special example by Fig. 4, this optimum amplitude is characterized by a minimum of the difference between the maximum (Z_{max}) and the average track depth (D) and can be easily determined by means of GEOPT from the condition that the depth in the track center (Z_0) is equal to the average track depth D.

The finite oscillation frequency f (usually f $\gtrsim$ 200 Hz) causes fluctuations of the surface temperature which decay rapidly below the surface. Their depth of penetration is about s $\approx \sqrt{2} * a / f$ ($\approx$ 0.25 mm for steel and 200 Hz). Mostly this is much less than the track depth. Hence the frequency influences strongly the surface temperature and slightly the width, but hardly the depth of the laser track, provided that melting and any damage of the absorption coating may be avoided. The optimum amplitude, therefore, is almost independent on the (not too small) frequency. This is confirmed by our experimental results. Further details on the influence of processing parameters on temperature field and track geometry in the case of beam oscillation are to be published elsewhere.

6. Summary

An integral expression for the laser induced temperature field caused by an arbitrary intensity distribution has been derived. The numerical results, which have been stored in a data base, are to be applied to the optimization of laser hardened tracks by means of the personal computer program GEOPT.

7. References

[1] D. Burger, Dissertation Universität Stuttgart, IFSW (1988).
[2] D. Lepski, W. Reitzenstein, Härterei-Tech. Mitt. 46 (1991) 178.
[3] E. Geissler, H. W. Bergmann, Opto Elektronik Mag. Heft 4 (1988).
[4] Th. Rudlaff, F. Dausinger, Proc. ECLAT'90 (1990) 251.
[5] D. Rosenthal, Trans. ASME 48 (1946) 848.

Characteristic Features of Laser-Induced Hardening
in Maraging Steels

I.L. Yakovleva, T.I. Tabatchikova, Institute of Metal Physics
Ural Division of Russian Academy of Sciences

Introduction

The characteristic features of structural transformations occurring as a result of laser treatment in constructional and tool
steels with various degrees of alloying have been examined in
detail in many publications. The main object of investigation
in this work has been the structure of maraging steels immediately following laser quenching, as well as after additional
ageing. It has been necessary to determine, whether it is possible in machine parts, which have undergone optimum heat treatment, to increase their surface hardness in comparison with
standart heat treatment without lowering the machine part core
properties. The second object has been the investigation of the
recrystallization process in martensite-ageing steels during
laser heating.

Materials

Industrial and laboratory-melted martensite-ageing steels were
used in these experiments. The composition of these steels is
presented in Table 1.

Table 1.

Chemical Composition of Investigated Steels

Steel grade	Content of main elements, wt%					
	C	Kh	Ni	Mo	Co	Ti
OKh11Ni10M2T	0,027	10,82	9,45	1,98	–	0,95
OKh11Ni10M2	0,027	11,60	9,72	2,05	–	–
OKh11Ni9M2	0,030	11,17	8,6	2,59	–	–
ONi18K9M5T	0,063	–	18,0	4,8	9,0	0,7
ONi18K9M5	0,103	–	17,9	5,0	9,0	–

The laser treatment was carried out with the help of a 1kW type
LT-1-2 laser unit. The specimen velocity under the laser beam
was 60 m/h. The metallographic specimens had surface parallel
and perpendicular to the laser-treated surface. A JEM-200 CX
electron microscope was to exemine thin foil specimens cut out
parallel to the laser-treated surface.

Experimental Results and Discussion

The laser-induced recrystallization was studied on specimens of
grade steel. Streaks of metal rapidly heated to the melting
point and than rapidly crystallized appeared on the specimen
surface as a result of laser irradiation. A zone laser-quenched
from the solid state was observed near the remelted zone. The
heating temperature there was somewhat higher than A_{c3},

but lower than the melting point. The zone of short-time ageing, where the temperature was less than A_{c1}, was closer to the initial structure.

It was noted that during recrystallization of martensite-ageing steels the most important result of laser heating of grade OKh11Ni10M2T steel prequenched from 1200°C was that the size and shape of the initial coarse grain were restored. No grain boundary effect was observed. It is significant that with sufficiently rapid laser heating the fine-grained strip adjacent to the remelted zone, the result of austenite recrystallization is eliminated (Fig.1). Thus, it became possible to overheat the restored and non-recrystallized austenite up to the melting point. Naturally, in this case the melted zone crystallizes on cooling on a coarse-grained substrate of restored austenite.

The preageing of quenched grade OKh11Ni10M2T steel at 300-600°C for 3 h had no effect on the process of recrystallization during the subsequent laser heating. That is, the initial austenite coarse grain is restored and there is no streak with recrystallized structure.

Thus, the maraging OKh11Ni10M2T steel shows a tendency to structural heredity and the recrystallization of austenite due to phase cold-work hardening is completely suppressed. It should be noted that recrystallization, a thermally activated process, requires a certain time interval to build up. Apparently, during laser heating recrystallization has no time to develop because of the high speed of laser treatment. It should be stressed that structural heredity was observed on all the investigated steels.

Microhardness in the laser-treated zone was measured in all cases. As a result of laser treatment of quenched specimens there emerges a zone having a microhardness very near to that of the initial quenched state (Fig.2, curve 1). It may be concluded that there is little difference between the hardness of a laser-quenched martensite-ageing steel and one which has undergone ordinary bulk quenching.

If the laser treatment is conducted after quenching and ageing, the hardness in the quenched zone decreases (Fig.2, curve 2). Such a result is to be expected, since martensite-ageing steels have after quenching a lower hardness than after quenching and ageing.

The loss of hardness by the surface layer subjected to laser irradiation has necessitated additional ageing. The effect of ageing temperature on the surface layer hardness was studied. The ageing was conducted in the intensive precipitation - hardening temperature range of 390-580°C. Such a treatment raises considerably the hardness of the zone laser-quenched from the solid state (Fig.2, curve 3). It should be noted that ageing at a temperature lower than 580°C has no effect on the properties of the matrix. The hardness of OKh11Ni10M2T steel following laser quenching and subsequent ageing at all temperatures is much higher than after bulk quenching from 980°C in water with subsequent ageing at 580°C (Fig.3, curve 1 and 2). Since the machine part core is usually heated to temperatures which ensure some over-ageing (560-580°C), the difference in hardness between laser-quenched layer and the core may be much higher (about 130 points). Depending on the heating method one observes a slight difference in the ageing kinetics.

Thus, the hardness maximum in case of bulk quenching lies within the temperature interval 480-500°C, while that for laser quenching is at 460°C. Such measurements of hardness were carried out on all the investigated steels. Table 2 shows maximum hardness values obtained after laser quenching and ageing, and after usual quenching, optimum ageing temperatures for both kindes of heat treatment.

The following regularities are observed for all the investigated steels.

Laser quenching with subsequent ageing ensures a higher value of hardness than one obtained after bulk quenching and ageing. The greatest effect is observed for titanium-containing OKh11Ni10M2T and ONi18K9M5T steels. In case of OKh11Ni10M2, OKh11Ni9M2 and ONi18K9M5 steels which do not contain titanium the hardening increment due to laser treatment and ageing is not significant (only 10-30 points) as compared with the usual heat treatment (see Table 2).

Table 2

Dependence of Maximum Value of Hardness on Ageing and of Optimum Ageing Temperature on the Composition and Method of Quenching of Maraging Steels

Steel grade	Bulk quenching		Laser quenching	
	HVmax	Ta opt	HVmax	Ta opt
OKh11Ni10M2T	460-470	490-500	490-510	460-470
OKh11Ni10M2	360-370	440-450	380-390	400-420
OKh11Ni9M2	370-380	-	360-380	-
ONi18K9M5T	530-540	480-500	580-590	460-470
ONi18K9M5	510-520	430-440	540-560	420-430

Thus, laser quenching has some specific features (as compared to the usual bulk quenching). The quenching method has little effect on the microhardness of hardened steels and, on the contrary, produces a substantial effect on the microhardness of aged steels. Steels containing titanium and laser-quenched on ageing even at lower temperatures become harder than when quenched in the usual way.

Electron Microscopy investigations of the laser-quenched zone structure were carried out on specimens of OKh11Ni10M2T steel. This grade showed maximum hardening effect following laser quenching and ageing. Prior to the laser quenching the steel was subjected to the usual heat treatment - quenching in water from 980°C and ageing at 580°C. The initial structure was composed of alpha-crystals in which were observed particles of Two specimens - χ-phase particles of rounded shape and η-phase fine needle-like particles.

After laser quenching (from the solid state) there was formed a structure consisting of martensite, a small amount of retained austenite and coarse χ-phase particles of the Fe-Cr-Mo type. The presence of η-phase particles having the composition $(Fe,Ni)_{34}Cr_{10}Mo_6Ti_8$ is explained by that they were found in the initial structure prior to laser treatment and had no time to get dissloved in the process of short-time laser heating. The subsequent ageing at 470°C ensuring maximum hardening of the laser-quenched steel (Fig.3) brings about the following changes in the structure.

The particles of the hardening phase decorate the dislocations and create the characteristic contrast in the electron micrographs (Fig.4). The size of the precipitate does not exceed 3,0-5,0 nm, however, its volume fraction is quite large. This makes it possible to observe in the micrographs not only alpha-phase reflections, but also those of the hardening phase. Ahalysis has shown that they pertain to type Ni_3Ti η -phase having an hcp lattice. Thus, with laser quenching with subsequent ageing the hardening is brought about by the precipitation of the Ni_3Ti hardening phase in a very dispersed form. When the ageing temperature is raised to 500°C the η -phase becomes coarser and acquires a needle-like shape. One can also see oval-shaped patches 0.1μm in size (Fig.5). Dark-field analysis has shown that these belong to austenite. Since at a lower ageing temperature (470°C) there was no austenite, one may conclude that the austenite was formed at 500°C and did not dissociate on cooling down to room temperature. This is due to that it is enriched with alloying elements which enhance its stability. Thus, in laser-quenched OKh11Ni10M2T steel it was possible to observe the formation of austenite at 500°C, while under the usual conditions the A_{c1} temperature of OKh11Ni10M2T steel is equal to 560°C.

As has been pointed out above, in case of laser quenching there remain undissolved coarse particles of χ -phase which has a high (as compared to the mean value in steel) content of ferrite-forming elements and an insignificant amount of nickel. For this reason, austenite on heating and martensite after laser quenching have a depleted amount of ferrite-forming elements and, on the contrary, a high content of nickel. This leads to a displacement of point A_{c1} in the direction of lower temperatures. One can expect that with ageing the valume fraction of the Ni_3Ti hardening phase will grow, as it would occur in steel with a high nickel content. The presence of defects in the crystalline structure after high-speed laser quenching facilitates the nucleation of the hardening phase at lower temperatures, since they serve as sites of preferential nucleation and enhance the hardening effect. Ageing in the low temperature range leads to the precipitation of fine η -phase particles, this also having a considerable effect on hardening with ageing.

All this factors explain why OKh11Ni10M2T steel when laser-quenched and aged has a higher hardness, than when subjected to the usual heat treatment.

Laser quenching of maraging steels produced on machine parts surfaces a layer having a hardness close to that of steels which have undergone bulk quenching. The high value of hardness of the surface layer is attained following additional ageing at 460-480°C. The hardness attained after laser quenching and ageing is greater than that of steels subjected to standart heat treatment. Steels containing titanium exhibit the greatest hardening effect after laser quenching and subsequent ageing. The optimum ageing temperature of laser-quenched layer is 20-40°C

lower than that of bulk-quenched steel. This makes it possible,
the core metal properies being the same, to obtain a much higher
surface hardness after laser treatment.

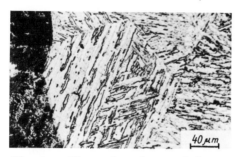

Fig.1 : Microstructure of laser-
treated zone in OKh11Ni10M2T steel
preliminary quenched in water
from 1200°C

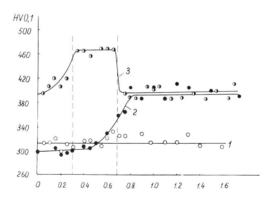

Distance, mm

Fig.2 : Distribution of microhardness values in
laser-treated zones in OKh11Ni10M2T steel after:
1 - double quenching in water from 980°C and
    laser treatment;
2 - double quenching in water from 980°C, 3h -
    ageing at 560°C, laser treatment;
3 - double quenching in water from 980°C, 3h-ageing
    at 560°C, laser treatment, 3h-ageing at 500°C

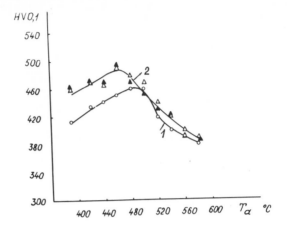

Fig.3 : Ageing temperature dependence of microhardness in OKh11Ni10M2T steel prequenched twice in water from 980°C(1) and by laser treatment (2).

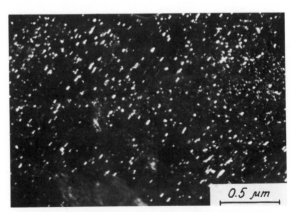

Fig.4 : Structure of OKh11Ni10M2T steel after quenching in water from 980°C, 3h-ageing at 560°C, laser quenching and 3h-ageing at 500°C. Dark-field image in Ni_3Ti reflection.

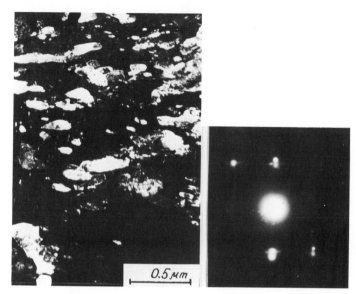

Fig.5 : Structure of 0Kh11Ni10M2T steel
after laser quenching and 3h - ageing
at 500°C. Dark-field image in austenite
(200) reflection.

Laser material processing with 18kW using a variable beam profile achieved with a deformable optic

H.W. Bergmann, T. Endres, R. Anders, B. Freisleben[*], D. Müller

Forschungsverbund Lasertechnologie Erlangen (FLE)
Lehrstuhl Werkstoffwissenschaften 2, Metalle
Universität Erlangen-Nürnberg
Martensstr. 5, W-8520 Erlangen

[*] Fa. Diehl GmbH & Co
Fischbachstr.16
W-8505 Röthenbach a.d. Pegnitz

1. Abstract

Laser solid state transformation hardening and surface remelting is well established in application investigations especially due to localized heating patterns. At present treatments have been restricted by limitation in availability of suitable optics. Complex geometries or variable component design require specific intensity profiles to obtain uniform and flexible processing results.

A possibility to influence the beam intensity profile are mirrors with a surface deformable by piezocrystals. The presented paper demonstrates the capabilities of such an optic, manufactured by Diehl GmbH & Co, Röthenbach. The evaluation includes HeNe-images, intensity profiles obtained with a high power CO_2-laser beam and microstructures of treated steel samples.

2. Introduction

The surface treatment of metals using CO_2-lasers is well established in research activities but it is limited in its industrial applications up to now. The potential of this technique is local treatment of components creating a wide variety of high quality surface properties. Therefore it is possible to increase the overall performance of components. If the laser surface treatment allows using inexpensive material or less complex production techniques for the rest of the component, a reduction of the costs can be achieved. In industrial applications the time needed for the laser surface treatment is a critical aspect. When using CO_2-laser with high output power, larger surface areas can be processed within a definite time which means increasing the processing speed. At surface treatments where large beamsizes are used, it is desireable to adapt the beamprofile and the geometry of the beam to the requirements of the process and the component to be treated. For this purpose different types of optics have been developed /1/ /2/.

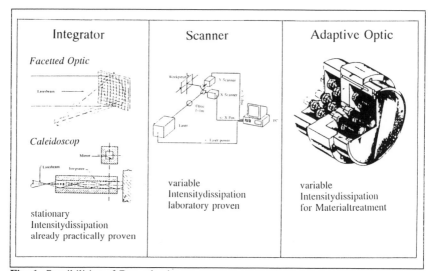

Fig. 1 Possibilities of Beamshaping

Integrating mirrors create a rectangular beam of constant size at a given working distance. This is realized in facetted mirrors or the caleidoscop optics. Scanning optics have very good capabilities in beamshaping and can create variable beam sizes and profiles /3/. These scanning systems are limited to lasers with a beampower up to 5 kW. Adaptive optics with flat mirror surfaces being deformed for example by piezocrystals combine extensive beamshaping performance with the capability of high power CO_2-laser application /4/.

3. Experimental Setup

Subject of the investigations done at the Forschungsverbund Lasertechnologie Erlangen (FLE) is the 19-channel adaptive mirror developed and manufactured by the company Diehl GmbH & Co, Röthenbach. This mirror uses 19 PC-controlled piezocrystals to deform a watercooled plane copperplate. The mirror is integrated into a 5-axis Messer-Griesheim beam handling system. For imaging purposes it is necessary to set up an additional spherical mirror behind the adaptive optics. This results in a working distance of 2 to 4 metres from the adaptive mirror to the workpiece. For the evaluation of the beamshaping

Fig. 2 Adaptive Mirror

photography was applicated using a HeNe-laser. When using a high power 18kW CO_2-laser (UTIL), results can be observed by burn prints or achieved processing results. Thermo-camera imaging is helpful to ensure the quality of the intensity profiles which have been created.

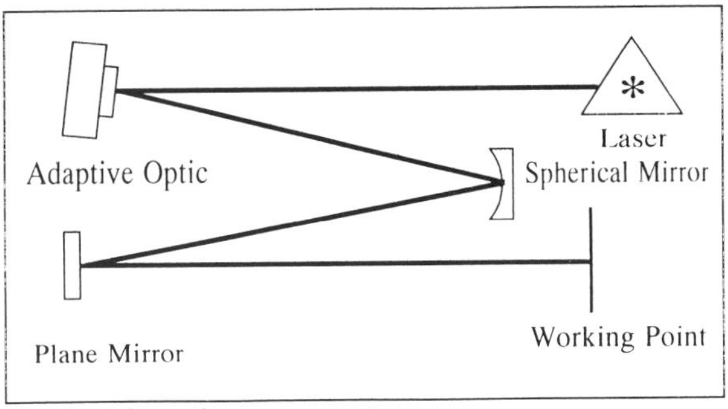

Fig. 3 Schematic arrangement

Adaptive Optic

Plane Mirror

Laser
Spherical Mirror

Working Point

4. Results

The combined use of the adaptive mirror with the spherical mirror (with a focuslength of 2,5 metres) opens a wide variety for the creation of beamprofiles. This is possible with the HeNe-laser as well as with the CO_2-laserbeam of the high power laser with unstable resonator. The optical setup is mechanically stable and the cooling capabilities are sufficient to withstand laser powers up to 20kW. The positioning of the piezocrystals is reliable and reproducible.

Using the same raw beam the adaptive mirror creates completely different beamshapes by changing the surface topography in the range of some 10 μm. Figure 4 demonstrates three

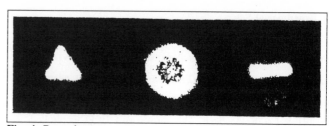

Fig. 4 Beamshapes

different beam-profiles using the identical gaussian HeNe-laser raw beam. The time which is necessary to switch between these profiles is only limited by the speed of the controlling system and the mechanical indolency of the mirror movement /5/. These profiles appear at the identical distance from the adaptive mirror. Therefore is not necessary to adapt the positon of the handling system relative to the workpiece. By adjusting the piezos slightly the size of the image changes in some cases, but the geometry of the intensity profile does not differ. In this way it is possible to change the dimension of a choosen intensity profile at a given working distance. On the other hand it is possible to create an identical beamsize at different positons of the beampath. So the working distance can be changed electronically while preserving the beamqualities in an improved way. In principle there is the capability of compensating different lenghts in the beampath of the beam handling system. It is therefore possible to install the adaptive mirror between laser and beam handling system. This combines the advantage of a safe mirror position

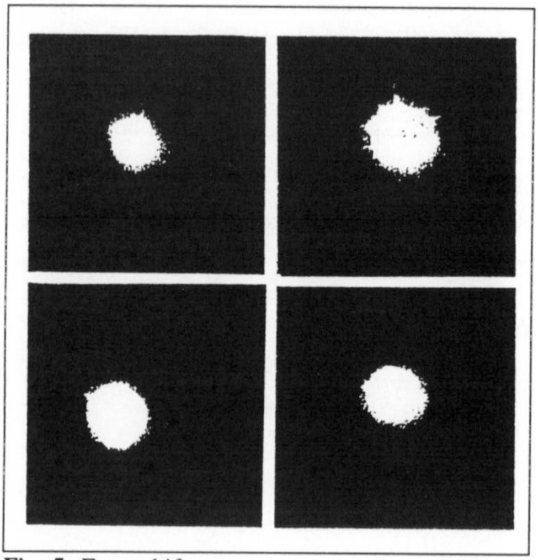

Fig. 5 Focusshift

from the processing point with the capabilities of influencing the beam. The evaluation of the intensity profiles was done with the HeNe-laser in the first place, followed by exposure with CO_2-laser radiation if necessary. Figure 6 demonstrates the correlation between the photography of a HeNe-laser image and the processing results of a cast iron sample. It is possible with both wavelenghts to create a rectangular beamprofile. Adjusting the surface of the mirror the lenght of the rectangle can be changed, leaving the width uninfluenced. Processing in the direction of the image width, the size of the processed track changes, but the interaction time remains the same at a constant feedrate. Processing in direction of the length the interaction time can be changed by the position of the piezocrystals.

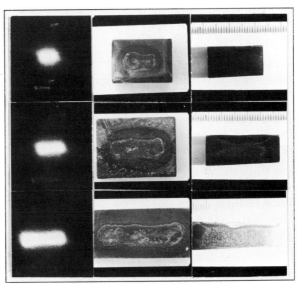

Fig. 6 Comparison between HeNe-Laserimage and CO_2-laser burn prints

5. Conclusion

The 19-channel adaptive mirror is capable of creating a wide variety of different intensity profiles using a high power laserbeam. Power levels up to 20kW do not cause mechanical or thermal problems. Technical relevant profiles like rings or rectangles are achieved. The adaptive optics has the potential for adjusting profiles electronically controlled to the requirements of the process. It is necessary to evaluate the influence of technical relevant parameters like the quality of the raw beam of the CO_2-laser, variation of the incident angle and the adjustment of the laserbeam relative to the adaptiv optics in detail. This will be subject to further investigations.

6. References

/1/ H. Hügel:
Strahlwerkzeug Laser
Teubner Studienbücher; B.G. Teubner, Stuttgart 1992

/2/ W.P.O. Jüptner, R. Becker, G. Sepolt:
Surface Treatments with High Power CO_2-Lasers Using Different Mirror Optics
SPIE, Volume 957 (1988) p. 82-85

/3/ Th. Rudlaff, F. Dausinger:
Hardening with variable intensity distribution
Eclat'90, Erlangen, Germany,17-19.Sept. 1990, Volume 1, p 251

/4/ R.K. Tyson :
Principles of Adaptive Optics
Academic Press, Inc ; Chapter 6, pages 185 - 212

/5/ V.V. Apollonov, S.A. Chetkin :
New High-Precision Displacement Devices Based Magnetostriction Effects for Adaptive
Optical Elements
Contributed Paper at the Ernst Abbe Conference, Jena, GDR

Laser Cladding of Paste Bound Carbides

E. Lugscheider, H. Bolender, Material Science Institute,
Technical University of Aachen, Aachen, FRG
H. Krappitz, Degussa AG, Hanau, FRG

Introduction

The process of surface cladding by laser beams offers a lot of new and interesting possibilities in a wide field of technical applications when e.g. wear and corrosion protection is needed (1-3). A new method is the laser cladding by an one-step-process, using paste-bound hardfacing materials (4).

The advantage of this process is the possibility of an exact positioning of the hardfacing alloy shortly in front of the laser beam, to receive defined melting conditions of the added material. Furthermore this process offers the possibility of prealloying the added hardfacing material by further elements (metals or metalloids) or by hard particles (e.g. carbides). By binding the material into a paste the decomposition of the blended hardfacing alloy is reduced , compared to the one-step laser-cladding process with powder injection.

Principle of laser processing paste bound materials

In the one-step-process with paste bound materials the material to be clad is fed to the place where the laser radiation interacts with the surface. This is accomplished by preplacing the paste on the substrate, directly in front of the laser interaction zone (Fig. 1).

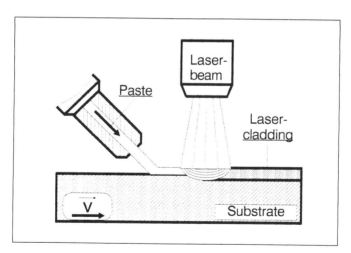

Fig. 1: Scheme of the one-step laser cladding process with paste bound cladding material

The substrate traverse conveys the paste to the laser radiation which melts the material continuously to produce a cladding layer on the surface of the substrate. By overlapping several single tracks, homogeneous hardfacings can be produced.
The rapid heating of the paste up to the melting temperature of the added material and the thermal emission of the heated sample demands optimisation of the paste, the feeding system and the laser cladding process.
The paste mainly consists of the hardfacing powder and a suitable binder. The binder must be able to dry within a short time while keeping the hardfacing material in a compact form, otherwise the powder particles are blown away by the shielding gas. The heating process must allow the evaporation of the binder and its decomposition products.

Materials
In these investigations hardfacings, made of nickel coated titanium carbide and cobalt coated tungsten carbide, additionally blended 20 wt.-% nickel or cobalt based hardalloys (Tab. 1 and 2), were produced by this one step laser cladding process.

Carbide	Matrix-alloy I	Matrix-alloy II
TiC/Ni 80/20	20 % Co-Cr-W-C	20 % Ni-Cr-B-Si (60HRC)
WC/Co 80/20	20 % Co-Cr-W-C	20 % Ni-Cr-B-Si (60HRC)

Tab. 1: Investigated carbide/hardalloy compositions [m%]

Added matrix material	B [%]	C [%]	Cr [%]	Fe [%]	Si [%]	W [%]	Co [%]	Ni [%]
NiCrBSi	3.0	0.75	13.5	4.75	4.25	--	--	bal.
CoCrWC	0.1	1.15	27.7	<3	1.15	4.5	bal.	<3

Tab. 2 : Chemical composition of the added matrix materials

Experimental
The laser cladding tests and the laser remelting of thermal sprayed coatings have been performed at the Fraunhofer Institut für Lasertechnik, Aachen, using a Heareus C76 - CO_2- Laser and at the Fraunhofer Institut für Produktionstechnologie, Aachen, using a Rofin Sinar RS 5000 CO_2-Laser.
The best results were achieved using the Heareus laser with a line optic (150 mm focus), a power output of 2 kW and a traverse speed between 200 and 300 mm/min.
After the laser treatment the coatings were investigated concerning their hardness, abrasive wear resistance and their microstructure. The behaviour against abrasive wear was tested in a

pin-on-disk test, using SiC paper (400 mesh) as abrasive. The specimens were superposed by a load of 0.05 MNm^{-2}. The results are compared to those of plasma sprayed hardfacing coatings of the same composition, in the as sprayed and in the laser remelted condition.

Metallurgical aspects

The coatings made of TiC/Ni and WC/CO with 20% NiCrBSi as additional matrix material is shown in Fig. 2. A good metallurgical bonding to the substrate was achieved, and the original TiC powder particles with randomly distributed smaller TiC particles are visible in the coating (Fig. 2a). Track heights between 400 and 650 microns were reached with pastes containing TiC and between 300 and 620 with those containing WC.

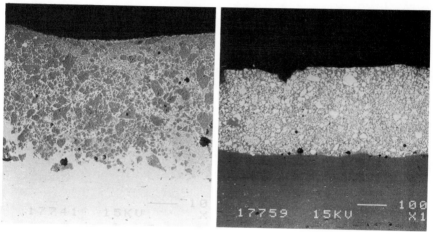

a)   b) $100\mu m$ ⊢——⊣

<u>Fig. 2:</u> Cross section of TiC and WC claddings
       a) TiC/Ni 80/20, b) WC/Co +20 % Ni-Cr-B-Si

A greater amount of TiC was dissolved in the melting during the cladding process and precipitated on cooling. The influence of the different matrix material is shown in Fig. 3 a-c). Claddings made of TiC/Ni and TiC/Ni with NiCrBSi showed higher precipitation rates of TiC (Fig. 3a,3b) than those with CoCrWC as additional matrix material (Fig. 3c).

As the melting point of CoCrWC is about 120°K above the one of NiCrBSi, this matrix material had a significant influence on the dilution of the cladding with the substrate and also on the formation of hard phases. A dilution of 22 % was measured in the shown micrograph, which enriched the matrix to 63 at.-% Fe and inhibited the precipitation of further TiC phases. The use of a matrix material, containing nickel, leads to an improvement of the coating quality if carbide precipitation and matrix hardness is taken into consideration.

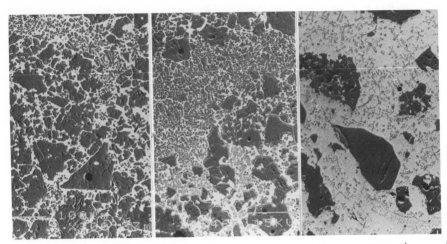

a)   b)   c)   50μm ├──────┤
<u>Fig. 3:</u> Micrograph of the different TiC / matrix compositions,
        a) TiC/Ni 80/20, b) TiC/Ni +20% NiCrBSi, c) TiC/Ni +20 %
        CoCrWC

Fig. 4 shows the micrographs of the laser cladded WC/Co 80/20
and WC/Co +20% NiCrBSi layers with an homogeneous distribution of
tungsten carbide in the cross section. The formation of pure
tungsten carbide within the tungsten enriched matrix is achieved
using the cobalt coated WC powders (Fig. 4a).

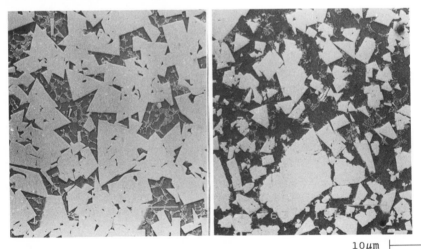

                                                      10μm ├──────┤
    a) WC/Co 80/20   b) WC/Co 80/20 +20 % NiCrBSi
<u>Fig. 4:</u> Microstructure of laser cladded WC/Co 80/20 coatings with
        different matrix material

In the darker phases 85 at.-% cobalt were detected, the tungsten/
cobalt ratio in the eutectic phases was measured with 64/36 at.-%
with a much higher carbon content. Fig. 4b clarifies the solution
of tungsten carbides with their later precipitation by the formation of pure tungsten carbide as well as complex tungsten carbides as matrix, when using NiCrBSi as additional matrix material. The matrix material is enriched with 14 at.-% tungsten in the
eutectic and 6.2 at.-% tungsten in the darker phases.

Fig. 5 shows the matrix hardness of the different carbide/matrix
combinations produced by this one step cladding process. The
WC/Co claddings reached the highest hardness values between 1000
and 1200 HV 0.1. It was decreased by the addition of NiCrBSi to
an average of 900 HV and to 800 HV, when CoCrWC was used. TiC and
its matrix materials tended similar, but CoCrWC decreased the
hardness between 400 and 600 HV 0.1.

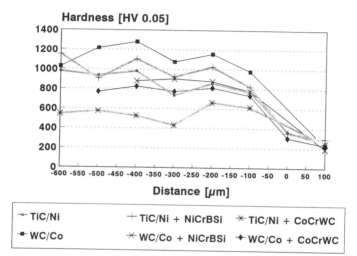

Fig. 5: Matrix hardness of the different carbide claddings

The formation of pores was one of the significant problems when
processing the paste bound materials. This phenomenon also occurs
during the laser remelting of plasma sprayed coatings and the one
step powder injection process; the reason for the formation of
pores is not jet clarified. The evaporation of the binder material is only one aspect. Crack formation occurred but was not identified as the outstanding problem.

Results of wear tests
The laser cladded and remelted hardalloy/carbide coatings were
tested concerning their abrasive wear resistance. The results
were standardized on plasma sprayed coatings which were used as
reference specimens (Fig. 6). The resistance against abrasive
wear was better than the one of the untreated thermal sprayed
conditions. For both carbide materials the addition of Co-Cr-W-C

374

reduced the wear resistance significantly by the formation of complex carbides with a lower hardness.

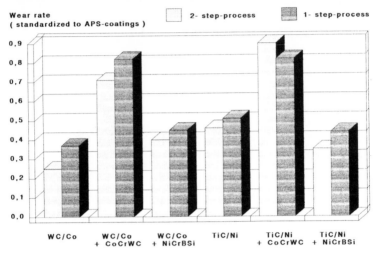

Fig. 6: Wear rates of the laser cladded and remelted coatings

A break out of bigger carbide particles is obviously responsible for the higher wear rates of the one step claddings, better results might be achieved using finer powder particles as cladding material.

Acknowledgement
Special thanks are directed to the German Ministry for Research and Technologie (BMFT) for financial support (No. 13 N 5592), to the company Degussa AG, Hanau for the production of pastes, to the ELGA GmbH, St. Ingbert for support with the production of the paste feeding system and to both the Fraunhofer Institut für Produktionstechnologie (FhG-IPT Aachen) and the Fraunhofer Institut für Lasertechnologie (FhG-ILT Aachen) for assisting with the laser processing.

References
(1) König, W. u.a., Oberflächenveredeln mit Laserstrahlung, Laser u. Optoelektronik, 2 (1988), 74-77
(2) E.Lugscheider, B.C.Oberländer und H. Meinhardt, Laser cladding and surfacing remelting of nickel-base alloys, Conf.Proceed. ECLAT '90, Ed. H.W. Bergmann, 1990, S. 555-568
(3) E.Lugscheider, H.Bolender, Laser cladding for wear and corrosion protection, Conf.Proceed. ECLAT '90, Ed. H.W. Bergmann, 1990, S. 111-121
(4) E.Lugscheider, H. Bolender und H. Krappitz, Laser Cladding of Paste Bound Hardfacing Alloys, Surface Engeneering 1991, Vol.7, No.4, S.341-344

Power absorption during the laser cladding process

C.F. Marsden, A. Frenk and J.-D. Wagnière

Swiss Federal Institute of Technology Lausanne
Centre de Traitement des Matériaux par Laser
1015 Lausanne, Switzerland

Introduction

Laser cladding by the blown powder method [1] involves injecting powder, of the alloy to be deposited, into a shallow molten pool produced by the laser on the substrate surface. Only powder impinging upon the molten pool will adhere to the substrate, generating a build up of dense material; powder particles striking solid material will ricochet and be lost. The substrate is then displaced relative to the laser and a single track clad is formed. In order to cover wider surfaces, it is necessary to lay down partially overlapping tracks. The inter-track advance between them is defined as the "step". Under correct processing conditions laser cladding is capable of producing dense, low dilution coatings, metallurgically bonded to their substrates.

Cladding is similar to other surface treatments using high power CO_2 lasers, in that the absorption and therefore the energy entering the treated specimen is unknown and changes with both substrate and surface finish. During the cladding process, the absorbed laser energy is used to melt part of the preceding track, in order to ensure a metallurgical bond between one track and the next, as well as the incoming powder and a thin layer of the substrate. If the quantity of powder adhering to the substrate is insufficient, with respect to the energy being absorbed, large quantities of the substrate will melt, leading to large values of dilution. In contrast, if insufficient energy is absorbed, then there is the risk of not generating a fusion bond between the substrate and the clad. The contribution made by the powder to the absorption process is unknown but is certainly dependant upon the powder feed rate, the powder efficiency and the absorption of the powder. It is therefore extremely difficult to make predictions about the relative influences of changing the processing parameters on a clad (in terms of its height and dilution) because of the complex inter-relationships existing between the processing parameters, the powder efficiency and the absorption. Thus, a great deal of empirical knowledge is required to produce low dilution clads of given heights for a particular materials couple. Even when using "computer aided laser cladding", Li et al [2] found it necessary to have a large data bank of previously acquired knowledge.

In this paper, expressions are presented for the power absorption during the laser cladding process and the attenuation of the incident power by the powder stream. Absorption experiments were carried out using a Stellite 6 powder and their results are discussed in the light of these expressions.

Factors related to the global power absorption

The global power absorption may be defined as the ratio of the total power absorbed during the process by the treated specimen to the power delivered by the laser. The total power absorbed is the sum of the power absorbed by the clad/substrate interaction zone (defined as the region of clad and substrate material illuminated by the laser), plus the power absorbed during flight by those powder particles that adhere to the substrate. For a given set of processing conditions, the global absorption β [-], may be written as :

$$\beta = \frac{\beta_{cs} \cdot (P - P_{at})}{P} + \eta \cdot \beta_p \left\{ \frac{P_{at} + (1 - \beta_{cs}) \cdot (P - P_{at}) \cdot (\frac{P_{at}}{P})}{P} \right\} \tag{1}$$

where P [W] is the laser power delivered and P_{at} [W] is the power that does not reach the substrate due to attenuation by the powder stream, while β_{cs} [-] is the absorption of the clad/substrate interaction zone and β_p [-] is the absorption of the powder. The powder efficiency, η [-], is defined as the ratio of the powder that adheres to the substrate to the powder delivered.

The first term on the right hand side of Equ.(1) refers to the power absorbed at the clad/substrate interaction zone. Before the laser energy reaches the interaction zone, it must first pass through the powder stream. As the powder is opaque, a certain proportion of the laser energy is absorbed and scattered. This results in a lower incident power, $P - P_{at}$, at the the clad/substrate interaction zone. The second term on the right hand side of Equ.(1) refers to the power absorbed, by those powder particles that adhere to the substrate, from both the incident beam and the power reflected from the molten pool. It should be noted that if the back reflection is not taken into account, then the value of β cannot exceed the larger of the two values, β_{cs} and β_p.

In order to quantify the power attenuation, it is necessary to know the volumetric fraction of powder in the powder stream, f_p [-], which is defined by:

$$f_p = \frac{\dot{m}_p}{\rho_p \cdot V_p \cdot S} \tag{2}$$

where $\dot{m}_p$ [kg/s] is the powder feed rate, ρ_p [kg/m^3] is the powder density, V_p [m/s] is the powder velocity and S [m^2] the cross sectional area of the powder stream.

The volume of powder stream in the laser beam can be determined by simple geometry, assuming that the laser beam and powder stream are two intersecting cylinders. If the shadowing of one powder particle in the laser beam by another is negligible (reasonable for the experimental conditions investigated, which resulted in volumetric fractions of powder of less than 1%), the power attenuated by the powder stream, P_{at}, due to the projected area of the particles within this volume, may be given for the two possible conditions as follows :

When the diameter of the powder stream, $\emptyset_p$ [m], is smaller than the diameter of the laser beam, $\emptyset_b$ [m], :

$$P_{at} \approx \frac{3\dot{m}_p \cdot P}{\pi \cdot r \cdot \emptyset_b \cdot \rho_p \cdot V_p \cdot \cos\theta} \tag{4}$$

When the diameter of the laser beam is smaller than that of the powder stream :

$$P_{at} \approx \frac{3\dot{m}_p \cdot P}{\pi \cdot r \cdot \emptyset_p \cdot \rho_p \cdot V_p \cdot \cos\theta} \tag{5}$$

where r [m] is the radius of the powder particles and θ [°] is the injection angle with respect to the horizontal. It is interesting to note that the two equations indicate that the power attenuated is proportional to both the laser power delivered and the powder feed rate, and inversely proportional to the radius and speed of the powder particles. When the diameter of the powder jet is smaller than that of the laser beam, the value of P_{at} is inversely proportional to the laser beam diameter, Equ.(4). In the opposite case, P_{at} is inversely proportional to the diameter of the powder jet, Equ.(5).

The attenuated power was determined for a Stellite 6 powder (30-60 μm) by comparing the power that passed through the powder stream with the laser power delivered. The experimental set up was similar to that used during cladding, as described in [4] with θ equal to 55°, except that no substrate was used. The exit diameter of the nozzle was 2 mm, but due to divergence of the powder stream, its diameter was measured to be 4 mm at the point of intersection with the laser beam. Powder feed rates between 0.05 and 0.4 g/s were used, with laser beam diameters between 1 and 5 mm and delivered powers between 400 and 1400 W. The power measurements were carried out using a commercially available calorimeter, giving a relative precision of ±5%. It was placed under the beam, well below the intersection point of the laser beam and powder stream. Figures 1a and 1b show the power attenuation, expressed as a percentage of the power delivered, as a function of the powder feed rate. Figure 1a shows the effect of varying the power delivered for a fixed beam diameter of 2 mm, and Figure 1b the effect of varying the beam diameter for a fixed power.

Both Figures show that the power attenuation increases linearly with the powder feed rate and is independent of the power delivered. The best fit of the data points shown in Figure 1a using Equ.(5) (Equ.(5) because the beam diameter was smaller than that of the powder stream), with a mean particle diameter of 45 μm and a value of 8400 Kg/m^3 for the density of Stellite 6 [3], led to a particle speed of 2.5 m/s. This speed is considered reasonable as it is lower than that of the transport gas and similar to values reported elsewhere [1]. The calculated power attenuation using this value is also shown in Figure 1a.

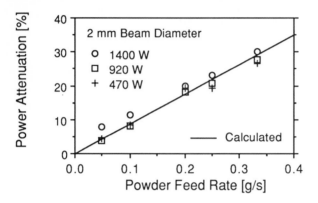

Figure 1a: Power Attenuation against Powder Feed Rate for different powers.

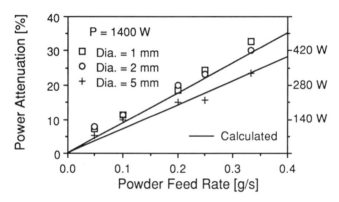

Figure 1b: Power Attenuation against Powder Feed Rate for different beam diameters.

Figure 1b shows how the power attenuation is independent of the beam size when the diameter of the powder stream is greater than that of the beam. However, the power attenuated decreases when the diameter of the laser beam is larger than that of the powder stream. The predicted power attenuations, when using a mean particle diameter of 45 µm and a particle velocity of 2.5 m/s in Equ.(4) and (5), are also plotted. The good correlation between the predicted and measured values, indicates that the assumptions and estimations made earlier seem reasonable.

However, during the actual cladding process no part of the laser beam traverses the entire powder stream before coming into contact with the molten pool. Therefore the results presented above, as well as the predictions, are over-estimations of the power attenuations occurring during cladding.

In order to determine the global absorption during the cladding process, a series of clads was produced on substrates of St 37 using the Stellite 6 powder mentioned above. The measurements were made using a calorimetric technique [5]. The only processing parameter altered was the powder feed rate, varied between 0.08 and 0.14 g/s (values towards the lower end of the range over which P_{at} was determine); the power, beam diameter, scanning speed and step remained constant at 1400 W, 2 mm, 500 mm/min and 0.66 mm respectively. The powder efficiencies, determined by weighing the specimen before and after cladding, increased from 39 to 51% within the range of feed rates investigated. The surface temperatures of the molten pools, measured by optical pyrometry, were all the same at approximately 1760°C. All the clads produced had less than 5% dilution with the maximum recorded depth melted into the substrate being 30 µm. As can be seen Figure 2, the global absorption increases with the powder feed rate. For powder feed rates above approximately 0.2 g/s, it was observed that no clad was produced. This occurred because the power attenuation by the powder stream became so great that the direct laser energy incident upon the substrate was insufficient to generate a molten pool and thus initiate the cladding process.

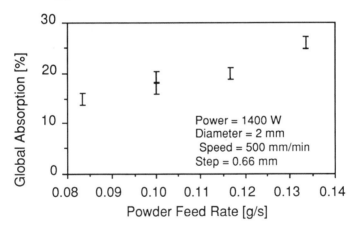

Figure 2: Global absorption as a function of powder feed rate for Stellite 6.

From Equ.(1) it can be seen that for the global absorption to increase with powder feed rate, the decrease in the direct laser power absorbed at the clad/substrate interaction zone must be more than compensated for by the extra power delivered by the powder adhering to the substrate. For this to occur, assuming β_p and β_{cs} remain constant, the value of β_p has to be much greater than that of β_{cs}. This is considered unlikely as the absorption of 10.6 µm radiation by metallic materials is known to be lower in the solid state than in the liquid [5-7]. However, as there is no reason why the value of β_p should change with powder feed rate, it is proposed that β_{cs} must have varied. Values of β_{cs} were calculated using Equ.(1), with the measured values of the global absorption and powder efficiency, together with the attenuated powers as predicted by Equ.(5) and a fixed value of β_p. It was assumed that the powder did not melt before arriving at the molten pool and had an absorption similar to that of the substrate material in the solid state; a value for β_p of 8% was found in the literature [5]. Values of β_{cs} between 16 and 28% were obtained, which being greater that the absorption of the substrate in the solid state, are considered reasonable. The proposed increase in β_{cs}, with powder feed rate, is interpreted as being due to an increase in the projected area of the molten pool with respect to the laser beam.

Conclusions

During laser cladding, part of the delivered power is attenuated by the powder stream. Expressions have been derived indicating that this attenuation is proportional to the powder feed rate and delivered power, and inversely proportional to the particle velocity and powder size. The first two points were verified experimentally.

The power absorbed has been expressed as the sum of two separate contributions; the power absorbed at the clad/substrate interaction zone and the power absorbed by the powder that adheres to the substrate, thus forming the clad. Experimental results showed that the global absorption increased with powder feed rate. These results appeared to be related to an increase in the absorption of the clad/substrate interaction zone with the powder feed rate, and were interpreted as being due to an increase in the projected area of the molten pool with respect to the laser beam.

Acknowledgements

The authors would like to thank M. Picasso and M. Rappaz for helpful and stimulating comments, together with the "Commission pour l'Encouragement de la Recherche Scientifique", Berne and Sulzer Innotec, Winterthur for financial support.

References

[1] Weerasinghe V.M., Steen W.M., *Lasers in materials processing*, ed. Metzbower, E.A., American Society of Metals, Ohio, 1983, pp166-174.
[2] Li L., Steen W.M., Hibberd R.B., *ECLAT'90*, ed. Bergmann H.W. and Kupfer. R., Erlangen, Germany, Sept 17-19, 1990, pp355-369.
[3] Haynes Stellite 6 catalogue, 1975, pp4.
[4] Marsden C.F., Frenk A., Wagnière J.-D., Dekumbis R., *ECLAT'90*, ed. Bergmann H.W. and Kupfer. R., Erlangen, Germany, Sept 17-19, 1990, pp535-542.
[5] Frenk A., Hoadley A.F.A.,Wagnière J.-D., Metall. Trans: B. Vol 22B, Feb 1991, pp139-141.
[6] Dekumbis R., Frenk A., *ECLAT'88*, Bad Nauheim, Germany, Oct. 13-14, 1988, pp134-137.
[7] Brückner M., Schaefer J.H., Uhlenbusch J., J. Appl. Phys., 66, 1989, pp1326-1332

Laser Cladding of Ni-Al Bronze on Cast Al Alloy AA333

Y. Liu and J. Mazumder, Center for Laser Aided Material Processing , Department of Mechanical and Industrial Engineering, University of Illinois at Urbana-Champaign, 1206 W. Green St., Urbana, Illinois, 61801, USA
K. Shibata, Nissan Motor Co. LTD., Nissan Research Center, 1, Natsushima-cho, Yokosuka, 237 JAPAN

Introduction

Surface treatment by energetic beams opens a new category of processing opportunities in composite material formation. Laser cladding is a process where a molten pool on the surface of a substrate metal is generated by laser interaction with simultaneous injection of a powder stream. The solidified clad track generally has very close composition to the powder used with fine microstructure resulting from high cooling rate. The application of laser cladding in material processing can be considered in two aspects. First, the effect of high cooling rate associated with laser cladding may be used to produce extremely fine microstructure with extended solubility of some desired elements (1,2). Second, the ability to metallurgically combine two kinds of materials possessing substantially different properties without limitation of equilibrium phase diagrams. For example, the wear resistance of Fe-Cr-Mn-C alloys can be improved by laser cladding (3). In either case, small dilution from the substrate is desirable. Ni-Al bronze has good wear resistance and thermal conductivity while Al alloy has low density. Laser cladding of Ni-Al bronze on Al alloys can extend application of Al alloys as mechanical components. The present paper summarizes the results of laser cladding Ni-Al bronze on cast Al alloy AA333 including processing and structure characterization.

Experimental Procedure

An AVCO HPL 10 kW continuous wave CO_2 laser was used as the energy source. A computer controlled stepping motor driven translation table was used to provide traverse motion. A detailed description of the cladding system can be found in Ref. (2). The compositions of the cladding powder and substrate are shown in Table 1.

Results

1. Effects of processing parameters on clad formation

The laser spot size was kept at 4.5mm in diameter which gives a reasonable power density. Good clad tracks with 2mm thickness were obtained in the traverse speed range of 8.6mm/sec ~ 25.0mm/sec. Traverse speed lower than 8.6mm/sec tends to produce large dilution while a larger traverse speed than 25mm/sec requires high laser power. Laser power has a substantial effect on clad formation. Insufficient laser power results in incomplete fusion in the interface, while excessive

	Cu	Si	Mg	Zn	Fe	Ni	Al	(O)
Bronze	Bal	---	---	---	2.19	5.27	9.79	0.008
AA333	3.14	8.92	0.298	0.103	0.486	0.013	Bal	---

Table 1. Chemical analysis of the clad Ni-Al bronze and substrate Al alloy AA333 (wt%)

laser power introduces dilution. Powder feed rate has a pronounced effect on clad formation. Very thin layer clad, less than 1mm, cannot be achieved without using assist gas He because of a strong tendency toward large dilution. Increasing Helium flow helps reduce dilution with the same other processing parameters being the constant. Eventually, clad tracks 0.7mm thick with good interface can be made with large Helium flow. However, large Helium flow introduces porosity as an unfavorable side effect. A critical step toward successful cladding of bronze on Al alloy has been the use of a polished surface substrate instead of sand blasted surface substrate. The clue to doing this was the idea that to avoid large dilution the laser energy must be put into the molten pool and must not be absorbed by the substrate surface right in the molten pool front. The mechanism of surface condition on clad formation will be further discussed later. With proper control of the processing parameters mentioned above, good crack-free interface clad tracks were achieved. One example of the composition profile obtained from SEM EDX is shown in Fig. 1. The measurement was started from the top of the clad and advanced until the unaffected substrate region was reached. In a not-too-large laser power range the composition distribution in the clad region is quite uniform and the content of each element is very near to that of bronze powder. A drastic change of composition occurs in the interface region which is usually about 100 μm. In the molten zone of substrate the composition of each element fluctuates in a certain range depending on which phase the electron beam has hit.

2. Overlapping clad

Overlapping clad is made by putting one clad track over another with a displacement in their position. Two parameters are used to describe the overlapping clad process. One is overlapping percentage, which is defined as the overlapping portion over clad width, and the other is aspect ratio defined by single clad width over single clad height. The laser power suitable for single clad is also suitable for overlapping. A good quality overlapping clad needs a proper combination of aspect ratio and overlapping percentage. One unfavorable phenomenon associated with overlapping is the interrun porosity which is due to a purely geometry effect, as described by Steen, et al. (4). Such interrun porosity occurs in combination of small aspect ratio and large overlapping percentage. When

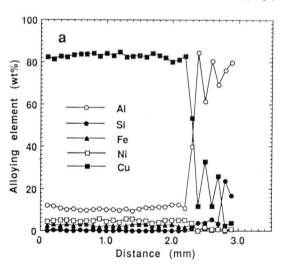

Fig. 1. SEM EDX composition profile of the clad.

overlapping percentage is smaller than 0.35 the surface of overlapping clad shows wave-like appearance. Best overlapping clads were achieved with combination of aspect ratio 3 ~ 3.7 and overlapping percentage 40% ~ 55%. One example is shown in Fig. 2. There is almost no sign of overlapping in the clad as polished and therefore good fusion between interruns has been achieved.

3. Microstructure characterization

Figure 3 is an optical micrograph of the clad on Al alloy substrate. The as-clad structure is featured by three regions, i.e., the clad region which is composed of a mixture of columnar and equiaxial grains, the interface region which is dark in the optical micrograph, and the molten zone which has a much finer structure compared with the unmelted substrate. The crystal structures of the phases observed in the clad region and interface were studied by Convergent Beam Electron Diffraction (CBED) and Selected Area Diffraction (SAD) techniques. The crystal structures and composition ranges identified are shown in Table 2. The α and β are martensites (5). The α martensite is located only in some keyhole area near the interface. The β martensite is the typical structure of the whole clad region. The γ_1 and θ phases are in the interface region. Al solid solution is in the molten zone. The lattice parameters are based on the SAD patterns. The point group and space group of β is based on CBED analysis which suggests $\bar{3}m$ symmetry. The point group and space group of other phases were established by earlier work (6-9) and confirmed in the present study. Each phase is presented by two compositions in which Al contents are two extremes: the minimum and the maximum. Since the cooling rate is high in laser cladding, this data may depend on the process parameters.

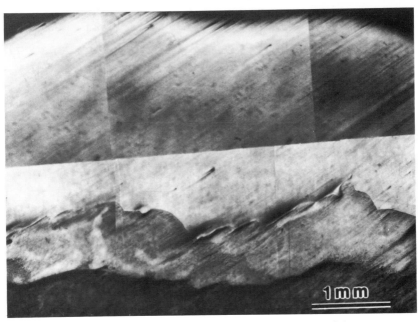

Fig. 2. Optical micrograph of cross-section of a overlapping clad.

Fig. 3 Optical micrograph of the cross-section of the clad.

Phase	Point Group	Space Group	Lattice Parameters Å	Composition Al	Si	Fe	Ni	Cu
α	m3m	Fm3m	a = 3.75	8.9 ~10.6	0.11 ~0.6	2.4 ~2.1	4.9 ~5.6	83.5 ~81.2
β	3̄m1	R3̄m1	a = 2.39 c = 17.74	10.0 ~12.6	0 ~0.17	2.4 ~3.8	6.3 ~8.1	79.8 ~75.3
γ₁	4̄3m	P4̄3m	a = 8.89 ~ 9.41	21.5 ~31.6	0.8 ~1.8	1.7 ~4.0	2.0 ~13.8	62.4 ~53.5
θ	4/mmm	I4/mcm	a = 5.93 ~ 6.18 c = 4.59 ~ 5.00	40.0 ~67.9	1.3 ~2.7	1.7 ~0.92	1.1 ~0.97	55.2 ~27.2
Al	m3m	Fm3m	a = 4.08 ~ 4.12	88.2 ~96.0	2.6 ~0.4	0.16 ~0.22	0.13 ~0.12	8.5 ~2.4

Table 2. Summation of structures and compositions of phases in Ni-Al bronze clad

Discussion

The effects of laser power, spot size, traverse speed are quite obvious because they can usually be incorporated in the overall energy balance equation (4). The effects of assisting gas He flow and substrate surface condition, however, have been less realized in the past. A laser generated molten pool model has been developed to explain the effect of assist gas He flow and substrate surface condition. Figure 4 shows schematic illustration of the molten pool model. The feature of the model is that the molten pool is behind the laser beam. The factors affecting dilution which can be considered are: substrate temperature, molten pool temperature and interaction time. There is a large difference of melting point (1063° - 577° =) 486° between the present cladding and substrate materials. If the interface thickness is about 0.1mm, the temperature gradient in the interface region is estimated as 486/0.1 = 4860°/mm. Such a large temperature gradient requires high molten pool temperature and low substrate temperature. Also, proper time is needed to achieve enough diffusion between the molten pool and the substrate. Qualitatively, it may be said that the lower the temperature of the substrate right before the molten pool the smaller the dilution that could be achieved. Based on the above argument, the sand paper polished surface has much higher reflectivity (0.81) than the sand blasted surface (0.20) and therefore absorbs less laser energy which results in low substrate temperature and therefore smaller dilution. The same argument also holds true for large Helium flow which has a significant cooling effect on the substrate under laser beam.

There are three phases, γ_1, θ and Al solid solution, in the interface. The γ_1 phase has a very large unit cell which contains 52 atoms. Generally the Burger's vectors in such a large unit cell are long and hence the Peierls-Nabarro stress necessary to move the superdislocation is expected to be high. It is then conceivable that the γ_1 phase is brittle. The θ phase is a tetragonal structure. The symmetry is lower than cubic structure which suggests the θ phase could be brittle. Therefore, regions composed of single γ_1 phase, single θ phase, and a mixture of γ_1 and θ phases should be avoided.

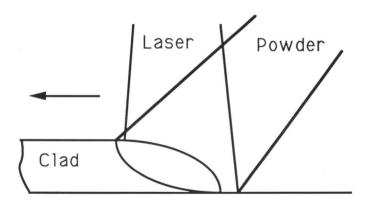

Fig. 4. The proposed molten pool model for the cladding of Ni-Al bronze on Al alloy AA333.

Acknowledgement

The authors would like to thank Mr. J. Koch for co-operation in the laser cladding work of the present materials. This work was made possible by a grant from Nissan Motor Co., Ltd.

References

(1) S. Sircar, C. Ribaudo, J. Mazumder: Metall. Trans. 20A (1989) 2267.
(2) J. Singh, J. Mazumder: Acta Metall. 35 (1987) 1995.
(3) J. Singh, J. Mazumder: Metall. Trans. 18A (1987) 313.
(4) W. M. Steen, V. M. Weerasinghe, P. Monson: "High Power Lasers and Their Industrial Aplications," SPIE. 650 (1986) 226.
(5) Y. Liu, J. Mazumder: Report to Nissan, February 1992.
(6) A. J. Bradley, P. J. Jones: Inst. Metals. 51 (1933) 131.
(7) A. J. Bradley, H. J. Goldschmidt, H. J. Lipson: Inst. Metals. 63 (1938) 149.
(8) J. B. Friauf: J. Amer. Chem. Soc. 49 (1927) 3107.
(9) Y. Liu, J. Mazumder: Report to Nissan, March 1991.

Laser cladding with amorphous hardfacing alloys

V. Fux, A. Luft, D. Pollack, St. Nowotny
Fraunhofer-Einrichtung für Werkstoffphysik und Schichttechnologie, Dresden (FRG)
W. Schwarz
Institut für Festkörper- und Werkstofforschung, Dresden (FRG)

1. Introduction

Due to their high alloy content and structure rich in hard material, cobalt- and nickel-based hardfacing alloys show a very insignificant ductility or no ductility at all. For this reason, they are generally not available as welding filler metals in the form of homogeneous wires or ribbons. They are fed into the hardfacing process as cast rods, powder or flux-cored wires.

One possibility of still achieving a higher ductility with such alloys is to make use of the amorphous condition. This method of having metal materials solidify amorphously from the molten mass has been used in international industry since the 1970s (1).

Classical field of application of these metallic glasses are chiefly magnetic materials and eutectic solder alloys. Essential to the production of the amorphous state are a high cooling rate during the weld freezing process ($>10^6$ K/s) and a retardation of crystallization caused by the technical conditions of the alloying process. According to this, amorphous metals always have a very small thickness (30 µm, as an example), on the one hand, and on the other, they generally contain metalloid additions (such as P, B, C and Si) having a crystallization inhibiting effect, thus supporting the glass forming process.

2. Welding filler metals

Since, so far, there were no suitable amorphous ribbons available for wear resisting surfacing, several ribbon materials were developed on the basis of CoCrWC-alloys (2). This type of alloy provides for a favourable starting position due to its natural metalloid contents.

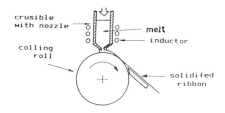

For the production of the ribbons, the "Planar Flow casting" process shown in the schematic representation of Fig 1 was used. According to this process, the molten metal is pressed onto a high-speed cooling roller through a slotted nozzle. After passing a few centimeters of the

Fig. 1: Schematic representation of the "Planar Flow Casting" - process

cooling path, it has already become a completely solidified bright metal ribbon. The melt solidifies in the same width as produced by the nozzle, so that the material is produced directly from the molten material and does not require any reworking. It is possible to produce ribbons of up to a few centimeters' width. The ribbon thickness of the alloys used was about 30 µm.

Amorphous metal ribbons have the advantage of distributing the alloy elements in an very homogeneous way cross-sectionally as well as lengthwise. The simplicity of adjustment of the ribbons' width allows for their adaption to the conditions of the process or the component, respectivily.

Test preparation and data

The tests were carried out on a strip feeding apparatus specifically designed for the purpose. The beam was produced by a cross-flow CO_2-laser with a maximum capacity of 4 kW. A very thin graphite layer was applied to the samples. In addition to the increased laser beam input originally envisaged there was a simultaneous partial improvment in the molten material's wetting to be observed on the substrate. The focal lengths of the focussing mirrors were approximately 200 to 400 mm.

Since the strip width preferred was 10 mm, nearly exclusive use was made of an oscillating beam. The frequencies used were ranging between 50 and 200 Hz. The laser capacity being constant across the width scanned, there were higher energy inputs at the deviation points. For test experiments, a line focus system of 200 mm of focal length was used in addition.

Further test parameters are outlined as follows:
mean radiation capacity   2.0 - 3.3 kW
(in the machinig place)
rate of feed v_v 50 - 400 mm/min
                                                (1400)
strip feeding speed v_f 1.5 - 10.5 m/min
average power density E_w 3   - 10 x 10^3 W/cm^2
irradiation angle   50 - 90°
strip feeding angle   (0)30 - 60°

3. Results

Fig. 2a shows the basic principle of ribbon surfacing. Since the significance of the direction of movement has become evident it is recommendable to classify ribbon surfacing according to two basic process alternatives, the motions of which can be derived from the partial representations 2b and 2c:

A   "Dragging" movement, i.e., the strip runs off in parallel with the feeding direction of the workpiece (Fig. 2b).
B   "Stabbing" movement, i.e., the strip runs off against the feeding direction of the workpiece (Fig. 2c)

In case of "dragging" strip surfacing, the surface of the workpiece is always covered by the hardfacing material, the molten bath and the ribbon. The laser beam can not act directly upon the surface of the workpiece. Thus, the required preheating of the sample's surface, which is to be wetted immediately before the material is molten, can be achieved only by heat conduction. This relationship is also the upper limiting factor for the feeding rate. With the laser capacities usable, velocities of only 150 mm/min were achieved (width and height of layer were 11.5 and 1.5 mm, resp.). Moreover, beads became increasingly narrower until the wetting failed completely and the liquid welding filler metal beaded off. With regard to this process alternative, the influence of the strip feeding angle is of minor importance. The ribbon may either be drained above the molten bath, or else dipped in the same. As a matter of fact, the latter is made possible only due to its small thickness of 30 μm. The cross-section of a surfacing bead typical of the process alternative is shown in Fig. 3a. The bead heights possible to be produced were between 0.9 - 3.0 mm. The best surfacing beads were found with layer thicknesses of 1,2 - 1.7 mm. Admixtures could be kept at a low level in this area, as can be seen from Fig. 4. Lower layer thicknesses led to increased admixtures. With higher bead thicknesses, lack of fusion was partially observed in the layer centre.

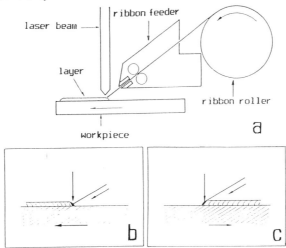

Fig. 2: a) Basic principle of laser cladding with amorphous hard-facing ribbons, b) the motion of "Dragging" movement, c) the motion of "Stabbing" movement

Different from the "dragging" movement, the laser beam gets through to the sample's surface partially in case of the "stabbing" surfacing process, thus having a direct influence on the wetting process of filler metal-to-substrate. This way, the surfacing speed is more intensively influenced by the capacity available, which may result in higher speed values. An inadequate capacity is evidenced by narrower and lower-quality surfacing beads.

Another benefit of a dirct contact between the laser beam and the surface of the substrate is shown in partial representations 3 b and 3 c. The increased energy input at the deviation points of the oscillation beam offers the possibility of influencing the formation of the surfacing bead. This envolves an alignment and maintenance of the welding filler metal melt close to the deviation points, which may lead to the formation of a flat-top layer. These may be provided with different types of side walls, in function of the wetting behaviour of the welding filler metal and further parameters. A more extensive fusion of the base material can be avoided similar to alternative A. The best-developed beads were found in a thickness range of about 1 mm. In the circumstances given, the flat-top shape disappeared with higher thicknesses. A reduced ribbon feed results in an arching of the welding filler metal edges, sometimes causing them to break in the middle. with this alternative, the influence of the strip feeding angle is of more significance.

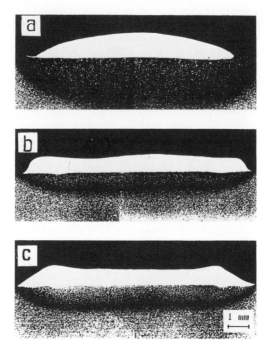

Fig. 3: Cross-sections of surfacing beads, a) cladded by "Dragging" movement, b) and c) cladded by "Stabbing" movement

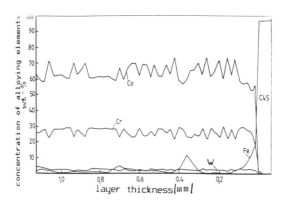

Fig. 4: Distribution of alloying elements in a surfacing bead

4. Characterization of layers

due to the fact the welding filler metal is fed in a compact way, there were comparatively few problems of voids and lack of fusion in the welds. The elements B and Si contained in these alloys are, at the same time, conductive to the formation of good beads. The amorphous state of the ribbons used for the surfacing process only serves to the purpose of facilitating its feedability. The surfacings are, in any case, of crystalline nature. From, this also derives that foil development should aim at producing the amorphous condition with a minimum of metalloid contents in order to reduce the negative effect of these elements on the ductility of the crystalline layers to the lowest possible level. The negative effect of these elements' increase in number mainly consists in the displacement of solidification morphology in the direction of eutectic conversion. This may be charakterized by the increasing share of an eutectic element in the structure as well as by the occurence of several such elements. This is accompanied by an inferior percentage of pre-eutectic mixed crystal which, though reducing wear resistance, is essential to a certain minimum ductility of the layers of hardfacing alloys. This minimum ductility, however, is predominantly required to take up shrinkage stress. This is all the more significant if the workpieces are not preheated before machining. Alloys such as stellites 6 and F are structured in the way mentioned and have been accepted not only in laser surfacing.

Compared to the CoCrWSiBC surfacing bead shown in Fig. 3b, the hardfacing material CoCrNiWSiBC of Fig. 3c has an optimized mealloid content and can be surface-welded as a single bead crack-free and without preheating. The structural composition of this layer clearly shows a large percentage of mixed crystals in the form of dendrites (3). The area of layer-to-substrate transition shows good metallurgical bonding. Above the dendritically solidified area lies a compact grained globularly solidified zone. Its extremously fine-grained structure is due to a spontaneous nucleation of the strongly undercooled molten material. This fine-grained character also leads to a higher homogeneity of properties as can be derived from the example of microhardness shown in Fig. 5. The flat-top shape of the surfacing bead and the reduced amount of reworking it envolves, make the partial preservation of this fine-grained zone possible.

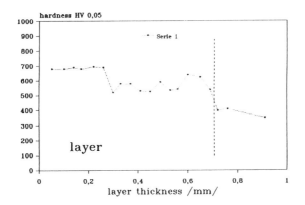

Fig. 5: Hardness curves of a CoCrNiWSiBC - layer

5. Concluding remarks

By previous work at this topic it could be shown that crack-free hardfacing layers can be produced with the help of specific amorphous metal ribbons. In this process, the welding filler metal is utilized to 100%. Under certain conditions, influence can be exercised on the formation of the geometrical design of the welding beads, allowing for the production of broad beads with low reworking requirements. For this reason, this procedure is most beneficial when applying surfacing beads to surfaces subjected to wear.

However, this process also offers a good chance for the production of layer thicknesses lower than 0.5 mm. In this case, a line focussing system was used. In test experiments using a model alloy, for instance, it was possible to achieve a very smooth , homogeneous layer thickness of 200 µm at a feeding speed of 1400 mm/min (layer width: 10 mm). This may open up new possibilities for even faster surfacing, with the componente being subjected to very low levels of thermal stress.

6. References

(1) U. Gonser, H.-G. Wagner: Metall 36(1982) 212
(2) W. Schwarz, V. Fux, St. Nowotny: Schweißen und Schneiden 44(1992), shortly
(3) A. Luft, W. Löschau, R. Gaßmann, V. Fux, W. Reitzenstein: "ECLAT '92", Proceedings

CAD/CAM Supported Laser Surface Treatment of Complex Geometry

W. König, D. Scheller, P. Kirner
Fraunhofer-Institut for Production Technology - IPT, Aachen, FRG

Introduction

Hot forming tools are subjected, in the course of some forming operations to extreme mechanical and thermal stress levels. Conventional heat treatment methods frequently fail to match up fully to these stresses and strains. The thermochemical variations of laser beam surface treatment, laser alloying and cladding, offer new ways of producing wear resistant, application-oriented coatings.

The deployment of laser beam surface treatment has resulted in increases exceeding 100% in tool life quantity and, in some cases, up to 500% as compared with tools which have received conventional heat treatment. The majority of the parts which have been refined until now have 2-D or $2\frac{1}{2}$-D geometries. Tools with complex 3-D geometries represent a special field of application for which the time required to write the NC-program is far in excess of the actual machining time.

Current state of technical development

The low level of automation of laser technology has greatly hampered its acceptance by industry. There is a lack both of computer-aided systems with which to plan operation-oriented machining paths for parts with complex geometries and of computer-aided process planning which would increase manufacturing quality and reliability.

The teach-in method is currently the one most frequently used to generate paths for laser beam surface treatment (**Fig. 1**). This entails the use of a homing laser to designate points on the contour which are subsequently used as a framework for the NC control. Very simple geometries can be machined rapidly and economically using teach-in. The machining of more complex parts however, necessitates a considerable number of support points, i.e. the amount of programming required increases out of all proportion. The treatment of such geometries quickly becomes uneconomical due to the relatively high hourly machine rate of the laser stations.

path generation options			

	Teach-In	2D-CAD/CAM	3D-CAD/CAM
machine time required	high on - line	low off - line	low off - line
geometrical complexity achievable	2D/3D-limited	2D-parts	3D-parts
availability	in existance	limited availability	non-existant

Fig. 1:   Methods of generating a path

Off-line programming systems based on simple CAD programs, but with additional special modules with which to generate automatically the path of the contours and surfaces to be treated, are currently available for complex 2D-geometries. Only level surfaces, however, can be tackled in this way. Complex 3D-geometries require sophisticated 3D-CAD/CAM systems with a specially adapted laser-specific technology module.

Problems surrounding the treatment of complex parts

A major obstacle to industrial acceptance of laser beam surface refinement as an innovative technology lies in the low level of automation which goes with it. The protracted machine idling times occurring during on-line generation of travel paths using teach-in, are extremely costly and make laser technology uneconomical (**Fig. 2**). The aim must now be, removed from the production machine, to make off-line programming accessible to laser technology. CAD/CAM systems capable of machining even complex 3D-geometries are, therefore, required.

CAD systems currently used in milling applications, permit sculptured surfaces to be treated and provide an interface to NC-machines via postprocessors. The laser technology which would permit the generation of laser-specific paths and the adaptation of process parameters to meet laser requirements has not, unfortunately, yet been integrated into these programs. This would require a computer-assisted technological knowledge base in the form of intelligent data bases to minimise the thermal strain on the part and to ensure that there was no occurrence of melting around the edges.

surface refinement of parts with
complex geometries
– problem –

□ long idling-times due to on-line keyboarding

□ lacking integration of laser technology in the
CAD/CAM-systems

□ complex 3D-machining requires 5-axial-movement

□ computer-aided technical knowledge not yet available

□ insufficient data preparation in the
digitization of sculptured surfaces

□

Fig. 2: Surface refinement of complex parts - difficulties

A 5-axial handling system, with timing accuracy which only high performance control systems can offer, is a further requirement for the machining of complex parts. Care must also be taken to ensure that the surface element being treated is always held in a horizontal position as the molten material otherwise runs, thereby producing an unwanted change in the geometry.

Approaches to the treatment of parts with complex geometries

CA engineering can bring about substantial reductions in the time required to write programs for the travel geometry and excessively long throughput times can be minimised. In a 3-D CAD system, the areas generated following digitisation, or the areas left after the cutting operation in the manufacturing process of the part to be treated, are analysed and prepared (**Fig. 3**). In this process, only those segments of the total

area which are subjected to severe stress when in operation
and which have been designated for laser refinement are selected. Using the "laser" as a tool, the individual segments are
marked out, in accordance with practise-oriented strategies,
by making a series of individual tracks which are subsequently
verified by a collision detection routine. This entails the
use of milling routines used in 5-axial machining operations,
in which the operator can change and optimise the paths generated via an interactive operator intervention option. The
paths are converted into a machine-oriented programming langu-
age by a postprocessor adapted to the laser and can be processed by the CNC-laser-machine.

surface processing of complex parts
- solution -

☐   CAD/CAM pre-processing of data relating to the segment in question

☐   design of computer-aided strategies for laser-specific path planning

☐   development of a technology module for the manipulation
    of process variables

☐   simulation and monitoring of machining process via the CAD/CAM-system

☐

Fig. 3:   Surface refinement of parts with complex geometries
          - solution

Fig. 4 shows the individual steps which together make up a
typical machining sequence of CAD- assisted treatment of a
connecting rod die, from the CAD model through to the final
result obtained after machining. The execution of this sequence permits complex part geometries to be machined without
tying up an excessive amount of machine capacity. Prerequisi-
tes for this are the availability of the geometrical information in a CAD system along with technological knowledge adapted to the requirements of laser machining permitting the
generation of laser-specific paths and the integration of
process variables adapted to the operation in hand.

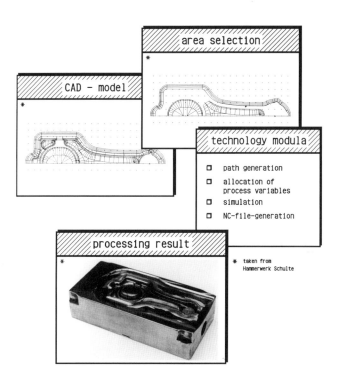

Fig. 4: Machining sequence of CAD-assisted laser beam treatment

Summary

The deployment of CAD/CAM techniques is the only way to make the generation of paths less time-consuming, thereby reducing long machine idling times. The integration of technological knowledge permits reliable process control and, at the same time, promotes the industrial viability of laser beam surface refinement.

Producing Hardfacing Layers on Aluminium Alloys by High Power CO_2-Laser

R. Volz, DLR, Institut für Technische Physik, Stuttgart, FRG

Introduction

Many methods for the laser surface treatment are already known. Most different materials are used as base materials. One of this base materials are aluminium alloys, they get more and more important as structural material. Aluminium alloys are used in aeronautics and automobile constructions, due to their low density and favourable recycling potential. For many applications however it is necessary to protect the surface of the aluminium substrate against wear. Many of the wear-resistant layers that have been developped so far have a disadvantage; they do not have the required bearing strength and thus fail in applications. A sufficient bearing strength and hardness of such a coating would increase the possibilities of applications enormously.

Besides thermal sprayed layers etc., laser cladding and alloying with CO_2-lasers (one-step process) are very interesting methods to produce wear-resistant layers on aluminium. The cladding/alloying powder is transported by a powder feeding system (the same which is used for the convential coating processes) through an own developped homogenizer directly to the laser processing point. With regard to the used powder and shielding gas as well as the powder feeding, coatings with a high bearing strength and without cracks and pores (single trace and overlapped) are produced.

Besides the basics like influence of shielding gas, powder, laser power, traverse speed, etc. on the hardness and the kind of coating; it is the aim of the paper to show new fields of applications resulting from this new method.

Experimental

For the investigations two types of CO_2-laser in reference to the following table were used:

Laser	RS 2500	10kW DLR
power [kW]	3,0	10,0
exited	hf	hf
gas flow	fast axial	fast transvers
resonator	stable	instable M = 2
mode	$TEM_{20}*$	ring

The beam focussing at both laser systems results of a mirror optic with 150 mm focal length. The processing head and the powder feeding system are distributed in two parts, shown in picture 1.

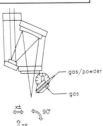

Picture 1: processing head, powder feeding

This system has the advantage, that it contains three axis of freedom in order to achieve high flexibility for laser surface treatment. The coating powder is transported under shielding gas argon over a homogenizer with a joined powder feeding nozzle with enclosed protective gas direct to the laser processing point. The protective gas and the powder feeding position have an important influence on the results of the one-step process. A selffluxing Ni-base alloy with the composition of 72.9 Ni, 15.5 Cr, 3.1 B, 4.3 Si and 0.7 C was used as coating powder (particle size 40-106 μm). Several cast aluminium alloys are cladded or alloyed by the one-step laser-process; their element distributions are listed in the following table:

	nominal composition %					
	Si	Fe	Cu	Mg	Ni	Al
Veral Si12CuNiMg	12	0,3	1,1	1,1	1,1	Bal.
Veral Si18CuNiMg	18	0,3	1,1	1,1	1,1	Bal.
Pantal 7	7			0,4		Bal.
Silumin Beta	9,5			0,4		Bal.

The laser treated samples were ground and polished. The melt width, the melt depth and the hardness profiles of the cladded/alloyed traces were determined. The etchings for the metallurgical estimations were carried out by the etching following Fuss (100 ml H_2O, 2.5 ml HCl, 0.8 ml HNO_3, 0.7 ml HF).

Results and Discussion

Different cast aluminium alloys are alloyed or cladded with Nicrobor 60 by one-step process. The results are controlled by the type and quantity of the shielding gas, powder quantity and feeding position, laser power, powder nozzle diameter and the processing speed. With optimated parameters pores and cracks are located neither in the single traces nor in the overlapping ones. Some applications shall show the potential of this one-step process in the field of laser surface treatment.

processing speed, powder nozzle diameter
The powder nozzle diameter influences the powder overspray, the trace depth and

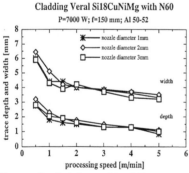

Picture 2: cladding with different nozzle diameter

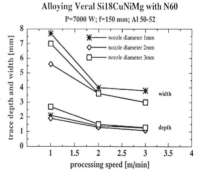

Picture 3: alloying with different nozzle diameter

width of the cladded or alloyed layer. In the case of cladding Veral Si18CuNiMg the best results regarding trace depth and width were produced by the powder nozzle 2 mm in diameter (picture 2). Alloying the same base material the least trace depth and width were achieved by a nozzle diameter of 2 mm, shown in picture 3. The cladding results are produced with a defocussing of 22 mm and the alloying ones with a defocussing of 15 mm. Picture 4 shows for 1 mm nozzle diameter the hardness profiles of cladded traces, produced by different processing speeds. With increasing processing speed the surface region with high hardness becomes smaller and smaller and the highest hardness regarding to the increasing processing speed decreases.

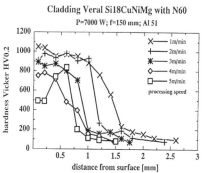

Picture 4: hardness profiles of cladded aluminium Veral Si18CuNiMg

beam defocussing
The beam defocussing regarding the position of the focal point above the base material influences the surface treatment result (alloying or cladding). Picture 5 shows the trace depth and width values for different positions of the focal point.

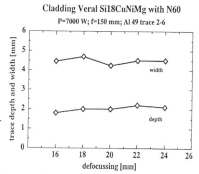

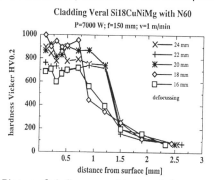

Picture 5: influence of beam defocussing on the trace depth
and width

Picture 6: influence of beam defocussing on the hardness
profiles

Hardness values of 700-900 HV0.2 up to 1.2 mm depth are achieved by a beam defocussing of 20, 22 and 24 mm (picture 6). The highest hardness (950 HV0.2) up to 0.7 mm depth was produced by a defocussing of 18 mm, this trace has a subsequent part of 1 mm depth with a hardness of 200-400 HV0.2 (alloying). The lowest hardness were achieved by 16 mm defocussing; many cladding powder particles were alloyed with the base material aluminium during laser processing.

shielding gas

Different shielding gases or mixtures of shielding gases strongly influence the geometry and hardness profile of the alloyed or cladded layer. With the use of nitrogen the deepest melted zones are achieved; the melt depth decreases with the use of argon and helium. Compared to argon and nitrogen, helium prevents a too deep penetration of the cladding/alloying powder into the base material aluminium. The hardest traces without cracks and pores and tubshaped surface-near layers are achieved by pure helium. Picture 7 and 8 show the influence of pure shielding gas He and CO_2 (different flow rates) on the trace depth and width. With the use of several mixtures of He and CO_2 with Ar, N_2 and He the trace depth and width mostly decrease.

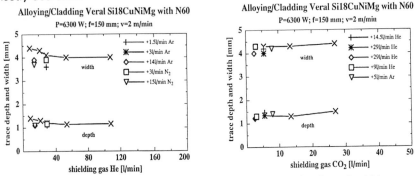

Picture 7 and 8: influence of shielding gas on the trace depth and width

applications

The production of load bearing layers on aluminium base alloys with the developped powder- and shielding gas feeding system is a new hardfacing technique for all aluminium users. Two significant applications shall show the potential of this laser cladding and alloying technique. Picture 9 shows the production of a spiral cladding layer on a circular workpiece. A very hard working surface, for example of

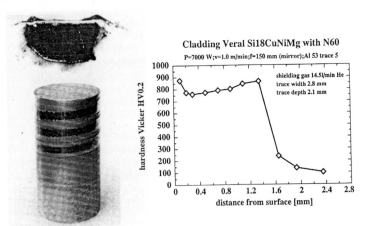

Picture 9: spiral cladded layer

cylinders, can be accomplished. The cladded traces placed side by side show a difference in height of about 0.2 mm between the cladded layer and the base material aluminium. A further application is the production of a hardfaced piston-ring groove. In a first step you have to mill a groove, this groove is alloyed or cladded with a hardfacing powder by laser treatment and after this procedure the geometry of the piston-ring groove is milled or ground into the laser treated trace (picture 10). A hard layer with an intermediate hardness of 700 HV0.2 up to 1 mm depth can be produced.

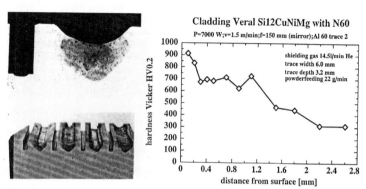

Picture 10: milled laser treated groove

summary

The one-step process for cladding and alloying of cast aluminium base alloys in the field of CO_2 laser surface treatment with hardfacing layers is a very interesting and potential application. A sufficient bearing strength and hardness can be achieved at utilization of Ni-base alloying/cladding-powders, thereby obtainable hardness values of the cladded or alloyed layers are at about 500-900 HV0.2. In connection with the developped powder- and shielding gas feeding system load bearing hardfaced layers up to 1 mm depth can be produced only with 3.0 kW laser power.

references

1. Animesh, K., Kulkarni, V.N., Sood, D.K. (1981). Laser treatment of chromium films on aluminium at high power densities. Thin Solid Films 86: 1-9.
2. Ayers, J.D., Schaefer, R.J. Robey, W.P. (August, 1981). A laser processing technique for improving the wear resistance of metals. Journal of Metals Aug. 81: 19-23.
3. Ayers, J.D. (1984). Wear behavior of carbide-injected titanium and aluminum alloys. Wear 97: 249-266.
4. Antona, P.L., Appiano, S., Moschini, R. (1986). Laser surface remelting and alloying of aluminium alloys. ECLAT 1986: 133-145.
5. Coquerelle, G., Fachinetti, J.L. (1986). Friction and wear of laser treated aluminium-silicon alloys. ECLAT 1986: 171-178.
6. Ferraro, F., Nannetti, C.A., Campello, M., Senin, A. (1986). Aluminium alloy surface hardening by laser treatment. LIM 1986: 233-243.
7. Luft, U., Bergmann, H.W., Mordike, B.L. (1986). Laser surface melting of aluminium alloys. ECLAT 1986: 147-162.
8. Bergmann, H.W., Gundel, P,H., Kalinitchenko, A.S. (1988). Laser surface alloying of AlSi12 and AlCr2,5Zr2,5 with transition metals and Si. Opto Elektronik Magazin Vol. 4, No. 5: 510-517.

9. Mordike, S. (1989). Umschmelzen und Umschmelzlegieren von Al und AlSi-Legierungen mit Laserstrahlung. Surtec 1989: 571-578.
10. Crooks, A., Hornbogen, E. (Okt., 1989). Silicon carbide impregnated aluminium alloys by laser surface treatment. Metall 43. Jahrg., No. 10: 957-959.
11. Crooks, A., Hornbogen, E. (Okt., 1989). An exploratory study of the effect of laser treatments on coated aluminium alloys. Metall 43. Jahrg., No.10: 954-956.
12. Gaffet, E., Pelletier, J.M., Bonnet-Jobez, S. (1989). Laser surface alloying on Ni film on Al-based alloy. Acta Metall Vol. 37, No. 12: 3205-3215.
13. Hagans, P.L., Yates, R.L. (1989). Laser treatment of thin molybdenum and chromium films on aluminium for enhanced corrosion resistance. Environmental degradation of ion and laser beam treated surfaces (ed. Was, G.S., Grabowski, K.S., The Minerals, Metals and Material Society, 1989): 215-236.
14. Bergmann, H.W., Endres, Th., Juckenath, B. (1989). Remelting and alloying of Al-Si-alloys. L.I.A. Vol. 69 ICALEO 1989: 239-250.
15. Pierantoni, M., Wagniere, J.D., Blank, E. (1989). Improvement in the surface properties of Al-Si cast alloys by laser surface alloying. Materials Science and Engineering A110 (1989): L17-L21.
16. Löwenberg, K., Arnesen, T. (1990). Surface modification of aluminium. Report from Center for Industrial Research Oslo, Norway.
17. Wagner, D. (1990). Laserumschmelzen von plasmagespritzten Hartstoffschichten auf Substraten mit niedrigem Schmelzpunkt. Steel and Metals Magazine, Vol. 28, No.9: 520-525.
18. Volz, R. (1991). Surface Treatment of Aluminium with a 10 kW CO_2 Laser. ICALEO 1991.
19. Kawasaki, M., Kato, S., Takase, K., Nakagawa, M., Mori, K., Nemoto, M., Takagi, S., Sugimoto, H., (1992). The Laser Cladding Application for Automotive Parts in Toyota. ISATA 1992: 293-300.
20. Volz, R. (1992). Surface Treatment of Aluminium Alloys with High Power CO_2-Laser. ISATA 1992: 301-310.

Adhesion and Microstructure of Laser Cladded Coatings

R. Gassmann, A. Luft, S. Schädlich, A. Techel, Fraunhofer-Einrichtung für Werkstoffphysik und Schichttechnologie, Dresden, FRG

Introduction

The application of a new cladding technique requires in addition to wear resistant alloy a good coating adhesion on the base material. Laser cladding is as far as his character is concerned a welding process, but the cooling times are short and the base material dilution is very low. Laser cladded cams of a heavy diesel engine (Fig. 1a) had a very low distortion and less wear than the hardened new ones. Due to the enlargement of the surface pressure ten times the normal size the coatings removed from the base metal and broke in pieces. Figure 1b shows the advanced crack along the nearly 10 μm thick transition layer between the deposited and the base material. The results of this study were published in 1989 /1/. Until now, the related coating adhesion problem did not receive the necessary attention. This is surprising because complete systems to laser clad engine valves are offered. In the present investigation the adhesion of various Co- and Ni-base alloys laser cladded onto austenitic and ferritic/perlitic steels was measured by a special adhesion tensile test. This was especially done to find out if there is a relation between the metallographic visible transition layer at the clad/substrate interface and the adhesion of the coating.

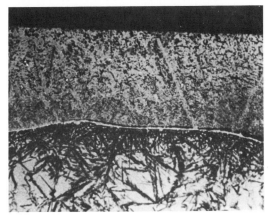

Fig. 1a: Laser cladded cam of a heavy diesel engine

Fig. 1b: Crack in the surface ground after the roll wear test

Testing Method

There are several methods to test the coat bonding on the base material, e.g. the micro bending test, the Oillard test, the scratch test and the adhesion tensile test. All these methods have in common that they have been invented to examine special layer systems and only to those they can reasonably applied. The DIN 50 160 provides the glue adhesion tensile test (Fig. 2a) to judge the influence of process parameter variation to the bonding of thermal sprayed coatings. In this experiment the coating is sprayed on body 1 to a stipulated sectional area, and then body 2 is glued to the coating. After hardening of glue,

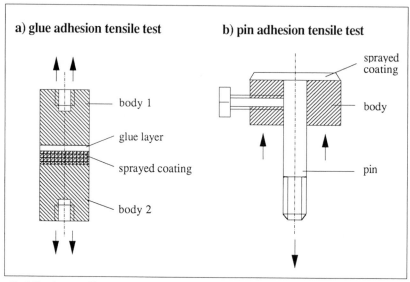

Fig. 2: Adhesion tensile test methods

the bodies are pulled apart to separate the coating from the base material. By reason of the equal sectional areas a pure tensile stress is assumed as long as elastic deformation holds. In case of a fracture in the surface ground the tensile power related to the sectional area is regarded as the adhesion strength. The application possibilities of the method are determind by the glue strength. On the basis of preexperiments /1/ adhesion strengths of laser cladded coatings about 200 MPa were registered. As this value, however, corresponds to the highest glue strengths a modified principle was used to evade the stick in the force path (Fig. 2b). A fitting pin was fixated in a body of same material. The even surface of both was cladded with a single track to avoid influences of track overlapping to the adhesion. The maximum pin diameter, which can be used, has to be in accordance with the sectional track dimensions. The maximum pin diameter amounts to 3 mm for track widths of 6 mm and track hights of 1.2 mm attainable by 5 kW laser power /2/. Figure 3 shows the used specimen geometry.

Both methods are distinguished by superposition of the load stresses and the shrinkage stresses due to the track cooling. In the pin adhesion tensile test additional shearing stresses due to the pin extraction are produced. Therefore in the case of crack shape in the coating the measured area load can not be equated with the tensile strength of the coating material.

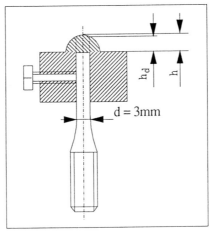

Fig. 3: Specimen geometry

Experiments

The cladding experiments were carried out with a focused TEM_{20}-CO_2-laser of 5.0 to 5.3 kW beam power. The distance between the focus and the working area was varied in 3 steps. This corresponded with medium intensities of 9.8, 14.2 and $17.4 * 10^3$ W/cm². Welding speed was between 100 and 600 mm/min. The powder feed rate was set to get the necessary track dimensions. The austenitic steel X 8 CrNi 18.10 and the ferritic/perlitic steel C 45 were used as base materials. 3 to 5 specimens of both base materials having one process parameter setting were cladded alternately to guarantee equal experimental conditions. According to the initial program the experiments were started with the Co-base alloys Stellite F* and Stellite 157* and the Ni-base alloy Deloro 40* as added materials. But these alloys cracked at the joint of pin and body even having low welding speeds. In spite of a fit free from play there was a notch point between pin and body and the absent bonding to the base material has led to track cracking due to the shrinkage stresses. By reason of the crack problem the ductile alloys Stellite 6* and Deloro 22* were added to the experiments.

Table 1: Chemical composition of the used materials

	C	Co	Cr	Ni	Fe	W	Mn	B	Si
C 45	0.45	-	-	-	base	-	0.65	-	<0.4
X8CrNi18.10	0.1	-	18.0	10.5	base	-	2.0	-	<0.8
Stellite 157	0.1	base	22.0	-	-	4.5	-	2.4	1.6
Stellite F	1.8	base	26.0	22.8	0.9	12.5	0.04	-	1.1
Stellite 6	1.1	base	28.0	-	-	4.0	-	-	1.0
Deloro 40	0.3	-	7.5	base	<2.0	-	-	1.7	3.5
Deloro 22	0.15	-	<1.0	base	<1.0	-	-	1.5	2.8

408

Results

Depending on the base metals, tracks with different sectional dimensions were generated by cladding with identical process parameters. Tracks cladded with welding speed of 100 to 150 mm/min on austenitic steel were on average 0.5 mm wider and 0.2 mm higher than tracks cladded on the ferritic/perlitic steel, in relation to a medium track width of 6 to 7 mm. This visible feature idicates a phenomenon being of fundamental importance to the interpretation of the tensile test results. Because of the more than factor 2 smaller heat conductivity of the austenitic steel ($\lambda = 0.18$ W/cm K) compared with C 45 ($\lambda = 0.5$ W/cm K) very different temperature distributions are created resulting in different sectional track dimensions. A metallographically proved finer dendritic microstructure and a higher hardness of tracks deposited on C 45 are the consequences of just as different cooling rates. Figure 4 shows a hardness diagramm of Stellite 157 that is representative for all used added materials. For a welding speed of 150 mm/min the track cladded on C 45 is on average 200 MPa harder than the track deposited on the X-steel. The effect that with increasing welding speed hardness is increasing too, equally applies to all added materials.

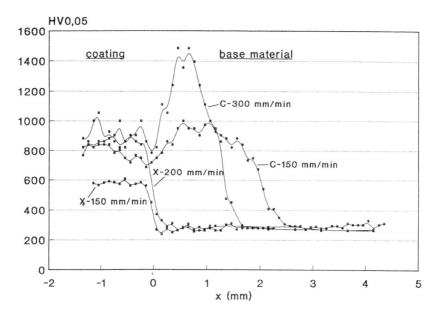

Fig. 4: Hardness of Stellite 157 laser cladded on C 45 and X 8 CrNi 18.10

In the tensile tests bearing load stresses between 300 and about 750 MPa were measured. Macroscopically 4 typical kinds of fractures could be observed: First, the fracture in the coating, second, the crack path along the surface ground, third, the ductile fracture of the X 8 CrNi 18.10-pins including the deep pin necking below the coating and, fourth, the brittle fracture of the C 45-pins below the heat affected zone. The fracture kinds 3 and 4 mean that the ultimate tensile strengths of the coating alloys were higher than the tensile strengths of the base metals. Bearing load stresses about 600 MPa were especially measured in the fourth group. Typical exponents were Stellite 157, Stellite 6, Stellite F and Deloro 40 cladded on C 45 having welding speeds below 300 mm/min. In case of the fracture in the coating, the coating tensile strength was the weakest point of the strength range. Those brittle fractures were found for Stellite 157 and Deloro 40 cladded on the X-steel, and less brittle for Deloro 22 also cladded on the X-steel. If comparing groups 1 and 4 it must be noticed that the same added materials cladded by unchanged process parameters on C 45 bore area loads of about 600 MPa, but cladded on X 8 CrNi 18.10, broke already with loads of about 300 MPa. This effect was probably caused by the deep plastic deformation of the austenitic pin material, which had induced considerable shearing stresses in the coating.

The metallographic examination of fractures along the surface ground (group 2) showed that a coating separation, as assumed initially doesn't take place. Figure 5 indicates that the crack passes above a light transition layer in the coating. The transition layer is a locally epitactically grown mixed crystal zone of nearly 10 μm thickness, which is planarly solidified. Only above this zone, if sufficient constitutional supercooling is reached, dendritic mixed crystal growth with subsequent eutectic solidification of the carbon rich interdendritic remaining melt starts. The thesis, that the transition layer being a weak point of laser cladded coatings, could be definitely refuted. In relation to the dendritic solidified coating material, the transition layer is distinguished by sufficiently high strength and ductility to stand all area loads, which have been generated in the experiments.

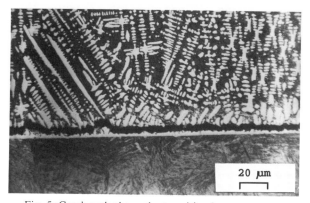

20 μm

Fig. 5: Crack path above the transition layer

Figure 6 shows the typical graphs of the measured fracture loads (fractures of group 1 and 2) versus the welding speed for Deloro 22. All measured fracture loads are not adhesion strengths according to the DIN-standard by reason of the fracture in the coating. They can be regarded as specific strengths of the coatings under the given test conditions. The diagramm shows 2 tendencies: Having low welding speeds, the bearing load of tracks cladded on C 45 is about 25 % higher than tracks cladded on X 8 CrNi 18.10. The bearing load also decreases in a linear way with increasing welding speed. If a welding speed of 600 mm/min is exceeded the difference of fracture loads of tracks cladded on C 45 and X 8 CrNi 18.10 is no longer significant. The described phenomena give the reasons therefore.

The difference in the fracture loads at a low welding speed could have been caused by additional shearing stresses due to the deeper plastic deformation of the X-steel compared with C 45. The fracture load decreasing versus the welding speed could be explained by a

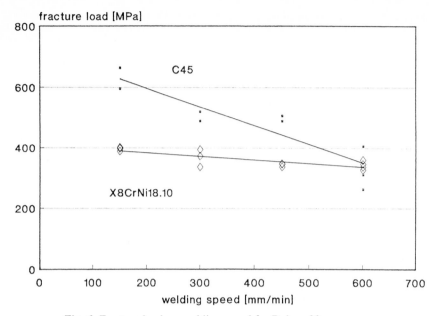

Fig. 6: Fracture load vs. welding speed for Deloro 22

decreasing deformation capacity of the coating material. With increasing welding speed the coating hardness increases too. Invers to this increase the deformation capacity decreases and the notch effect of the pin/body joint reinforces. At a welding speed of 600 mm/min a significant difference in the coating hardness depending on the base material couldn't be noted anymore, i.e. bearing loads become more and more similar.

Conclusions

The planarly solidified mixed crystal layers being created at the interface by laser cladding of dendritic Co- and Ni- hard alloys on steel substrates are not weak points of the coating-substrate-compound under mechanical load. On the contrary, because of their ductile mixed crystal structure they are well-suited to compensate internal stresses and to transmit surface loads from the hard coating to the usually less hard base material. In the case of laser cladded cams a roll wear test with extremely reinforced load had been made. The calculation of stress distribution showed, that the shearing stress maximum of 500 MPa, due to the rolling load, was exactly in the clad/substrate interface. It leads to the conclusion that dynamic shearing stresses in the interface are more likely to cause a damage to the coating than static stresses, because of the big difference in the properties of both materials.

References
(1) R. Gassmann: A-thesis, Hochschule für Verkehrswesen Dresden (1989)
(2) A. Techel: unpublished report (1992)

* Trademarks of the THERMADYNE Deloro Stellite, Inc.

Laser cladding with a heterogeneous powder mixture of WC/Co and NiCrBSi

B.Grünenwald[1,2], J.Shen[1], F.Dausinger[1], St.Nowotny[3]
[1]Institut für Strahlwerkzeuge (IFSW), Universität Stuttgart, FRG
[2]Max-Planck-Institut für Metallforschung, Institut für Werkstoffwissenschaft, Stuttgart
[3]Institut für Werkstoffphysik und Schichttechnologie, Dresden (FRG)

1. Introduction

Among the techniques for processing metal surfaces with laser, the cladding process with heterogeneous powder mixtures has shown potential for becoming an important technique for surface modification. By this process ceramic/metal composite layers can be obtained on different substrates like steel [1,2], Inconel [3,4], Ti- and Al-alloys [5,6]. Besides hardening the surface, incorporation of hard particles in a ductile metallic matrix markley improves the resistance to abrasive and erosion wear [5].

In this study a heterogeneous powder mixture of WC and NiCrBSi was injected directly into the melt pool during the laser irradiation. Depending upon the processing parameters (laser intensity, transverse speed of samples and powder feeding rate) different kinds of composite layers were produced. In addition to the track performance microstructural aspects such as dilution with the substrate, amount of non-melted tungsten carbide particles, composite hardness, microstructure of the matrix and the occurence of pores and cracks are discussed. A direct connection between processing parameters, microstructure and mechanical properties is given.

2. Experimental Procedure

The experiments were performed on a low alloyed steel (16MnCrS5) using a 5 kW CO_2-laser at a constant output power of 3500 W. The heterogeneous powder mixture consisted of 60 wt.% WC/Co (88/12) and 40 wt.% NiCrBSi with a median particle size between 45 and 90 µm. The chemical compostion of both powders is shown in tab.1.

WC/Co (88/12)			
Co [wt.%]	C_{total}	C_{free}	WC
10.5-13.5	5.1-5.7	< 2.0	rest
NiCrBSi			
Si [wt.%]	Ni	B	Fe
2.3	rest	1.3	0.5

Table 1: Chemical compostion of the used powder mixture.

A series of single tracks was produced using the following range of laser processing parameters:

      laser power $P = 3500$ W

      beam diameter $d = 3.0\text{-} 9.4$ mm

      traverse speed $v = 0.1\text{-} 0.3$ m/min

      powder feed rate $m = 6.3\text{-}17.8$ g/min

For a detailed investigation of the influence of traverse speed and powder feed rate on the track performance and the microstructure a beam diameter of 4.2 mm resulting in a intensity of 25 kW/cm² was used. After the metallographic preparation of the cross-sectioned specimen the shape and the microstructure of the laser tracks was characterized by optical and scanning electron microscopy. The track performance was described by the parameters of width, thickness, build-up and melted depth. The element distribution was determined by electron probe microanalysis (WDS).

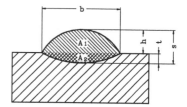

Fig. 1:   Schematic diagram of shape and dimension of a laser cladded track

        b = width

        h = build-up

        s = thickness

        t = melted depth

3. Results and Discussion

High intensities (> 30 kW/cm²) result in a keyhole shape of the processed zone with many pores and low carbide contents [7]. For lower intensities the zone shape shows a conduction limited profile and an optimal performance. Due to the reduced remelting of the substrate an intensity of 25 kW/cm² was selected for the following investigation.

3.1 Influence of the traverse speed on the track performance

Fig. 2 and 3 show the relationship between track performance and traverse speed at four different powder feed rates. Track width and thickness decrease with increasing traverse speed. The buildup decreases parallel to the thickness of the tracks. The principal shape of the tracks concerning the dilution with the substrate is independent upon the traverse speed.

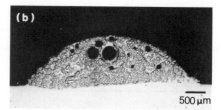

500 μm

Fig. 2:   Cross sectional macrographs for different traverse speeds ($P = 3500$ W, $d = 4.2$ mm, $m = 12.6$ g/min) **(a)** $v = 0.15$ m/min **(b)** $v = 0.3$ m/min

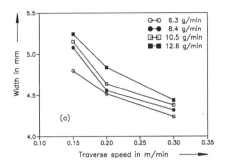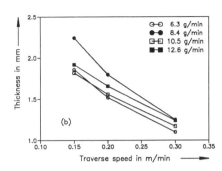

Fig. 3: Track width **(a)** and thickness **(b)** as a function of the traverse speed for different powder feed rates (P = 3500 W, d = 4.2 mm)

For high speeds (> 0.2 m/min) the tracks often contain pores formed by gas precipitation in the liquid state (fig. 2). This is due to the reduced time for the dissolved gases to evade by the surface at high traverse speeds.

3.2 Influence of the powder feed rate on dilution, microstructure and hardness

The amount of dilution of the substrate material with the clad layer at increasing powder feed rate is illustrated in fig. 4a. The amount of dilution is defined as follows (fig. 1):

$$Dilution = \frac{A_2}{A_1 + A_2} \cdot 100\%$$

For feed rates higher than 8.4 g/min the dilution is reduced to values less than 10%. In this case a great part of the laser power is absorbed by the powder and only a small part reaches the substrate. For lower feed rates more laser power is absorbed directly by the substrate and therefore the amount of dilution increases. A similar behaviour as the dilution with the substrate is shown by the iron content of the tracks (fig. 4b).

The shielding of the substrate by high powder feed rates also influences the temperature of the melt pool. High feed rates result in low melt pool temperatures and therefore most of the dispersed WC-particle remain unmelted during the cladding process. For low feed rates high melt pool temperatures are achieved resulting in a high dissolution of the carbides. Due to the excessive dissolution of the carbides the amount of graphite in the melt pool rises leading to an extensive porosity by vaporisation and combustion [8].

In fig. 5 a complex array of microstructures is shown for different powder feed rates. Within this powder feeding range besides the varying degree of dissolution of the WC-

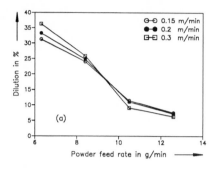 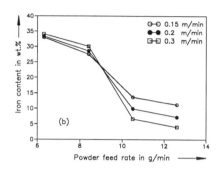

Fig. 4: Amount of dilution **(a)** and iron content **(b)** versus powder feed rate for different powder feed rates (P = 3500 W, d = 4.2 mm)

particles a significant difference in the resulting microstructure of the metallic matrix is obtained. For a feed rate of 6.3 g/min most of the WC-particles are dissolved completely and resolidified again in a dendritic shape (fig. 5a). The small black points between the precipitated carbides can be identified as grahite spheres due to the high

carbon content in the melt pool for low powder feed rates. For 10.5 g/min the microstructure of the matrix shows another morphology consisting of script-type eutectic carbides in the interdendritic regions of the Ni-rich dendrites (fig. 5b). Due to the lower dilution with the substrate and the limited dissolution of the injected WC-particles the chemical composition of the metallic matrix and the shape of the precipitated tungsten carbides changes [9,10,11]. For 12.6 g/min the microstructure only consists of a great amount of unmelted WC-particles and a ductile, Ni-rich metall matrix with some interdendritic Ni$_3$B (fig. 5c) [12].

Fig. 5: Optical micrographs showing the microstructure of the metallic matrix (d = 4.2 mm, v = 0.15 m/min):

(a) 6.4 g/min
(b) 10.5 g/min
(c) 12.6 g/min

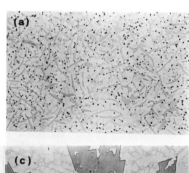

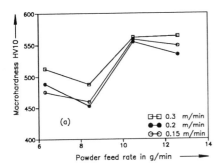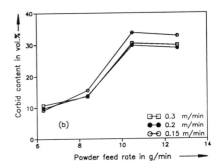

Fig. 6: Composite hardness **(a)** and carbide content **(b)** depending upon the powder feed rate for different traverse speeds (P = 3500 W, d = 4.2 mm)

For high powder feed rates (> 10 g/min) a closer look at the interface between melt zone and heat-affected zone revealed an intermediate layer between the cladding and the substrate. This layer mainly consists of a brittle meta-stable phase with high hardness and affects the adhesive strength adversely (fig. 7). Lower feed rates are characterized by a fairly homogeneous distributed microstructure throughout the cladding layer because of the higher convection in the melt pool.

The macrohardness (HV10) of the ceramic/metall composite as a function of the powder feed rate is plotted in fig. 6a. Due to the small carbide content for low powder feed rates (fig. 6b) the macrohardness strongly depends upon the microstructure of the metallic matrix. For low powder feed rates the matrix microstructure is very hard and brittle because of the resolidified carbides. Therefore cracks occur perpendicular to the melt pass and the macrohardness of the composite layer is relatively high (500 HV10). High feed rates are characterized by an high amount of non-dissolved WC-particles. This is shown in fig. 6b. High carbide contents result in a maximum composite hardness and a ductile metallic matrix without pores and cracks.

Fig. 7: Microhardness of the metallic matrix as a function of depth from surface for 12.6 g/min powder feed rate (P = 3500 W, d = 4.2 mm, v = 0.3 mm/min)

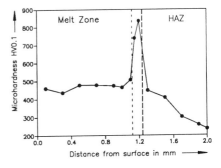

4. Conclusions

The results of the investigation show that laser cladding with a heterogeneous powder mixture of WC/Co and NiCrBSi can be used to produce hard and wear resistant layers on a ductile substrate. Due to the strong dependence of dilution, melt pool temperature, carbide content, microstructure, hardness and wear resistance an optimized combination of the processing parameters (intensity, traverse speed and powder feed rate) is necessary.

For low intensities (< 30 kW/cm^2), low traverse speeds (< 0.3 m/min) and high powder feed rates (> 10 g/min) the cladded layers are characterized by a high amount of homogeneously distributed WC-particles and a ductile metallic matrix without pores and cracks. The composite hardness of such layers reaches values up to 550 HV10 and the resistance to abrasive sliding wear is compatible to sintered WC hard metalls [13].

This report is based on the EUREKA project EU 194 FKZ 13EU0062 funded by the Bundesminister für Forschung und Technologie. The author is responsible for the content of the paper.

5. References

[1]   J.D. Ayers, T.R. Tucker: Thin Solid Films 73 (1980), pp. 201-207.
[2]   J.D. Ayers: "Lasers in Metallurgy", AIME (1981), pp. 115-125.
[3]   K.P. Cooper: J. Vac. Sci. Technol. A4, No.6 (1986), pp. 2857-2861.
[4]   K.P. Cooper, P. Slebodnick: "Laser Materials Processing", Proc. of ICALEO'88, Springer-Verlag (1989), pp. 3-16.
[5]   J.D. Ayers: Wear 97 (1984), pp. 249-266.
[6]   A. Gasser, E.W. Kreuz, W. Krönert, K. Lohmann, K. Wissenbach, C. Zografou: Proc. of ECLAT'90 (1990), pp. 651-660.
[7]   J. Shen, St. Nowotny, F. Dausinger, H. Hügel: Proc of LAMP'92, pp. 755-760.
[8]   R. Vilar, J. Salgado Figueira, R. Sabino: Proc. of ECLAT '90, Erlangen, pp. 593-604.
[9]   M. Fiedler, H.H. Stadelmaier: Z. Metallkde 66 (1975) No. 7, pp. 402-404.
[10] H.H. Stadelmaier, C. Suchjakul: Z. Metallkde 76 (1985) No. 3, pp. 157-161.
[11] A. Merz, L. Illgen: Neue Hütte 9 (1964) No. 12, pp. 733-736.
[12] O. Knotek, E. Lugscheider, H. Reimann: J. Vac. Sci. Technol. 12 (1975) No. 4, pp. 770-772.
[13] J. Shen, F. Dausinger, B. Grünenwald, St. Nowotny: Laser und Optoelektronik 23 (1991) Nr. 6, pp. 41-49.

IV.
Laser Application
in the
Automotive Industry

Vortrag: "Anwendung der Lasertechnik im Automobilbau"
von Dr.-Ing. E. h. W. Kirmse am 14. Oktober 1992 in Göttingen

4th European Conference on Laser Treatment of Materials "Eclat 92"

Der Wettbewerb auf dem Automobilsektor zeigt, daß sich die Produkte hinsichtlich ihrer Funktionalität und ihrer Qualität immer mehr einander nähern. Diese Produkte sind jedem zugänglich. Neuentwicklungen sind mit dem Erscheinen eines neuen Produktes sichtbar, die Konkurrenz lernt voneinander. Beträchtliche Unterschiede ergeben sich jedoch bei den Herstellungskosten der Produkte.

Wo liegen nun die Resourcen, die wir nutzen können, um die Kosten unserer Produkte zu reduzieren?

Wenn wir auf unsere bisherigen alten Erfolgsfaktoren, wie Innovation und Qualität schauen, auf Zeitvorsprung und Flexibilität, auch auf Produktivität und Kostensenkung, dann werden wir uns aber auch auf die Nutzung unserer eigenen Stärken, auf die Nutzung unseres ganz spezifischen Know-hows zu besinnen haben. Sicher liegen große Kostenreduzierungsresourcen auf dem organisatorischen Gebiet unserer Fabriken, aber auch bei den Produktionsmethoden, wie z. B. bei intelligentem Einsatz der Laserbearbeitungsverfahren. Und diese optimal eingesetzten Produktionsmethoden bringen Kostenvorteile, die die Konkurrenzfähigkeit sichern.

Die Anwendung der Lasertechnik sollte daher nicht allein im Hinblick auf die optimale Lösung von Bearbeitungsproblemen betrachtet werden, sondern auch vor dem Hintergrund der angespannten Situation in der deutschen Werkzeugmaschinenindustrie.

So stellt sich für Automobilhersteller, Zulieferer und Hersteller von Bearbeitungslasern die Frage, wo in Zukunft Anwendungsschwerpunkte liegen werden.

Ausgehend von der Situation im Werk Untertürkheim der Mercedes-Benz AG ist hier an erster Stelle das Laserschweißen, gefolgt vom Laserschneiden und für die Zukunft auch das Beschichten mit Laser zu nennen. Diese Laserbearbeitungen bieten, eingebunden in das gesamte Fertigungskonzept, gegenüber herkömmlichen Verfahren häufig deutliche Vorteile bezüglich Qualität und Kosten.

Geht man über zu einer Betrachtung des Lasereinsatzes, aufgeschlüsselt nach den drei großen Wirtschaftsräumen, so ergeben sich weitere Hinweise für einen künftigen Einsatz der Laser in der Fertigung.

	Hersteller	Zulieferer	Summe
Europa	104	40	144
USA	131	117	248
Japan	105	410	515
Gesamt	340	567	907

	Schneiden	Schweißen	Oberflächen-Bearbeitung
Europa	54 %	42 %	4 %
USA	23 %	64 %	13 %
Japan	91 %	8 %	1 %

Bild 1: Laser im Produktionseinsatz bei Automobilherstellern/-zulieferern
Quelle: Rofin-Sinar

Aus dem oberen Teil des Bildes ist zu ersehen, daß in Japan die Zulieferer der Automobilhersteller mit 410 installierten Lasern fast 4 mal soviel Einheiten installiert haben, wie die Automobilhersteller. In Europa haben die Zulieferer mit 40 Lasern knapp 40 % installiert im Vergleich zu den Herstellern von Automobilen.

In der unteren Hälfte des Bildes ist sichtbar, daß für die Bearbeitungsverfahren Schneiden und Schweißen in Europa größenordnungsmäßig etwa gleich viel Einheiten eingesetzt werden. Es verhält sich Schneiden zu Schweißen, wie 54 zu 42 %. In Japan hingegen werden 91 % der Laser für das Schneiden eingesetzt und nur 8 % für das Schweißen.

Die große Zahl der in Japan installierten Laser und der hohe Anteil des Laserschneidens daran zeigen uns, daß wir neue Wege überlegen müssen, um künftig die Vorteile der Lasertechnik noch besser nutzen zu können.

Ich möchte nun anhand einiger Beispiele aufzeigen, wie wir bei Mercedes-Benz, insbesondere auch bei der Einführung der Laser-Technik, vorgehen und was wir bereits realisiert haben.

Informationen über den Einsatz des Lasers zum Randschichthärten von Zylinderlaufbahnen in USA waren bereits 1976 Anlaß für uns, die Möglichkeiten des thermischen Werkzeuges LASER auch für unsere Bearbeitungsaufgaben in der Aggregatefertigung zu beleuchten. Eine durchgeführte Analyse der amerikanischen Laser-Hersteller und Anwender brachte uns bereits damals überraschend viele Anbieter und Einsatzfälle, die uns veranlaßten diese Technologie intensiv zu untersuchen. Mit der Einrichtung eines Laser-Labors im Jahr 1981 und der Aufgabenstellung, dieses neue thermische Werkzeug hinsichtlich der Einsatzmöglichkeiten für unsere Produktion abzugrenzen, begann dann unsere aktive Beschäftigung mit der Laser-Technologie.

Die Leistung unserer CO_2 Laserlaboranlage wurde von zunächst installierten 2,5 kW vor 5 Jahren mit einem neuen, hochfrequenzangeregten Laser auf 5 kW erhöht. Auf 3 Arbeitsstationen werden die heutigen Schwerpunktthemen **Trennen, Schweißen, Randschichtbehandeln und Beschichten** bearbeitet und die Ergebnisse technisch/wirtschaftlich den Alternativverfahren gegenübergestellt.

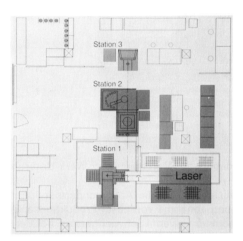

<u>Bild 2:</u> **Grundriß Laser-Labor**
Daten: 5 kW-CO2-Laser, Station 1: 4-Achsen-NC-Anlage.
Station 2: Planetenträger-Schweißen; Station 3: Rohre trennen

Neben diesen Aufgabenstellungen, die damit im Rahmen von Aggregateoptimierungen, Grundsatzentwicklungen für laufende und neue Automobilteile und Versuchsteilefertigungen in Angriff genommen werden, ist die ganz wichtige Aufgabe dieses Labors und der hier tätigen Mannschaft darin zu sehen, die neue Technologie gegenüber unseren Entwicklungs-, Konstruktions-, Planungs- und auch Produktionsbereichen bekannt zu machen.

Dieses Laserlabor ist eines der anwendungstechnischen Labors der Verfahrensentwicklung im Werk Untertürkheim der Mercedes-Benz AG. In diesem Werk werden sämtliche Motoren, Getriebe und Achsen für unsere PKWs und kleine Nutzfahrzeuge hergestellt.

Neben Prototypteilen werden auch Kleinserienteile mit dem Laser in diesem Labor bearbeitet sowie Serienanlauffertigungen durchgeführt, um nach Freigabe der Versuchsteile bis zur Lieferung der Produktionsanlage den Serienanlauf sicherzustellen.

Beratung der Konstruktions-bereiche zur Bauteilgestaltung 1	Bauteilversuche, Entwicklung bis zur Produktionsreife 4	Grundsatz-untersuchungen Schweißen, Schneiden und Oberflächen-behandeln 7
Ansprechpartner für Institute und Industriebetriebe 2		Inbetriebnahme von Produktions-anlagen 8
Produktions-betreuung und Schulung 3	Kooperation mit Produktionsmittel-Fertigung (Anlagenkonzept) 6	Serienanlauf-fertigung und Kleinserien-fertigung im Labor 9

Bild 3:   Aufgaben des Laserlabors bei Mercedes-Benz Werk Untertürkheim

Die Fachleute dieses Labors sind aber auch die Ansprechpartner für Institute und Industriebetriebe und mit der Inbetriebnahme von Laser-Produktionsanlagen befaßt. Sie betreuen außerdem die Produktion, wenn Schwierigkeiten bei der Serienbearbeitung auftreten. Dieses Labor ist ein operativer Bereich, der die neuesten wissenschaftlichen Erkenntnisse schnell umzusetzen vermag.

Im Folgenden werden zunächst ausschnittweise Beispiele aus der Motoren- und Getriebeteilefertigung vorgestellt.

Laser-Schweißen

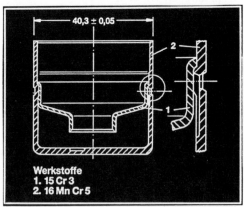

Bild 4:   Stößelgehäuse

Einleitend soll nur kurz auf die bekannten Schweißbeispiele "Stößelgehäuse" und "Vorkammer für PKW-Dieselmotoren" hingewiesen werden. Von den Stößelgehäusen wurden seit 1984 ca. 8 Mio. Stück in lasergeschweißter Ausführung gefertigt. Bei diesem Stößelgehäuse werden zwei ineinandergepreßte, topfförmige Blechteile mittels Laserschweißung verbunden.

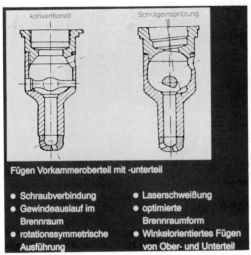

Bild 5: **Laserstrahlschweißen Vorkammer, PKW-Dieselmotor**

Bei der Vorkammer wurde die Schraubverbindung zwischen dem Zwischenteil und dem Endstück in Verbindung mit Abgasverbesserungsmaßnahmen durch eine Laserschweißung ersetzt (im Bild durch Pfeil markiert).

Neben diesen Beispielen des Laserschweißens von Motorenteilen nun weitere Beispiele aus dem automatischen Getriebe.

Für den Planetenträger eines 5-Gang-Automatik-Getriebes und eines verstärkten Vierganggetriebes wird das Laserschweißen jetzt serienmäßig für das Fügen der Blechteile eingesetzt.

Dazu werden die vier Segmente des kaltumgeformten Blechteils mit einer ebenen Platine verbunden. Beide Teile bestehen aus einem mikrolegierten Stahl und haben eine Materialstärke von 5 mm bzw. 4,5 mm.

424

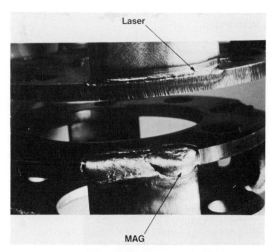

Bild 6: Planetenträger für automatische Getriebe;
Vergleich Laser/MAG-Schweißung

Wie im Bild zu erkennen ist, wird durch Laser-Schweißen gegenüber dem konventionellen MAG-Schweißen die Wärmeeinbringung in das Bauteil reduziert. Es entstehen kleinere geometrische Verzüge, die ein manuelles Richten nicht mehr erforderlich machen, bzw. den Aufwand für die Spannvorrichtung in der Schweißmaschine reduzieren. Ebenso ergibt sich ein geringerer Nacharbeitsaufwand zur Beseitigung des Spritzerauswurfes. Sowohl beim Laserschweißen wie auch beim Elektronenstrahlschweißen entsteht ein zwar feinerer, aber nicht zu unterschätzender Spritzerauswurf, insbesondere im Bereich des Strahlaustritts. Durch entsprechende Maßnahmen muß dieser Spritzerablagerung vorgebeugt werden. Die vergleichende Erprobung der Laser- und MAG-geschweißten Planetenträger hat aufgrund der möglichen vollen Verbindung der Wandstärke eine höhere Festigkeit und Steifigkeit des fertigen Bauteils ergeben.

Geschweißt wird mit einer Laser-Leistung von 4,5 kW und einer Schweißgeschwindigkeit von 2,0 m/min. unter einem Schutzgasmantel aus Argon.

<u>Bild 7:</u>   Teile für Automatik-Getriebe

Die Vorteile des Lasers gegenüber dem Elektronenstrahl, die im Wesentlichen auf
Wegfall der Vakuumkammer und auf der Möglichkeit des Mehrstationenbetriebes
beruhen, sind Veranlassung, das Laserschweißen verstärkt auch für bisher elektro-
nenstrahlgeschweißte Getriebebauteile einzusetzen.

<u>Bild 8:</u>   Laserschweißanlage für Getriebeteile; Strahlführung

Das Bild zeigt die Anlage auf der Podestebene über den drei Arbeitsstationen mit
dem modernen hochfrequenzangeregten 5-kW-CO2-Laser und die externe Strahl-
führung, die mit gefilterter Luft geflutet wird. Außerdem sind die Strahlweichen für
die drei Arbeitsstationen zu erkennen.

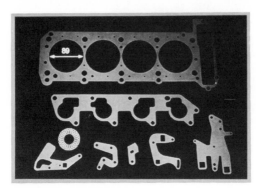

Bild 9:   Laser-Trennen, Motoren- und Prototypteile

Laser-Trennen

Unter der Überschrift Lasertrennen sind zunächst einmal die auf der eingangs gezeigten Anlage in unserem Labor ständig durchzuführenden Trennaufgaben an Prototypteilen und Serienanlaufteilen - wie im Bild gezeigt - zu nennen.

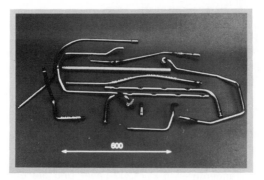

Bild 10:   Laser-Trennen, Bohrungen in Rohrleitungen

Unter der Überschrift "Trennen" ist auch das Bohren mittels Laser einzuordnen.
Dazu zeigt das Bild Beispiele aus unserer Motorenteilefertigung.
In dieser Fertigung fällt ein erheblicher Fertigungsaufwand für Rohrleitungen an,
die für den Transport von verschiedenen Medien, wie z. B. Kraftstoffen, Kühlflüs-
sigkeiten, Bremsflüssigkeiten, Ölen, Abgasen benötigt werden. Das Bild zeigt eine
Auswahl dieser Rohrformen. Diese Rohrleitungen müssen mit Bohrungen für seitli-
chen Zu- und Abführungen versehen werden.

Die für das "Bohren" bisher zur Verfügung stehenden Fertigungsverfahren, wie
mechanisches Bohren oder EC-Senken, ergeben Schwierigkeiten z. B. beim mecha-
nischen Bohren hinsichtlich der Innengratbeseitigung und der Standzeit der Bohrer.

Als Alternative wurde das "Bohren" mit CO2-Laser untersucht und aufgrund tech-
nischer und kostenmäßiger Vorteile vorgesehen. Die im Bild gezeigten, vorwiegend
aus X 5 Cr Ni 18 10 bzw. X 10 CrNi 18 9 gefertigten Rohre müssen somit mittels
diesem Verfahren mit bis zu 18 Bohrungen versehen werden. Die Rohrdurchmesser
liegen zwischen 12 und 28 mm, die Wandstärken zwischen 1,0 und 1,5 mm und
die einzubringenden Bohrungs-Durchmesser zwischen 1 und 12 mm.

Zum Laser-Bohren auf der Produktionsanlage wird der Bearbeitungskopf entspre-
chend der Bohrung über 3 CNC-Achsen bewegt (flying optics). Mit konzentrisch
zum Laserstrahl durch die Arbeitsdüse zugeführtem Sauerstoff wird das heiße Material nach innen geblasen. Dabei ergaben sich bei den ersten Versuchen Probleme
mit der Entsorgung der Schlacke und der Verschmutzung der Rohrinnenseite durch
die heiße, ausgeblasene Schlacke. In verschiedenen Patenten sind bezüglich dieser
Probleme Vorschläge gemacht worden. zufriedenstellende Ergebnisse für die hier
vorliegenden Aufgabenstellungen, bei den hier vorliegenden Rohrgeometrien konn-
ten aber damit noch nicht erreicht werden

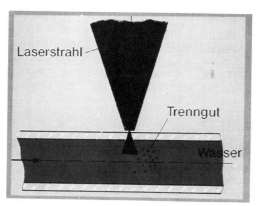

Bild 11: Laser-Trennen mit Wasserspülung, Prinzip

428

Eine Lösung wurde schließlich nach dem im Bild gezeigten Prinzip gefunden. Während des Laser-Trennens wird das Rohr innen mit Wasser durchspült, das die heiße, ausgeblasene Schlacke abkühlt, die erstarrte Schlacke abtransportiert und die Rohrinnenseite von Rückständen freihält. Gleichzeitig wird dadurch - in Verbindung mit den entsprechenden Laser-Parametern - erreicht, daß die Bohrungskanten frei von Grat und Rücksständen sind und keine Nacharbeit mehr erforderlich ist.

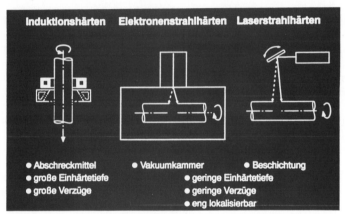

Bild 12: Thermophysikalische Randschicht-Härteverfahren

Laser-Härten

Ein weiteres, mögliches Laser-Einsatzgebiet ist das Randschichthärten. Das Bild zeigt thermophysikalische Randschichthärteverfahren, die in Konkurrenz untereinander, aber auch insgesamt in Konkurrenz zu thermochemischen Härteverfahren, wie Einsatz- und Nitrierhärten zu untersuchen sind. Das Randschichthärten ist sicher eine der schwierigsten Aufgabenstellungen, die mit dem Laser aufzugreifen ist, zumal hier auch meistens das Induktionshärten - und für spezielle Aufgaben, bei denen das Induktionshärten, z. B. aus geometrischen Gründen schon nicht mehr möglich ist - auch das Elektronenstrahlhärten eine Konkurrenz darstellen.

Außerdem sind die hohen Taktzeiten und die hohen Kosten ein Hindernis bei der Einführung der Strahltechnik im Vergleich zu der in der Serie bewährten Induktionshärtung.

Besonders interessant wird das Strahlhärten für Härteaufgaben, bei denen sehr schlecht zugängliche Bereiche behandelt werden müssen, eine Ankopplung von Induktoren somit nicht möglich ist und deshalb bisher z. B. das Einsatzhärten oder Nitrieren praktiziert wird - mit allen Nachteilen, wie schlechter Wirkungsgrad, Zwischentransport, Verzug und auch größerer Umweltbelastung. Hier bietet sich das Strahlhärten an. Dabei konkurrieren der Elektronenstrahl und der Laserstrahl.

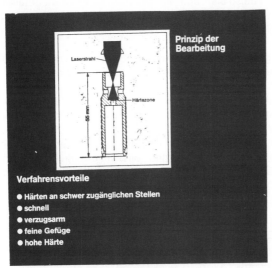

Bild 13:   Laserhärten von Druckbolzen zur Verschleißreduzierung, Prinzip

Als Beispiel für einen Lasereinsatz wird im Bild die partielle Härtung des Kugelsitzes eines Druckbolzens für einen Kettenspanner gezeigt. Mit einem defokussierten Strahl - der Fokuspunkt liegt 70 mm über der Bearbeitungsfläche - wird hier der im Werkstück weit zurückliegende Kugelsitz erreicht und auf Umwandlungstemperatur erwärmt. Ein "Coating" der zu behandelnden Fläche ist bei diesem Beispiel nicht erforderlich, da die aus dem vorangeschalteten Arbeitsgang "Vergüten" resultierende dünne Oxidschicht für eine ausreichende Absorption sorgt.

Laser-Trennen/-Schweißen in der Karosseriefertigung

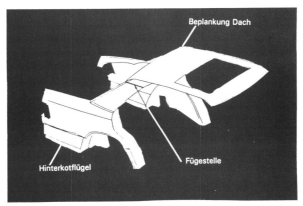

Bild 14:   Zusammenbau Hinterkotflügel mit Beplankung Dach

Nach diesen Laser-Einsatzbeispielen aus der Motoren- und Getriebeteilefertigung abschließend zwei Anwendungsfälle aus der Karosseriefertigung, die in unserem Werk Sindelfingen realisiert wurden.

Mit Anlauf der Fertigungsanlagen für die neue S-Klasse kam hier erstmals das Laserschweißen in der Karosseriefertigung serienmäßig zum Einsatz.

Anstoß für diese Entwicklung war bei dem ersten Einsatzfall der Verbindung Dach-Hinterkotflügel - die hohe Nacharbeit infolge der großen Schweißraupe und der Verformung im Fügebereich aufgrund der eingebrachten Wärmemenge. Mit dem Einsatz des Lasers wird zum einen eine geringere Oberraupe erzeugt, zum anderen weniger Wärme eingebracht. Hierzu wurden wissenschaftliche Vorarbeiten von einem Laserinstitut geleistet.

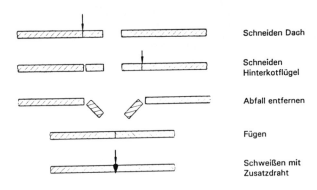

	Schneiden Dach
	Schneiden Hinterkotflügel
	Abfall entfernen
	Fügen
	Schweißen mit Zusatzdraht

Bild 15:   Schematische Verfahrensfolge Laserstrahlschneiden und -schweißen

Zur Reduzierung der manuellen Nacharbeit infolge von Wärmeverzug werden, wie im Bild gezeigt, mittels des sogenannten Offsetschneid-Verfahrens die beiden Bauteile "Dach" und "Hinterkotflügel" präzise besäumt und anschließend mit Zusatzwerkstoff verschweißt. Bei der installierten Anlage werden die Bauteile zur Erzeugung eines möglichst engen Fügespaltes zunächst in der Vorrichtung nacheinander lasergeschnitten und anschließend unter Zuführung eines Zusatzdrahtes miteinander verschweißt. Während des gesamten Schneid- und Schweißvorganges bleiben die Bauteile in derselben Vorrichtung gespannt.

Die Bearbeitung erfolgt mit einem 5achsigen CNC-Portalsystem und einem 2,5 kW-CO_2-Laser. Die Schneid- und Schweißgeschwindigkeiten liegen bei ca. 6 bzw. 3 m/min. Der max. überbrückbare Fügespalt beträgt 0,2 mm; die Nahtlänge ca. 350 mm pro Seite.

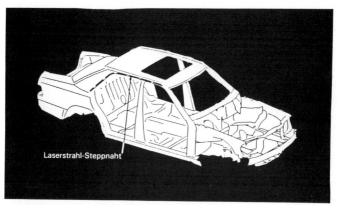

Laserstrahl-Steppnaht

Bild 16:   Laserstrahlschweißen Dach/Seitenwand

Beim zweiten Einsatzfall in der Karosseriefertigung - der Verbindung Dach-
Seitenwand - ließ die stilistische Lösung unter Serienbedingungen keine Punkt-
schweißverbindung zu. Es wird deshalb eine Überlapp-Lasernaht in Form von
8 mm langen Strichnähten auf einer Länge von ca. 1,80 m eingebracht. In dem
Fügebereich kommt verzinktes Blech zum Einsatz, was besondere Maßnahmen,
wie Abstandhalter (Noppen) von 0,1 bis 0,2 mm Höhe erfordert, um ein Entgasen
des Zinkdampfes aus der Fügeebene zu ermöglichen.

Mit diesen Beispielen aus den Bereichen Schweißen, Trennen bzw. Bohren und Randschichthärten ist die Vielseitigkeit des thermischen Werkzeugs LASER, in diesen Fällen zunächst einmal des CO_2-Lasers, für unseren Automobilbau demonstriert. Weitere Einsatzfälle sind in Vorbereitung. Die hierbei zu verfolgenden Ziele der Laserbearbeitungstechnik und deren Umsetzung in Produkte sind zusammenfassend für uns:

1 Die "lasergerechte" Konstruktion von Bauteilen muß im frühen Entwicklungsstadium wertanalytisch beeinflußt werden.

2 Durch Laser-Bearbeitungsversuche in den anwendungstechnischen Labors der Wissenschaft und Industrie sind kostengünstige Produktions-Lösungen zu finden.

3 Eine konsequentere Umsetzung neuer wissenschaftlicher Erkenntnisse in Produkte ist durchzuführen.

4 Konzentration und Abstimmung der Laserkapazitäten in Wissenschaft und Forschung sind unabdingbar.

5 Die flächendeckende Information der Entwicklungs-, Planungs- und Produktionsbereiche der Industrie einerseits und der Institutionen der Lehre andererseits über den Stand der Laserbearbeitungstechnik ist zu verbessern.

6 Weiterführung und Intensivierung des Erfahrungsaustausches zwischen Wissenschaft und Industrie ist erforderlich.

Wissenschaft und Forschung müssen der Industrie neue Erkenntnisse der Lasertechnik aufzeigen, welche die Industrie konsequent in Produkte umsetzt, die uns in diesem Land den Lebensstandard weiterhin sichern sollen.

Recent Trends in Laser Material Processing in the Japanese Automotive Industry

K. Shibata, Materials Research Laboratory, Nissan Research Center, Nissan Motor Co., Ltd., Yokosuka, JAPAN

1. Introduction

A review of the progress achieved in laser processing technology in Japan shows three distinct stages. The first stage, during the 1970's, was a period of searching for possible laser applications. In this period, very few companies had laser machines. In the second stage, during the 1980's, great strides were made as a result of a national project sponsored by the Ministry of International Trade and Industry (MITI). Now, in the third stage, a lot of companies are installing laser machines for use in manufacturing operations and are developing them in-house. Laser technology in the Japanese automotive industry was reviewed by the author in 1990 (1). Since then, a number of new applications have been developed, and laser technology seems to be finding more active use. This paper presents an updated picture of laser material processing in automobile manufacturing in Japan. Subjects covered include cumulative sales of laser systems in the Japanese automotive industry to date and a survey of their uses, recent applications in the areas of cutting, welding and surfacing, a discussion of why laser technology has been implemented, and the tasks that should be addressed to promote further progress of laser technology.

2. Recent picture of laser processing in automobile manufacturing in Japan

2.1. Unit sales of laser machines

Figure 1 shows the cumulative unit sales of laser machines in the Japanese automotive industry from 1982 to 1991. Nearly 2,000 CO_2 and Nd:YAG laser machines are estimated to have been installed in the last decade. In particular, since 1987, more than 200 CO_2 laser systems and nearly 150 Nd:YAG laser systems have been installed in the industry every year. As a result, cumulative unit sales of laser machines rose from a little over 500 in 1987 to more than 1,600 systems in 1990, more than a threefold increase. After the annual sales growth rate for both CO_2 and Nd:YAG laser machines reached a peak in 1989, the rate seems to have dropped off slightly. However, a favorable increase in sales is predicted. According to a report by the Yano Research Institute (4), CO_2 laser machines delivered to the automotive industry were estimated to account for 23% of all units sold in Japan in 1990.

434

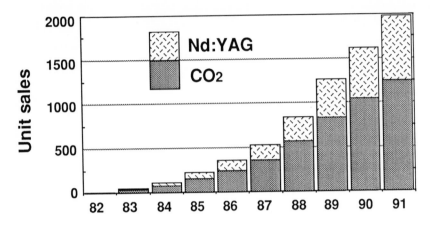

Fig. 1: Cumulative unit sales of laser machines in the Japanese automotive industry between 1982 and 1991

2.2. Processing applications of laser machines

Figure 2 shows a breakdown by processing application of the laser machines installed in the Japanese automotive industry to date. Breakdowns are also given for the respective uses of CO2 and Nd:YAG lasers. As a whole, 60% of the laser machines are used for cutting and drilling. Marking and welding account for 20% and 11%, respectively. These figures are considerably different from laser applications in North America and Europe, where welding is the main use and is estimated to account for more than 35% of all laser machine installations (2). The percentage of

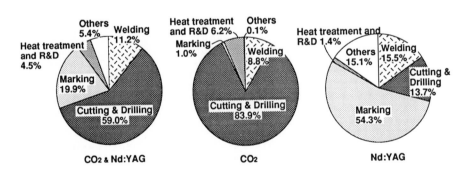

Fig. 2: Laser processing applications in the Japanese automotive industry

laser machines in Japan being used for welding seems to have increased slightly over the 1990 survey result (1), but the figure is still lower than that in Europe and North America.

The main applications of CO2 lasers are for cutting and drilling, accounting for 84% of all such machines in use. Less than 9% are used for welding. In contrast, 54% of all Nd:YAG lasers are used for marking, 16% for welding and 14% for cutting and drilling. Marking is by far the main application for Nd:YAG lasers. Other applications of Nd:YAG lasers include trimming and processing of IC components. Compared with the last survey (1), the proportion of Nd:YAG lasers used in welding applications increased to 16% from 11%.

Surface treatment applications do not seem to be so common in Japan as they are in Europe and North America. However, as will be shown in a later section, some interesting applications have been developed recently.

2.3. Power level

The power levels of laser systems installed in the industry in 1989 and 1990 are discussed for various applications. In the case of CO2 laser machines used for welding, 50% were in a power range of 1 to 1.5 kW and 50% had an output of 3 kW or greater. For cutting and drilling, nearly all machines had a power output of 700 W to 1.5 kW. For surface treatment, laser systems with a 3-kW or larger resonator were installed. In the case of Nd:YAG lasers for welding applications, 100-W to 600-W resonators were installed, but most had a 400-W resonator. For cutting and drilling, 100-W to 600-W resonators were implemented, but the majority were in a power range of 200 to 300 W. For marking applications, 50-W resonators were dominant, but some 200-W units were also installed. For scribing and trimming, lasers with an output of under 10 W were implemented.

2.4. Price change of high-power CO2 lasers

The going price of the laser resonator has been decreasing. According to reference 5, CO2 lasers are priced at under 10,000 yen per watt. A certain manufacturer was reported to be selling a 1500-watt CO2 laser for 10,000,000 yen. Figure 3 shows the change in the relative price of 5-kW CO2 laser resonators and the unit sales of high-power machines with an output of more than 2 kW. Following the implementation of a national project by the MITI between 1977 and 1984, the price of laser resonators decreased considerably. Since then, the price has been decreasing at a rate of 6 to 7% per year due to increasing sales and technical developments. The rate may be higher if resonators below 2 kW are included.

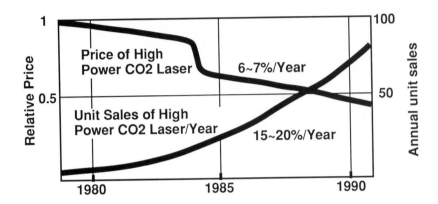

Fig. 3: Relative price of high-power CO2 laser resonators and unit sales of high-power laser machines

3. Applications

Laser processing applications in the automotive industry are discussed in the proceedings of the seminars and conferences, referred to in references 1-3, 6-10, 12-16 and 18. In this paper, recent trends and applications of this technology in automobile manufacturing in Japan will be described.

3.1. Cutting and drilling

At present, laser cutting is an indispensable technique for the trimming and piercing of panel components for trial and small-lot production. Laser cutting is used for body panel parts, such as bumpers, interior trim parts and outer body panels made not only of steel but also of many types of plastics. However, because the cutting speed is so much slower than that of press machines, laser cutting is not used for medium- or large-lot production. An on-line laser cutting process was recently installed by Ford and others for cutting holes in body panels for signal lamps, mirror and antennas, some of which vary to meet different regulations in other countries or customer demands. This system is expected to be used increasingly in Japan in the shift toward flexible manufacturing systems that can produce small lots of a larger variety of models. A Nd:YAG laser robot with an optical fiber system is thought to be suitable for this laser cutting process.

For some high-performance engines, an exhaust manifold made of stainless steel pipe has been adopted in place of a cast iron unit. Laser processing is used for cutting holes in the main port of the manifold. Figure 4 shows the laser machine used at Nissan and the processed pipe. This part is

300 to 500 mm in length and requires a three-dimensional
cutting process. A Nd:YAG laser with an optical fiber
system has been installed for cutting holes and provides a
relatively low-cost three-dimensional processing system.

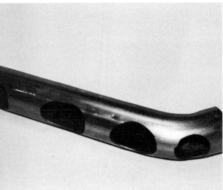

Fig. 4: Laser cutting system for exhaust manifold and
laser cut manifold

3.2. Welding

The situation for laser welding in the Japanese automotive
industry is different from that of the European and
American industries. In Europe and the U.S., laser welding
is considered to be superior to electron beam welding for
manufacturing transmission parts (10). Laser welding is
also used for manufacturing transmission parts in Japan,
but electron beam welding has been more common up to now.
In this paper, recent developments for laser welding of
parts made of sheet metal will be described.

Figure 5 shows a metal catalyst support used at Nissan. A
ceramic honeycomb made of cordierite has been the major
catalyst support used for automotive engines. A metal
catalyst support makes it possible to increase the contact
surface with the exhaust gas and to reduce gas flow
resistance and catalyst size, thanks to its thinner walls.
As a result, the maximum power of the engine can be
increased. Because of these advantages, production of metal
catalyst supports is increasing. A metal catalyst support
consists of a laminated structure of corrugated steel
sheets. A laser is used to weld the edges of the steel
sheets. Laser welding makes it possible not only to
increase the rigidity of the structure, but also to boost
productivity.

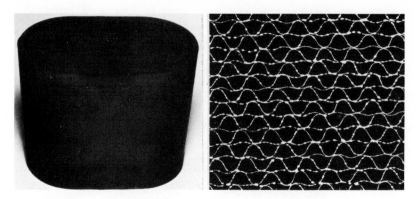

Fig. 5: Metal catalyst support

Recently, laser welding of body panel metals has drawn
more attention. Toyota has developed a laser welding
process for the metal sheets used to make body panels (11).
Metal sheets of varying thicknesses, gages and surface
coatings are welded into one piece before stamping. The
tailor-made panel sheet thereby provides the right
combination of metals needed for the body structure and
helps to reduce the weight of the car body. Most car
companies are developing this technology. Honda has
introduced laser welding systems using 1.4-kW CO_2 lasers to
hem hood panels (12). These technologies for welding car
bodies are expected to advance greatly in the future
because they contribute directly to the attainment of
lighter vehicles and greater design flexibility.

3.3. Surface modification

In spite of the successful application of laser hardening
to gear housings by General Motors in 1974, very few other
surface modification applications have been reported in
comparison with those for cutting and welding. In Japan,
the use of lasers in the production line was reported for
hardening of the grooves of piston rings for diesel engines
(13), cladding of stellite alloys on exhaust valve facings
for gasoline engines (14), and hardening of the trimming
edges of dies (8). Toyota has recently developed a laser-
based process for cladding the valve seat insert metal on
the cylinder head (15).

One unique application for surface modification is laser
textured steel sheet for outer body panels, which was
developed through collaboration between Kawasaki Steel Co.
and Nissan (16). The original concept of laser texturing
was developed by the CRM laboratory in Belgium (17). Steel
sheets for making body panels are given a dull surface

texture to prevent galling during stamping and to ensure good paint adherence. The dull surface pattern is transcribed from the surface of the roller used in temper rolling. The dull pattern on the roller surface was formerly produced by shot blasting or electrical discharge machining. However, the pattern produced with these methods was irregular. The irregularity caused surface undulations in the steel, which remained after painting. As shown in figure 6, a pulsed laser can produce a regular dull pattern on the surface of the temper roller, which works to increase the image clarity of painted outer body panels. This steel sheet has been used for the outer body panels of commercial vehicles at Nissan since 1987.

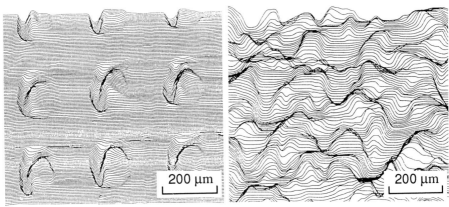

Laser dull steel sheet   Shot blasted dull steel sheet

Fig. 6: Three-dimensional surface roughness of dull steel sheet for outer body panels

Another application related to painting is the use of lasers for repairing paint defects on the car body surface, a technique that is under development (18). Small paint defects can be ablated by irradiation with a Nd:YAG laser with an output power of 200 mW that is transmitted through an optical fiber. The ablated holes are then filled with a compound and polished. The use of this technique is expected to make it possible to shorten the repair time after painting.

4. Why laser processing

There are few examples of applications where laser processing is the only means of obtaining added value. From the point of view of quality and/or cost, in some cases, previous processing methods are superior to laser processing. For example, electron beam welding provides

better quality than laser welding. However, the main reason
for choosing laser processing is freedom from the
restrictions imposed by the vacuum chamber, which is
necessary in the case of electron beam welding. With
respect to cost, laser welding is no match for spot
welding. However, the main reason for selecting laser
processing is the flexibility it offers for forming various
types of joints. One common characteristic of these laser
application examples is that laser processing reduces the
restrictions found in the previous methods and allows
greater flexibility in design and production. Seen from
another perspective, this means that the benefits of laser
processing can not be fully exploited by simply
substituting it for conventional methods. Not only the
production line, but also component designs should be
modified to make the best use of the flexibility and
controllability provided by laser processing .

5. Further tasks in laser processing

As mentioned earlier, many laser machines have been
introduced in the Japanese automotive industry. However,
the main applications of high power lasers are for cutting
and drilling. Relatively few lasers are used for welding or
surface modification. The use of lasers for welding in
particular has not increased as much as was expected. There
are many reasons for this. However, from the standpoint of
laser users, the following tasks are considered to be
important issues for the future progress of laser
technology in the future.

(1) Assurance of product quality in the production line:
In implementing a new technology in the production line,
one of the key factors is the question of how to assure
stable product quality in mass production operations. If
there is any doubt about quality assurance, production
engineers will be reluctant to adopt the technology. In
fact, it is often observed that production engineers are
concerned about problems caused by the instability of the
beam power, mode or orientation, or focus point. A process
monitoring system, for example, would help to resolve much
of the concern about the assurance of stable quality.

(2) Necessity for the accumulation of laser processing
data:
In order to use laser welding, part designs should be
premised on the conditions characteristic of laser
processing. A designer who lacks sufficient knowledge of
laser welding can hardly be expected to create a design
compatible with laser technology. Therefore, data should be
accumulated in databases regarding the features of
different laser machines, process factors in laser
processing, the mechanical properties of welded materials,
different joint structures and other related
characteristics.

(3) Need for lower cost laser systems:
This task is actually a request to laser machine builders.
In general, laser welding is more expensive than the
conventional methods such as spot welding and electron beam
welding. Even when the advantages of laser welding are
taken into account, laser processing should be more cost
competitive in order to promote greater use of laser
systems. Although the price of the resonator itself has
been decreasing every year, operating costs for optics,
gases and maintenance have not changed so much. The cost of
laser systems should be reduced further to promote greater
penetration.

6. Conclusion

In order to study laser processing trends in the Japanese
automotive industry, a survey was made of laser sales and
applications in the industry. It was found that nearly 200
CO2 laser machines and 150 Nd:YAG laser machines have been
installed in the industry annually in recent years. These
figures indicate that laser processing has been making
great progress in this industry recently. Examples of
recent laser applications were reviewed. The dominant
factors determining the implementation of laser processing
were discussed along with future tasks for promoting
further progress in laser technology.

Acknowledgments

This survey was conducted with the collaboration of many
Japanese manufacturers and venders of laser systems for
material processing. The author wishes to thank them for
their kind disclosure of this information. The author also
wishes to thank laser processing engineers at Nissan for
their valuable discussions.

References

(1)   K. Shibata, "Laser Applications in the Japanese
      Automotive Industry," Proc. of Laser-Material-
      Bearbeitung für den Automobilbau, 2nd Int.
      Anwenderforum, Bremen, Germany (15-16, Nov.,1990).
(2)   D. M. Roessler, "Laser Processing: Global Overview
      and Future Trends," Paper No.89541, Proc. of 21st
      Int. Symposium on Automotive Technology and
      Automation (ISATA), Wiesbaden, West Germany (6-10
      Nov.,1989).
(3)   D. A. Belforte, "Laser Applications in the Auto
      Industry," Proc. of 5th Int. Conf. on Lasers in
      Manufacturing (LIM-5), Stuttgart, West Germany, (13-
      14 Sept., 1988) H. Hügel ed., pp. 61-70.
(4)   Report of Yano Research Institute Ltd., (1991).
(5)   J. Nagasawa, Laser Focus World, May 1990, pp. 185-
      189.

(6)   Y. Iwai, et al., "Application of Laser Processing for Automotive Parts Manufacturing," Proc. of LAMP'87, Osaka, Japan (21-23 May, 1987), pp. 517-522.

(7)   M. Itoh, et al., "Application of Laser Processing for Automotive Manufacturing in Japan," Proc. of Laser Processing: Fundamentals, Applications and Systems Engineering (1986) pp. 201.

(8)   K. Shibata, "Laser Applications at Nissan," Int. Seminar on Applications of Laser Processing in Automobile Fabrication and Related Industries, The Welding Institute, Abington, Cambridge, U.K. (3-4 Dec., 1987).

(9)   K. Shibata et al., "CO2 and Nd:YAG Laser Cutting of Steel Panels and Plastic Components," Paper No. 89536, Proc. of 21st Int. Symposium on Automotive Technology and Automation (ISATA), Wiesbaden, West Germany (6-10 Nov., 1989).

(10)   M. Ogel and D. Guastaferri, "Why Laser," Int. Seminar on Applications of Laser Processing in Automobile Fabrication and Related Industries, The Welding Institute, Abington, Cambridge, U.K. (3-4 Dec., 1987)

(11)   Manual of Toyota Cercio, ed. by Toyota Motor Corp., (Nov., 1989).

(12)   K. Fukuda,"Laser Welding Applications at Honda," (in Japanese), Sectional Meeting for Automotive Engineering, The Japan Welding Engineering Society (5 Nov., 1991).

(13)   Y. Asaka, et al., "Laser Heat Treatment of Piston Ring Groove," Proc. of LAMP'87, Osaka, Japan (21-23 May, 1987) pp. 555-560.

(14)   K. Mori, et al, " Application of Laser Cladding for Engine Valve," Paper No. 885181, Proc. of 22nd FISITA Congress (1988).

(15)   K. Kawasaki, et al.,"Development of Engine Valve Seat Directly Deposited on Aluminum Cylinder Head by Laser Cladding Process," SAE Paper No. 920571, (24-28 Feb., 1992).

(16)   M. Imanaka, et al., "Development of High Image Clarity Steel Sheet by the Application of Laser Texturing," Proc. of 15th Biennal Congress, Controlling Sheet Metal Forming Processes, Dearborn, MI, U.S.A. (16-18 May, 1988).

(17)   J. Crahay, et al.,"Present State of Development of the Lasertex Process," Fachberichte Huttenpraxis Metallweiterarbeitung, Vol. 23, No.10 (1985) pp. 968-975.

(18)   H. Ito, et al., "Application of FHG Nd:YAG Laser Transmitted Through an Optical Fiber," Proc. of Laser'91, San Diego, U.S.A. (8-15 Dec., 1991).

V.
Thin Film Technology

Laser Ablation and Arc Evaporation

W. Pompe, B. Schultrich, H.-J. Scheibe, P. Siemroth, H.-J. Weiß
Fraunhofer-Einrichtung für Werkstoffphysik und Schichttechnologie Dresden

Introduction

The deposition of coatings by beam sustained PVD methods involves the following phenomena:

*   energy input into the target surface region by high power irradiation,
*   ablation of target material,
*   transport of the ablated material to the substrate,
*   condensation on the substrate.

Laser Pulse Vapor Deposition (LPVD) is based on laser beam power (Fig. 1). It distinguishes itself from nearly all other PVD methods by the degree of lateral concentration of the energy input and by its contraction to a short pulse leading to high instantaneous ablation rate and high excitation of the ablated material and the emission of micrometer-sized droplets. The latter is unacceptable for many applications. There is only one other PVD method, the Vacuum Arc Deposition (VAD), where related phenomena appear: Even under dc regime the elementary events are governed by localized and short-lived cathode spots which act as intense plasma sources. Again there is emission of droplets. In the following, LPVD and VAD are compared with respect to their deposition conditions, their advantages and their problems. It will be shown that a suitable combination of both, the LASER-ARC process, may open new possiblities for deposition of high quality films at high rate.

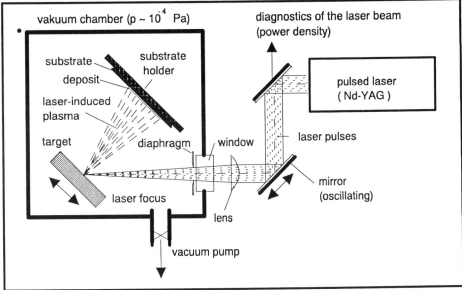

Fig. 1:   Scheme of the LPVD process

Laser Pulse Vapor Deposition

The laser sustained deposition process is usually based on short pulse lasers, especially excimer lasers, Q-switched Nd lasers and TEA-CO_2 lasers with typical pulse duration of some ten nanoseconds, pulse energies of some Joule and pulse power of some ten MW.

The mechanisms involved in the laser induced ablation process can be summarized as follows:
* Absorption of the laser radiation by excitation of electrons in a thin surface region (for metals < 100 nm)
* Transfer to thermal lattice energy
* Heat flow from the absorption region
* Evaporation-like or (at high intensities) detonation-like material ablation
* Ionization of the vapor by the incident laser radiation leading to increasing absorption of laser energy by the plasma cloud :
- in the case of energetic coupling of plasma and surface: large increase of the energy transfer into the surface (especially for the highly reflecting metals)
- in the case of weak coupling: shielding of the surface from the laser beam
* Emission of droplets by the effect of vapor (or plasma) pressure on the melt.

No other conditions than sufficient absorptivity of the target material are required. The process works in vacuum and in inert or reactive atmosphere.

As an essential advantage over charged particle beams, the laser beam can be easily focussed to a narrow spot on the target also in the case of high intensity. Together with the small pulse duration, this leads to extreme power densities of 10^7 - 10^{11} W/cm^2. (For other applications as laser induced nuclear fusion, values of 10^{14} W/cm^2 and more are realized.) Under such conditions an intense flow of highly excited particles is ablated from the target. This combination of high particle flux (up to 10^{19} particles/cm^2 ·s) and high particle energy (up to some 100 eV) determines the unique position of the LPVD amid the PVD processes. The corresponding high instantaneous deposition rate (up to 100 µm/s) combined with the high energy carried by the particles effects the activation of the growing layer, which results in the special qualities of LPVD films: dense, fine-grained, often amorphous coatings, low substrate temperature, smooth surface on the nanometer-level. Also there are the possiblities of nearly stoichiometric transfer of complex target compositions because of the athermal detonation-like ablation at higher intensities and of good control of the pulse-for-pulse deposition, allowing a well-defined production of multilayers with thicknesses down to some nanometers.

The short pulse duration τ necessary to assure the special advantages of LPVD restrict its application: Because of the limited repetition rate f (usually between 1 and 100 Hz) the averaged laser power and deposition rates are lower by a factor $\tau \cdot f < 10^{-5}$. Hence, the LPVD process in its current version is especially suited for the deposition of coatings with special qualities as X-ray mirrors or high-temperature superconductors where the rate and effectivity are of less importance. But on the other hand, the droplet emission is especially critical in this case.

Because of the strong dependence of the ablation process on energy density, pulse duration and laser wavelength besides the material parameters, characteristic values for the ablation can only be given as a crude orientation: The laser energy input of about 10^4 kJ/cm^3 necessary for ablation overcomes the heat of evaporation of about 50 kJ/cm^3 by a factor 200. From this an instantaneous ablation rate of 1 cm^3/s and an averaged of 10^{-5} cm^3/s follow, corresponding to deposition rates of 100 μm/s (instantaneous) and 1 nm/s (averaged) for a 1 dm^2 substrate.

The droplet emission is apart from inhomogeneities in the target material caused by the action of the plasma pressure on the molten superficial layer splashing it away. These effects are more pronounced for higher power intensities and for lower melting target materials. The droplet emission can amount to more than 50 % of the total ablation yield, depending on experimental parameters. A variety of approaches has been proposed for reducing the splashing of particulates besides the trivial ones of low intensity and increased target homogeneity: Mechanical, pulse synchronized filters, centrifugal effect of high speed rotating targets or the collision of two intersecting ablation clouds. They all use with some success the lower velocity and the higher stability of the trajectories of the heavier droplets. The additionally required equipment and the lower yield of the thus modified deposition techniques restrict their wider application so far.

Vacuum Arc Deposition

The basic equipment consists of the target as cathode in a low voltage circuit, arranged in a vacuum chamber representing the anode (Fig. 2). Working in an inert or reactive atmosphere is possible. The arc is burning in the vapor of the evaporated target material. Hence, the process must be ignited by an initial evaporation pulse realized by mechanically contacting the cathode with an ignition electrode or by a high-voltage pulse.

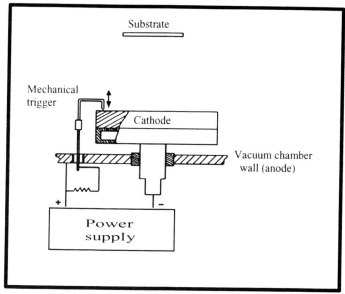

Fig. 2:   Scheme of the VAD process

The mechanisms involved in the arc induced ablation process can be summarized as follows:
* Ionization of the vapor by the electrons
* Acceleration of positive ions of the vapor cloud towards the cathode
* Pinching of the arc current by magnetic forces to small spots on the cathode
* Heating of the cathode spots
* Thermal release of electrons, ions and neutrals from the cathode
* Ionization of the vapor by the electrons
* Burning of the elementary arcs for about ten nanoseconds each in rapid random succession

* Random movement of the cathode spots over the target surface superposed by a
  deterministic motion controlled by the magnetic field.

Typical working conditions are voltages of some 10 V and currents between some ten and some 100 A . Considering the small dimensions of the cathode spots of about 10 μm, this corresponds to local current densities of about 10^8 A/cm^2 and local power densities of about 10^9 W/cm^2. That means that the duration and power density are similar to those of LPVD. Differences exist with respect to the succession and to the instantaneous size of the events (Tab. 1).

Table 1: Ablation conditions for LPVD and VAD process

	LPVD	VAD
Event duration / ns	10-100	10-50
Event energy / J	0,1-10	10^{-6}-10^{-5}
Event power / MW	10-100	10^{-3}
Spot dimension / mm	0.1-10	0.01
Local energy flux density / W/cm^2	10^7-10^{10}	10^9
Event repetition rate / Hz	1-100	$>2 \cdot 10^7$
Mean power / W	< 100	500-5000

The vapor has a high percentage of ions (50 - 100 %) with kinetic energies in the range 10 to 100 eV (Table 2). This energy range is optimal for most deposition processes. This is illustrated by the fact that with LPVD and VAD the amorphous carbon coatings (DLC) with the highest similarity to diamond can be produced. With bias assisted Vacuum Arc Deposition, DLC coatings with nearly complete sp^3-diamond bonds are reported.

It turns out that the single cathode spots carry a definite current so that higher current means a larger number of cathode spots whereas the processes in the single spots are not essentially effected. Hence, the ablated material per charge is nearly independent on current, it amounts to 50 ... 500 μg/C for metals. This means the energetic efficiency is lower by a factor 10 ... 100 compared with evaporation. The growth rate can only be discussed in relation to a certain substrate area, because of the source is of the point type as in the LPVD process. Typical substrates of 1 dm^2 correspond to 100 nm/s.

The comparison of the ablation conditions for the LPVD and the arc process shows that the single events on the target are comparable with respect to their duration and energy flux density, whereas the spot dimensions are much smaller and the event repetition rates are much higher for the arc process. Much higher average power inputs are realized for the arc leading to much higher average ablation and deposition rates. Essential differences concern the variability of the processes:Contrary to laser ablation, the arc can only be varied over a restricted range of intensities and the target erosion is difficult to control because of the statistical nature of the events. In summary: The arc process is more determined but less deterministic.

Table 2: Deposition conditions for LPVD and VAD process

	LPVD	VAD
Source type	point-like	point-like
Deposition regime	pulsed	(mostly) continuous
Particle energy / eV	10-1000	10-100
Ionization	0-100 %	50-100 %

The droplet emission is caused by phenomena similar to the related LPVD processes (Table 3): The melt of the cathode spot is splashed away by the gas pressure produced by the localized arc discharges of short duration. Again these effects are more distinct for lower melting materials. But compared with LPVD, there are important pecularities as the random motion of the spot and the high degree of ionization of the ablated material. Both items are used for reducing the droplet emission apart from the trivial way of geometrical shielding of the substrate, which suppresses besides the droplets the high energetic particles, too, necessary for high quality deposition: The spot can be fastly moved over the target surface in a controlled manner by an external magnetic field ("steered arc"), thus avoiding the burning in at a fixed place. More effective, but also more expensive is the installation of a mass filter, which separates the ions for deposition by a magnetic field.

Table 3: Conditions for reduced droplet emission in LPVD and VAD processes

	LPVD	VAD
Target materials	high melting point	
Energy flux density	low or very high	(not controllable)
Event duration	short	(not controllable)
Duration on same target area	short	

LASER-ARC process

The new method of laser-assisted vacuum arc deposition (LASER-ARC) intends to combine the high efficiency of arc deposition with the easy controllability of laser beam processes in order to overcome the complementary deficiencies of both. In the LASER-ARC, the repetitive radiation of a small pulsed laser is focussed on a target biased as a cathode (Fig. 3). By each of the laser pulses a small amount of plasma is induced which initiates a vacuum arc discharge of limited duration determined by the current source. Because of the limitation of the arc burning time, the cathode material is eroded and evaporated only in the vicinity of the ignition point. By moving the target or the laser beam, the location of the laser focus and of the erosion area can be positioned in such a way that the cathode is evenly eroded in the process of repeated initiation and extinction of the arc. Limitation of the arc burning time avoids the formation of a large melt pool, consequently droplet emission will be essentially reduced. In this way, high quality films of diamond-like carbon have been produced for optical and tribological applications with the high rate typically for arc deposition.

Thickness of the deposit is easily controlled by counting the pulses. In this way, complex layer stacks can be obtained by using several targets alternately. This potential of the LASER-ARC method was successfully demonstrated by the deposition of multilayer systems with thicknesses of the single layers in the ten nanometer range.

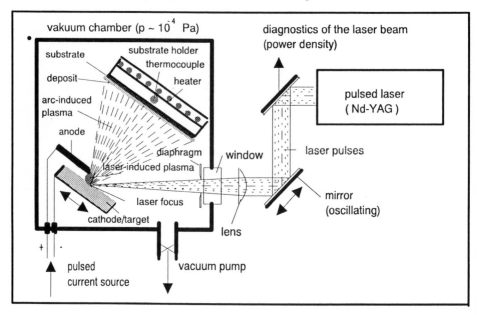

Fig. 3:   Scheme of the LASER-ARC process

Characterization and Applications of PLD Oxide Ceramic Films

M. Alunovic, J. Funken, E.W. Kreutz, H. Sung, A. Voss, Lehrstuhl für Lasertechnik, Rheinisch-Westfälische Technische Hochschule Aachen, Aachen, FRG
G. Erkens, O. Lemmer, T. Leyendecker, CemeCoat GmbH, Aachen, FRG

Introduction

The pulse laser deposition (PLD) exhibits various consecutive processes (1,2) separated in space and time such as the energy coupling into the target material, the removal of the material from the target, the transfer of the target material as vapour and/or plasma to the substrate via the gas phase and the growth of thin films onto the substrate. The properties of the deposited films strongly depend on their growth conditions, which are highly influenced by the surface temperature and the mobility of the particles at the surface either of the substrate or of the films deposited.

PLD includes several convincing arguments for technical applications: First, the deposition of various kinds of materials is possible since the processes are not limited by the charging either of the target or of the substrate. Secondly, the method is capable for the stoichiometric transfer of the target material. Third, the deposition rate is scaled up with a decrease in processing time depending on laser parameters and processing variables. As a consequence PLD is used as a technique for the deposition of films for a variety of applications such as optical, electrical, mechanical and tribological coatings. The latest trends include semiconductor oxides (3), ferroelectric materials (4-6) and photorefractive materials (7).

The pulse laser deposition generally involves a vapour and/or plasma close to the target within the interaction zone. The nature, the speed and the charge of the spezies present in the gas phase substantially affect the properties of the deposited films via the deposited energy from the gas phase.

The formation of Al_2O_3 and ZrO_2 films with laser radiation was investigated as a function of laser parameters (beam geometry, fluence, power density distribution) and processing variables (rf bias, geometric arrangement of target and substrate). Structure and properties of the films were investigated with SEM, X-ray diffraction, EDX and mechanical testing by the ball grindig method, the scratch test and the measurement of current density potential curves. The different film structures ob-

tained by matching the laser parameters and processing variables are discussed in view of applications in particular to
wear, corrosion and thermal protection of material surfaces.

Experimental

PLD is performed in a high vacuum chamber (base pressure 6 x
10^{-5} mbar). The radiation provided by different laser radiation
sources (Table 1) is incidenting through a window onto the target.

laser	wavelength [μm]	average power [W]	repetition rate [Hz]	mode of operation	pulse length [nsec]
TEA CO_2	10,6	26	10	pulsed wave	200; 3500
CO_2	10,6	750	-	continuous wave	-
Nd:YAG	1,06	20,5	5000	pulsed wave	50
Excimer	0,308	35	20	pulsed wave	20

Table 1: Laser parameters.

The targets are sintered discs of ceramics (Al_2O_3, ZrO_2) with a
purity of about 99%. The removed material is deposited onto the
substrate parallel to the target. Substrates are   plates
(10 x 10 mm^2) of stainless steel or hard metal (WC grains embedded in Co).

The distance between substrate and target is adjustable in the
range of 0.5 - 4 cm allowing the deposition with the luminous
laser induced plasma either in contact or in non-contact with
the substrate. The substrate holder is electrically insulated
against the chamber and a rf generator is connected to the substrate for the cleaning of the surface before deposition and
for the use as bias during deposition. Thin films are prepared
either for matched (alternating phases of electrical and optical field) or for unmatched (phase of superposed electrical and
optical field followed by fieldfree phases) conditions. The
sputter cleaning of the surface induces a heating of the substrate up to 550 K, temperatures up to 1100 K are obtainable by
resistance heating.

For the analysis of the deposited thin films the substrates are
cutted on the backside to obtain a predetermined breaking line.
After deposition the plates and films are cracked yielding a
cross-section through the film for SEM analysis. Structure, topography and stoichiometry are investigated with x-ray diffrac-

tion, SEM and EDX. Mechanical properties of the films are tested with the ball grinding method (wear) current density-potential curve (corrosion) and scratch test (adhesion).

Results

At small fluences the weight loss of the target (Al_2O_3, ZrO_2) increases with fluence and then it decreases at high fluences to a nearly constant value after having passed a maximum (2, 8). The weight loss of the target (Al_2O_3, ZrO_2) as a function of processing gas pressure also exhibits a maximum at a certain gas pressure with its position depending on the nature of processing gas (2, 8).

Fig. 1 shows the influence of fluence on the properties of the deposited films. For small fluences the films are less dense with a lot of pores (Fig. 1a), for high fluences the films are dense with a glassy structure (Fig. 1b) giving no indication of pores. X-ray diffraction (2) yields an amorphous structure.

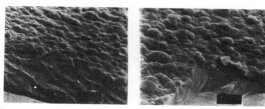

4µm   2µm

Fig. 1: Morphology of Al_2O_3 deposited on stainless steel with pulsed CO_2 laser radiation for low (10 J/cm^2, left hand side) and high (27 J/cm^2, right hand side) fluence.

Within the experimental resolution EDX measurements reveal that the deposited films are stoichiometrically transferred from the target.

The influence of the rf bias is demonstrated in Fig. 2. Without bias the film structures are equal to those obtained for various fluences (see above and Fig. 1).

454

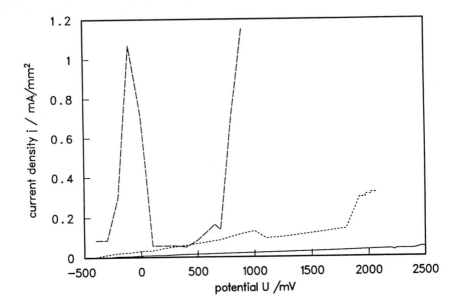

Fig. 2: Current density potential measurement of Al_2O_3 deposited on stainless steel with pulsed CO_2 laser radiation (—— uncoated substrate ---- deposition with bias ······ deposition without bias).

As evaluated from SEM investigations, the films prepared under matched conditions are dense with a glassy structure (Fig. 2b), those under unmatched conditions give evidence for large grains (Fig. 2c).

The film thickness in the range of 2-200 μm, as obtained from the ball grinding method (2), strongly depends on the growing conditions governed by the laser parameters and processing variables. For extended deposition times the cross-section of the calottes is regular; for short deposition times the cross-section of the calottes is irregular as seen from optical microscopy indicating the adhesion at the substrate surface. The adhesion, for example, of Al_2O_3 on stainless steel or hard metal is generally poor but improves with the use of a bias (2).

The current density-potential curves of coated substrates with matched and without bias are shown in Fig. 3.

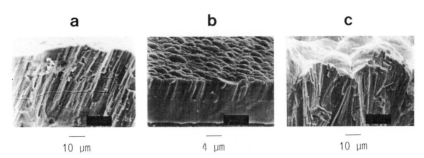

a	b	c
10 μm	4 μm	10 μm

<u>Fig. 3</u>: Morphology of Al_2O_3 deposited on hard metal with pulsed CO_2 laser radiation at high (27 J/cm^2) fluence (a) without bias b) matched bias c) unmatched bias).

The film deposited without bias shows a stronger increase of the current density at high positive potentials than the film which was deposited with rf bias. The latter shows no indication of dielectric breakdown up to a potential of 2500 mV.

The volume loss of an Al_2O_3 film deposited under matched bias conditions is in the range of 1.46 x 10^{-3} mm^3/min, which is nearly the wear value of the polycrystalline target material of 1.56 x 10^{-3} mm^3/min. The amorphous films always achieved lower values than the polycrystalline target.

Similar results are obtained by varying the processing variables such as the distance between target and substrate allowing the deposition with the luminous laser-induced plasma either in contact or in non-contact with the substrate. Al_2O_3 films deposited on cold substrates far away from the target without plasma contact are amorphous or consist of nanocrystallites with less than 10 nm diameter, because there is no indication for x-ray diffraction (2). Al_2O_3 films deposited on hot substrates very close to the target with plasma contact show various structural properties according to the local deposition conditions. The film in the center of the plasma zone consists of small grains and a smooth surface. At the edge of the plasma zone the dimension of the grains increases with increasing surface corrugation. Outside the plasma zone the film shows less uniformity. Outgrowth suppresses the growth of the surrounded cristallites. In further distance from the plasma zone the cristallites grow columnar. The structures observed for ZrO_2 films are comparable to those of Al_2O_3 films. The regimes of the corresponding structures are shifted towards the center of the plasma zone. The stucture with fine grains and a smooth surface is missing.

Discussion

During PLD the energy coupling into the target depends on its optical constants (refractive and absorption index) at laser wavelength. For ceramic target materials the use of CO_2 laser radiation is suitable because of the low reflectivity. The removal of the ceramic target materials as a function of fluence is dominated by two regimes as discussed in detail recently (2): a vapour dominated one with increasing weight loss and a plasma dominated one with decreasing weight loss as a function of fluence, respectively. The transfer of the removed ceramic target material as a function of processing gas pressure also shows two regimes (2): a plasma dominated one with increasing weight loss and a collision dominated one with decreasing weight loss, both of them depending on the composition of the gas. The physics and the chemistry of the gas phase highly influence the removal of the target material and the film deposition by the interaction with the gaseous particles, which originates in the transfer of particle energy onto the target or the film and in chemical reactions at the surface of the target or the film.

Various structures of the deposited films thus arise from the laser parameters (wavelength, power density distribution etc.) and processing variables (target substrate distance, substrate temperature etc.). At laser fluences above the threshold of evaporation irradiance the dominant mechanism is vaporization, at higher laser fluences the plasma formation (2). As a consequence the transfer of target material from the target to the substrate plays the major role giving clear evidence for the need to get information on the transfer processes with high temporal and spatial resolution.

Deposited particles have a certain surface mobility, increasing with increasing particle energy and/or increasing substrate temperature. By surface diffusion the particles loose energy until they have reached stable positions in equilibrium at surface temperature. The degree of relaxation is ruled by the ratio between deposition rate and surface diffusion. The influence of bulk diffusion in the film becomes increasingly important with growing deposition rate. For deposition rates in the range of one monolayer per pulse (repetition rate 10 Hz) with substrate temperatures below 500 K, the relaxation is determined by the surface diffusion. The bulk diffusion within the film deposited is still negligible because of the higher activation energy.

The variation of the fluence on the target enables the variation of the energy of the removed particles. Evaporated particles with energies just above the evaporation energy have low velocities so that the mobility on the surface is low with a lot of particles stuck on the surface before having reached a stable site. At high fluences with the generation of a material vapour/plasma the energies of the removed particles are high so that the mobility on the surface is higher than at low fluences resulting in a pronounced surface motion.

The variation of processing variables enables several possibilities to influence the relaxation of particle energy such as variation of fluence, substrate temperature and target substrate distance in combination with the superimposed rf-field. Fig. 2 exemplarily shows the schematic representation of the structures of deposited Al_2O_3 film grown at various bias modes.

Under matched bias conditions deposition and sputtering cycle with the processing gas is alternating (100 msec Ar^+ sputtering, 6 μsec deposition without bias). After deposition the atoms and molecules relaxate. The following sputtering of the Ar^+ ions disturbs the relaxated arrangement by collisions and the generation of grain boundaries is avoided. The film grows in a glassy structure, since the growth conditions are essentially determined by the sputtering.

Under unmatched bias conditions the power is coupled into the evaporating material vapour / plasma and accelerates the particles. The deposited atoms and molecules on the surface have a higher mobility than particles which are deposited without or with matched bias. The high degree of relaxation allows the growth of large grains. After the breakdown of the laser-induced plasma the rf power becomes reflected (6 μsec deposition with rf bias, 100 msec pause). The processing gas is not excited. No sputtering of the surface with processing gas ions occurs. The grain boundaries are not destroyed growing unhampered with each deposition cycle, the growth conditions are governed by the deposition.

For films deposited at short distance between substrate and target with high energies of the deposited particles the deposition occured under non-equilibrium conditions before the cooling by expansion via particle transfer was finished. The deposited particles consequently exhibit a very high mobility, which is comparable to a high substrate temperature. According to the Thornton model (9), describing the structures of sputtered films grown on heated substrates, the structure of the Al_2O_3 film in the center of the plasma zone corresponds to higher substrate temperatures ($T > 0.8\ T_m$). With increasing distance from the center of the plasma zone the particles in the vapour / plasma reach a higher degree of cooling before deposition. Hence, these films exhibit structures similar to those of films, which were deposited at lower temperatures ($T < 0.8\ T_m$). In case of ZrO_2, the $0.8\ T_m$ value is higher than the one of Al_2O_3 because the melting temperature strongly differs (T_m $(Al_2O_3)=2305$ K, $T_m(ZrO_2)=2895$ K).

To a large extent the mechanical properties of the deposited films depend on their structure. The resistivity against corrosion is improved by depositing the film under matched bias conditions, since glassy structures have less surface for a chemical attack and there are no grain boundaries throughout the film. In films with grain boundaries deposited without bias the corrosive medium may diffuse along the grain boundaries allowing corrosion on the film substrate interface. The wear against diamond also is lower for amorphous material than the wear of crystalline Al_2O_3 because of the higher ductility of the amorphous material. The crystalline material is more easily abrased by the diamond powder.

Conclusion

Depending on the laser parameters and processing variables, PLD-films of Al_2O_3 and ZrO_2 show a broad variety of different structures quite similar to sputtered films on heated substrates.

PLD-films with glassy structure exhibit good mechanical properties. Especially if the temperature of the substrate must be kept below 500 K, the application as protective film against corrosion is promising. The use of amorphous ceramic films is also promising for combined resistance to wear and corrosion. A possible technical application is the protection against erosion in wet tribological systems. PLD-films with columnar structures, which were deposited at 500 K with plasma contact, are interesting for heat protection. A high thermal gradient may be compensated by elastic expansion or contraction of the columns. The application as thermal shock protection for steel or other temperature sensitive materials like alloys with age hardening at about 1000 to 1100 K is obvious. The solution is a short deposition time or deposition at low temperature. The PLD fulfils both possibilities.

Acknowledgement

The authors are very indebted to the Ministry for Research and

Technology of the Federal Republic of Germany for financial support within the framework of the project 'PVD at low temperatures and new coatings' (13 N 5611).

References

(1)   J.T. Cheung, H. Sankur: CRC Critical Reviews in Solid State and Material Science 15 (1988) 63.

(2)   J. Funken, E.W. Kreutz, M. Krösche, H. Sung, A. Voss, G. Erkens, O. Lemmer, T. Leyendecker: Surface and Coatings Technology 52 (1992) 221.

(3)   A. Slaoui, E. Fogarrasy, C. Fuchs, P. Siffert: J. Appl. Phys. 71 (1992) 590.

(4)   H. Tabota, T. Kawai, S. Kawai, O. Murata, J. Fujioka, S. Minakata: Appl. Phys. Lett. 59 (1991) 2354.

(5)   D. Roy, S.B. Krupanidhi, J.P. Dougherty: J. Appl. Phys. 69 (1991) 7930.

(6)   N. Maffei, S.B. Krupanidhi: Appl. Phys. Lett. 60 (1992) 781.

(7)   F.E. Youden, R.W. Eason, M.C. Gower, N.A. Vainos: Appl. Phys. Lett. 59 (1991) 1929.

(8)   J. Funken, E.W. Kreutz, M. Krösche, H. Sung, A. Voss, G. Erkens, O. Lemmer, T. Leyendecker: Appl. Surf. Science 54 (1992) 141.

(9)   J.A. Thornton: Ann. Rev. Material Science 7 (1977) 239.

Preparation of Thin Films by Laser Induced Deposition (Laser-PVD)

H. Haferkamp, C. Möhlmann, Laser Zentrum Hannover e.V., FRG

Introduction

Since powerful lasers were developed it has been possible to deposit thin films by ablation respectively evaporation [1,2]. At this rather young thin film deposition method a high power laser, e.g. an UV-excimer laser, is focussed onto the target material which shall be evaporated. An additional gas can be let into the vacuum chamber in order to influence the chemical composition of the film. The target material is evaporated and condenses on a nearby substrate with a defined thickness.

A specific change in the process parameters will lead to a wide range control of thickness, stoichiometry and structure of the films. This method makes it possible to synthesize films, especially of compound materials, and film structures like multilayers in the range up to a few micrometers with unique characteristics, that can hardly be reached by other methods.

This method has already been used successful to prepare thin films of superconducting materials like $YBa_2Cu_3O_{6+\delta}$ [3,4,5]. It became obvious that especially Laser-PVD is able to keep the correct composition of such multinary compounds. This property is very important to ensure the functional characteristic like the superconductivity in the case of $YBa_2Cu_3O_{6+\delta}$. Thermal evaporation of such compounds would lead to heterogeneous composition over the film thickness.

In this paper we describe the typical process characteristics of Laser-PVD. Different materials were used for evaporation respectively deposition. A new attempt was made to prepare molybdenum thin films for use in optical parts.

Description of the Laser-PVD Process

Evaporation and Expansion

Usually short laser pulses in the ns-range, as supplied by UV-excimer lasers, with powers of about 10^7 W are applied on material surfaces. Cw laser irradiation would create a larger melt bath on the surface of the target. This way energy consumption for melting will be higher, thus reducing the efficiency for the evaporation process. Also the creation of larger droplets will disturb the film homogeneity. Focussed laser pulses reach a power density up to 10^{10} W/cm^2. Such intense photon fluxes will induce several absorption and evaporation effects on the target surface and in the created vapour respectively plasma plume. Different approaches have been made to model this process. At dense materials like metals the radiation is absorbed only in a surface layer smaller than the light wavelength, e.g. for 248 nm on copper an optical penetration depth of 11 nm was calculated [6]. The absorbed light energy changes into thermal energy. The thermal diffusion length δ during a laser pulse is given by [7]

$$\delta = 2 \sqrt{\varkappa \, t_p}$$

with the thermal diffusivity $\varkappa$ and the laser pulse length t_p. For copper δ is 2,2 µm, therefore a magnitude larger than the wavelength. This layer will at least be molten and partially be evaporated.

In order to show the degree of material removal, a solid titanium target was irradiated by 308 nm excimer laser pulses (50 ns, 1000 pulses) normal to the target surface in a vacuum chamber (p=10⁻²mbar). The results are depicted in Fig.1. The volume was calculated from the measured profile of the ablated spots, either by conical, ellipsoidical or spherical approximation. At small energy densities (<3J/cm²) the ablated volume increases and reaches a maximum between 3 and 4 J/cm² due to increased energy coupling into the surface. Towards larger energy densities the ablated volume decreases because of energy absorption in the expanding plasma plume and a geometrical effect, i.e. the hole is drilled fastly, the area of the hole walls is enlarged and the light has a larger angle of incidence so that the effective intensity decreases.

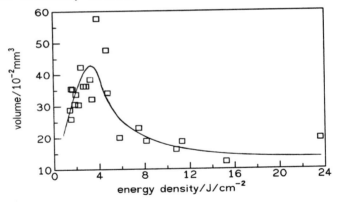

Fig.1: ablated volume of a Ti surface versus energy density of each 1000 excimer laser pulses (308 nm)

Since enough laser energy is converted into thermal energy, the evaporation process starts. Low laser intensities produce only thermal evaporation and no interaction between the material vapour and the laser pulse occurs. The flight velocity of the target atoms increases with the logarithm of the laser intensity [8]. Beyond a first threshold intensity the interaction between the laser photons and the vapour atoms start. Via inverse bremsstrahlung and multiple-photon absorption the material vapour is heated and a plasma is created. Because of the hypersonic velocity a shock wave drives the front of the expanding vapour. This is the so called Laser Supported Combustion domain. Beyond a second threshold at about 10^9 W/cm^2 (dependent on the wavelength) a Laser Supported Detonation wave ($v \geq 10^6$ cm/s) drives the atoms. The laser photons are strongly absorbed in the dense detonation wave. According to an one dimensional, hydrodynamic model [9] the velocity of this detonation wave is proportional to $I^{1/3}$.

In the beginning phase of the laser induced plasma the inner temperature will be in the range of 10^5 K. This produces an optical broad band emission in the XUV range with maxima of a few ten nanometers [10]. According to the Saha-equation the electron and ion density can be calculated. For electron densities of a laser induced copper plasma we got as results $1,93 \cdot 10^{14}$ cm^{-3} ($I=10^7$ W/cm^2) that increased to $1,40 \cdot 10^{19}$ cm^{-3} ($I=10^{10}$ W/cm^2).

During the initial, dense stage of the plasma the absorption via inverse bremsstrahlung is dominant while in later, diluted stages the multi-photon absorption becomes dominant for UV radiation.

Optical Time-Of-Flight Measurements

For the determination of the particles velocity in the expanding plasma time-of-flight measurements were carried out. In this case two fluorescence lines of cadmium (Cd, neutral atom: 508,6 nm; Cd$^+$, single charged ion: 538 nm) were observed time resolved at various distances to the target by means of plasma image formation and a monochromator (Fig.2). Laser irradiation of 248 nm and $2,8 \cdot 10^{10}$ W/cm^2 was used and vacuum pressure was 10^{-2} mbar.

At a distance of about 12 mm the Cd$^+$-signals diminished and disappeared. The signals from the neutrals were observed farther to a distance of about 20 mm. The inner, higher excited part of the plasma can be distinguished this way. In general, the ions have a higher velocity than the neutrals. This can be explained with the occurence of temporal electric fields in the plasma that leads to higher acceleration. The expansion velocity at the first four millimeters was determined to $1,7 \cdot 10^6$ cm/s (Cd) and $2,5 \cdot 10^6$ cm/s (Cd$^+$) and is in good agreement with theoretical predictions. This is equivalent to a kinetic energy of 168 eV respectively 364 eV. These energies are one to two orders higher than electron or ion beam deposition methods. At the first ten millimeters there is a considerable decrease in velocity. It results from collisions between the particles. Beyond 10 mm distance from the target the velocity stabilizes. The plasma density is now low enough so that collisions are strongly reduced.

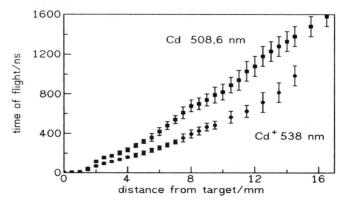

Fig.2: time-of-flight signals of Cd (508,6nm, neutral atom) and Cd⁺ (538nm) in a laser induced plasma

Deposition Process

At a distance of a few centimeters the material vapour reaches the substrate. Thin film growth is mainly influenced by deposition rate, substrate temperature, vacuum pressure, and surface morphology of the substrate. These factors determine microstructure and adhesion of the films. Laser-PVD can be applied for film thicknesses from a few atomic layers up to a few micrometers. The influence of laser energy density in case of depositing titanium nitride shows an almost linear dependence (Fig.3). This is in accordance to the ablation measurements (Fig.1). A maximum seems not to be reached as seen in Fig.1 at about 4 W/cm^2 because the target rotated during ablation in contrast to the measurement of the ablated volume.

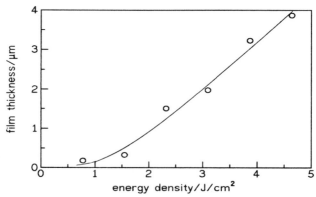

Fig.3: Influence of laser energy density on film thicc- ness in case of TiN (excimer laser 308 nm, 100-600 mJ, 5000 pulses, distance target-substrate 30 mm, N_2 as reactive gas)

The lateral distribution of the film thickness was investigated for three samples of TiN/TiO$_2$. The Ti-target was irradiated under N$_2$/O$_2$ reactive gas. An ellipsometric measurement of the partly transparent film revealed a refractive index of n=1,58. The visible interference color rings were counted under normal incident light that was reflected (n·2d=k·λ, d: film thickness, k: order of ring). The thickness pofiles are given in Fig.4. They show that most material (half maximum thickness) is ejected from the target in a full angle of 38 degrees. There is no qualitative change with varying laser energy density. The lowest profile in Fig.4 is described by a function of A·cos^3(x/B)+C with x as distance from the middle of the film.

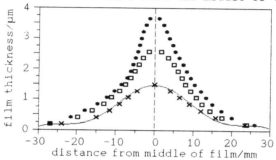

Fig.4: film thickness profile of TiN/TiO$_2$ films (5000 pulses, 308 nm excimer laser, different energy density, distance target-substrate 30 mm, N$_2$/O$_2$ reactive gas, pressure 2 mbar)

Molybdenum Films and Their Optical Properties

Thin metal films with thicknesses smaller than the wavelength of the optical radiation can be used as anti-interference coatings, as absorbers for the infrared, as semitransparent mirrors in interferometric devices and beam splitters, as mirrors [11], and in the case of Mo as resistive coating for IR-components. Also multi-layers in combination with Si or B$_4$C can be used as mirrors for soft-X-rays. Another application of Mo-films is known as temperature sensor. Typically, evaporated films of refractory metals are fine grained and porous with lower reflectivity than bulk material.

In the present work, a new approach was made to get sufficiently dense Mo-films. The target was irradiated with 308 nm excimer laser pulses (50 ns) and intensities up to 1,6·10^{8}W/cm^2. Film thicknesses varied between 100 and 400 nm. SEM observations showed no evidence for a micro-structure or grains larger than a few ten nm. Results from XRD-measurements can later be reported, TEM measurements for a better determination of the micro-structure are still to be done. Mo-films produced by magnetron sputtering in an Mo-Si-multilayer system [12] were polycrystalline with a [110] texture in growth direction. Each layer had a thickness of 2,6 nm.

Substrates of Cu, Si, and polycrystalline Ge were used in the present work. Optical microscopy shows the occurence of μm-size Mo-droplets in the films. They can be minimized by using lower laser intensities. The adhesion of the Mo-film on Cu was not sufficient. The results of optical measurements are given in Table 1.

Table 1: optical properties of Mo-films on different substrates

	R	A	S
Mo/Si	0,79	0,14	12 ppm
Mo/Ge	0,91	≤0,09	593 ppm
Mo/Cu	0,93	≤0,07	1100 ppm
Mo bulk	0.98	≤0,02	

R: reflectivity at 10,6 μm, reference: Au-mirror
A: absorption at 10,6 μm by adiabatic laser calorimeter,
   Mo/Ge, Mo/Cu: estimated
S: scattering at 633 nm

The relatively high absorption is probably due to the water va-pour content in the films that depends on the vacuum pressure. An increase from $\cdot 10^{-5}$ to $\cdot 10^{-2}$ mbar reveals the double absorption value. Progress can also be achieved by a better pretreatment of the substrates. Best Mo-films prepared by other thin film deposition methods have absorptivities between 2 and 4 %. The scattering of the film is on the Si-substrate rather low and comparable to values of the substrate surface itself. Reflecti-vity of 93 % could be achieved on Cu. Recent works from Hoffman et al. [13] showed a maximum reflectivity of 97,8 % by using arc evaporation with a substrate bias. Polished Mo bulk material shows the highest reflectivity of 98,1 %. This is due to the large grain size of up to 10 μm, while the grain size of the films is only a few ten nm.

References

[1] H. Sankur, R. Hall: Applied Optics, Vol.24 No.20(1985), 3343-3347
[2] S. Metev, G. Sepold: Laser und Optoelektronik 21 (3),1989, 74-78
[3] D. Bäuerle: Appl. Phys. A 48 (1989), 527-542
[4] B. Roas, L. Schultz: J. of the Less-Common Metals 151 (1989), 413-418
[5] T. Venkatesan, X. Wu, A. Inam, Ch.C. Chang, M.S. Hegde, B. Dutta: IEEE J. Quant. Electr. vol.25 no.11 (1989) p.2388
[6] G.M. Holtmeier, D.R. Alexander, J.P. Barton: J. Appl. Phys. 71 (2) 1992 p.557
[7] T.R. Jervis, M. Nastasi, T.G. Zocco: Mat. Res. Soc. Symp. Proc. 100, 621 (1988)
[8] R. Poprawe: Materialabtragung und Plasmaformation im Strahlungsfeld von UV-Lasern, Ph.D. thesis, Darmstadt 1984
[9] Ya.B. Zel'dovich, Yu.P. Raizer: Physics of Shock Waves and High-Temperature Hydrodynamic Phenomena, Vol.1 a. 2, Academic Press, 1966
[10] J.M. Bridges, C.L. Cromer, T.J. McIlrath: Appl. Opt. vol.25 no.13 (1986) p.2208
[11] S. Bauer: Am. J. Phys. 60 (3) 1992 p.257
[12] D.G. Stearns, R.S. Rosen, S.P. Vernon: J. Vac. Sci. Technol. A 9 (5) 1991 p.2662
[13] R.A. Hoffman, J.C. Lin, J.P.Chambers: Thin Solid Films 206 (1991) 230-235

In Situ Laser Assisted PVD and Characterization of Y-Ba-Cu-O Layers inside a Laser Coupled Scanning Electron Microscope

K. Wetzig, S. Menzel
Institut für Festkörper- und Werkstofforschung Dresden e. V.,
Dresden, Bundesrepublik Deutschland

1. Introduction

Laser assisted PVD is a successful technology for preparation of Y-Ba-Cu-O layers [1]. Such HTSC layers possess an excellent conductivity and a high T_c value. For requirements of their electronic parameters an optimization of processing conditions effecting an excellent film quality is necessary. A primary cause reducing the film quality is the existence of particle defects during plasma condensation.

HTSC layers were generated in situ inside a scanning electron microscope by irradiation of a $YBa_2Cu_3O_{7-x}$-target with a Nd-glass-laser ($\lambda = 1,06$ μm) (see Fig. 1). The experimental parameters are summarized in Tab. 1. The laser was combined with a SEM, and the beam was focussed on the target with a spot size of $0,01 \ldots 4$ mm^2 [2]. Positionable substrate holders permit both an x-y-adjustment and a face overturn so that alternating sequences of in situ coating and electron optical observation were possible.

The experiments were carried out in order to minimize the density of failure particles - both droplets and indented big particles - in layers for an increasing film quality.

Tab. 1: Experimental Parameters

a) Laser		b) SEM	
wave length, λ/μm	1,06	pressure, p/Pa	$< 10^{-3}$
pulse time, τ/ms	10	solution,s/nm (d=8 mm)	7 (slow scan)
pulse energy, E/J	≤ 2	voltage, U_B/kV	5 - 50
spot area, A_{sp}/mm^2	$0,01 - 4$	working distance, d/mm	6 - 60
angle of incident, $\alpha/1°$	45	mode	SE/BSE
pulse number, N	1,2,10,100...		
beam diameter, d_B/mm	6		

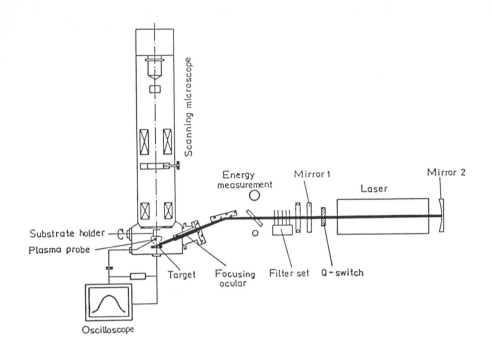

Fig. 1: Schematic section through a laser coupled SEM

2. Results and Discussion

Target and substrate are arranged parallel in a distance of about 10 mm. Failure particle defects produced in layers after several laser pulses were measured in dependence on the angle ϕ between the target normal in the spot centre and the observed substrate microarea normal.

By image analysis methods area fraction A_A, object density N_A and mean area $\bar{A}$ *) were determined from SEM micrographs. An (100) oriented B-doped silicon substrate was used and experiments were carried out at room temperature with target material from pressed $YBa_2Cu_3O_{7-x}$ compound. The target position to the incident

---

*) $A_A = \sum_i A_i/A_M$ A_i measured area of particle i

    $N_A = N/A_M$ A_M measured layer area

    $\bar{A} = A_A/N_A$ N   measured particles

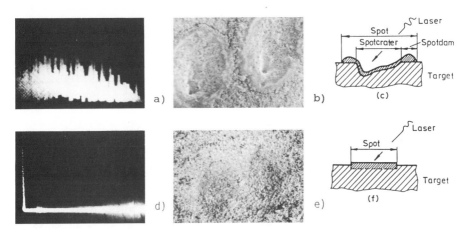

Fig. 2: Laser pulse course demonstrated by (a) oscillographs, (b) SEM spot images and (c) models for the spot cross sections for different pulse lengths. (a-c) τ = 10 ms; (d-f) τ = 15 ns.

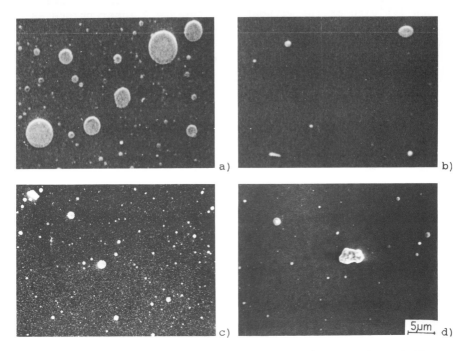

Fig. 3: Typical SEM images of LPVD layers with or without the Q-switch for different position angles. (a) φ=0°, τ=10 ms; (b) φ=30°, τ=10 ms; (c) φ=0°, τ=15 ns; (d) φ=30°, τ=15 ns.

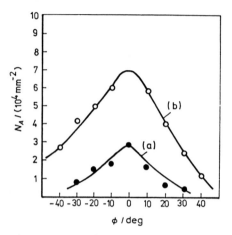

Fig. 4: Particle density N_A in LPVD layers after $N_P=6$ laser pulses (a) with or (b) without Q-switch as a function of position angle ϕ.

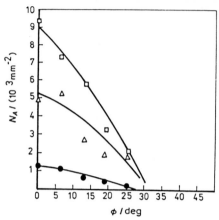

Fig. 5: Particle density N_A in LPVD layers without Q-switch after $N_P=30$ laser pulses as a function of position angle ϕ. □ , 1.9 J; △ , 1.0 J; ●, 0.45 J.

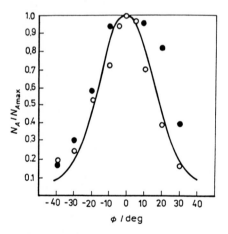

Fig. 6: Normalized particle density N_A in LPVD layers without Q-switch after $N_P=3$ (o) or $N_P=6$ (●) pulses as a function of position angle ϕ

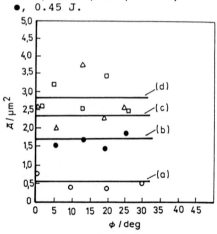

Fig. 7: Mean particle area $\bar{A}$ (a) with ($N_P=6$ pulses) or (b-d) without ($N_P=30$ pulses) Q-switch as a function of position angle ϕ. (a) E=0.15 J; (b) E=0.45 J; (c) E=1.0 J; (d) E=1.9 J.

laser beam was changed from spot to spot in order to avoid crater formation. Fundamental laser-target processes are described in (3) as a function of process parameters. Material transfer can be caused by ablation and evaporation. If the unswitched laser pulse ($\tau \approx 10$ ms) energy E' is high enough, parts of surface material will be evaporated. An important parameter is $E'/\tau^{1/2}$ (4). Q-switched laser pulses ($\tau \approx 15$ ns) also effect ablation processes for materials transfer. Fig. 2 shows typical laser spots on the target for free-running and Q-switched mode with a power increase from ~ 100 W to ~ 10 MW. Because of the short irradiation time the surface target erosion is reduced and typical spot craters are not visible. The spot crater cross section is related to the laser beam energy profile. In the LPVD layers two different kinds of defects exist: there are circular, flat size particles generated from the liquid state, and besides crystalline particles with an irregular size ablated in the solid state from the target surface can be detected (Fig. 3). In the free-running mode all detected particle defects are droplets. The droplet density along the direction transverse to the laser beam is symmetric in regard of the spot centre (see Fig. 4). The particle density for the Q-switched mode is about 50 % lower than that for the free running mode. In the Q-switched mode both droplets and irregular particles which are smaller than droplets are obtained on substrates (Fig. 3 c, d). In the free running mode droplet density increases proportional with the laser energy for energy values > 1 J. For energies lower than 1 J the droplet density is much more diminished (see Fig. 5). In that case the resulting temperature at the target surface ist not sufficient for evaporation processes.

Normalized particle density N_A/N_{Amax} registrated in Fig. 6 corresponds quite well with the normalized thickness of a Sm-Ba-Cu-O layer as a function of the position angle ϕ. It can be nearly related by a relation $d \sim \cos^n\phi$ (n = 10) described elsewhere (5). The conclusion is that the particle density in the plasma seems to possess the same solid angle distribution as the plasma itself. It should be noted that the mean particle area $\bar{A}$ is quite well constant in relation to position angle ϕ as demonstrated in Fig. 7. Differences for higher pulse energies may be particularly caused by a particle overlap (see Fig. 3).

3. Summary and conclusions

1. In the result of laser PVD processes for production of HTSC-layers two kinds of defects - droplets from liquid particles and irregular target particles by ablation in the solid state - are incorporated in $YBa_2Cu_3O_{7-x}$-layers.

2. It was proved by EPMA results that irregular particles possess the same elemental concentrations as the target does.

3. The defect particle density in the layers shows the same dependence on the position angle ϕ as layer thickness does.

4. The mean area of particles is independent of the angle ϕ.

5. Density and mean area of irregular particles are lower than those of the droplets.

6. Both the deposition mechanism and the relation between density of droplets and irregular particles are mainly caused by the laser pulse length τ.

References:
(1) X. D. Wu et al.: Appl. Phys. Lett. 54 (1989) 179.
(2) K. Wetzig et al.: Scanning 9 (1987) 99.
(3) J. Narayan et al.: Mat. Res. Soc. Symp. Proc., Vol. 13, New York (1983).
(4) R. Benz: J. Nucl. Mater. 150 (1987) 128.
(5) R. A. Neifeld et al.: Appl. Phys. Lett. 53 (1988) 8.

Growth Kinetics of Thin Films Deposited by Laser Ablation

A. Richter[1], G. Keßler[2], and B. Militzer[3]

[1]Fachhochschule Ostfriesland, Fachbereich Naturwissenschaftliche Technik,
Constantiaplatz 4, D-W-2970 Emden, F. R. G.
[2]Frauenhofer-Institut für Werkstoffphysik und Schichttechnologie, Helmholtzstr. 20,
D-O-8027 Dresden, F. R. G.
[3]Humboldt-Universität, Fachbereich Physik, Invalidenstr. 110, D-O-1040 Berlin,
F. R. G.

The pulsed laser ablation method has recently become an important technique for deposition of a wide range of materials (1). At this process, the interaction of the laser beam with a solid target results in the generation of a plasma which condenses as a thin film on the substrate. Advantages of the laser ablation technique for in-situ thin film growth are:
- stoichiometric non-thermal material removal from a solid multi-component target with the same composition as the desired film,
- energetic particles and clusters within the plasma,
- presence of reactive gases within the deposition chamber to allow chemical reactions within the plasma and on the film surface,
- high deposition rates of $R = 10^{18} - 10^{20}$ cm^{-2}s^{-1} with definite material supply by pulse-like material removal and detailed time stages as the length of the laser pulse, time of flight of the plasma and the repetition frequency of the laser pulses.

A disadvantage of the laser ablation technique is the generation of particles and droplets on the top of the film surface. The number of particles and droplets can be substantially reduced by use of a target of high density. Another very new method (2) uses the interaction of two plasma plumes which are deflected from their initial direction in such a way that the ablated material does not hit the substrate directly. Additionally, metallic screens retaine particles and droplets, thus smooth films without droplets could be grown.

Thin film formation is a complex process of nucleation, growth, and coalescence of islands to reach continuity. Growth mode, that is three-dimensional (3D) or two-dimensional (2D) layer growth, and structure of the deposit mainly depend on the mobility of particles on the substrate, the microstructure of substrate and interface as well as the supersaturation determined by the material transport of the laser plasma (3, 4). An analysis of film growth with respect to the mentioned peculiarities necessarily requires the investigation of transient nucleation, kinetics of layer growth, and a bond energy description (5). Basic processes on the substrate are shown in fig. 1. The growth mode is basically determined by the atomic bond energies between deposit atoms ε_{dd} and bonds between film and substrate ε_{ds} related to the thermal energy kT and the effective coordination number.

Additionally, the particle energy of the laser plasma increases the mobility and fluidity of the material after impact upon the surface. Part of the plasma energy E_{cl} is transformed into kinetic energy of the atoms with the velocity $v = b\sqrt{E_{cl}}$ where b is an effective factor for transformation into kinetic energy relative to thermal energy. Due to an additional mobility of atoms the growth mode is preferably changed in the direction of 2D growth. The substrate temperature T_s is a measure for surface diffusion of atoms D_a. With the desorption time τ the mean diffusion length is given by $<l^2>^{1/2} = \sqrt{D_a \tau}$. The desorption rate $R_{des} = Z_1/\tau$ depends on the concentration of monomers Z_1 and the desorption time τ, because only desorption of single atoms is considered.

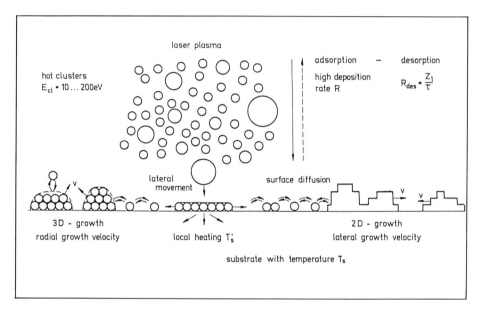

Fig. 1: Basic processes on the substrate during laser plasma deposition of thin films.

Within a computer simulation film growth and layer formation is considered. The basic equation for nucleation and growth on the substrate consists of a set of coupled differential equations:

$$\frac{dZ(i,t)}{dt} = D_{i-1}^+(t)\, Z(i-1,t) - D_i^-(t)\, Z(i,t) - D_i^+(t)\, Z(i,t) + D_{i+1}^-(t)\, Z(i+1,t) \qquad (1)$$

$$= I_{i-1}(t) - I_i(t).$$

The cluster densities $Z(i,t)$ depend on time t and cluster size i, D_i^+ and D_i^- are the frequencies of a single atom attachment to and detachment from a cluster, respectively. Taking into account a starting distribution of cluster densities $Z(i,t_0)$ known from plasma investigations (6, 7) and a determination of D_i^+ and D_i^- by surface diffusion, capture number, concentration of adatoms and specific bond energies, a numerical simulation of time dependent film growth is possible:

$$\Delta Z(i,t) = (I_{i-1}(t) - I_i(t))\ \Delta t.$$

This expression does not include the alteration in cluster population during the sufficiently small time intervall Δt and is true only in the limit $\Delta t \rightarrow 0$.

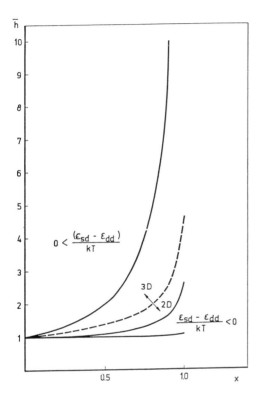

Fig. 2: Mean film height $\bar{h}$ in atomic layers versus layer coverage x for 3D and 2D growth mode without consideration of particle energy.

The height function h (x, y, t) reflects the surface profile at the location (x, y) on the substrate at time t. Three-dimensional layer growth is related to a large mean thickness of the film if complete continuity is reached. Fig. 2 shows the mean film height for complete layer coverage at different growth modes for deposition of tungsten (3D) and diamond-like carbon (2D) on silicon without consideration of the plasma energy. The 3D growth mode corresponds to a larger surface roughness than the 2D growth mode.

Particle energy and relaxation time t_c are responsible for changes in the 3D growth mode and flatten the film surface. The additional kinetic energy allows the atoms to fill free space in the layer below. Enough time between the laser pulses at sufficiantly high substrates temperatures leads to similar results, because the atoms move along the surface due to thermal diffusion. Fig. 3 shows two examples for full and partial relaxation for 3D and 2D growth due to different repetition frequencies of the the laser pulses.

476

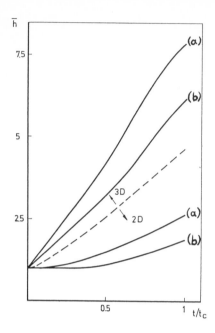

Fig. 3: Mean film height h versus time
relation t/t_c for different relaxation:
(a) partial, (b) complete relaxation

Snapshots of a part of the growing film are shown in figs. 4 and 5 illustrating layer
growth at different time stages for 3D and 2D growth mode, respectively. Generally an
area of 150 x 150 atoms with periodic boundaries on the substrate and 20 atomic layers
in height can be investigated. The parameters correspond to deposition of tungsten layers
(fig. 4) and diamond-like layers (fig.5) on silicon with a plasma energy of about 18 eV
characterized by the electron temperature. With a repetition frequency of 30 Hz the time
constants are large enough for complete relaxation. It is obvious that tungsten shows a 3D
growth mode whereas carbon has a tendency to a 2D growth mode.

These considerations are of practical importance for deposition of multilayer systems as
quantum well superlattices and x-ray mirrors (8, 9).

References

(1) G.K. Hubler et al., MRS Bulletin, February 1992, p. 26 - 58.
(2) E.V. Pechen et al., phys. stat. sol. (a) 131 (1992).
(3) J.A. Venables et al., Rep. Prog. Phys. 47 (1984) 399.
(4) A. Richter, ZfI Mitteilungen 134 (1987) 139.
(5) B. Lewis, Thin Solid Films 7 (1967) 179.
(6) A. Richter, Energy Pulse and Particle Beam Modifications of Materials, ed. K.
    Hennig, Akademie Verlag 1988, p. 405.
(7) S. Becker et al., ZfI Mitteilungen 134 (1987) 1.
(8) J.J. Dubowski, Chemtronics 3 (1988) 66.
(9) W. Pompe et al., Laser-Jahrbuch, ed. H. Kohler, Vulkan Verlag Essen 1990, p. 300.

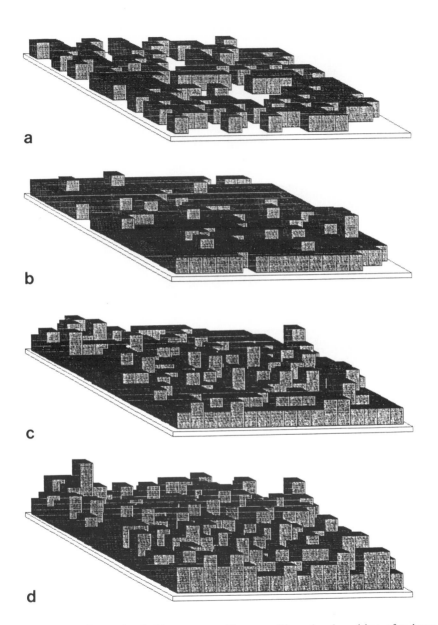

Fig. 4: Snapshots of growth of thin tungsten films on silicon by deposition of a laser plasma with mean energy of 18 eV, 3D layer growth after 5 (a), 10 (b), 15 (c), and 20 (d) laser pulses.

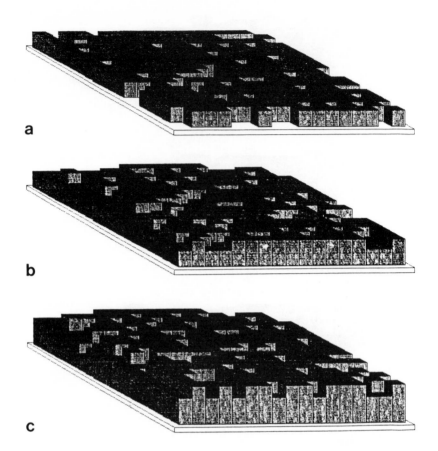

a

b

c

Fig. 5: Snapshots of growth of thin diamond-like carbon films on silicon by deposition of a laser plasma with mean energy of 18 eV, layer growth after 5 (a), 10 (b), and 15 (c) laser pulses.

Laser Pulse Vapour Deposition of Metal-Carbon Superlattices for Soft X-Ray Mirror Applications

R. Dietsch, H. Mai, W. Pompe, B. Schöneich, S. Völlmar
Fraunhofer-Einrichtung IWS, Dresden, FRG
S. Hopfe, R. Scholz, Max-Planck-Institut für
Mikrostrukturphysik, Halle, FRG
B. Wehner, Technische Universität Dresden, FRG
P. Weißbrodt, Jenoptik GmbH, Jena, FRG
H. Wendrock, IFW Dresden, FRG

Introduction

The application of particular techniques of physical vapour deposition of thin solid films (e.g. thermal and electron beam evaporation, Knudsen cell evaporation, ion beam coating, laser pulse vapour deposition) is often limited, when the coating of large sized substrates is required.
A typical emission characteristic of the vapour source covering merely a restricted solid angle is the reason why homogeneous thickness profiles at moderate deposition rates are only achieved within the central part of the substrate area. Thus larger source to substrate separations have to be chosen for the preparation of larger thin film specimens and this is correlated with a serious decrease of the deposition rate.
Other means to increase the area portion that is homogeneous in film thickness are substrate motion perpendicular to the particle beam (e.g. substrate rotation). But, this will also often not give a satisfactory solution. In particular in LPVD (1), where the particle ejecting plasma plume FWHM is concentrated within a solid angle of typically 1/3 sr, for a small target/substrate separation (typically 30 mm) even when substrate rotation is applied, the useful area having an adequate thickness homogeneity is of the order of magnitude of one cm^2. Therefore it has been tried to realize a new idea for the LPVD of thin films that is based on the particular plume behaviour, i.e. that maximum particle fluence is always observed close to the normal of the target surface.

Experimental

The conventional arrangement of LPVD for the synthesis of thin solid films is characterized by the following features (fig. 1 a):

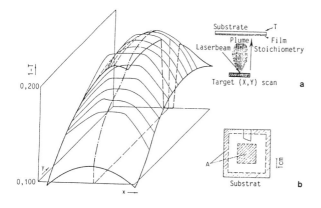

Substrate ⌐T
Plume ⌐Film
Laserbeam ⬛ ⌐Stoichiometry

Target (X,Y) scan a

A⌐

Substrat b

Fig. 1: Schematic diagrams of conventional LPVD-process
a) – experimental arrangement
b) – lateral thickness distribution across substrate surface
(T-optical transparency of thin solid film)

480

a) a focused laser beam is directed at the target surface to induce material ablation,
b) a planar target is rotated or x,y -scanned in the focal plane of the laser beam to achieve a stationary ablation rate,
c) a planar substrate located at a typical separation from the target is held stationary or is rotated for homogenization of the deposition rate.
A typical pattern of the optical density distribution (1-T) as an equivalent to the layer thickness profile of a tungsten thin film prepared in a conventional system is shown in fig. 1 b.
The microphotometer traces obtained from the hatched areas (A) of the substrate (inset at right hand side) indicate a maximum deviation of a factor of two in layer thickness. At the most an area of typically 1 cm^2 provides a useful thickness homogeneity (target/substrate separation d = 50 mm).
Now, to increase this area portion, a target/substrate arrangement has been proposed that will provide substrate motion, target motion and waging of the plasma plume.
To achieve a homogeneous thickness distribution of the prepared films it is only necessary to scan with the plume axis across the total substrate surface of arbitrary shape by waging the target surface normal. Thus an actual projection of a particle emitting extended area at the target onto the substrate surface is achieved by means of an area scan and the appropriate magnification process. In the arrangement provided this effect is achieved solely by the application of cylindrical targets.

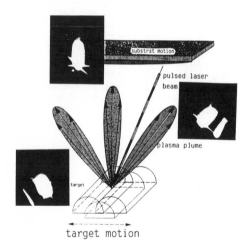

target motion

Fig. 2: Schematic diagram of LPVD target/substrate handling with plume direction control

The schematic diagram of the new LPVD-arrangement is shown in fig. 2. The features of major importance that can be found are the target that has the shape of a half cylinder, the substrate with linear motion and a target position control that allows x,y-scan and the adjustment of target position in vertical direction to compensate for lense to target distance changes.
For three different target positions the resulting declination of the plume is indicated and appropriate insets show photographs of the actual plume behaviour inside the ablation reactor.
The resulting thickness profiles for various handling regimes have been calculated. For target in motion and stationary substrate (fig. 3), the

thickness profiles show a substantial thickness homogeneization as compared to fig. 1.
At present the objective of major importance is the preparation of nm-layers having an optimum thickness homogeneity of the order of ± 1 %.
Theoretical and experimental efforts gave up to now results that are quite close to these limits. For a more technical solution the appropriate experiments for deposition rate optimization are now in progress.

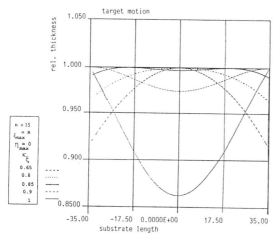

Fig. 3: Distribution of relative film thickness observed for various sin-function powers during target motion

The schematic diagram of the apparatus for LPVD of nm-thin films is shown in fig. 4. Basic sections are:
- UHV-chamber and pumping system
- residual gas pressure control
- target/substrat-manipulators
- Nd-YAG-laser and laser beam handling
- PC-control of laser, quarz oscillators, (ellipsometer) and stepper motors of the target/substrat-manipulators.

The characteristic features of the LPVD process and its interaction with the residual gas atmosphere are explained in fig. 5. The main components of particle fluxes involved in the synthesis of thin solid films are caused by
* laser pulse ablation,
* residual gas condensation,
* wall desorption followed by condensation and
* reaction of substrate adsorbates.

For the preparation of high quality films, i.e. specimens of a good physical perfection and high chemical purity, not only the film morphology but also the film composition have to be optimized.

Thus, for deposition rates of the order of $0,1 \text{ nms}^{-1}$, which is equivalent to the condensation of typically $3 \cdot 10^{14}$ atoms $\text{cm}^{-2}\text{s}^{-1}$, if the residual gas contribution should remain distinctly below 0,1 %, a base pressure of the order of magnitude of $p = n \cdot 10^{-10}$ mbar (n = 1...5) has to be guaranteed (for equivalent sticking coefficients).

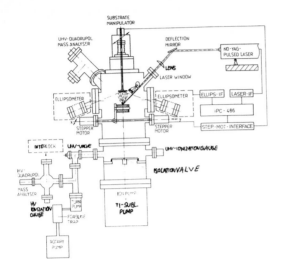

Fig. 4: Schematic diagram of the LPVD thin film reactor

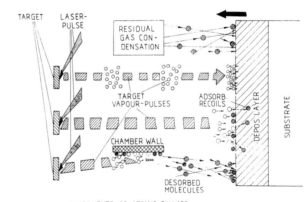

Fig. 5: Schematic explanation of particle flux components involved in thin film preparation

COMPONENTS OF ATOMIC FLUXES STRIKING THE SUBSTRATE

Selected results of film preparation

Various specimens of thin solid films (C, Ni) and C/Ni-multilayers have been prepared by the arrangement described above. Thin film characterization has been carried out by ellipsometry, X-ray diffraction, cross sectioning TEM and SNMS.
Selected results for the layer thickness homogeneity achieved in our initial experiments are the following:
The variation of the thickness d of a C-layer measured by ellipsometry for the whole substrate length has a standard deviation $\sigma = 1{,}6$ %. Phase shifts and polarization changes measured along a multilayer specimen gives fig. 6. Here only the overall effect of the multilayers (20 double layers of Ni/C and a C-coverage, Si-substrate) on the reflected ellipsometer beam can be taken for a homogeneity estimate.

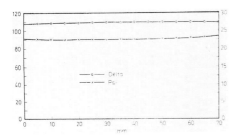

Results of thin film
characterization by
ellipsometry
- phase shift and polarization
change across the surface of
a multilayer
Si - 20x(Ni/C) - C

Thus local deviations over a 70 mm measuring path remain very close to 1 %. The
appropriate micro-homogeneity of the layer stack has been determined by TEM
cross section imaging. Fig. 7 shows the regular arrangement of the individual
layers. From the high-magnification micrograph a mean thickness for Ni-layers
of 25 Å and for C-layers of 32,5 Å can be evaluated. The interface roughness
is clearly below $\sigma = 2$ Å. Thus basic requirements for the preparation of X-ray
monochromators have been realized.

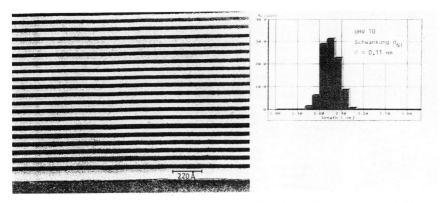

Fig. 7: TEM-micrograph of C/Ni-multilayer (20periods, C-coverage and Si-substrate), determination of interface roughness by image processing of HRTEM
micrographs (magn. $4 \cdot 10^6$), interface roughness $\sigma_{IF} = 0,08$ nm, inset-standard
deviation of Ni-layer thickness $\sigma_{Ni} = 0,11$ nm

Typical depth profiles obtained from the same specimen by SNMS are shown in
fig. 8. A good reproducibility of the Ni-peak heights and a depth resolution
very close to the theoretical limit (Δz typically 10 Å) are the confirmation
of a very regular stack morphology even in macroscopic dimensions (SNMS-probe
area typically 10...30 mm^2). Concentrations of nitrogen and oxygen remained
below 1 % - at.
Finally an example for a X-ray optical response of a 30 period layer stack has
been verified by grazing incidence X-ray diffraction (fig. 9).

484

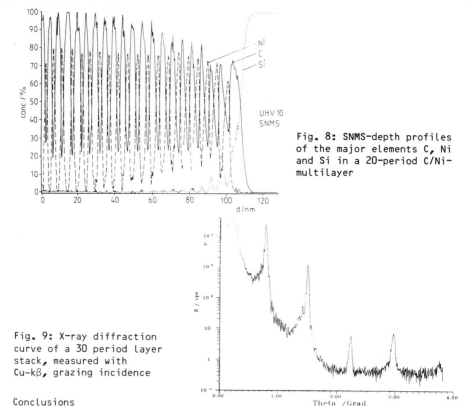

Fig. 8: SNMS-depth profiles
of the major elements C, Ni
and Si in a 20-period C/Ni-
multilayer

Fig. 9: X-ray diffraction
curve of a 30 period layer
stack, measured with
Cu-kβ, grazing incidence

Conclusions

First tests of a newly designed and constructed LPVD-reactor for the preparation of thin solid films and nm-layer stacks show that high quality specimens can be prepared.
Regular morphology of multilayers and desired thickness homogeneity of individual layers and atomically flat interfaces have confirmed the applicability
for X-ray monochromator preparation. Chemical purity of the films can be
realized owing to residual gas pressures of the order of 10^{-10} mbar.

References

(1) T. Venkatesan, Q. Li and X. X. Xi, Laser Ablation,
    Mechanisms and Applications, J. C. Miller and R. F. Haglund (Eds.)
    Springer Verlag Berlin, 1991, p. 12

Financial support of this research work has been provided by the Ministry of
Research and Technology of the Federal Republik of Germany under Contract-
No.: 13 N 5945.

Laser-CVD on High-Tensile Ceramic Fibres

V. Hopfe, A. Tehel and S. Böhm
Technische Universität Chemnitz, FB Chemie, Germany

Introduction

Fibre reinforced composites are now widely accepted materials having superior mechanical behavior at low specific weight (Fig. 1). Carbon fibres embedded into ceramic matrices e.g., are hopeful new materials suitable for high temperature applications in the 2000K-region. They have a good chance to bridge the gap between metallic superalloys with their superior mechanical properties but very restricted thermal stability and, ceramics with their extremely high thermal stability but intrinsic brittleness. At higher temperatures, carbon fibres must be protected against chemical matrix-attack or atmospheric corrosion by a dense coating. In addition to their ability to act as a diffusion and reaction barrier, the coatings must meet a complex set of chemical and mechanical requirements (1), above all, they should not diminish the intrinsically high tensile strength of the fibres. Recently we have introduced laser-driven chemical vapour deposition (LCVD) as a method for coating carbon fibres (2). In comparison with hot-wall thermal CVD, which is now well established in the coating of fibres, the LCVD method is characterized by several significant advantages: high deposition rates (roughly 1μm/s), small-volume cold-wall reactors, thermal reaction paths supported by photochemical channels, almost no restrictions on deposition temperatures.

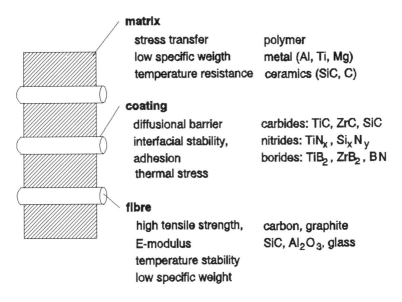

matrix

stress transfer	polymer
low specific weigth	metal (Al, Ti, Mg)
temperature resistance	ceramics (SiC, C)

coating

diffusional barrier	carbides: TiC, ZrC, SiC
interfacial stability,	nitrides: TiN_x, Si_xN_y
adhesion	borides: TiB_2, ZrB_2, BN
thermal stress	

fibre

high tensile strength,	carbon, graphite
E-modulus	SiC, Al_2O_3, glass
temperature stability	
low specific weight	

Fig. 1: Fibre reinforced composites

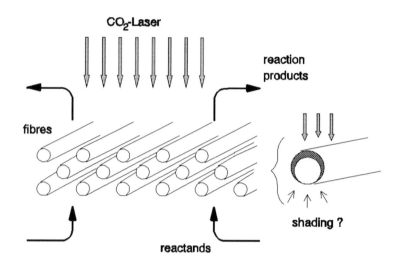

<u>Fig. 2:</u>   Principle of laser CVD reaction cell

<u>The deposition process</u>

Obtaining good-quality and low- price coatings seems to be one of
the key problems in producing fibre reinforced ceramics.
In principle, we need high deposition rates together with a wide-
area deposition zone. But in 'conventional' LCVD, e.g. in 'direct
writing', the situation seems to be completely different. High
deposition rates on compact substrates can only be achieved by
focussed laser beams, due to the kinetics of quasi
three-dimensional gas-phase transport of the reactants inside the
localized microreaction region (3). In coating fibres or other
dispersed materials, we are faced with an inverse geometry (see
Fig. 2). Micro-scaled particles are irradiated in a widened laser
beam and these quasi zero-dimensional heat-sources now achieve
high transport rates in the mass-flow controlled kinetic regime.
The productivity of the process may be upscaled by enlarging the
reaction zone, which means increasing the power of the CW laser
source at a predetermined level of power-density, needed for the
reaction desired.
The asymmetrical irradiation of the fibres suggests that shading
on the surface of the fibres may occur resulting in unwanted
thickness profiling of the layers around and across the fibres.
Surprisingly, most of our experiments did not give any evidence of
shading (see Fig. 4). This phenomenon may be interpreted by the
high thermal conductivity of the fibres which, together with their
short characteristic heating time, prevents an appreciable
temperature difference of the irradiated and the reverse side of
the fibre (2). Using the 10 μm radiation of the CO_2 laser, strong
scattering may occur within the fibre roving as the wavelength
matches the diameter of the 7μm fibres. On the other hand, the

layer homogeneity may be increased using a optically driven 'self stop mechanism' of the deposition rate within the transition region in between kinetically and mass-flow controlled regimes. Depositing highly reflecting layers, e.g. the majority of transition-metal carbides, nitrides and borides, onto the surface of the initially strong absorbing (graphitic) carbon fibres, the temperature of the fibres goes down with increasing layer thickness. This is accompanied by an appreciable decrease in the deposition rate at layer thicknesses of typically 50 nm ... 100 nm, which, in turn, results in layers with optimal dimensions for the subsequent fabrication of composite materials.

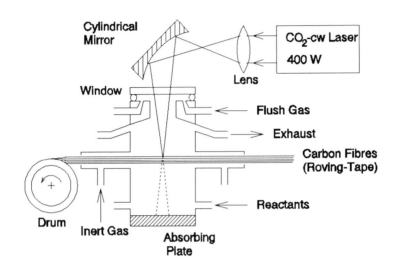

Fig. 3: Apparatus for continous LCVD on fibre rovings

Laser-CVD apparatus

The scheme of the LCVD reactor for continuous deposition at normal pressure is sketched in Fig. 3. A 6k-roving, which has been flattened out into a tape, is transported continuously through the reaction zone via two Ar-flushed 'gas-windows'. The beam of a technological multi-wavelength CO_2-cw laser (400 W, TEM00) has been shaped into a line focus perpendicular to the fibre axis, which irradiates the moving fibre tape in full width. The power density may be controlled by beam-focussing from 0.1 to 10 W/mm^2. Parameters to be optimized in the LCVD experiments are: power density, fibre velocity, selection of precursors, flow rate and concentration of reactants. The temperature of the fibres has not yet been measured in the experiments reported here. Very recently, using a two-wavelength pyrometer, the temperature has been measured on a reactor configuration similiar to the one described above, yielding temperatures in the region at 1300 K to 2700 K. Commercially available high tensile carbon fibres (T800, TORAY) have been pretreated prior to LCVD by thermal outgassing at 660 °C in flowing N_2.

The morphology, layer thickness and the structure of the layers have been analysed by SEM/EDX, TEM/electron diffraction and by X-ray diffraction, sometimes supplemented by FTIR, XPS and GDOS.

Deposition of refractory coatings

The parameters of optimized LCVD experiments and some typical results of the extensive deposition tests are summarized in Tab. 1. The titanium systems are characterized by a pronounced decrease in the deposition rate caused by the increasing reflectivity of the deposit. In SiC_x, SiN_x and pyro-C depositions however, an increase or even oscillation due to interference, is to be expected since the optical absorptivity of these layers exceeds that of carbon fibres. Therefore, the deposition rates, given in Tab. 1, are only mean values and may be higher in the initial stage of layer-growth. All experimental data summarized in Tab. 1 have been carried out at normal total pressure, using partial pressures of the precursors which prevent homogeneous nucleation in the gas phase.

Some preliminary experiments at reduced total pressure did not give any evidence of higher deposition rates or other advantages. The highest deposition rates have been found in the TiB_2-system. They are situated in the 2...10 μm/s region, which may be an indication of photolytic enhancement of the mainly photothermal LCVD process. This is caused by the BCl_3 molecule which may undergo multiphoton dissociation even at rather low power densities (5). The deposition rate of pure photothermal LCVD processes, on the other hand, may be influencend by thermodiffusion and by Stefan mass-flow effects, which have been shown in depositing TiC_xN_y layers with selected carbon precursors (2). In contrast to CVD-grown layers, which normally show a rugged nodular surface, LCVD layers are normally smother and denser. The layers exactly copy the underlying filaments of the fibre and have quite a high degree of adhesion to the substrate, see Fig.4. Despite the fairly high deposition rate, the layers consiste of well-crystallized phases. In the titanium systems the mean grain size of the crystallites is comparable to the layer thickness suggesting a dense 'cobble-stone pattern'. After the deposition of refractory coatings tensile strength of the fibres decreases more or less pronounced (Tab.1). The influence of the deposition parameters on tensile strength is rather complex, but power density of the radiation and fibre velocity rank uppermost (4).

Conclusions

The laser-CVD method developed for continuous coating of rovings holds great promise of extending the application area of conventional deposition processes, particularly with regard to high deposition rates and fine-grained but well-crystallized layer structures. A greater variety of precursors as well as coatings can be deposited including the high temperature region and photolytical reaction channels.

Considering the variety of parameters, the LCVD process is more difficult to optimize with respect to certain properties and high tensile strength of the coated fibres.

Table 1
Typical parameters and results of laser CVD experiments

Layer	Reactants	Partial pressure of precursor [kPa]	Fibre velocity [mm/s]	Power density [W/mm^2]	Layer thickness [nm]	Tensile strength [MPa]	Deposition rate [µm/s]
TiC$_x$	TiCl$_4$ / H$_2$	2,3	5 ... 1	5 ... 10	50 ... 100	1800 ... 1300	0,1
TiC$_x$	TiCl$_4$ / C$_6$H$_6$ / H$_2$	2,0 / 1,0	1	5 ... 10	50 ... 200	2900 ... 2300	0,1
TiB$_2$	TiCl$_4$ / BCl$_3$ / H$_2$	2,8 / 8,0	20 ... 5	5 ... 10	80 ... 120	2800 ... 900	2,0
SiC	CH$_3$SiCl$_3$ / H$_2$	10 ... 5	20 ... 5	1 ... 5	50 ... 100	2500 ... 900	0,8
pyro-C	C$_2$H$_2$ / Ar	30	30 ... 5	5 ... 10	-	4000 ... 3400	0,7
(Uncoated fibre)	(H$_2$)	-	1	10	-	3500	-

490

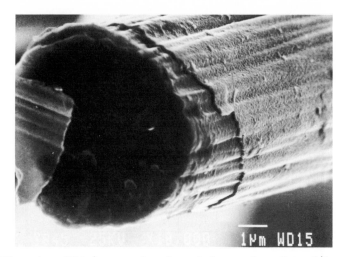

Fig. 4:   SEM image of cut end-face of carbon fibres coated
          by laser CVD with a layer of TiC

Besides their superior efficiency, performance and reliability,
the advantageous use of technological cw-CO_2 layers is enhanced by
the strong absorptivity of commercial high tensile ceramic fibres
and their diameter, which causes a rather homogenous
electromagnetic field inside the roving by scattering. By
enlarging the reaction zone, high productivity may be obtained by
technological multi-kilowatt lasers. The principle of the method
may be easily transfered from the deposition of rovings to the
coating of highly dispersed particles, thus opening up exciting
fields in materials science.

Acknowledgements

The authors would like to thank Dr. Eva Kieselstein for scanning
the electron micrographs as well as Mrs. Goldmann and Mr.
Froehnert for their technical assistance though out this research
project.

References

[1]   J.R. Strife and J.E. Sheehan
      Ceramic Bull. 67 (1988), 369
[2]   V. Hopfe, A. Tehel, A. Baier and J. Scharsig
      Appl. Surf. Sci. 54 (1992), 78-83
[3]   D.J. Ehrlich and J.Y. Tsao
      J. Vac. Sci. Technol. 81 (1983), 969
[4]   V. Hopfe, S. Böhm, G. Wieghardt and A. Schulze
      Appl. Surf. Sci. (accepted)
[5]   N.V. Karlov
      Zh. Exp. Teor. Fiz. 14 (1971), 214
      Appl. Opt. 13 (1974), 301-9

Laser Induced Chemical Vapour Deposition of Conductive and Insulating Thin Films

R. Ebert, U. Illmann, G. Reisse, A. Fischer, F. Gaensicke,
Ingenieurhochschule Mittweida, Mittweida, FRG

Introduction

In future microsystem technology the preparation of conductive and insulating thin films with laser methods could play an important role (1). We investigated the laser induced chemical vapour deposition of Co and Ti conductive films using a direct writing method and, moreover, the large area deposition of SiO_2 films in an excimer laser parallel beam configuration. To better understand the deposition process we calculated laser induced temperature fields by means of a computer program developed in our research group.

Experimental equipment

The apparatus for the direct writing experiments consists of the following components: a 10 W argon ion laser, a gas handling system for the supply of parent and carrier gases, a vacuum pumping unit, a stainless steel reaction chamber with substrate heating facility and total pressure measurement and a computer controlled x-y table (Fig. 1). Additionally it is possible to heat the reac-

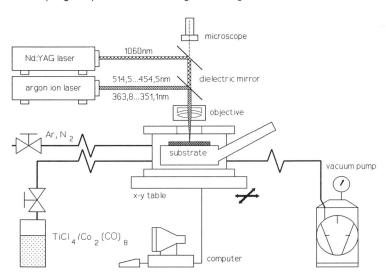

Fig. 1: Equipment for direct writing laser-CVD

tion zone with a Nd:YAG laser, specially for the deposition at UV-wavelengths. The computer program allows to deposit complex structures, such as spirals at continuously varied scan velocity.

The second apparatus for area and structured deposition consists
of gas handling, pumping and scanning facilities and a reaction
chamber (Fig. 2). An excimer laser beam is directed parallel to
the substrate surface in order to activate a gas phase reaction.
The special substrate holder allows to preclean the substrate
surface by a rf-plasma process before the deposition runs and to
heat or to cool the substrates during film deposition. With an
additional argon ion or Nd:YAG laser the local heating of the
growing film is possible.

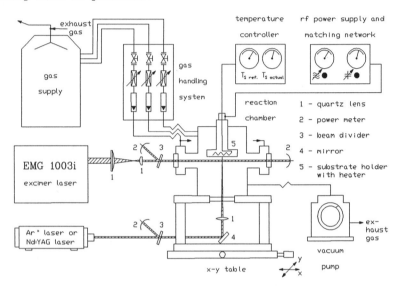

<u>Fig. 2:</u> Equipment for area and structured laser-CVD

Experimental results

First experiments were carried out with $Co_2(CO)_8$ as parent gas for
the argon ion laser induced cobalt deposition on semiconducting
(c-Si) and insulating (SiO_2, glazed Al_2O_3 ceramic) substrates. The
deposition from this precursor compound is characterized by an
initial nonthermal photolytic stage followed by a combined photolytic / pyrolytic reaction. This behaviour was also investigated
by Ehrlich (2). Adhesive and scratch resistant Co films could be
deposited on SiO_2 substrates at typical process parameters of a
Gaussian Radius $r_G = 4 \mu m$,   a scan velocity of $v_s = 0.2$ mm/s ,
a laser power of $P_L = 0.2$ W and a precursor partial pressure of
$p = 13$ Pa at a substrate temperature of 300 K. These parameters
result in a deposition rate of $R = 5 \mu m/s$ and a minimum specific
film resistivity of $\rho = 20 \mu\Omega cm$. The film deposition on ceramic
substrates is accompanied by a lateral growth outside the laser
spot. This effect can be reduced by addition of Ar as buffer gas.
After nucleation and with the generation of the first Co layers on
the SiO_2 or the Al_2O_3 surface the contribution of the pyrolytic

growth is large compared to the photolytic growth. This is because of the fact that the first Co layers are absorbant for the incident visible argon ion light causing the surface temperature to rise supported by the low thermal conductivity of the SiO_2 or Al_2O_3 substrate. The deposition of Co on c-Si substrates is characterized by lower deposition rates of R = 0.75 μm/s even at higher laser power of P_L = 0.65 W and lower scan velocities of v_s = 0.02 mm/s. One possible reason is the lower temperature rise on c-Si caused by the higher thermal conductivity and smaller differences in optical and thermal properties between the film material and the substrate in comparison with the Co deposition on SiO_2 substrates. Co films deposited on c-Si are smooth at laser power of P_L = 0.5 W or lower. When the laser power is increased an inhomogeneous growth occurs at the center of the line. This "cluster-like" growth is probably a temperature dependent phenomenon. The dependence of the line width on laser power is shown in Fig. 3.

Further experiments were carried out with $TiCl_4$ as parent gas for direct writing of titanium lines on c-Si and SiO_2. First runs were performed using H_2 as carrier gas for the $TiCl_4$ transport into the reaction chamber. Despite of the parameter variation the experimental conditions generally led to the etching of the silicon surface and possibly to a substitution reaction between the $TiCl_4$ * H_2 reactant vapour and the silicon surface. Therefore it was not possible to deposit Ti-lines on silicon under these conditions. Ensuing described investigations showed, that the Ti-deposition on stainless steel substrate can be assisted by multiphoton processes. Because of this fact it was possible to deposit Titanium on SiO_2 (being transparent at the deposition wavelength) with the following typical parameters: r_G = 4 μm, v_s = 0.1 mm/s, P_L = 0.7 W, T = 300 K and a $TiCl_4$ partial pressure of about p = 1.3 kPa. Etching of the substrate surface was not observed. With further improvement of the film properties laser deposited cobalt and titanium should be a suitable alternative for interconnects or in the case of cobalt for thin ferromagnetic films in microsystem technology.

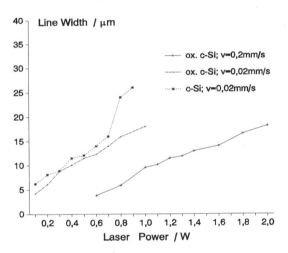

Fig. 3: Dependence of line width on the laser power

A last series of experiments was directed to the large area deposition of SiO_2 films from SiH_4, N_2O and Ar as buffer gas on c-Si and SiO_2 substrates using an ArF excimer laser (3). The laser

beam passes the surface parallel in a small distance of about
0,3 - 1 mm. Typical deposition parameters were: pulse energy
E = 5 mJ, beam cross section A = 3*8 mm^2, pulse repetition rate
f = 100 Hz, total pressure p = 1.3 kPa, scan velocity v_s = 0.35
mm/s, a substrate temperature T = 520 K and a N_2O / SiH_4 flow
ratio of 110. A typical value for the deposition rate at these
parameters was R = 15 nm/min. The refractive index of n = 1.48 was
determined by ellipsometric analysis. The films were well adherent
on SiO_2 (adhesion strength > 16 MPa) and on c-Si (> 9 MPa), where
the adhesion strength of the films decreased with decreasing
substrate temperature. Simultaneously, films deposited at
temperatures lower than 470 K were milky and scratch-sensitive.
The breakdown field strength was measured to be 0.1-0.5 MV/cm
(ρ_s = 10^9-10^{10} Ωcm).
Additional investigations for structured deposition of SiO_2 are
planned.

Temperature field calculations

For the calculation of the laser induced temperature field the
heat conduction equation is transformed into a nonlinear first
order differential equation system by subdividing the space into
hollow cylinders and by integration over the volume of these
elements (4). This differential equation system is numerically
solved according to a predictor-corrector procedure by the method
of Gear with an automatic control of time step and the order of
the polynominal statement (5). The complete computer program
allows the calculation considering phase changes, mechanical
stresses and gas phase diffusion processes (6).

For the laser - CVD
direct writing process
the Gaussian Radius is
the parameter, which
determines the propor-
tion of substrate
temperature (being
decisive for pyrolytic
deposition) and photon
density (being deci-
sive for multiphoton
induced deposition).
By means of the
computer program we
have calculated the
temperature at the
center of a laser spot
on a stainless steel
surface with different
objectives (Fig. 4).
As can be seen the
double Gaussian Radius
requires approximately
the double laser power
to obtain the same
surface temperature.

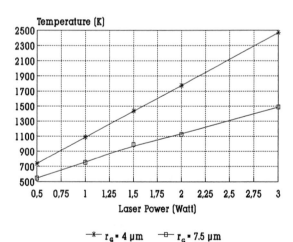

Fig. 4: Calculated maximum temperatures
(in K) on a stainless steel surface as
function of the laser power (λ=500nm) at
two different Gaussian Radii.

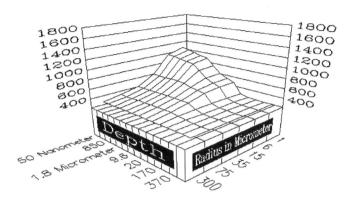

Fig. 5: Calculated laser induced temperature field (in K) in stainless steel without layer after reaching steady state (t=15ms, λ=500nm, r_G=4μm and P_L=1 W)

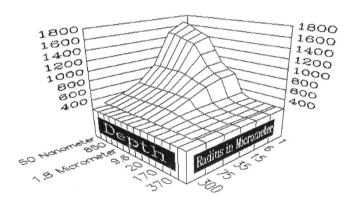

Fig. 6: Calculated laser induced temperature field (in K) in stainless steel with a 100nm TiN-layer after reaching steady state (t=15ms, λ=500nm, r_G=4μm and P_L=1 Watt)

On the other handside because of the reciprocal quadratic dependence of the laser intensity on the Gaussian Radius the fourfold laser power is necessary to get the same laser intensity and therefore the same photon density at a double Gaussian Radius. Hence, the variation of the Gaussian Radius and the comparison of the related deposition threshold powers is a suitable method to distinguish between pyrolytic and photolytic processes. In order to investigate this we deposited titanium dots on stainless steel from precursor TiCl$_4$ at a pressure of 1.3 kPa with an argon ion laser. Two different objectives were used to adjust the Gaussian Radii. The details are content of Table 1.

experiment	1	2	3	4
r_G [μm]	4	7.5	7.5	7.5
P_L [W]	0.3	0.53	1.0	1.8
T_S [K]	620	620	810	1070
S_0 [W/cm^2]	$6*10^5$	$3*10^5$	$6*10^5$	$1*10^6$
result	deposition	no dep.	no dep.	deposition

Table 1: Experimental details of Ti dot deposition

In experiment 1 a deposition could be observed at a calculated surface temperature of 620 K and a laser intensity of $6*10^5$ W/cm^2. At the same surface temperature and the half laser intensity in experiment 2 a deposition did not start. This indicates that the deposition at this temperature is photolytic (induced by multi-photon processes) in nature. Experiment 3 shows that this photo-lytic process is dependent on substrate temperature since with the same photon density as used in experiment 1 no deposition could be observed at a surface temperature of 810 K. A possible reason of this result could be a smaller dwelling time and thus a lower density of adsorbed molecules on the substrate surface at higher temperatures. In experiment 4 the sixfold laser power was required for the deposition to start, and so the high surface temperature indicates that the deposition at these conditions is rather pyro-lytically induced. These experiments demonstrate that the Gaussian Radius influences directly the deposition mechanism.

In forthcoming experiments the deposition of TiN with the direct writing equipment and the argon ion laser is planned. To this end further temperature field calculations were carried out (Fig. 6 and 7). As can be seen the difference in the maximum temperature is as large as 600 K for the blank stainless steel surface and the covered surface respectively at the same parameters. This result is important for the direct writing process of areas because the laser spot can overlap with the as-deposited lines, where the sur-face temperature would consequently rise to higher values.

References

(1) D. Bäuerle: "Chemical Processing with Lasers", Springer
     Berlin (1986).
(2) D.J. Ehrlich: EPA 0241190 (1987).
(3) M. Tsuji, N. Itoh, Y. Nishimura: Jpn.J.Appl.Phys.
     30 (1991) 2868.
(4) W. Schwarzott: Forsch.Ing.-Wes. 38 (1972) 165.
(5) C.W. Gear: "Numerical Initial Value Problems in Ordinary
     Differential Systems", Prentice-Hall (1971).
(6) A. Fischer, G. Reisse, F. Gänsicke, K. Zimmer:
     Appl.Surf.Sci. 54 (1992) 41.

Film thickness variation of pulsed laser deposited carbon

B. Keiper, G. Reisse, A. Fischer,
Ingenieurhochschule Mittweida, Mittweida, FRG

Introduction

In recent years pulsed laser deposition has been increasingly used to produce thin films. In comparison to the conventional thermal evaporation the advantages of this technique are the better film purity due to the absence of heating elements in the vacuum chamber, the higher energies of the film forming particles, the high deposition rates, the possibility to work in reactive atmospheres and the possibility to evaporate materials with high evaporation temperature. Moreover, because of the high temperature rising velocity, materials with complex stoichiometric composition can be deposited with the film composition being almost the same as that of the target (1). Problems are the evaporation of the target at only one spot in the laser focus, which can lead to a deep hole in the surface of the target and the extremely forward directed expansion of the evaporated material, which complicates the production of homogeneous films. The aim of our investigations was to optimize the conditions for the deposition of homogeneous films over the whole substrate area.

Experimental apparatus

For the deposition of thin films by the pulsed laser deposition method we constructed a special vacuum apparatus (see Fig.1) con-

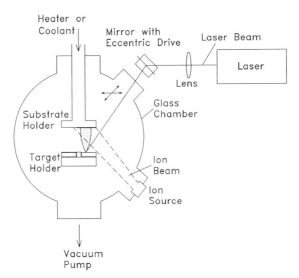

Fig.1: Experimental setup for pulsed laser deposition of thin films

sisting of a glass chamber that can be evacuated by an oil diffusion pump to a base pressure of $6.7*10^{-4}$Pa. The water cooled target holder can carry four targets which are exchangeable during the deposition. The substrate holder can be both cooled using liquid N_2 or heated from the inside of the holder. The temperature of the substrate was measured by means of a thermocouple. In our experiments a Lambda Physics excimer laser EMG 1003i (248nm, 20ns, 200Hz, 200mJ) was used. The laser beam was focused through a quartz glass window onto the target surface. It can be scanned with an eccentric drive across a target area of 2cm*2cm. This drive consists of a laser mirror tiltable in X- and Y-direction by two motor driven eccenters.

The laser beam scanning should result in a homogeneous target ablation and should prevent the development of a deep hole in the target, which would change the deposition conditions during the experiment. Additionally, the laser scanning should enlarge the size of the homogeneous film area. By means of an ion beam directed onto the substrate we can clean the substrate surface before each deposition as well as modify the film properties during deposition.

Thin film deposition

In order to investigate the film thickness variation of pulsed laser deposited films we prepared amorphous carbon films on Si-substrates by excimer laser evaporation of a graphit target. The experiments were carried out in the above explained apparatus, where the substrate-target distance was 28mm. To determine the film thickness variation caused by a single laser pulse we directed the laser beam to a fixed spot and moved the target surface relative to the laser beam. With this arrangement the ablation took place from a smooth surface and the laser couldn't produce deep holes in the target, which would have changed the film thickness variation. The film was actually deposited using a large number of laser pulses but was considered as the sum of subfilms caused by single laser pulses. Films with a maximum thickness of 100nm to 180nm were deposited.

Furthermore, we produced films applying the above explained laser beam scanning with eccentric drive. The size of the scanned area was limited by the target size and the geometrical dimensions of the vacuum chamber. In our experiments we used a size of 13mm*20mm or 19mm*25mm.

Determination of the film thickness variation

The film thickness variation was determined by means of the color chart for diamondlike carbon films on silicon (2). We found a film thickness variation differing in X- and Y-direction with a nearly elliptical shape. Fig.2 shows the film thickness variation of both major axis obtained from a typical film compared to the cosinus distribution of the conventional thermal evaporation process.

Similar results were obtained by R. K. Singh et. al.(3), who investigated the film thickness variation of pulsed laser deposited (XeCl excimer laser, 308nm, 45ns) high temperature superconductor thin films ($YBa_2Cu_3O_7$). In this work the spatial thickness as well as compositional variations were determined by Rutherford backscattering spectrometry (RBS).

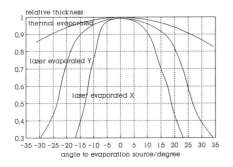

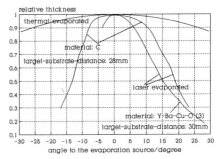

Fig.2: Film thickness variation of pulsed laser deposited carbon films using fixed laser beam

Fig.3: Film thickness variation of ref.3 in comparison to our results

According to our experiments the laser beam scanning size of 13mm*20mm or 19mm*25mm was too small to lead to the expected distinct increase of the homogenous film area. Fig.4 shows the film thickness variation of such a film compared to the film deposited without laser scanning.

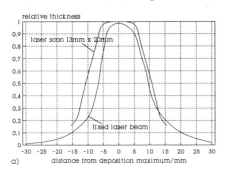

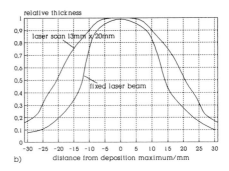

Fig.4: Film thickness variation of pulsed laser deposited carbon films using both fixed and scanned laser beam (a)X- and (b)Y-direction)

Calculation of the film thickness variations resulting from laser scanning

In order to determine conditions, that should be convenient for the production of homogeneous films over the whole substrate area we calculated the film thickness variations that result from various methods of laser beam scanning. For this purpose we developed a computer program. The thickness variation obtained from a single laser pulse was assumed as being the same for each pulse corresponding to Fig.2. The values of the film thickness variation were input into the computer program for both major axis. The lines with equal thickness were assumed to be of elliptical shape and the

points between the values were approximated by a curve. To calculate the film thickness variation for film deposition with laser
beam scanning we added up the film thickness variations of each
laser pulse considering the motion of the laser beam between the
pulses. Our computer program allows the calculation of the
thickness variation in dependence of the size of the scanned area
in X- and Y-direction, the deposition time and the laser pulse
repetition rate.
In the following we present the results of our calculations, that
were performed for two deposition methods. Method 1: Laser beam
scanning with constant velocity, every point of the scanned area
will be hit with the same frequency. Here we were able to calculate
the thickness variation with different scanning velocity in X- and
Y-direction. The results of an calculated example are shown in
Fig.5.

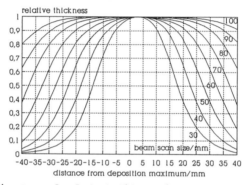

Fig.5: Calculated film thickness variation
corresponding to method 1 (X-direction)

Method 2: Laser beam scanning with eccentric drive, so that the
edges of the scanned area will be hit more frequently. Different
speeds of both motors of the drive can be adjusted. Fig.6 and Fig.7

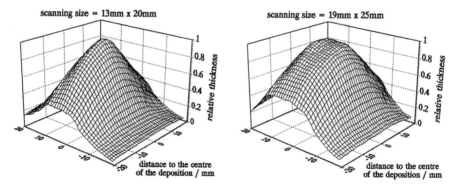

Fig.6: Calculated results with the parameters used in our experiments

show the calculated results of method 2. The laser pulse repetition rate, the scanning velocity and the motor speeds respectively don't affect the film thickness variation noticeably if these parameters are suitably chosen, i.e. the laser pulses must be distributed over the whole scanned area.

As can be seen from both the experimental results (Fig.4) and the calculated results (Fig.6), the scanning sizes of 13mm*20mm and 19mm*25mm, as used in the experiments, don't lead to a distinct increase of the homogeneous film area, what can only be achieved by a further enlargement of the scanning size.

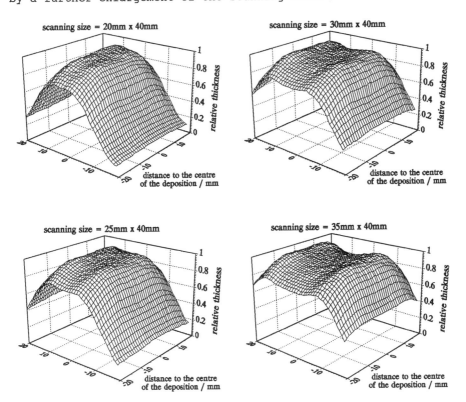

Fig.7: Calculated film thickness variations corresponding to method 2

In general, method 1 requires a relatively larger scanning size than method 2 and thus a larger target area, too. A disadvantage of method 2 is that for the used target-substrate distance only one optimum size of the laser beam scanning exists and so the size of the homogeneous film area is not variable. In the calculated example, which was adapted to the conditions of the experiment with the eccentric drive, the optimum scanning size was 25mm*40mm, where the film thickness divergence of a film area of 16mm*18mm was as low as 6%. Method 1 requires for the same homogeneous area 50mm*50mm scanning size. The results of our calculations lead to the conclusion that our experimental setup demands a scanning size of 25mm*40mm to get sufficient homogeneous films. Nevertheless, instead of using so large scanning sizes it is possible to control the laser beam scanning and the laser pulse rate in such a way, that the laser will hit the target at previously calculated points, so that the overlay of the successive partial depositions lead to the desired homogeneous films. For this purpose we developed a new deposition system with computer controlled laser beam scanning and laser pulse triggering.

References

(1)   S. Metev, G. Sepold: Laser und Optoelektronik 21 (1989) 3, 74-78 .
(2)   T. J. Moravec: Thin Solid Films 70 (1980) L9-L10
(3)   R. K. Singh, O. W. Holland, J. Narayan: J. App. Phys. 68 (1990) 1, 233-247

Deposition of Hard Coatings by Laser Induced PVD Technique

F. Müller and K. Mann, Laser Laboratorioum Göttingen e.V., Göttingen, FRG

Introduction

The laser induced PVD process has already become state of the art for the deposition of high T_c superconducting films (1). Due to the substantial particle energies within the laser generated plasma, the LPVD process offers great advantages also for the deposition of other novel materials, as there are hard coatings like TiN, SiC, and also amorphous diamond-like-carbon (DLC) films (2-5). The latter are of increasing interest as protective coatings for IR optical applications, in particular due to their high and broad-band optical transparency, high hardness, thermal conductivity and chemical inertness against any solvent (6,7).

The main advantage of the LPVD process with respect to DLC deposition is the possibility of film growth at low substrate temperatures and at the absence of aggressive hydrogen. Earlier investigations have also revealed larger portions of diamond bonds than in the case of other deposition techniques; the corresponding sp^3/sp^2 ratio strongly depends on the power density of the laser used for target ablation (8). In this work we have investigated the energy and mass distribution of excimer laser ablated target particles, using Ti and graphite targets. The influence of these quantities on the properties and morphology of the deposited films gives insight into the fundamental mechanism of film growth in the LPVD process. In particular with respect to the formation of transparent diamond-like-carbon coatings a strong correlation between laser fluence, particle energies and film properties is observed.

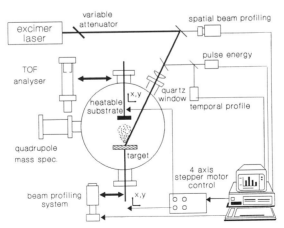

Fig. 1: Computer controlled experimental set-up for LPVD process.

Experimental Set-up

A schematic representation of the experimental set-up is shown in Fig. 1. The deposition is performed in an UHV chamber pumped by a wide range turbo molecular pump (Balzers TPH 450). The vacuum system has a base pressure of $< 1 \times 10^{-7}$ mbar; the residual gas composition is controlled by a quadrupole mass spectrometer (Leybold PGA 100). For thin film deposition targets are ablated with an excimer laser (Lambda Physik EMG 202) of 248nm wavelength and 30ns duration at a repetition rate of 10Hz. The beam is imaged by a spherical lens through an UHV window (fused silica) onto the target, using an angle of incidence of 45deg. Opposite to the target a heatable substrate is mounted on a manipulator at a distance of typically 8cm. During deposition, both target and substrate to be coated can be rotated by computer controlled stepper motors, allowing efficient use of the target material as well as a homogeneous film thickness on the substrate.

In order to determine the energy density of the laser radiation in the target plane, the 2D intensity profile is measured by a UV camera in combination with a PC based laser beam profiling system (9). For this purpose the camera is brought into the chamber instead of the target manipulator. Corresponding measurements reveal a relatively smooth profile without any hot spots, with an irradiated target area of 6.4×10^{-3} cm^2. The overall energy is measured by a pyroelectric energy monitor.

For an investigation of the energy and mass distributions of laser ablated ionic species, a time-of-flight (TOF) spectrometer is installed perpendicular to the target surface instead of the substrate manipulator. The arrangement of the TOF spectrometer is shown in Fig. 2. After target ablation the particles enter a field-free drift tube of length s=60cm; positive ions are detected at the end of this tube by an electron multiplier. After amplification and shaping, the ion pulses are displayed temporally resolved on a fast digital storage oscilloscope, which is triggered by the ablating laser pulse. The trace obtained for a single pulse is transfered via IEEE interface to a PC, which integrates the spectrum for a preselected number of pulses.

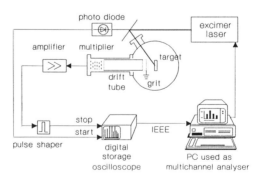

Fig. 2: TOF arrangement for plasma analysis.

For mass analysis the target is biased against a grit to a voltage U=+2.5kV. Mass separation is possible, since the corresponding energy of the accelerated ions of mass m and charge q is much higher than the particle energy E_{abl} obtained during the ablation process. The kinetic energy is given by:

$$E_{kin} = q\,U + E_{abl} = m/2\ (s/t)^2 \qquad \text{Eq. 1}$$

Having identified the masses of the ablated ions, this set-up can also be used for investigation of the initial particle energies. For these measurements the target is grounded and the ions enter the drift tube with an energy given only by the laser ablation process.

Plasma analysis by TOF measurements

Plots of the count rate of laser ablated carbon ions versus time of flight are presented in Fig. 3. In any case the spectra are divided into two temporal regions, one of fast atomic or small molecular ions and a broad distribution at higher flight times, which can be attributed to formation of large carbon clusters. As seen from Fig. 3, the data show a strong correlation between laser fluence and the corresponding particle velocities. While the peaks of the small particles are getting sharper with increasing laser fluence, the broad cluster distributions are shifted towards higher flight times (lower velocities). Obviously the cluster size is getting larger with increasing portion of absorbed energy. On the other hand, the peaked velocity distributions of the smaller ions observed above -10J/cm² are shifted to the left for increasing fluence (cf. Fig. 3), indicating an enhancement of the kinetic energies of the ablated particles (see below). As a further result, we observe a strongly increased portion of carbon clusters compared to smaller ions with increasing fluence. A similar behaviour is seen during ablation of Ti and Cu targets, while Al, due to its low absorption coefficient, does not show any cluster formation in the investigated fluence range.

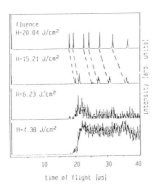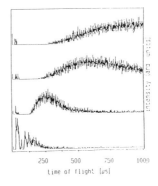

Fig. 3: TOF spectra of positive carbon ions ablated with different laser fluences at 248nm (cf. text).

In order to identify the smaller molecular masses corresponding to the peaks in the TOF spectra, the target is biased with an increasing voltage, and the shift of the peaks is monitored until the energy E_{abl} from the ablation process can be neglected (cf. Eq. 1). This evaluation leads to the result that below 9J/cm^2 atomic carbon is formed in the plasma, while at higher fluences only molecules of three and more atoms are observed as smallest particles. Having identified the masses, we can calculate the kinetic energies E_{abl} from the flight times of the corresponding ions. In Fig. 4 these energies are compiled for C^+ and C_3^+ as a function of the laser fluence, which has been varied from 0.5-23J/cm^2. We observe an almost linear relation, with maximum kinetic energies as high as 220eV at 23J/cm^2. Obviously, the energy of the small molecular ions is independent of the mass and only a function of the laser fluence. Absolute values of the kinetic energy are in agreement with Ref. (5).

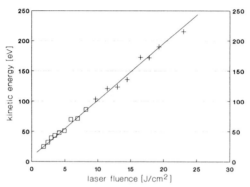

Fig. 4: Energy of ablated carbon ions as a function of incident laser energy density ($\square = C^+$, $+ = C_3^+$)

Film deposition and characterization

Thin films of TiN and amorphous diamond-like-carbon (DLC) have been deposited on polished stainless steel and fused silica substrates of 20mm diameter. For deposition of TiN a titanium foil is used as target material. The ablation and subsequent nitridation on the substrate takes place under a N_2 working gas pressure of 1×10^{-3}mbar. The obtained films show a golden colour well known for stochiometric TiN coatings, while the films deposited at lower pressures have a metallic appearence. With increasing laser fluence the roughness of the films increases as a consequence of droplet formation.

For carbon film growth sintered graphite targets are irradiated with 20000 laser pulses for each substrate. A new target site is exposed every 2000 pulses. Using a conical nozzle, an argon gas stream from the UHV window onto the target is applied in order to avoid deposition of ablated particles on the window. The corresponding pressure during ablation is about 1×10^{-3}mbar.

Carbon film deposition has been carried out on fused silica substrates at different substrate temperatures and different laser fluences. The films deposited at high substrate temperatures (> 200C) and lower fluence values are opaque and conducting, indicating a graphite bond character. The adherence to the substrate and scratch resistance are bad.

On the other hand, transparent films with good substrate adhesion are obtained at low deposition temperatures <u>and</u> at high laser fluences (>8J/cm^2). At these conditions the coatings are electrical insulating and have a scratch resistance higher than that of conventional optical coatings. In particular when a substrate bias voltage is used, the films are very homogenous and smooth, with rms-roughness values equal to the uncoated substrate. From x-ray diffraction analysis we know that the films are amorphous with a small crystalline portion, which can partially be attributed to microcrystalline diamonds, as seen in SEM micrographs. The number of microcrystals is reduced by use of the bias voltage. These observations, which are consistent with the results of other groups (10,11), lead to the assumption that a relatively large portion of diamond bonds is deposited. The influence of substrate temperature and laser fluence on the transmission characteristics of the coatings, as measured with an UV-VIS-NIR spectrometer (Lambda 19, Perkin-Elmer), is displayed in Figs. 5 and 6 in comparison to the uncoated substrate.

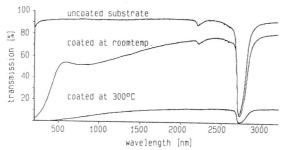

<u>Fig. 5:</u> Transmission spectra of carbon films deposited at different substrate temperatures (laser fluence H=12J/cm^2)

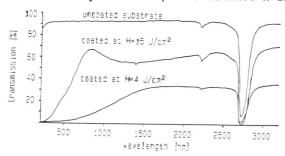

<u>Fig. 6:</u> Transmission spectra of carbon films deposited with different laser fluences at room temperature.

The diamond-like films show a high transmission particularly in the IR spectral range, in accordance with DLC coatings deposited with conventional CVD techniques (12). Interference maxima are observed in the visible region, indicating film thicknesses corresponding to this wavelength range. This is confirmed by measurements with a mechanical profilometer.

Conclusion

In this paper we have investigated the two seperated process steps of the LPVD process, i.e. laser induced emission of target particles and subsequent film deposition on a substrate. By help of TOF measurements, a correlation between laser fluence and particle emission characteristics on the one hand, as well as film properties on the other hand could be obtained. In particular, we observe an almost linear increase of the kinetic energies of ablated carbon ions with increasing laser fluence; above a threshold of about $10J/cm^2$, where the width of the energy distributions is getting very small, the deposited carbon films are found to be diamond-like. This justifies the assumption, that energetic particles are responsible for diamond bond formation, possibly by preferential sputtering of sp^2 bonded sites (13).

With increasing laser fluence, TiN as well as DLC films show an enhanced surface roughness due to droplet and microcrystal formation. The corresponding TOF spectra clearly indicate, that this is caused by an increasing portion of large particle clusters in the plasma plume, which obviously has to be avoided for high quality, low absorption optical thin films. Our experiments show that this can be partially achieved by use of a substrate bias voltage. Further improvements are expected in the near future from laser beam homogenizing optics, which will guarantee an extremely uniform target ablation (9). Moreover, the use of a short-pulse (fs) excimer laser is planned, from which a high degree of atomization can be expected.

References

(1) G. Endres et al.: Opt. Elektr. Mag. 6 (1990) 66.
(2) E. D'Anna et al.: J. Appl. Phys. 69 (3) (1991) 1687.
(3) M. Balooch et al.: Appl. Phys. Lett. 57(15) (1990) 1540.
(4) S.S. Wagal et al.: Appl. Phys. Lett. 53(3) (1988) 187.
(5) J.J. Cuomo et al.: J. Appl. Phys. 70(3) (1991) 1706.
(6) A. Bubenzer, B. Dischler, G. Brandt, P. Koidl: Opt. Eng. 23(2) (1984) 153.
(7) P.K. Bachmann et al.: Ber. Bunsenges. Phys. Chem. No.11 (1991) 95.
(8) F. Davanloo, T.J. Lee, D.R. Jander, H.Park, J.H. You, C.B. Collins: J.Appl.Phys. 71(3) (1992) 1446.
(9) K. Mann et al.: Laser und Optoelektronik 24 (1) (1992) 42.
(10) J. Krishnaswamy, A. Rengan, J. Narayan, K. Vedam, C.J. McHargue: Appl. Phys. Lett. 54 (24) (1989) 2455.
(11) T. Sato, S. Furuno, S. Iguchi, M. Hanabusa: Appl.Phys.A 45 (1988) 355.
(12) W. Müller-Sebert et al.: Mat. Sci. Eng. B11 (1992) 173.
(13) T. Miyazawa et al.: J. Appl. Phys. 55 (1984) 188

Photophysical and Photochemical Processes in Excimer Laser Deposition of Titanium from TiCl4

P. Kubát and P. Engst, J.Heyrovský Institute of Physical Chemistry and Electrochemistry, Czechoslovak Academy of Sciences, Prague, Czechoslovakia

Introduction

The laser deposition of titanium for fabrication of optical waveguides is of considerable technological interest (1). While the UV laser pyrolysis of $TiCl_4$ on a substrate surface yields a titanium film (2 - 5), photolysis in the gas phase results in the removal of a Cl atom and the formation of $TiCl_3$ particles. Two growth regimes have been proposed (4). A titanium film is formed by photolysis of $TiCl_4$ during an initiation step. In a second step, deposition occurs as a consequence of the laser heating of the previously deposited film. In this paper, we describe a study of the mechanism of laser photolysis of $TiCl_4$ using time - resolved emission and absorption spectroscopy.

Experimental

The experimental setup has been described in detail previously (6). XeCl excimer laser used for irradiation operated at 308 nm. The time - resolved emission and absorption spectra were measured on an Applied Photophysics kinetic spectrometer in the static regime. The recorded radiation is separated by a grating monochromator and detected by a Hamamatsu R928 photomultiplier. The measuring, evaluation and storage of the spectra was controlled by an Acorn Archimedes computer. The IR spectra were measured on a Philips PU 9800 FTIR spectrometer, and the UV spectra on a Hewlett Packard 9153C UV spectrometer. $TiCl_4$ (International Excines), H_2 and Cl_2 were the commercial products.

Results and Discussion

The time - resolved emission is dependent on the radiation wavelength (Fig. 1a) and laser fluence (Fig. 2). The broad band (Fig. 1, dashed line) corresponds to transition of $TiCl_4$ to the ground state ($3a_1 \longrightarrow 3t_2$). The bands at 383, 402, 418, 430, 440 and 454 nm correspond to the emission of energy-rich Ti and $TiCl_n$ (n = 1 - 3) species. There were two intense lines in the emission spectra of titanium generated by XeCl laser ablation of metallic titanium (Fig. 1b) - 430 nm ($x^5D_0^0 \longrightarrow a^5F_1$) and 454 nm ($y^5F_5 \longrightarrow a^5F_5^0$)(7). The production of titanium atoms in the laser photolysis of $TiCl_4$ monitored at 430 nm is preferred at laser fluences greater than 2 $J.cm^{-2}$(Fig. 2d). Dissociation of all the chlorine atoms requires an energy of 1700 $kJ.mol^{-1}$ (8). Therefore the most probable process of direct formation of titanium atoms is multiphoton dissociation (9), but a step-by-step process on a ns time scale or less is also possible. The calculated mean lifetime of the excited titanium atom ($x^5D_0^0$) is 196 ns. The bands at 383, 402 and 418 nm correspond to the $TiCl_n$ (probably

510

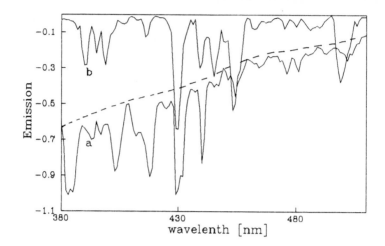

<u>Fig.1.</u> The emission spectra of TiCl$_4$ irradiated by an XeCl laser, laser fluence 2180 mJ.cm^{-2}(a) and emission spectra of excited titanium atoms generated by laser ablation of metallic titanium (b), the broken line corresponds to the transition of TiCl$_4$ to the ground state

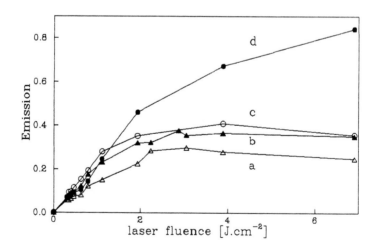

<u>Fig.2.</u> The dependence of the emission of the products at 20 ns after a laser pulse on the laser fluence monitored at 403 (a), 418 (b), 383 (c) and 430 nm (d).

TiCl₃) particles. Their mean lifetime is less than 30 ns and these particles are generated at low laser fluences (Fig.2a,b,c).
UV spectrum of the yellow deposit obtained at low laser fluences (150 mJ.cm⁻²) is characterized by a broad band at 434 nm (Fig. 3b).

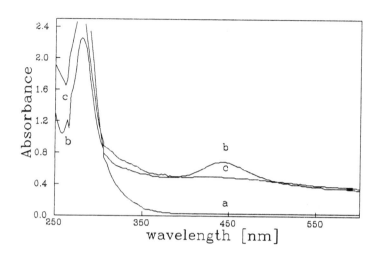

Fig. 3. The UV spectra of titanium tetrachloride (a), solid deposit at the reactor window after 10000 pulses (laser fluence 150 mJ.cm⁻²) (b) and solid deposit in vacuum after 100 pulses (c)

The deposit contains adsorbed titanium tetrachloride (band at 280 nm, Fig. 3a), adsorbed chlorine (shoulder at 330 nm) and TiClₙ(n=2,3) (IR spectra (10)). Irradiation of this deposit in a vacuum yields a thermodynamically stable species - a gray deposit of metallic titanium, titanium tetrachloride and a small amount of molecular chlorine (Fig. 3c). Direct formation of titanium atoms was observed at high laser fluences (>2 J.cm⁻², Fig.2d).
In the presence of molecular hydrogen, hydrogen chloride was also identified in the gas phase (IR spectra) and the concentration of chlorine, monitored by absorption at 330 nm (¹Σg⁺--> ¹π(1u) transition (12)), decreased (Fig. 4a,b).
Molecular hydrogen removes the reactive Cl atoms and molecules from the reaction mixture (11) and prevents the reverse reaction of solid titanium and titanium subchlorides to titanium tetrachloride (Fig. 4 a,b). Chlorine molecules are formed mainly by photolysis in the adsorbed state. Decomposition of TiCl₄ in the gas phase (1 s after cleaning substrate surface by a laser pulse) yields only a small amount of molecular chlorine (Fig. 4c).

512

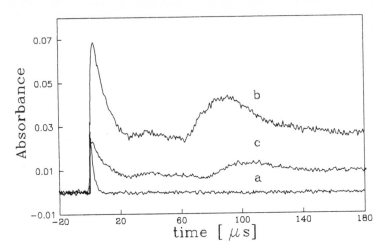

Fig.4. The time dependence of the absorbance monitored at 330 nm after irradiation of TiCl₄ by an XeCl laser without (a), in the presence of 34.9 kPa H₂ (b) and 1 s after cleaning of surface substrate by a laser pulse without H₂ (c), laser fluence 2150 mJ.cm⁻²

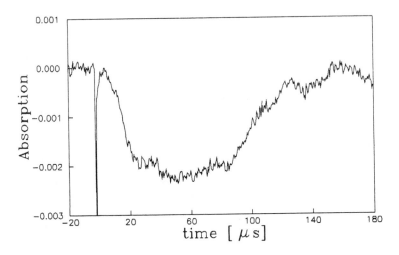

Fig.5. The time dependence of the absorption monitored at 500 nm after irradiation of TiCl₄ by an XeCl laser (average from 100 traces), laser fluence 150 mJ.cm⁻²

An attempt was made to demonstrate the presence of Cl atoms.
Cl atoms react with I_2 and the decrease in the I_2
concentration was monitored at 500 nm ($^1\Sigma_g^+ \text{-->} B^3\pi(O_u^+)$)
transition (12)). This decrease in the I_2 concentration (Fig.
5) is probably a consequence of reaction with Cl atoms but
energy transfer from the excited $TiCl_4$ and subsequent
decomposition of energy-rich I_2 cannot be excluded. The change
in the absorption of reaction mixture in time 100 µs is
probably caused by desorption and convection phenomena. I_2
molecules desorb from the reactor walls and their
concentration in the gas phase which is monitored, increase.
Following process is convection of energy rich species out
from the monitoring beam. The same effect can be demonstrated
for desorption of $TiCl_4$ (Fig. 4a, time 80 µs). The high heat
conductivity of hydrogen eliminates the temperature gradients
in the reaction mixture and prevents convection of $TiCl_4$
molecules desorbed from reactor walls into the monitoring beam
(Fig. 4b).

Conclusion
$TiCl_3$ particles are formed by photolysis of $TiCl_4$ mainly in
the adsorbed state in the first step of laser deposition of
titanium. $TiCl_3$ absorbs subsequent laser photons and decompose
to titanium, titanium tetrachloride and chlorine. Titanium
atoms can also be produced by laser pyrolysis of $TiCl_4$ on
a hot $Ti/TiCl_3$ surface or at high laser fluences (>2 J.cm^{-2})
directly by photolysis of $TiCl_4$.

References
(1) J.Y.Tsao,R.A.Becker,D.J.Ehrlich,F.J.Leonberger: Appl.
    Phys. Letters 42 (1983) 559.
(2) J.Y.Tsao,D.J.Ehrlich: Mat.Res.Soc.Proc. 29 (1984) 115.
(3) Z.Izquirdo,C.Lavoie,M.Meunier: Appl. Phys. Letters 57
    (1990) 647.
(4) C.Lavoie,M.Meunier,Z.Izquierdo,S.Boivin,P.J.Desjardins:
    Appl. Phys. A53 (1991) 339.
(5) C.Lavoie,M.Meunier,S.Boivin,Z.Izquierdo,P.J.Desjardins:
    Appl. Phys. A53 (1991) 339.
(6) P.Kubát,P.Engst: J. Photochem. Photobiol. A 63 (1992) 257.
(7) Tables of Spectral Line Intensities (ed. W.F.Meggers,
    Ch.H.Corliss, B.F.Scribner). NBS, Washington 1975.
(8) V.I.Vedenyan, L.V.Gurvich, V.N.Kondraťyev, V.A.Medvedev,
    Ye.L.Frankevich: Bond energies, ionization potentials and
    electron affinities. Arnold, London 1966.
(9) A.Gedanken,M.B.Robin,N.A.Kuebler: J.Phys.Chem. 86 (1982)
    4096.
(10) J.W.Hastie,R.H.Hauge,J.L.Margrave: High. Temp. Sci. 3
     (1971) 257.
(11) S.W.Benson,F.R.Cruishank,R.Shaw: Int. J. Chem. Kinet. 1
     (1969) 29.
(12) Ch.Okabe: Photochemistry of small molecules. Moscow, Mir
     1981.

Formation and Spectroscopy of Thin Films Produced by Laser Vaporization of Ceramics

F.W. Froben, M. Ritz and H. Yu
Institut für Experimentalphysik, Freie Universität Berlin, Germany

Introduction

For the last two decades lasers have been used not only as a monochromatic light source but also for modification and vaporization of solid material and for the production of high-temperature and high density plasma (1-3). A large number of experimental methods like mass spectroscopy, electron microscopy and optical spectroscopy at different wavelength are involved and the areas of research include fusion, cluster physics and reactivity, catalysis and material science. Many results in these different research areas have been obtained, but very little is known on some of the basic questions like the characterization of the plasma and laser assisted thin film formation. The investigation of the plasma is also important to solve the question of cluster growth. Are only atoms evaporated? Where starts the clustering? The dynamical process especially after pulse vaporization and differences in thin film formation by laser sputtering or neutral cluster deposition have important applications (4-6) and some experiments in this context will be discussed in his contribution. The materials used belong to the technologically important new type of high temperature ceramics.

Experimental

The experimental set up is shown schematically in Fig. 1. Part of it has been described (7). In short, it consists of a freely rotating rod of the material to be

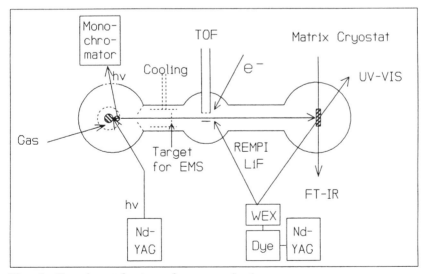

Fig. 1: Experimental set-up - laser vaporization

vaporized - a modification of the original Smalley design (8). A Nd-Yag laser is used, either the fundamental at 1.06 μ or the second harmonic. The effect of different wavelength is very small. The laser is focussed on the rod to a spot size of 0,2 mm^2 corresponding to a power between 10^8 and 10^{10}W/cm^2. The vaporization rate is monitored by a quarz microbalance. The rate of vaporization is between 10^{18} and 10^{20} atoms per second.

The ceramic materials are: YBa$_2$Cu$_3$ O$_7$-δ, the pellets are pressed together to form a rod. BN, Al$_2$O$_3$, SiC and ZrN are used as commercially available rods.

The measurements include: The energy contents of the plasma and the ion/neutral ratio, the time and spatially resolved emission spectra of the plasma, the mass distribution of the vaporized material by mass spectroscopy (TOF) and of the structure by electron microscopy of the deposit on a carbon film.

Several other parts of the experimental set-up are not used for the experiments reported here.

Results and Discussion

One of the first questions arising when a solid is exposed to a short pulse of high energy laser radiation is to estimate the energy content of the plasma produces and the type of particles evaporating. Typically the ratio of ions to neutrals is between 3 and 10 increasing with power and the energy of the charged particles is up to 300 eV. So it is really a hot plasma very far from equilibrium. In Fig. 2 the measurements of ions produced by vaporization of SiC are shown, the velocity of the particles is around 10^4m/sec. The early stages of evaporation can be described not only by the plasma eg temperature and density but also by observing the optical spectra with spatial and time resolution.

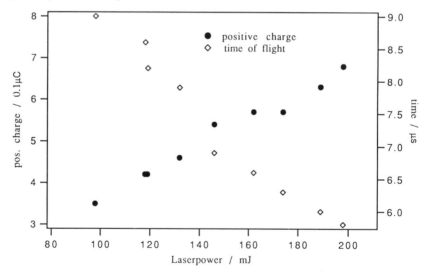

Fig. 2: Measurement of positive SiC ions

Boron nitride, for example, shows an extremely nonthermal distribution of the vibrational levels of the A-X transition of B_2 (9) during the first 10 μ sec after the pulse of the vaporizing laser. After 15 μsec the vibrational distribution is thermalysed. In addition to the spectra of atoms and ions also emission of the BN-species can be observed very close to the target. In this case it seems clear that at least diatomics leave already the surface.

In most of the experiments the emission can be observed up to a distance 20 cm downstream from the target and up to 50μ sec after the laser shot. In simple cases of metal alloys it is possible to use the emission of atomic lines to optimize laser power, pulse duration and focussing for laser treatment of the surface, cutting or welding. But the original hope to use also molecular or cluster emission to optimize thin film formation of more complex material like YBCO has not yet been fulfilled. The emission spectra from various research groups show very dissimilar spectra due to different experimental arrangements (7,10,11). Nevertheless it was shown that uniform and superconducting thin films were formed (12,13).

As an example Fig. 3 shows the emission of SiC, with mostly atomic features with different intensities for increasing distance from the target and different time structure depending on the emitting species. In addition to atoms and ions, also emissions from C_2 and SiC_2 has been observed.

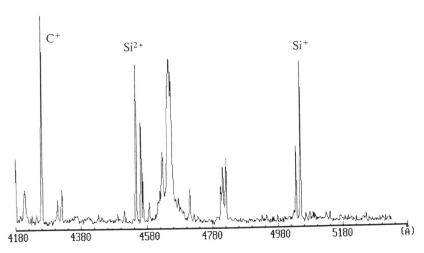

Fig. 3: Emission spectra of laser vaporized SiC

The optical measurements are supplemented with LiF in the beam, additional ionization with electrons or photons for mass spectroscopy and with spectroscopy in low temperature rare gas matrix (s. Fig. 1).
Additional information has been gotten by electron microscopy of the deposits. Small particles up to a diameter of 200 Å with 3 or 5 fold symmetry have been

detected and also amorphous films. Fig. 4 shows YBCO as an example, on the left side a 150 Å particle with structure and on the right side a 80 Å particle. The lattice distance observed corresponds to approximately 12 Å, very close to the c axis of YBCO. This is already observed for the deposition without further treatment. A 5 to 20% variation of the oxygen content of the carrier gas does not change the results. It is also possible to observe coagulation under the microscope for very small particles, so some precautions have to be taken to minimalize this effect.

The thin films from laser vaporization show different properties like colour and adhesive strength depending on the vaporization process. But it is still under investigation how the optical, mechanical and electrical properties of the thin films depend on all the experimental parameters. It is expected, that these films produced by laser vaporization either directly deposited or after cluster formation in the gas phase will show different properties and will allow the formation of new materials (14-16).

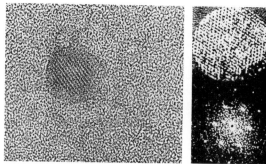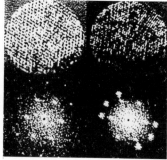

Fig. 4:   EMS of YBCO; 150 Å diameter cluster (left), 80 Å diameter cluster using computer assisted image reconstruction (right).

Acknowledgement
We thank B. Tesche and J. Urban from the Fritz Haber Institut of the Max Planck Society for the Electron Microscopy and the German Research Organization (DFG-Sfb 337) for financial support.

References
(1)   R.W. Ohse, Pure Appl. Chem. 60, (1988) 309
(2)   S.K. Gosh and T. Ericsson ed., Laser 4-High Power Lasers in Metal Processing, i.i.t.t. intern. Gournay - France 1989
(3)   J.W. Hastie ed., Materials Chemistry at High Temperatures Vol. 1,2 Humana Press, Clifton NJ. 1990

(4)   S.M. Rossnagel and J.J. Cuomo, MRS Bulletin, (2) (1987) 40

(5)   C.E. Morosana: Thin Films by Chemical Vapour Deposition, Elsevier 1990

(6)   R.K. Singh and J. Narayan, Phys. Rev. B 41, (1990) 8843

(7)   F.W. Froben, J. Kolenda and K. Möller, Z. Phys. D 12, (1989) 485

(8)   J.P. Hopkins, P.R.R.R. Langridge-Smith, M.D. Morse and R.E. Smalley,
       J. Chem. Phys. 78 (1983) 1627

(9)   F.W. Froben, J. Kolenda, High Temp. Science 28, (1990) 15

(10)   X.D. Wu, B. Dutta, M.S. Hegde, A. Inam, T. Venkatesan, E.W. Chase, C.C.
       Chang and R. Howard, Appl. Phys. L. 54, (1989) 179

(11)   T.J. Geyer and W.A. Weimer, Appl. Phys. L. 54, (1989) 469

(12)   L. Lynds, B.R. Weinberger, G.G. Peterson and H.A. Krasinski,
       Appl. Phys. L. 52 (1988) 320

(13)   A. Gupta and B.W. Hussey, Appl. Phys. L. 58 (1991) 1211

(14)   M.Y. Chen, P.T. Murray, J. Mat. Sci. 25, (1990) 4929

(15)   T.H. Allen, W.T. Beauchamp and B.P. Hichwa, Photonics Spectra
       (3) (1991) 103

(16)   M. Balooch, R.J. Tench, W.J. Siekhaus, M. J. Allen, A.L. Connor and D.R.
       Olander, Appl. Phys. L 57, (1990) 1540 .

Polishing of lead-crystal-glass using cw-CO_2-Lasers

A. Geith[1], C. Buerhop[1], R. Jaschek[2], R. Weißmann[1], A. Helebrant[1], H.W. Bergmann[3]

1) Institut für Werkstoffwissenschaften 3 (Glas und Keramik),
   Universität Erlangen-Nürnberg, Erlangen, Germany
2) Applikations- und Technikzentrum für Energieverfahrens-, Umwelt-
   und Strömungstechnik (ATZ-EVUS), Vilseck, Germany
3) Institut für Werkstoffwissenschaften 2, (Metalle),
   Universität Erlangen-Nürnberg, Erlangen, Germany

1 Introduction

In producing smooth and homogeneous glass surfaces, two methods are commonly used in the lead-crystal-industry [1]:

The first method is the fire-polishing technique, which takes advantage of the viscous flow at elevated temperatures. As the glass softens and begins to flow, rough surfaces become smoother. However, the reducing atmosphere of the gas burners may result in undesired metallic lead segregations.

The second method is the washing of glass in an acid solution (HF:H_2SO_4 = 1:6). After the solution has had time to dissolve a thin surface layer, the waste particles are washed from the glass. However, large amounts of lead-, flour-, arsenic-, and antimony-containing waste, as well as unused acids result. Their disposal is expensive and pollutes the environment.

Due to the disadvantages of both these methods, it is of technical, economic, and environmental interest to develop feasible and harmless alternative techniques for polishing lead-crystal-glass.

Several authors have reported CO_2-Laser polishing as an effective method for finishing of fused silica surfaces [2], [3]. The lack of success in applying lasers to other glasstypes may be due to their intrinsic stress and fracture characteristics when they are subjected to intense concentrated radiation. Thus, a successful laser treatment of these critical glasses must be performed at states where cracks cannot be sustained. This means that the temperature of the glass specimen should be in its annealing range [4]. The possibility of successful laser-polishing of lead-glass has been shown in former experiments [5], [6], [7].

2 Interaction between CO_2-Laser-radiation and glass

The CO_2-Laser emits radiation of 10.6 μm. In this region of the spectrum silicate glasses absorb strongly due to the broad Si-O-vibrational band between 8.5 μm and 11 μm [8]. Approximately 80 % of the applied laser energy is absorbed and converted to heat within a thin surface layer.

However, glass is a brittle material with small thermal expansion and low thermal conductivity. It therefore has little thermal shock resistance. For this reason, using a CO_2-Laser, the glass must be treated at elevated temperatures, in order to avoid critical thermal stresses. Preheating the glasses at temperatures just below the tranformation temperature T_g, minimizes these stresses, avoids slumping, and so a local heat treatment with a CO_2-Laser can be carried out. When the surface temperature in the irradiated area is high enough, the glass viscosity is lowered locally, and the glass in this region may flow. By adjusting the temperature, the surface may be polished homogeneously.

3 Experimental procedure

The experiments are performed using commercially cut lead-crystal glass. The experimental set-up is shown in fig. 1. The beam of a 4 kW cw-CO_2-Laser is directed using a mirror and a focussing optic. A facet optic is used, to give an homogeneous rectangular intensity distribution. The samples are preheated in a furnace to their annealing range, at $T \approx 490$ °C. During the laser treatment the glass is kept in an heat-isolating sample holder, which rotates as it moves across the beam. After the laser treatment the glass samples are annealed in order to eliminate residual stresses.

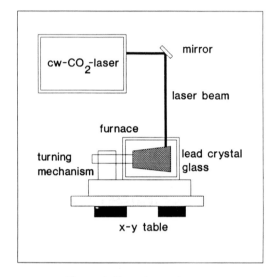

Figure 1: Experimental set-up

4 Experimental results

Laser polishing proceeds at intensity levels from 50 to 450 W/cm² depending on the feed speed. Fig. 2 exhibits the laser-polishing quality as a function of laser intensity and feed speed. Three sections with typical process features are remarkable:

Region A: The laser-treated surface appears rough and dull. The surface has not been smoothened perfectly because the interaction time between laser radiation and glass has been too short.

Region B: The surface appears homogeneously smooth and brilliant. The temperature in the irradiated area is high enough and reducing the feed speed, the optical appearance becomes even better.

Region C: The surface shows many bubbles. The the interaction is too long and the glass overheats. Depending on the vapour pressure of the components, various chemical reactions of physically solved gases in the glass matrix take place.

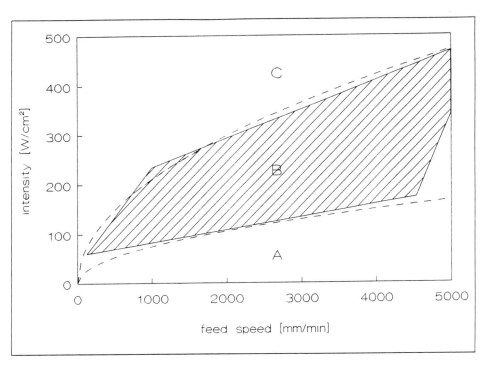

<u>Figure 2</u>: Optimal parameters for the laser-polishing process of lead-crystal-glass (experimentally determined area: solid line, calculated values: dashed line)

In comparing the realizable roughness using different polishing techniques, the mean roughness R_z of selected samples is measured. Tab. 1 shows that laser-polishing achieves fairly smooth surfaces.

Surface treatment	Roughness R_z
Cut (grain size 40 μm)	13.3 μm
Acid-polishing	3.3 μm
Fire-polishing	3.2 μm
Laser-polishing	2.5 μm

<u>Table 1</u>: Mean surface roughness after polishing

Fig. 3 shows SEM-photographs of a commercially cut (grain size 40 μm) and a laser-polished surface. After the laser treatment, the previously rough surface appears perfectly smooth. Fig. 4 illustrates the transition region between the ground and the polished regions.

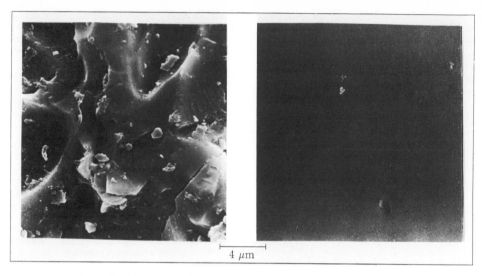

Figure 3: Original rough surface (left), laser-polished surface (right)

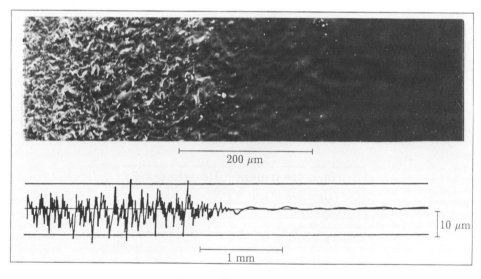

Figure 4: Transition region

In addition, the possibility of evaporation of components like lead-oxide during the treatment is of major importance. For this reason, the samples are carefully analysed using EDS-, SIMS (Secondary Ion Mass Spectroscopy)-analyses. These measurements show that concentration differences between irradiated and untreated glass do not exist. Therefore, no detectable amounts of PbO evaporate during the laser treatment. No environmental pollution occurs.

5 Mathematical simulation

Due to the high absorption coefficient, the absorption layer of the applied laser beam energy is negligible in comparison with the heat diffusion path in this particular case. The glass surface temperature after the interaction can be estimated by solving the heat flow equation for the case of a surface source, assuming temperature-independent properties. Under the presumption of one-dimensional heat flow the surface temperature T as a function of interaction time t can be calculated [9],

$$T - T_o = \frac{I(1-R)}{\delta \rho c_p \pi^{1/2}} (4\delta t)^{1/2} \tag{1}$$

where T_o is the temperature at t=0, R the reflectivity, ρ the glass density, c_p the heat capacity of the glass, δ the thermal conductivity, and I the laser beam intensity.

With the laser beam diameter d and feed speed v, the interaction time is given by

$$t = d/v \tag{2}$$

Rearranging (1) and (2), the relationship between I and v is given by

$$I = \frac{\rho c_p}{(1-R)} \left(\frac{\pi \delta}{4d} \right)^{1/2} (T - T_o) v^{1/2} \tag{3}$$

Considering the region of successful polishing is bounded by a certain viscosity interval, the influence of the laser parameters on the process temperature can be calculated. The viscosity is related to the temperature by the Vogel-Fulcher-Tammann equation

$$\log \eta = A + \frac{B}{T - C} \tag{4}$$

Then

$$I = \frac{\rho c_p}{(1-R)} \left(\frac{\pi \delta}{4d} \right)^{1/2} \left(\frac{B}{\log \eta - A} + C - T_o \right) v^{1/2} \tag{5}$$

The constants A, B, C are estimated by fitting the four experimetally determined viscosity points by equation (4): A=0.59, B=2078, C=625.5 for η in dPas and T in K.

By solving the one-dimensional heat diffusion equation, the dependence of between the laser intensity I on the feed speed v determined by equation (5) can be compared to the experimentally determined region, as shown in fig. 2. The upper limit is determined by $\eta = 10^2$ dPas (T $\approx$ 1800°C) and the lower limit by $\eta = 10^{4.2}$ dPas (T $\approx$ 930°C). For glass manufacturing these viscosity values are well known glass working temperatures: the melting point and the sink point. This calculated melting temperature is rather high because the simple one-dimensional model did not considere radial heat diffusion or heat loss due to radiation.

6 Conclusions

Experiments with the cw-CO_2-Laser shows that a crack-free heat treatment of lead-crystal-glass is possible. After the laser-treatment the glass surface appears smooth and brilliant. Entire glass articles have been successfully polished. Preheating the samples to temperatures just below T_g is necessary to avoid cracking while treating the glass. According to the mathematical simulation, laser polishing of lead crystal glass can be carried out successfully within a small viscosity range, implying a large temperature interval. This fact offers a wide range of possible laser parameters. EDS- and SIMS-analyses confirm that this technique is an alternative, especially considering the environmental load of traditional processes because no products like PbO are emitted into the atmosphere.

References

[1] A. Kaiser, H.Schmidt, and H. Scholze.
Untersuchungen zum Verfahren der Säurepolitur von Kristall- und Bleikristallgläsern.
Glastechn. Ber., 15 (1985) 7, 200-209.

[2] A.F. Stewart and A.H. Guenther.
Laser polished fused silica surfaces: absorption data.
NIST Spec. Publ. (Laser induced damage opt. Mater.: 1988), 775 (1989), 176-182.

[3] P.A. Temple, S.C. Seitel, and D.L. Cate.
Carbon dioxide laser polishing of fused silica: recent progress.
NBS Spec. Publ. (U.S.), 669 (1984), 130-137.

[4] G. K. Chui.
Laser cutting of hot glass.
Cer. Bull., 54 (1975) 5, 514-518.

[5] C. Buerhop, B. Blumenthal, and R. Weißmann.
Wechselwirkung von Hochleistungslasern mit Glasoberflächen.
64. Glastechn. Tagung der DGG, Düsseldorf, (1990) Kurzreferate, 29-31.

[6] C. Buerhop, B. Blumenthal, R. Weißmann, N. Lutz, and S. Biermann.
Glass Surface Treatmant with Excimer and CO_2 − Lasers.
Appl. Surf. Sc., 46 (1990), 430-434.

[7] A. Geith, R. Jaschek, and H.W. Bergmann.
Umweltfreundliche Politur von geschliffenen Bleikristallgläsern mit einem CO_2-Laser.
Sprechsaal, (1992), to be published.

[8] F. Geotti-Bianchini, L. De Riu, G. Gagliardi, M. Guglielmi, and C.G. Pantano.
New interpretation of the IR reflectance spectra of SiO_2-rich films on soda-lime glass.
Glastechn. Ber., 64 (1991) 8, 205-217.

[9] H.S. Carslaw and J.C. Jaeger.
Conduction of Heat in Solids.
Oxford at the Clarendon Press 2nd ed., Oxford, 1959.

Fundamental aspects of the damage free removal of hard coatings using excimer laser

R. Queitsch[1], H. Kukla[2], H.W. Bergmann[1], J. Naser[1], K. Schutte[1]

[1] Forschungsverbund Lasertechnologie Erlangen, Friedrich-Alexander-Universität, Lehrstuhl Werkstoffwissenschaften 2 (Metalle), Erlangen, FRG
[2] Thyssen Edelstahlwerke AG, HOT Nürnberg, FRG

Abstract

One method to generate protective wear resistant devices is the deposition of thin hard films. During lifetime of the component the protective layers are removed by abrasive wear. Especially for complex and expensive tools a redeposition would have economic advantages. The damage free material removal using an excimer laser is a possible application avoiding nasty chemical treatments.

First investigations were carried out with different samples geometries of TiN-surface layers on steel. Reported is the removal rate and the substrate damage as function of both the laser parameters (energy density, number of pulses and repetition rate) and processing parameters (geometry, processing atmosphere) applied.

Introduction

Hard coatings are of essential interest to improve the tribological behaviour (wear resistance, durability, corrosion resistance) of high-duty components for applications in tooling industry (1).

During lifetime of the parts the protective surfaces will be destroyed by abrasive wear. A recoating is carried out several times for expensive and large tools. In this case it is necessary to remove the old layers completely without damage of the surface (e.g. hardness and roughness) or the bulk material.

Nowadays cleaning and removing of thin, hard coatings from components for the tooling industry are exclusively performed by chemical etching techniques and / or mechanical treatments like sand blasting (2). Removing of PVD- and CVD-coatings by these treatments may lead to high production costs (con-

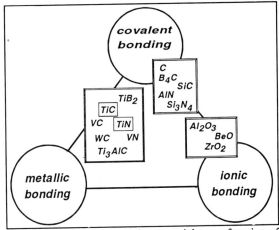

Fig. 1 Groups of hard coating materials as a function of different types of bonding (bonding triangle)

cerning the long time for the decoating process) and to pollution problems because of the aggressive chemicals used.

Figure 1 exhibits the different hard coating materials commonly used as protective layers. Due to the type of bonding it is possible to classify these materials (3). Materials properties (roughness, adherence, hardness, high temperature properties) change depending on the position within the bonding triangle. Titanium nitride (TiN) is a typical hard coating used as protective layer for tools or for decoration purposes because of its gold colour. The centred position of TiN in the bonding triangle shows a mixed bonding property with covalent, metallic and ionic character, see figure 1.

Alternatively to chemical treatments the authors examined a contact free, optical process using an excimer laser. Due to the small penetration depth of the UV-light and the intensive, short laser pulses an ablation of thin films can be achieved (4). The coatings should be removed step by step with ablation depths corresponding to laser wavelength depending on the other laser parameters.

Experimental Setup

The fundamental investigations were carried out using an high power excimer laser and two different processing gases (air and argon). Argon can be applied to the samples surface by a gas nozzle directed to the treated area. A schematic drawing of the experimental setup is shown in figure 2.

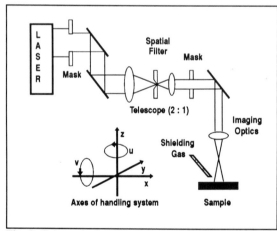

Fig. 2 Schematic drawing of the experimental setup

The applied beam source was a Siemens-type excimer laser XP2020 with an optical power of 2J/pulse, an average power of 40W using XeCl as laser active medium (wavelength 308nm, pulse duration 55ns, repetition rate 20Hz). The original cross section of the laser beam of about 45mm x 55mm has to be reduced to an area of 16mm x 19mm by a Kepler telescope. An imaging optic using a mask projection system was build up to illuminate the samples. The energy density applied to the specimen can be varied from 5 to 140mJ/mm^2. Changing the distance between sample and imaging optic or using other suitable optics will lead to a large variation of parameters.

The characterization of the laser treated samples was performed by auger electron spectroscopy (AES), roughness measurements and optical methods. Enhanced process security and quality is provided by means of spectral differential reflectometry (5). The reflection signal detected by a spectrometer can be used to terminate the laser illumination and to suppress an ablation of the ground material. Using the equipment described above the process could be controlled to ensure a complete removal of the TiN layer without a substrate damage.

Experimental Results

The material investigated was a vacuum-hardened high speed steel (St. 1.2379) coated with TiN. The deposition of the protective layers was carried out using a commercial plasma arc equipment (PVD), generating a layer thickness of 3μm ± 200nm. The influence of the laser parameters, surrounding atmosphere and pressure was examined to optimize the removal rate and surface quality of the treated samples.

The first experimental step was the characterization of surface modification and removal of the hard coatings in air at normal pressure. Applying one to ten pulses and low energy densities (from 15 up to 40mJ/mm²) only a small volume of the layer is modified. The spectral reflection of the illuminated areas changes significantly. AES scan and profile measurements prove that a layer with a thickness of about 200nm changes from TiN to TiN_xO_y, see figure 3.

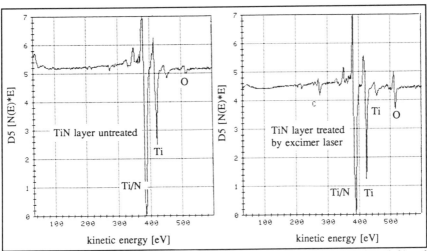

Fig. 3 AES-spectra (sputter depth: 50nm) of TiN coated high speed steel before (left) and after (right) an excimer laser treatment in air (1bar)

The changes in stoichiometry and the oxidation of the TiN layer could be suppressed using a shielding gas (e.g. argon) during the excimer laser treatment. No considerable removal rate can be detected varying the laser parameters within the limits specified above.

The next experimental steps gained in the determination of the influence of the processing parameters avoiding an ablation or roughening of the substrate material.

530

Figure 4 traces the maximum ablation depth versus the number of pulses using one specific energy density and different repetition rates. No improvement of the removal process concerning the applied laser frequency could be observed. The influence of changing the processing atmosphere can be easily seen. The removal rate using argon is increased twice to three times compared to the excimer laser treated targets in air.

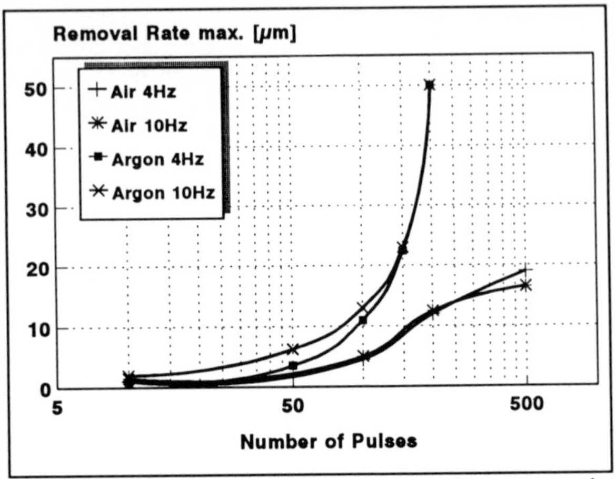

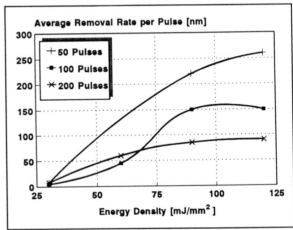

Fig. 4 Influence of processing gas and repetition rate on the maximum ablation depth; energy density 60 mJ/mm^2

Illuminating the targets surface a processing gas plasma of high ionized state is formed. A bombardment of the materials surface by the plasma particles occurs. The propagation is directed both to the surface and to the incident laser beam. The velocity of the expanding plasma plume is about 10 km/s, which can be proved by experimental investigations using a high speed diagnostic (6). Additionally the generating of a shock wave during the fast expansion of the plasma will support the surface removal by creating mechanical stresses in the layers structure. This kind of a mechanical destruction of the TiN layer could be in fact improve the ablation process.

The correlation between energy density and removal rate (average ablated depth per pulse) is shown in figure 5. Processing the samples with high energy densities (up to 120 mJ/mm^2) an increasing of the ablated layers volume is detected. Below a characteristic threshold the removal rate grows nearly linear with increasing energy density. Above this threshold value a saturation effect can be observed. The TiN layer is completely removed from the steel surface but simultaneously damaging the steel substrate (change of surface roughness and hardness).

Fig. 5 Average ablation depth per pulse in air as function of the energy density

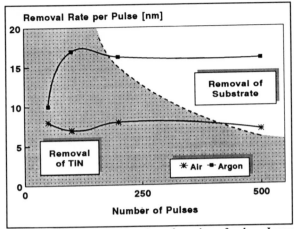

Fig. 6 Removal rate as function of number of pulses; Laser parameters : energy density 30 mJ/mm², repetition rate 4Hz

During the last experimental step the processing parameters has to be optimized.

The dashed curve in figure 6 connects the points where the product of the number of pulses and the achieved removal rate is equal to the layers thickness of 3 μm. On the left side of this line only a removal of the protective material without a damage of the bulk material occurs. On the right side the excimer laser irradiation causes an undesired ablation of the original steel surface.

The optimized removal rates can be obtained using the processing parameters summarized in the table below.

energy density	number of pulses	repetition rate	shielding gas
30 - 60 mJ/mm²	30 - 100	4 - 10 Hz	argon

Conclusions and Outlook

The work presented in this paper outlines the possible application of excimer laser surface treatment for typical components of the tooling industry. Using a fast, optical process it is possible to remove hard coatings from steel completely.

In future investigations will be carried out to optimize this process by applying other shielding gases (neon, helium, hydrogen, nitrogen) and a high vacuum system (changing the pressure). This may result in a higher removal rate and to an improved surface quality of the decoated materials.

References

(1)   P. Hedenqvist, M. Olsson, S. Söderberg: Surf.Eng. **5**, No. 2 (1989) p. 141
(2)   B. Gysen, M. Malik: Surf.Eng. **5**, No. 3 (1989) p. 208
(3)   H. Hollek: in "Surface Engineering & Heat Treatment", P.H. Morton ed., The Institute of Metals London (1991), p. 312

(4)   S. Rosiwal, H.W. Bergmann: Proc. ECLAT '90, H.W. Bergmann and R. Kupfer eds., Sprechsaal Publishing Group Coburg (1990) p. 895

(5)   H.W. Bergmann, E. Schubert: Proc. ECLAT '90, H.W. Bergmann and R. Kupfer eds., Sprechsaal Publishing Group Coburg (1990) p. 813

(6)   K. Schutte, R. Queitsch, H.W. Bergmann: Proc. ECLAT '92 (1992), in print

Paint stripping with short pulse lasers

H.W. Bergmann, R. Jaschek, G. Herrmann

Friedrich-Alexander-Universität Erlangen-Nürnberg, Forschungsverbund Lasertechnologie Erlangen (FLE), Institut für Werkstoffwissenschaften 2, Metalle, Martensstr. 5, 8520 Erlangen, FRG

1. Abstract

For many industrial applications paints are used with good adhesive strength. Using such kinds of paints difficulties may occur if it is necessary to remove the paint. Conventional methods are insufficient because either bulk material could be damaged or the waste disposal of chemicals is difficult, expensive or will harm the environment.

Different short pulse lasers (TEA-CO_2-laser, excimer laser, Nd:YAG-laser) have been tested for their ability of paint stripping. Short pulse lasers can remove layers of paint locally, in a short time and without damaging the substrate. The temperature within a thin surface layer is increased very quickly so that the paint evaporates faster then absorbed energy is transported into deeper layers. The TEA-CO_2-lasers seem to be favourable for stripping of painted steel because the substrate will not be damaged due to the high reflectivity in the infrared region. Excimer lasers seem to be favourable for paint stripping of fibre reinforced materials allowing a reduction of the heat transfer caused by the short wavelength and special photochemical reactions.

Long pulse and cw-lasers (Nd:YAG-laser, CO_2-laser) proved to be unsuitable because of their disadvantageous energy transfer causing melting of the substrate surface.

2. Introduction [1,2]

In many industrial branches different paints are used like topcoats, rain erosion coatings, textured anti-skid walkways, sealants, primers and other surface coatings. These paints can be subdivided by either their function or their structure and can be build up by different layers and with different materials like polyurethane, alkyl phenol resin or coatings based on an epoxy matrix, oil or water. The coatings are dried in different atmospheres, by heat treatment or by reaction between two components. Sometimes these coatings have to be renewed and therefore they must be removed all at once or layer by layer without damaging the bulk material or the coating below. For example all aircraft with painted surfaces must be stripped and repainted on an average four-year cycle. This process necessitates the use of toxic chemicals and considerable hand scraping, resulting in an enormous expenditure of personnel hours and the creation of serious air and water pollution. Waste disposal is another problem for which a solution has become increasingly important. Therefore a new process for paint stripping has to be developed which must be automated, environmentally acceptable and with considerably reduced costs. In early 1986 a survey of all potential stripping processes revealed that laser stripping had more potential savings than any other process. Laser paint stripping eliminates the environmental hazards associated with chemical paint stripping, reduces waste removal to only paint residue and allows stripping from a lot of substrates. Waste reduction by a factor of 100 over present chemical stripping techniques is expected with concentration and containment of the waste product in an environmentally acceptable form. The laser literally vaporizes the coatings to be removed, leaving only solid waste of about 10-50 % of the original paint volume with the paint volatiles removed by an exhauster and air filters. Other automated stripping methods such as aquastripping, dry ice crystals, granulated plastic and abrasive wheat germ pellets may cause surface deformation, surface hardening and other intrusive and impact related problems [1,2].

In this paper the potential of paint stripping with short pulse lasers will be discussed. Spectral difference reflectometry and measurement of the electrical resistance are two methods for process control.

3. Experimental setup

The investigations were carried out using commercial TEA-CO_2-laser, excimer laser and Nd:YAG-laser. Table 1 summarizes the characteristic parameters of these short pulse lasers.

Table 1: Specifications of the short pulse lasers used for the paint stripping experiments

laser	λ	P	P_{max}	repetition rate	FWHM	η
Nd:YAG-laser	1,06 µm	7 W	46 MW	10 Hz	15 ns	~2 %
Excimer laser	308 nm	40 W	40 MW	20 Hz	60 ns	~1 %
TEA-CO_2-laser	10,6 µm	75 W	15 MW	30 Hz	50 ns	~8 %

The laser beam is focused on commercially produced paint with a thickness of 20 µm on steel substrates. The paint stripping is carried out without moving the work piece. The important experimental laser parameters are the laser beam diameter, the repetition rate, the laser beam intensity and the homogeneity of the laser beam. These parameters have a significant influence on the stripping rate and the quality of the stripped surface. After the paint stripping the roughness of the surface and the microstructure of the substrate should not be changed and the paint should be stripped completely without remnants. Two methods are used for the detection of surface modifications like complete paint stripping or oxidation of the substrate. The quality of the paint stripping is first controlled by the electrical resistance between two different measuring points on the surface of the substrate. Ten measurements on each stripped area are averaged when the resistance is lower than 1 Ohm. A second technique allows a contactless process control by means of spectral difference reflectometry in the UV-VIS range because of its high flexibility and fast response (< 0,01 s) [3]. The reflected light from the surface of the work piece is recorded between the wavelengths of 320 nm and 720 nm in comparison to a standard (1 µm polished aluminum). This process controls firing of the laser, when a special paint is detected and to stop firing when the substrate is reached or when a layer below is reached which does not have to be stripped. This method increases the quality and the safety of the overall process.

4. Experimental results

4.1 Paint stripping with a Nd:YAG-laser

With the best parameters for paint stripping with a Nd:YAG-laser an area of 1 cm^2 can be stripped with 10 pulses during 16 s. Fig. 1 shows the quality of the stripped surface as a function of the number of laser pulses.

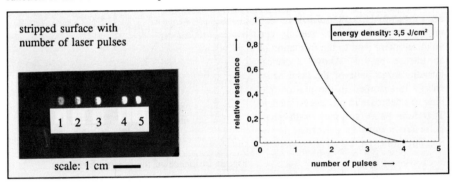

Fig. 1: Paint stripping with a Nd:YAG-laser

The quality of the stripped area is good. Sometimes remnants of paint are found which could be due to the inhomogeneity of the laser beam. The beam of a Nd:YAG-laser can be delivered through an optical glass fibre resulting in a high flexibility of this process. A higher stripping rate can be expected with higher repetition rates.

4.2 Paint stripping with an excimer laser

With the best parameters for paint stripping with an excimer laser an area of 1 cm^2 can be stripped with 100 pulses during 5 s. Fig. 2 shows the quality of the stripped surface as a function of the number of laser pulses.

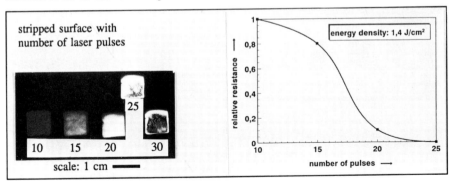

Fig. 2: Paint stripping with an excimer laser

The quality of the laser stripped surface is very good. Sometimes oxidation of the steel surface is observed due to a high number of laser pulses. Beside the stripping of painted steel excimer lasers seem also to be favourable for paint stripping of fibre reinforced materials allowing a reduction of the heat transfer caused by the wavelength in the UV-region and special photochemical reactions.

In the case of paint stripping with an excimer laser the influence of the plasma formation is demonstrated as seen in Fig. 3. Ionisation of the air and volatized paint lead to the formation of an intense plasma. Above a critical intensity large parts of the laser beam energy is absorbed in the plasma leading to a decrease in process efficiency. Therefore paint stripping with short pulse lasers must be performed below this threshold.

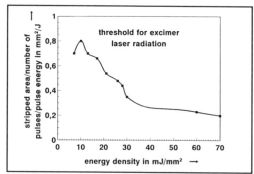

Fig. 3: Influence of the plasma formation

4.3 Paint stripping with a TEA-CO_2-laser

With the best parameters for paint stripping with a TEA-CO_2-laser an area of 1 cm^2 can be stripped with 30 pulses during 1 s. Fig. 4 shows the quality of the stripped surface as a function of the number of laser pulses.

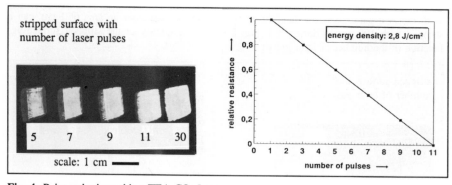

Fig. 4: Paint stripping with a TEA-CO_2-Laser

The quality of the stripped area is very good. The stripping rate for the TEA-CO_2-laser is expected to be the highest of the tested short pulse lasers due to the large beam diameter.

4.4 Process control by spectral difference reflectometry

An appropriate technique for online and contactless process control is possible by means of the spectral difference reflectometry. The reflected light from the surface of the work piece is recorded during the paint stripping with an excimer laser. Fig. 5 shows the obtained relative reflection as a function of wavelength. The original paint surface show low reflection due to its black colour. After 30 laser pulses the paint is stripped and a high reflection from the surface of the steel substrate can be detected.

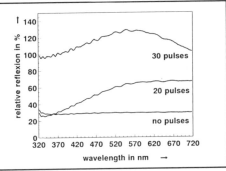

Fig. 5: Process control

5. Discussion and Conclusions

During this experimental work long pulse and cw-lasers (Nd:YAG laser, CO_2-laser) have also been tested but they proved to be unsuitable because of their disadvantageous energy transfer damaging the surface of the substrate by melting it.

Table 2 summarizes the results for paint stripping with short pulse lasers. Due to their individual wavelengths different absorption in the paint and the substrate takes place. With the TEA-CO_2-laser high temperatures can be reached for a short time leading to vaporization of the paint. The excimer laser emitting in the UV-region shows ablation of the paint. The stripping rates listed in Table 2 have to be considered with the used average laser power as shown in Fig. 6. The stripping rate with the TEA-CO_2-laser is high because the area of the laser beam is large in comparison to the area of Nd:YAG-laser beam. It would be desirable to increase the repetition rate of the TEA-CO_2-laser for even a higher process efficiency. For a Nd:YAG-laser a higher stripping rate is also expected when a laser with high repetition rates is used.

For future applications it is necessary that the laser beam will be scanned over a special frame to achieve large stripped areas [2]. Therefore a compromise has to be found between laser beam diameter, frequency and feed rate of the work piece. With a small beam diameter and a high frequency, e.g. for the Nd:YAG-laser, a high scanning velocity is necessary. But a moderate velocity with larger beam diameters will achieve similar stripping rates.

The investments for the laser equipment are also listed in Table 2. These averaged values are related to the optical laser power. Paint stripping with an excimer laser is expensive due to the high investment in comparison to a Nd:YAG-laser and TEA-CO_2-laser. Using the Nd:YAG-laser with an optical glass fibre for transmission of the laser beam a high flexibility is achieved and stripping of work pieces with a complex shape is improved. Even at high repetition rates and a high number of laser pulses a serious damage of the substrate surface could not be observed for all three short pulse lasers, especially for the TEA-CO_2-laser. Even at high repetition rates in the range of some kHz the time between the laser pulses are 10^3-10^4 higher than the laser pulse duration.

Table 2: Summary of the experimental results

laser	stripping rate	investment $/W	flexibility	substrate damage
Nd:YAG-laser	16 s/cm²	440	fibre delivery	low
Excimer laser	5 s/cm²	4700	mask projection	low
TEA-CO₂-laser	1 s/cm²	300	projection with mirrors	low

Fig. 6 shows a plot for stripping time as a function of the average laser power related to an area of 1 cm² and a coating thickness of 20 μm. The laser parameters for the best stripping results are extrapolated to other values of the average laser power. Higher stripping rates can be achieved using lasers with a higher repetition rate or a higher average power.

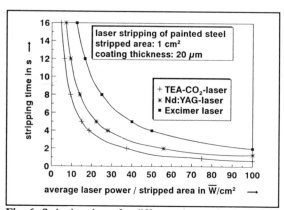

Fig. 6: Stripping times for different short pulse lasers

6. References

[1] J.S. Foley: "Laser paint stripping: An automated solution" in *Industrial Laser Review*, 8, 1991, p.4-9

[2] Development contract received for automated laser paint stripping system" in *The laser Edge,* Information folder from United Technologies Industrial Lasers, Hartford, USA, 7, 1990, p.1

[3] S. Rosiwal, H.W. Bergmann: "Surface treatments with excimer laser and quality control by means of difference reflectometry" in *Proc. of 3ʳᵈ ECLAT*, edt. by H.W. Bergmann and R. Kupfer, Sprechsaal Verlag, Coburg, 1990, p.895-904

Fretting Wear of PVD-TiN Coatings Oxidized with Excimer Laser Radiation

M. Franck, B. Blanpain, J.P. Celis, J.R. Roos, Katholieke Universiteit Leuven, Department of Metallurgy and Materials Engineering, Leuven, B
E.W. Kreutz, Rheinisch-Westfälische Technische Hochschule Aachen, Lehrstuhl für Lasertechnik, Aachen, FRG
M. Wehner, K. Wissenbach, Fraunhofer-Institut für Lasertechnik, Aachen, FRG

Introduction

Hard ceramic TiN coatings are used as wear resistant layers on cutting and forming tools in the metal and plastic working industry. Oxide compounds were found to play an important role during the wear of TiN coatings (1-2). It was also hypothesized by Gardos (3) that slightly oxygen-deficient rutile (TiO_{2-x}) behaves like a low shear strength, lubricous oxide, and TiN was suggested as the most suitable rutile-forming substrate. Stable oxides acting as solid lubricants are actually candidates for application as in-situ engineered oxide layers in tribosystems operating at high temperatures in oxygen containing atmospheres.

The use of Excimer laser radiation for advanced materials processing has received a lot of attention over the past few years. The UV wavelength and the short pulse duration make the Excimer laser radiation well suited for surface processing of thin films and microstructures in the submicron depth range (4-6). This paper is concerned with the post-deposition surface treatment of PVD TiN coatings with photon beams (7). The laser-induced surface oxide layers on TiN were characterized and tested in comparison to the as-deposited TiN coating.

Experimental procedure

The substrate material used was a high-speed steel type AISI M2 which was thermally hardened to 64-66 HRC and contained 0.9 %C, 4.1 %Cr, 5.0 %Mo, 1.9 %V and 6.4 %W. The flat subtrate specimens were first cut from a 1.4 mm thick sheet, then ground and mechanically polished. They were subsequently coated with stoichiometric TiN by Physical Vapour Deposition (PVD) using a triode ion-plating equipment (WTCM, Diepenbeek). The TiN coatings were about 3.0 μm thick and had a surface roughness of 0.05 μm R_a. After the specimen preparation, the TiN coatings were surface treated with the UV laser radiation. The laser radiation was provided by a short pulse (τ≈30 nsec) Excimer laser operating on a KrF gas mixture which emits at 248 nm. Rectangular apertures were used for obtaining a uniform distribution of the energy density over the irradiated area. A quartz imaging lens concentrated the energy of the laser beam on the surface in a rectangular spot of size 0.8 mm x 2.3 mm. The amount of incident energy per pulse could be changed by a rotating attenuator. The TiN-coated specimens were mounted on a CNC-controlled XY-table. The laser-induced oxidation experiments were carried out in ambient air at a pulse repetition rate of 20 Hz and an energy density of 0.9 J/cm².

The number of pulses was selected between 1 and 5,000. The large surface area needed for proper characterization of the oxidized TiN layers was produced by processing adjacent single spots with multiple pulses or by a unidirectional scanning in overlap mode.

The surface morphology was examined by scanning electron microscopy. Rutherford Backscattering Spectrometry (RBS) was used to investigate elemental depth profiles. RBS spectra were recorded with 2.0 MeV He$^+$ incident ions and a detection angle of 13°. Data analysis was done using RUMP (8) simulation software. The surface hardness was measured by nanoindentation (9), which probes the mechanical properties in the submicron depth range. The Nanoindenter was operated in the load controlled mode at a constant loading rate of 100 μN/sec, and different maximum loads ranging from 3 to 40 mN. The MTM fretting apparatus (10) was used for investigating the wear behaviour of the flat specimens subject to small oscillatory displacements. The contacting counterbodies used were hardened (63-64 HRC) chromium-steel (AISI D3) balls (diameter of 10 mm) with composition 85.3 %Fe, 12.0 %Cr, 2.1 %C, and a surface finish of 0.02 μm R_a. Flat specimens and balls were cleaned with acetone before testing. The fretting tests were performed in ambient air of 67-73% RH and 25°C under dry sliding conditions. A displacement stroke (A) of 100 μm was imposed, and a normal load (F_n) of 50 cN was applied giving mean hertzian contact stresses, neglecting the TiN contribution, of 0.16 GPa. The frequency (ν) was fixed at 10 Hz, and tests were run for 5,000 cycles. By sampling the displacement and tangential force, the normalized energy dissipation per cycle could be calculated. The wear surfaces were examined by light optical microscopy with Nomarski interference contrast, and by scanning electron microscopy with an energy dispersive X-ray attachment.

Results and discussion

A single Excimer laser pulse irradiation below the 1.3 J/cm^2 threshold for selective nitrogen evaporation, has been reported to give rise to a fast thermal transient in the solid state with preservation of the original TiN surface morphology (7). The multiple pulse irradiation investigated in this study, was causing the progressive solid-state oxidation of the TiN compound by repeatedly heating to high temperatures. Due to the short pulse duration, the heat affected zone was confined within the TiN coating itself, while the pulsed Excimer irradiation is considered to generate independent thermal cycles because of the low pulse repetition rate used. The thermal oxidation reaction thus proceeded under non-isothermal conditions, with strong temperature gradients which may have a substantial influence on the mass transport processes involved (11).

A distinct change in the surface morphology of TiN was observed after 1,000 laser irradiation pulses (Fig. 1). The as-deposited TiN coating consists of crater-like micro-roughnesses at the surface (Fig. 1a). This morphology results from the argon-ion etching step prior to the TiN deposition. The laser-oxidized TiN surface on the other hand, is uniformly covered with spherical protuberances of nearly equal size (Fig. 1b). A grain boundary structure containing a large number of small circular holes was observed around these protuberances. This typical boundary structure is believed to originate from different oxidation processes occurring in the as-deposited columnar TiN grains and along the defect-rich grain boundary areas. The pores were then probably formed by accumulation of the numerous defects like voids present in TiN.

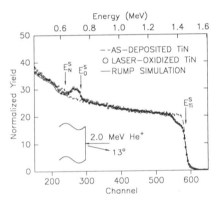

Fig. 1: SEM of morphology of TiN (a) as-deposited and (b) after Excimer irradiation with 1,000 pulses of 0.9 J/cm^2 in air.

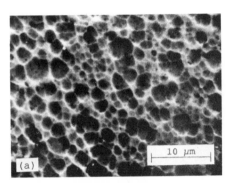

Fig. 2: RBS of as-deposited and laser-oxidized (500 pulses, 0.9 J/cm^2) TiN.

Fig. 3: Oxygen content by RBS of laser-oxidized TiN vs. number of pulses.

The existence of an oxidized toplayer on TiN after Excimer irradiation was confirmed by RBS. The RBS spectrum after irradiation with 500 pulses is shown in Fig. 2. By RUMP data analysis, a titanium oxide compound with stoichiometry between $Ti_3O_{4.6}$ and $Ti_3O_{5.1}$ was found for the different samples, excluding the samples after 100 and 150 pulses, where the oxide layer is too thin to determine the stoichiometry. No further oxidation to the thermodynamically more favourable rutile TiO_2 phase occurred for an increasing number of pulses, and the oxygen deficient titanium oxide remained structurally stable during subsequent laser irradiation. This behaviour is different from the isothermal furnace oxidation of TiN which leads to rutile formation (12). The transition between the surface oxide layer and the underlying

TiN, as observed in Fig.2, is not really sharp, probably as a result of the prevalent surface roughness.

The oxygen content *vs.* the number of pulses was determined by RUMP simulation of the corresponding RBS spectra (Fig. 3). The oxide layer thickness reached 201 nm after 5,000 pulses, assuming a Ti_3O_5 oxide density of 4.35 g/cm³. Initial oxidation kinetics are obviously linear (k_{lin}=2.6x10¹⁵ atoms/cm² per pulse), indicating that a surface or phase boundary process or reaction is rate determining. The oxidation process at this early stage could be limited by the oxygen supply at the surface or by the steady-state formation of the oxide at the oxide/TiN interface. After extended oxidation for more than 300 pulses, the linear oxidation rate quickly dropped off to an even slower than parabolic rate. Such kinetic behaviour could indicate that the oxidation process which at this stage could be expected to be controlled by the oxygen diffusion, is also affected by other factors like e.g. the non-isothermal conditions in the oxide layer (13).

The hardness of TiN in the near-surface region decreased after 200 and 500 Excimer pulses from 18-23 GPa to 12 GPa in a similar way (Fig. 4). Since the surface oxide layer was only about 100 nm thick after 200 pulses, the considerable hardness drop could only originate from a mechanical softening of the underlying TiN coating. The TiN subsurface is supposed to endure the thermal annealing effect of the pulsed laser heating, which was completed within the first 200 pulses. The heat affected zone is restricted in depth as suggested by the converging hardness curves of the as-deposited and oxidized TiN.

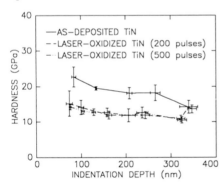

Fig. 4: Hardness *vs.* indentation depth corrected for elastic recovery for as-deposited and laser-oxidized (0.9 J/cm²) TiN.

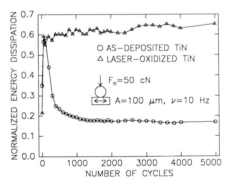

Fig. 5: Energy dissipation *vs.* number of cycles during fretting of as-deposited and laser-oxidized (1,000 pulses, 0.9 J/cm²) TiN against Cr-steel.

The friction and wear behaviour during fretting was investigated by monitoring the normalized energy dissipation per cycle. The normalized energy dissipation was calculated per unit stroke displacement and per unit normal load after the number of cycles run. The contact conditions used resulted in a gross slip displacement without any reversible elastic contribution. The normalized energy dissipation is therefore representing also the coefficient of friction. In case of as-deposited TiN (Fig. 5), the dissipated energy is typically displaying 2 stages, being the break-in transition followed by a steady-state regime (9). During the transition stage with a maximum energy dissipation around 0.6, strong adhesive interactions initiate a material transfer from the Cr-steel counterbody and a subsequent formation of tribo-oxidized wear particles. These hard abrasive particles started then to progressively scar the TiN surface (Fig. 6), finally leading to the local removal of the coating. A characteristic feature of the steady-state regime is the reproducible and low normalized energy dissipation of about 0.18, which could be attributed to the intermediate third body wear particles which prevent further adhesive interactions and easily shear off the low strength tribo-oxides from the surface.

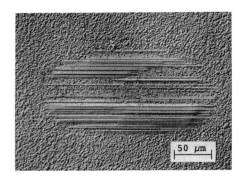

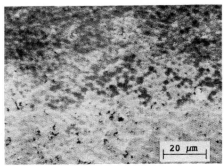

Fig. 6: Nomarski contrast of fretting scar on as-deposited TiN after 5,000 cycles against Cr-steel (A=100 µm, F_n=50 cN).

Fig. 7: SEM of transfer layer on laser-oxidized TiN (1,000 pulses, 0.9 J/cm²) after 5,000 cycles against Cr-steel (A=100 µm, F_n=50 cN).

Fretting of the laser-oxidized TiN was characterized by the immediate rise of the normalized energy dissipation to a steady-state value of approximately 0.6 (Fig. 5). This value is typical for the rubbing of a steel-steel material pair and was also observed during the break-in transition for as-deposited TiN. SEM of the wear surface of the laser-oxidized TiN (Fig. 7) demonstrates the built-up of a thin transfer layer coming from the Cr-steel ball and covering the entire contact area. EDX-analysis indicated the presence of iron in this compacted third body layer so that an iron(oxide)-iron(oxide) contact was governing the further wear behaviour. In contrast to as-deposited TiN, the laser-induced oxide layer on TiN was not systematically broken down during fretting under the same conditions, and thus exhibited a higher mechanical shear strength and load-bearing capacity.

Conclusions

Submicron surface oxide layers were formed on PVD-TiN by irradiating the coatings with Excimer laser radiation in air. The solid-state oxidation reaction of TiN proceeded under non-isothermal conditions. The surface morphology was modified during irradiation showing a distinct grain boundary structure containing pores. The laser-induced oxide layer is found to preserve a stoichiometry between $Ti_3O_{4.6}$ and $Ti_3O_{5.1}$ for an increasing number of pulses, while no rutile TiO_2 appeared as expected from isothermal furnace oxidation. Under laser irradiation, the oxidation kinetics slow down from a linear rate at the initial stage to a less than parabolic rate after an extended number of pulses. The mechanical softening of the TiN sublayers was caused by the laser-induced thermal cycles. The fretting wear behaviour of the laser-oxidized TiN slid against steel is characterized by the formation of a protective iron-based transfer layer, while as-deposited TiN is damaged by the abrasive action of the oxidized wear debris. The laser-induced oxide layer demonstrates a higher mechanical strength compared to the tribo-oxides on as-deposited TiN.

Acknowledgment

The authors gratefully acknowledge the support of WTCM-Diepenbeek for providing the TiN coatings. This work was supported by the Belgian Government and was carried out within the framework of the Interuniversity Networks for Fundamental Research (IUAP contract n°4).

References

(1) I.L. Singer, S. Fayeulle, P.D. Ehni: Wear 149 (1991) 375.
(2) K.H. Habig: Surf. Coat. Technol. 42 (1990) 133.
(3) M.N. Gardos: MRS Symp. Proc. 140 (1989) 325.
(4) T.R. Jervis, J.P. Hirvonen, M. Nastasi: Lubric. Eng. 2 (1992) 141.
(5) M. Wehner, G. Grötsch, E.W. Kreutz, W. Schulze, K. Wissenbach: ECLAT'90 Proc. (1990) 917.
(6) H.W. Bergmann, E. Schubert, K.J. Schmatjko, J. Dembowski: Thin Solid Films 174 (1989) 33.
(7) J.P. Celis, M. Franck, J.R. Roos, E.W. Kreutz, A. Gasser, M. Wehner, K. Wissenbach, H. Pattyn: Appl. Surf. Sci. 54 (1992) 322.
(8) L.R. Doolittle: Nucl. Instrum. Methods B 15 (1986) 227.
(9) J.P. Celis, J.R Roos, E. Vancoille, M. Franck: Wear of Materials'91 Proc. ASME (1991) 655.
(10) J.R. Roos, J.P. Celis, M. Franck, H. Pattyn: Surf. Coat. Technol. 45 (1991) 89.
(11) M. Wautelet: Appl. Phys. A 50 (1990) 131.
(12) H.G. Tompkins: J. Appl. Phys. 70 (1991) 3876.
(13) N. Birks, G.H. Meier: "Introduction to High Temperature Oxidation of Metals", Arnold Ltd. (1983) 60.

VI.
Precision Machining
and Structuring

Precision Drilling With Short Pulsed Copper Vapor Lasers

Roland Kupfer*, Hans W. Bergmann

Forschungsverbund Lasertechnologie Erlangen (FLE), Lehrstuhl Werkstoffwissenschaften 2, Metalle, Universität Erlangen-Nürnberg, Martensstr. 5, D-8520 Erlangen

* jetzt: FLUMESYS GmbH, Kropfersrichterstr. 6-8, D-8458 Sulzbach-Rosenberg

ABSTRACT

Laser beam drilling has been established in industry for a few years, in addition to mechanical, spark erosion and electro-chemical drilling techniques. Most of the laser drilling techniques use single pulse or percussion drilling (with milliseconds long pulse durations). The development of more powerful and reliable short pulsed lasers allows the evaporation drilling of materials more precisely and without detectable damage to the base material. Typically, lasers which could be used for drilling of metal materials are especially copper vapor (CVL), Nd:YAG, CO_2 and excimer lasers.
In this paper results are presented which demonstrate the possibilities of copper vapor lasers on the drilling velocity and the quality of drilled materials, like metals and semiconductors.

1. INTRODUCTION

Materials processing by means of lasers has readily been adopted by industrial manufacturers during the recent years. The broad range of possible tasks in industry is given by laser cutting and welding applications [1] as well as for the purpose of drilling [2]. Conventional lasers, e.g. CO_2 and Nd:YAG-lasers offer high efficiency due to their maximum power output, and various possibilities for the direction and the formation of the beam. In the case of short pulse lasers, the pulse duration, the pulse energy, the pulse repetition frequency, and the focusability of the beam have to be taken into consideration. The present elaboration deals with the application of CVLs for precision drilling tasks in various metals, alloys, and semiconductor materials.

2. COPPER VAPOR LASER TECHNOLOGY

Stimulated emission at two visible wavelengths (510.6 and 578 nm) can be obtained with a Copper Vapor Laser device by direct impact excitation of the neutral copper atoms. In view of the characteristics of the energy level, only short pulses within a range of 20 - 50 ns can be achieved [3]. In table 1, typical laser parameters are listed.

Table 1: Specifications of Copper Vapor Lasers

λ (nm)	P_{MEAN} (W)	f_p (kHz)	E_p (mJ)	t_p (ns)	P_p (kW)
510.6/578	max. 100	2-32	max. 20	20-50	max. 400

The beam divergence of a CVL is around 6 mrad when fitted with a plane-plane cavity. With this pair of parallel mirrors highest extraction efficiency is obtainable. When an unstable cavity is fitted, the beam divergence can be reduced by a factor of 10 and more. The near field beam profile shows in both cases a top hat intensity distribution. In the far field the beam profile tends to be annular for the plane-plane cavity. Using the unstable cavity the far field profile consists of a series of rings with progressively smaller diameter. The highest power is concentrated in the smallest diameter rings because of the low divergence of this part of the laser pulse [4].

3. EXPERIMENTAL SETUP

The experimental setup is shown as a schematic diagram in Fig. 1. The investigations were carried out using a Copper Vapor Laser CU40. The beam could be apertured in the near field.

A beam splitter offered the possibility to separate both wavelengths and to monitor the beam characteristics. Using biconvex lenses, the laser pulses were focussed onto the samples. A translation stage allowed the variation of the focal plane position and hence, the power flux density. The number of pulses for target exposures were varied by computer control. In order to estimate the drilling depth the material was slice wise removed and the development of the drilling depth as well as the aspect ratio record-

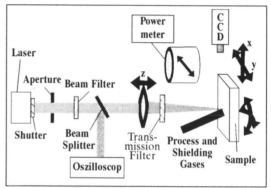

Fig. 1: *Experimental setup*

ed. SEM and light microscopy investi-gations were carried out to determine a heat affection of the base material. The investigated materials and some of their thermodynamic data are listed in the following table 2.

Table 2: *Materials data of the drilled samples*

Material	T_M (°C)	T_V (°C)	λ_W (W/cmK)
Cu	1083	2600	3,94
Ck45	1520	3000	0,5
X12CrNi17.7	1410	3000	0,15
Si	1410	2355	1,48
Ge	938	2830	0,6
InP	1062	-	0,68

T_M: melting point T_V: vaporization point λ_W: thermal conductivity

4. RESULTS

4.1 Drilling of metals and alloys

In Fig. 2 the drilling behaviour of different stainless steel materials is given. Fig. 2a shows to a 10 μm thick stainless steel foil which was exposed with CVL pulses. The non-circular shape and the second small hole derive from inaccuracies in the experimental setup.

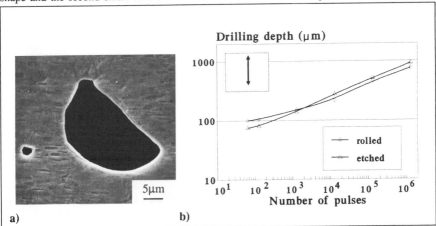

a) b)

Fig. 2: *CVL drilling of stainless steel: a) steel foil (10 μm thickness); b) bulk material*

In Fig. 2b the drill hole formation for bulk stainless steel is given. Two different surface conditions of the material were used to determine the influence of the surface. The variations observed have to be regarded as low taking into consideration that a very laborious technique of preparation had been used which made it very difficult to determine the exact depth of the drill hole.

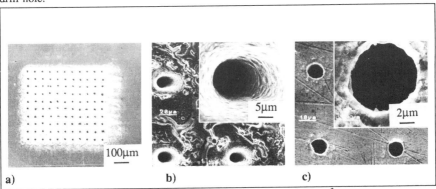

a) b) c)

Fig. 3: *CVL drilling of copper: a) top view on 100 holes at 0.25mm²; b) enlargement of Fig. 3a); c) bottom view*

In Fig. 3 SEM pictures show the maximum drill hole density in copper sheets. In the left picture 100 drill holes on an area of 0.25 mm² are shown viewed from above. The enlargement in Fig. 3b) indicates that the drill holes are nearly of circular shape with a wall roughness of less than 1 μm. The reproducable drill hole diameter is less than 15 μm on the beam entry side and less than 10 μm on the reverse side, see Fig. 3c). These results were obtained using an aperture in the near field which reduced the beam diameter to 15 mm. The average power was then less than 2 W.

The development of the aspect ratio for various cold rolled copper sheets shows Fig. 4a. Values of 40:1 can be received. As already reported elsewhere [5] aspect ratios of 90:1 are obtainable in copper materials. The lower curve indicates the ratios for the beam without an aperture in the near field. The upper curve gives the results for the already mentioned beam aperture. Obviously, the holes drilled with the full beam lead to enlarged diameters which reduce the aspect ratio for comparable sheet thicknesses. The cross-section in Fig. 4b) shows a 0.6 mm thick cold rolled copper sheet drilled with the "hot spot" arrangement. The diameter is around 15 μm.

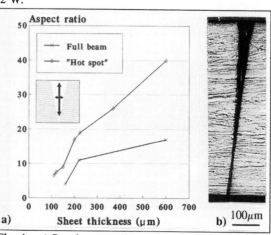

Fig. 4: *a) Development of the aspect ratio in copper; b) cross-section of a 0.6mm cold rolled copper sheet drilled by CVL*

Fig. 5: *Cross-section of a 1 mm thick copper sheet drilled by a CVL under 45° inclination*

Fig. 5 presents a cross-section of a drill hole in a 1 mm thick copper sheet, drilled under an angle of incidence of 45°. It is possible to incline up to 60°, as shown elsewhere [5]. The typical cone-like hole geometry lead to outlet diameters of 15-20 μm while the entry diameter is around 100 μm due to the larger area exposed with the laser pulses.

Determination of heat affected zones in mild steel and copper

As accurate measurements of the temperature inside the drill hole are difficult to accomplish, one may choose materials which act as temperature sensors themselves. This is possible for materials which undergo a solid state transformation or likewise as with highly deformed metals. Fig. 6a) presents a cross-section of drilled ferritic-pearlitic mild steel Ck45. From the SEM picture one can derive that the pearlitic lamellae remain unchanged right up to the drill hole without any indication of transformation. Obviously, no deterioration of the parent material occurs.

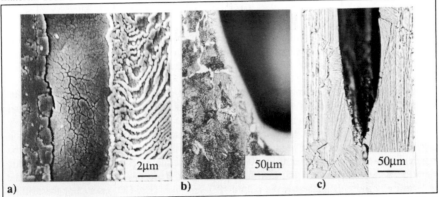

Fig. 6: *Cross-sections of different materials for the determination of HAZ: a) CVL drilled Ck45 (SEM picture); b) Nd:YAG drilled Ck45; c) CVL drilled cold rolled copper*

In Fig. 6c) the cross-section of the cold rolled copper is shown. Near the drill hole of this highly deformed material no transformation or recrystallisation zones are detectable. The above-mentioned result in a temperature gradient of values up to 10^3 K/μm. A comparative drilling in Ck45 using a high power Nd:YAG-laser led to a transition zone of 20-30 μm. This indicates a temperature gradient of around 50 K/μm, see Fig. 6b).

4.2 Drilling of semiconductor materials

In Fig. 7 the drilling results for various semiconductor materials are shown. Obviously, the material removal rate (around 3 μm per pulse) is 2 to 3 times higher for low number of pulses compared to drilling of metals, see Fig. 7a. This can also be seen from the low number of pulses

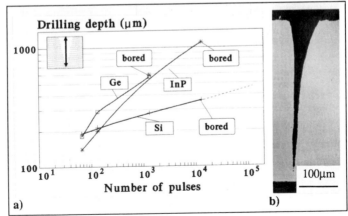

Fig. 7: *a) Drill hole development in semiconductor materials; b) cross-section of CVL drilled Si*

necessary for drilling through the specimen. In Fig. 7b a typical cross-section of CVL drilled Ge is given. Remarkable is the burr free drill edge, however, the broad opening diameter of 70-100 μm is not desired. Optimization of beam quality and optical setup will lead to better drilling geometries, as shown in Fig. 3 for metals.

5. DISCUSSION

The aforesaid results underline that CVLs are suitable tools for drilling in metals and semiconductor materials. Specifically the good beam quality, the high ablation rate which results from the high repetition rate and the lack of a detectable heat affected zones qualifies this laser.

The comparision with other short pulse lasers point out clearly the advantages of CVLs for drilling fine holes in metals. Excimer lasers are suitable tools for ablation and removal processes especially in ceramic and plastic materials. Due to the short wave lengths a very good focusability is achieved. Unfortunately, the restricted repetition rate limits the ablation rate. Nd:YAG-lasers are used in industry for various drilling applications. In most cases the drill hole diameters are larger than 100 μm. To generate finer drill holes single mode operation is necessary which reduces the efficiency of these lasers. The limitation in the focusability is a disadvantage for the pulsed IR-lasers like TEA-CO_2-laser. For the drilling process the room temperature absorbtion is not the decisive factor as the achievable power flux density always result in a massive ablation of the illuminated material. Though, the generated plasma leads to a shielding effect which arises for the IR-lasers at lower power flux densities as for the shorter wave length lasers. This reduces the efficiency of the drilling process. In the case of drilling with TEA-CO_2-lasers one may elude this disadvantage by prolonging the pulse duration. The reduced power flux density, if the pulse energy is considered to be constant, effects an increased ablation rate due to less reflection losses in the plasma plume. The influence of different gas mixtures on the pulse duration and hence, on the material processing is subject to current investigations.

6. ACKNOWLEDGEMENTS

The authors acknowledge Oxford Lasers Ltd., Oxford, for the financial and technical support in carrying out the experiments. Furthermore, they would like to thank M. Lingenauer and B. Aumüller for their helpful assistance in the practical work.

7. REFERENCES

[1]   M. Geiger, P. Hoffmann, G. Deinzer: *Combination of laser beam cutting and welding for car body manufacturing: welding with wire as filler.* Proc. Eclat'90, 689.

[2]   T.J. Rockstroh: *Application of Slab-based Nd:YAG Laser at GE Aircraft Engines.* Proc. SOIE 1601, (1990), 132

[3]   C.E. Webb: *Copper and Gold Lasers: Recent Advances and Applications.* Proc. Int. Conf. on Lasers '87, (1988), 276

[4]   N.N.: *Cavity Design for Metal Vapor Lasers.* Technical Note No.1, Oxford Lasers Ltd., Oxford, (1988)

[5]   R. Kupfer: *Laserstrahlbohren mit Kupferdampflasern.* Doctoral Thesis, Univ. Erlangen (1992), to be published

Surface Microstructural Changes on Alumina and Zirconia after Excimer Laser Treatment

A. Tsetsekou, Th. Zambetakis, C.J. Stournaras, CERECO S.A., Chalkida, GR
G. Hourdakis, E. Hontzopoulos, FORTH/IESL, Iraklion, Crete, GR

Introduction

The high energy and the high absorbency by most materials of the ultraviolet irradiation of excimer lasers render this type of laser a very useful tool, especially in the cases when very high precision micromachining (microelectronics industry) is required or surface treatment (modification of surface properties) is needed, due to the small absorption depth of the radiation and to the photochemical reactions that are likely to happen on the surface.

Over the last few years, there has been a growing interest in the study of excimer lasers influence on ceramics (4-8) and on ceramic coatings (8,9).

In the context of this study the effects of KrF (248 nm), ArF (193 nm) and XeCl (308 nm) excimer lasers on ceramics such as a-alumina and partially or fully yttria-stabilized zirconia are investigated.

The aim of this research effort is the investigation of the possibilities for improving the surface treatment of ceramics.

Experimental work

Samples of a-alumina and zirconia (partially or fully stabilized by yttria) were irradiated using KrF (248 nm), ArF (193 nm), (Lambda Physik, LPX 210) and XeCl (Lambda Physik, LPX 315) excimer lasers. The laser pulses were of 29, 17 and 30 ns duration respectively for the KrF, ArF and XeCl mixtures.

Al_2O_3 samples were prepared by the tape casting method, using a commercial powder (Mandoval AlM41). The sintering was performed at 1600°C for 4h. Zirconia samples were prepared by cold isostatic pressing (300 MPa) and sintering at 1500°C for 2h.

Commercial zirconia powders fully stabilized (8% per mole) (HSY-8 of Zirconia Sales) and partially stabilized by yttria (3% per mole) (TZ3 of TOSOH) were used. The fully stabilized zirconia samples consisted of pure cubic zirconia with no porosity, while the partially stabilized zirconia samples consisted of cubic and tetragonal zirconia with a porosity of about 14%.

The influence of laser radiation on the samples surface was studied using an optical microscope and a Scanning Electron microscope (JEOL 840). X-ray diffraction (SIEMENS D-500 Diffractometer) analysis was used for the investigation of the possible phases modifications. The treated surface and the effect of the irradiation parameters (wavelength, energy density, repetition rate, surrounding atmosphere) on the surface microstructure were studied.

Results and Discussion

i) Irradiation of alumina

The surface microstructure of the irradiated samples depends on the irradiation conditions. The energy density of the radiation is the most important parameter and its value may or may not influence the appearance of a surface melted layer and the grain size of the surface (8). The melting of the surface layer occurs when the energy density is greater than 2.5 J/cm^2. A completely melted layer, which is very smooth, appears when the energy density is 4.5 J/cm^2. When higher energy densities are used, ablation of the irradiated material occurs.

The repetition rate also influences the treatment results. As the value of this parameter increases while the other parameters are kept constant, the influence of the irradiation becomes more and more "thermal" resulting in growing grain size. The high repetition rate doesn't permit the material to be cooled between the pulses and consequently results in increased heating of the surface.

The application of scanning of the surface during treatment instead of localized irradiation (spot) has the same result. It also leads to the decrease of the melting threshold (4.5 J/cm^2 when scanning is applied, 4.5-6 J/cm^2 without scanning). The reason is once again the increased heating of the surface provoked by the repeated irradiation of a given surface point due to the overlapping of the steps during scanning . The increased heating enhances this phenomenon by provoking heat transport and additional heating to the surrounding area. When the energy density of the radiation exceeds the melting threshold, the application of scanning results in better surface quality of the treated area as well. The melted surface layer appears to have microcracks and 1-2 μm pores when the irradiation takes place without scanning, while when scanning is applied and the other parameters are kept constant, the number of microcracks and the porosity decrease greatly.

The improvement of the surface layer quality after the laser treatment can also be achieved by the increase of the imposed number of pulses per step (a surface layer having a great number of microcracks and a high porosity appears when 10 pulses per step are imposed, whereas a substantial reduction in the number of microcracks and of porosity occurs when 100 pulses per step are applied with the other parameters being wavelength 248 nm and energy density 7.5 J/cm^2). The melted layer depth has a mean value of about 10μm.

Summarizing the above observations, the resulting conclusion is that some optimum laser parameters do exist for the achievement of a high quality laser treated surface without microcracks and porosity. Such a surface results in improved wear resistance of the ceramic. When the used wavelength is 248 nm, these parameters have been found to be the following: energy density of about 6 J/cm^2, application of scanning using 60% overlap of the steps, number of imposed pulses per step 100 and repetition rate 100 Hz. Irradiation using less energy density, smaller pulse number per step and without scanning of the surface results in a less than satisfactory surface quality (microcracks, porosity) because the melted layer is still too thin (4.5 J/cm^2 is the threshold limit above which the melting is completed). Irradiation using greater values than those mentioned above for the energy density or the pulse number, leads to material removal with the consequence being the appearance of a thin molten layer with many microcracks and porosity. The repetition rate of 100 Hz is thought to be the most appropriate one, since when using this value, faster treatment is achieved and additionally, the melting is supported because of the increased heating of the sample.

The wavelength also influences the treatment quality. The less "thermal" effect of the smaller wavelength (193 nm) radiation requires more intensive irradiation conditions for the achievement of the same result (minimization of microcracks number and porosity) as the one which is achieved using the wavelength of 248 nm. The melted surface layer shows intensive secondary crystallization. On the contrary, the higher wavelength (308 nm) has more "thermal" effect leading to a glassy surface layer full of microcracks.

The treatment quality doesn't seem to be greatly influenced by the irradiation atmosphere (vacuum 10^{-8} mbar or mixture H$_2$, N$_2$ 10% per volume in H$_2$). However, in the case of the vacuum, a thermal influence of the surface up to a depth of 60 μm (wavelength 248 nm) and the appearance of orientation in the surface layer (X-ray diffraction analysis, **Fig. 1**) was observed. These phenomena were not observed when the irradiation was made in air atmosphere. The limited heat transport to the surrounding area leads to a more localized irradiation which is a possible explanation for these findings when a vacuum atmosphere is used.

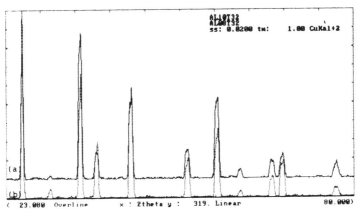

Fig. 1:   X-ray diffraction analysis of alumina a) untreated b) KrF laser treated at 6 J/cm^2 under vacuum (10^{-4} mbar)

X-ray diffraction analysis of the irradiated samples showed the presence of the metastable phase γ-alumina when the irradiation was made using an energy density greater than 6 J/cm^2 at 248 nm. The quantity of this phase seems to depend on the intensity of the irradiation conditions and especially the number of the imposed pulses per step (8).

The irradiation is always accompanied by colour changes on the sample surfaces (from white to brown). The intensity of the colour also depends on the intensity of the irradiation conditions. This colour change can be attributed to the creation of oxygen vacancies due to the influence of the irradiation.

ii) Irradiation of zirconia

The excimer laser radiation also influences the zirconia surface by a melting-recrystallization mechanism. However, the melting starts at a lower energy density than that of alumina (this energy density is about 1.2 J/cm^2) and is completed at 3 J/cm^2. A possible reason for this differentiation observed between the two ceramics is the smaller thermal conductivity of zirconia (0.47·10^{-3} cal/s/cm/°C) compared to the corresponding value of alumina (7.25·10^{-3} cal/s/cm/°C) which results in a lower energy loss due to heat transport. (However, zirconia has a much higher melting point than that of alumina, 2700° instead of 2050°C). Irradiation using an energy density greater than 3 J/cm^2 leads to the occurrence of laser ablation. At 248 nm and 7.5 J/cm^2 the material removal is too strong and carries off the layer influenced by the radiation.

The irradiation of zirconia (partially or fully stabilized by yttria) leads to grains merging, melting of the superficial layer and to almost complete elimination of porosity. However, the presence of a great number of microcracks after irradiation at 248 nm is remarkable. These microcracks continually become more intense (especially in the case of fully stabilized zirconia, **Fig. 2b**). The stronger partially stabilized zirconia seems to have a smaller number of microcracks after the irradiation and this number is reduced as the repetition rate increases. The depth of the influenced layer doesn't exceed 1μm.

A significant differentiation appears when the smaller wavelength of 193 nm is used. The zirconia grains merge and form greater ones while at the same time, intensive secondary recrystallization inside the grains occurs. This secondary recrystallization depends directly on the applied pulse number. When 6 J/cm^2 and 100 pulses per step are used, the surface layer is covered by these fine grains (size <0.1 μm) which form clusters with a mean size of 0.3 μm (**Fig. 2c**). The treatment using the higher wavelength of 308 nm results in a much less satisfactory surface quality. A glassy phase full of microcracks (many of them being 0.5 μm in width and 6 μm long) appears.

X-ray diffraction analysis of the samples after the laser irradiation didn't show any destabilization of zirconia by creation of the monoclinic phase (**Fig. 3**). This is an important fact indicating that the laser treatment can be used in zirconia samples without causing any destruction of the ceramic material.

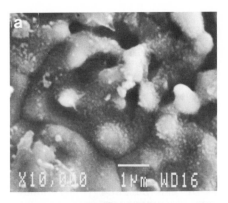 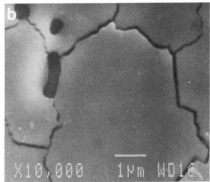

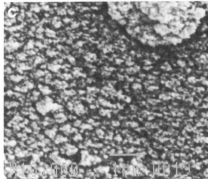

Fig.2 : Fully yttria-stabilized zirconia surface (a) untreated (b) KrF laser treated at 5 J/cm^2 (c) ArF laser treated at 3.5 J/cm^2.

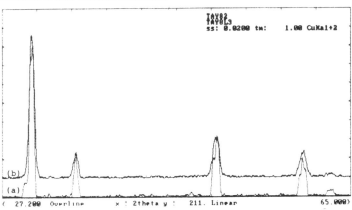

Fig. 3: X-ray diffraction analysis of fully yttria-stabilized zirconia (a) untreated (b) KrF laser treated at 6 J/cm^2

The irradiation is always accompanied by colour changes in the ceramic surfaces (from white to grey or black). The colour intensity depends on the irradiation conditions and it is less intense when the wavelength of 193 nm is used.

Conclusions

- Excimer lasers influence the surface of ceramics through a mechanism of melting-recrystallization. The melting threshold for alumina is 2.5 J/cm^2 at 248 nm while the corresponding value for zirconia is 1.2 J/cm^2.

- Irradiation of alumina using energy density greater than 6 J/cm^2 results in material removal. The corresponding value for zirconia is about 3 J/cm^2.

- The properties of the irradiated area (porosity, microcracks, microcrystalline structure) depend directly on the treatment conditions. More intense conditions with longer treatment duration lead to better results while irradiations without scanning of the surface (spot) under mild conditions lead to microcracks creation.

- The irradiation of alumina in a small quantity causes the formation of the metastable phase γ-alumina depending on the laser parameters while the irradiation of zirconia doesn't destabilize it.

- The quality of the treated area is not influenced significantly by the irradiation atmosphere but it has been observed in the case of alumina that the irradiation under vacuum causes a thermal effect up to a depth of about 60 μm and causes an orientation of the surface layer as well.

Acknowledgement

This work is supported by the Greek General Secretariat for Research and Technology under EUREKA, EUROLASER EU 205 programme.

References

(1) P. Helzer: Phototonics Spectra 1 (1989) 112.
(2) T.A. Znotins, D. Poulin and J. Reid: Laser Focus/Electro-Optics 5 (1987) 54.
(3) J.K. Wright: Industrial Laser Handbook, Annual Review of Laser Processing (1988) 40.
(4) K.J. Schmatjko, G. Endres, U. Schmidt and P.H. Banz: Proc. SPIE 957 (1988).
(5) K.J. Schmatjko, H. Durchholz and G. Endres: Proc. SPIE 1025 (1988).
(6) K.J. Schmatjko, G. Endres, and H. Durchholz: Zur Veroffentlichung Proc. LIM-5 (1988).
(7) E. Hontzopoulos and E. Damigos: Appl. Phys. A52 (1991) 421.
(8) G. Hourdakis, E. Hontzopoulos, A. Tsetsekou, Th. Zambetakis and C.J. Stournaras: Proc. SPIE 1503 (1991) 249.
(9) G. Gravanis, A. Tsetsekou, Th. Zambetakis, C.J. Stournaras and E. Hontzopoulos: Surface and Coatings Technology 45 (1991) 245.

Integration of Materials Processing with YAG-Lasers in a Turning Center

M. Wiedmaier[1], E. Meiners[1], T. Rudlaff[2], F. Dausinger[1], H. Hügel[1]
(1): Institut für Strahlwerkzeuge (IFSW), Universität Stuttgart, FRG
(2): Zentrum Fertigungstechnik Stuttgart (ZFS), FRG

Introduction

High-performance Nd-YAG lasers in the kilowatt range are available today with beam guiding through a thin (0.4 - 1 mm diameter) flexible glass fibre to the machining point. Integrated within the working space of cutting machine tools the potential of a laser beam can be used in processes being performed either subsequently or parallel to the machining operation. The laser optic can be integrated directly in the tool-changing system of a machining centre as an additional tool without any significant restrictions in cutting tool functions.

Along Fig. 1 "integration" is defined not as a linking of a laser machining system with other single-tasks systems in a production line, but as a machine tool with integrated laser with the possibility to combine metal cutting with laser welding, hardening, marking or laser caving. Among these combinations, laser processes may serve three different purposes:

o   First, they can be used as a tool "supporting" or "assisting" a cutting process to increase the process efficiency or just to enable a cutting process of materials with difficult cutting properties. Examples are chip breaking [1] or laser assisted machining.

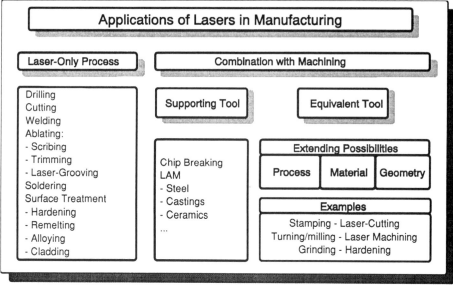

Fig. 1   Classification of the possibilities in materials processing integrating a laser in machine tools

o Second, the laser process can be integrated into a machine tool as a technique equivalent to a classic one. Stamping-cutting machines equipped with lasers represent this type of combination [2] to fulfill higher demands on flexibility, accuracy or quality. Technologies like laser caving or drilling are other examples to extend the machinable materials and geometries.

o Third, with welding [3] and surface treatment processes like hardening [2] generally seperated external production steps can be included in one production sequence of metal cutting and laser processing in the machining center. A complete machining in one setting can be performed with technical and economical advantages resulting from higher accurracies, a reduced flow of material or shorter production times.

Discussing one particular process - laser machining a ceramic thread bolt in a lathe - the principle feasibility of integrating laser processing into machining centres will be illustrated with the new technologie of laser caving.

2. Integration Concept of Laser Processing in a Lathe

2.1 Choice of laser type

Choosing the Nd:YAG-laser, a decisive point is the effort and space needed for the beam guiding into the working-space of the machine tool. The Nd:YAG-laser enables a beam transmission via flexible fibres and has advantages in laser processing concerning efficiency and quality resulting from the shorter wavelength compared to the CO_2-laser as proposed or just realized [4], [5].

For the experiments in laser caving presented in this paper, a pulsed Nd:YAG-laser with a maximum mean power of 600 W (*HAAS Laser GmbH*) was chosen. This laser offers a parameter range of frequencies up to 1000 Hz, pulse durations from 0,1 to 10 ms and a maximum peak power of 12 kW with a glass fibre of 0,6 mm diameter. In further experiments on complete machining - aiming for example on hardening - a cw Nd:YAG-laser with mean powers from 1000 - 2200 W related to fibre diameters of 0.4 - 1 mm can be applied as well.

2.2 Technical realization

The integration of the YAG-lasers has been realized in a small turning centre (INDEX GS30) with two turrets and five controlled axes, four of them linear and one as an rotational C-axis. The two turrets with cylinder shaft tool-holding mechanisms are positioned with the turret rotating axes parallel and vertical to the spindle, respectively. The cutting tools and the laser machining head are swivelled into position by rotating the turret. To ensure that the fibre does not require to be bent and, consequently, the working area and chip removal remain unimpeded, a rotary degree of freedom around the laser beam axis has to be provided. The fibre then can be held away from the working area by means of a tension spring.

The numerical control link between laser and lathe is realized by a programmable digital and analogue interface [6].

Fig. 2   General view of the working space of a lathe with integrated laser working head, left, and close view of laser working head in processing position.

Fig. 2 shows the setup for circumferential laser machining. In this setting, the second turret can be used parallel for machining with a cutting tool (provided that the machining speeds are compatible).

The optical head shown in Fig. 2 is an improved design of a focussing optic with an imaging scale of 1:1. The pipings for working gas and cooling water are integrated into the fibre outlet. The focal point can manually be shifted 4 mm.

3. Laser Caving as an Example for Integrated Laser Machining

Laser caving is described here as an example out of possible processes like welding, hardening, marking, drilling and others. Material removal can be realized by various means; the fundamental strategies for a general "laser machining" are discussed in [10]. The material removal by melting and vaporization (with or without supporting coaxial gas jet) only, will be named here as laser caving. The material's chemical composition and its properties together with the laser parameters determine the amount and composition of the removed material. A detailed discussion of laser caving several steels, metals and ceramics is given in [7], [8] and [9] with a summary in [10].

This investigations on laser caving of steel and ceramics have shown that its advantages have to be compared with cutting, grinding or eroding extending the machining possibilities. As a material of particular interest for laser caving silicon nitride (Si_3N_4) has been chosen. This extremely hard and electrically non-conductive material has a restricted machinability when shaping with diamond grinding, while electrical eroding is not possible. A shaping by form sintering calls for expensive and inflexible moulds and hinder therefore the evolution of prototypes.

Fig. 3 and Fig. 4 represent the experiments on laser caving Si_3N_4. With appropiate spot scanning patterns the surface roughness can be kept constant at a value of $R_z = 20\ \mu m$ while the laser parameters influencing the ablation rate or the ablated depth per pulse are

562

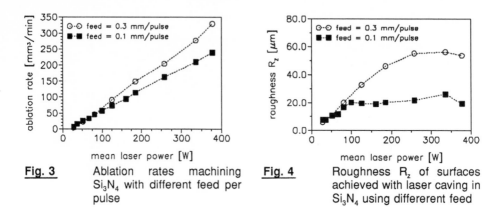

Fig. 3   Ablation rates machining Si_3N_4 with different feed per pulse

Fig. 4   Roughness R_z of surfaces achieved with laser caving in Si_3N_4 using differerent feed

varied. With reduced ablation rates a finishing can be performed to achieve a surface roughness below $R_z = 5$ µm.

These very good surface qualities result from a sublimation-like dissoziation of Si_3N_4 and vapor removal when irradiated with energies above a certain threshold level in contrast to a predominant melt removal, when metals are irradiated. Melt removal is ongoing with a deposition of oxides and melt reducing quality and accuracy.

Material removal by a process in its effect similar to sublimation, therefore, offers the possibility to achieve sufficient surface quality over a wide range of processing parameters. This allows a flexible process control and, in addition, a relatively simple yet reliable theoretical modelling. The excellent correspondence of qualitative results (roughness) and quantitative results (ablation rates) of simulation and experiment allows to develop and proove working strategies for three-dimensional shaping [11].

Fig. 5   In a lathe laser grooved and marked thread bolt made of Si_3N_4.

To demonstrate the feasibility of laser caving in the lathe supported by an optimization of working strategies by numerical modelling, a trapezoid thread bolt Tr 20x4 of Si_3N_4 was produced and inscribed. With the thread cutting function of the lathe the rough thread-shape has been produced by ablating trace to trace in four layers. Subsequently, finishing has been achieved by multiple precision machining coupled with measurements of the geometry by an optical sensor in the machine. At a mean power of 400 W the production of a 40 mm long thread and the following marking have taken 30 minutes. The "demonstration product" is shown in Fig. 5 and its final geometry, as measured by the optical sensor is presented in Fig. 6.

This process - in the two stages roughing and finishing linked by a measurement to an open loop control - allows to achieve sufficient accuracies when then ablation rates are reduced or if a material like Si_3N_4 shows good-natured characteristics. To optimize the accuracy as well as the machining times

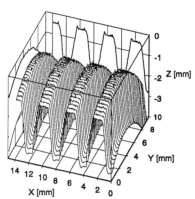

Fig. 6 Processing quality of the work piece shown in Fig. 5 as measured with an optical sensor

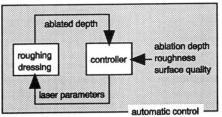

Fig. 7 Scheme on closed loop control with a measurement by optical sensor

especially for materials with melt removal (e. g. steel) an automatic on-line control of the ablated depth is needed. The concept shown in Fig. 6 can be realized with an optical sensor (a measurement is shown in 4) integrated in a focussing device [12], [13]. This device is recognized as a key part introducing this technologie in an automated industrial manufacturing [14].

It should be mentioned, that 1979 in [15], [16] the machining of a ceramic thread bolt is shown with a CO_2-Laser as first experiments on laser caving. The demands on "integration" as outlined in chapter 1 there could not be fulfilled with the realized beam guiding. Today, with an automatic control, the knowledge of many experimental data, CAM - possibilities and reliable lasers, laser caving can be introduced in manufacturing.

4. Summary

The integration of laser and machine tool as described here demonstrates the flexibility of the link via a glass fibre. Laser caving is shown as a new development opening up new possibilities for shaping. Because of the restricted conventional machining possibilities and the achievable machining quality ceramics, particularly Si_3N_4 appears interesting. The state of the art is described in the production of a ceramic thread bolt with open loop control. Developments will upgrade the process by an on-line control of the ablated depth per pulse, making it suitable for an automated industrial application.

The beam guiding via glass fibres to a focussing head integrated in the tool changing system allows the alternative use of pulsed and cw-lasers enabling cutting, joining and surface treatment with laser in combination or parallel to cutting operations of the lathe. The economic justification should be given by a complete machining in one setting including several production sequences and reducing flow of material.

References

[1] M. Wiedmaier et al.: "Materials processing with YAG-Lasers integrated in a turning center". ICALEO'92, Orlando Okt. 1992 - Orlando: LIA, to be published.

[2] N.N.: "LASERPRESS - Trumpf pulst sanft aufs Blech". maschine + werkzeug, 23/1979.

[3] N. N.: Stanzpaketieren von geblechten Teilen. Firmenschrift BRUDERER GmbH.

[4] F. Dausinger: Lasers with different wave length - implications for various applications. Proc. of 3. ECLAT, Vol. 1, Erlangen, 1990 - Sprechsal 1990.

[5] F. Dausinger: "Beam Matter Interaction in Laser Surface Modification". LAMP '92, Nagaoka, 1992, pp. 697.

[6] M. Wiedmaier, E. Meiners, R. Niethammer: "In Drehmaschinen integriertes Abtragen mit dem Laserstrahl". FTK '91, Stuttgart 1991 - Springer, 1991.

[7] E. Meiners, F. Dausinger, M. Wiedmaier: "Micro Machining of Ceramics by pulsed Nd:YAG Laser". Laser Materials Processing, ICALEO'91, San José 1991 - Orlando, Florida: LIA, 1992.

[8] J. Arnold, U. Stark, F. Dausinger: "Structuring of Metals by Ablation with Excimer Laser". Laser Materials Processing, ICALEO'91, San José 1991 - Orlando, Florida: LIA, 1992.

[9] E. Meiners, Wiedmaier, Dausinger, Krastel, Masek, Kessler: "Micro Machining of Hard Materials using Nd:YAG-Laser". LASER '91, München, 1991- im Druck.

[10] H. Hügel: "Integration of Laser Materials Processing in Metal Cutting Machine Tools". LAMP '92, Nagaoka, 1992.

[11] E. Meiners et al.: Approaches in Modelling of Evaporative and Melt Removal Processes in Micro Machining. ECLAT '92, Göttingen, 1992 - to be published.

[12] I. Masek: Integration eines optischen Tasters in eine YAG-Optik zur Tiefenmessung beim Abtragen. Diploma thesis 3/92 IFSW Uni Stuttgart, 1992.

[13] G. Eipper: Entwicklung einer Online-Tiefenregelung für den Abtragprozeß mit gepulsten YAG-Lasern. diploma thesis 92/4 IFSW Uni Stuttgart, 1992.

[14] G. Eberl, S. Sutor: "Abtragen mittels Laserstrahlung - Prozeßanalyse und Regelungskonzepte". LASER '91, München, 1991- im Druck.

[15] M. Copley, M. Bass: "Shaping Materials with a Continous Wave Carbon Dioxide Laser". Proc.: Appl. of Lasers in Materials Processing, Washington, DC 1979 - ASM, 1979.

[16] S. Copley, M. Bass, B. Jau, R. Wallace: "Shaping materials with lasers". In: Bass (Hrsg.): "Laser Materials Processing". Amsterdam: North Holland, 1983.

Acknowledgements

The project this report is based on is suported by the Bundesministerium für Forschung und Technik (BMFT) of the Federal Republic of Germany and recorded under code number 13N5747. The authors are responsible for the content of this publication.

Laser-Plasma Deposition of TiN Coatings

S. Metev, R. Becker, S. Burmester
BIAS - Bremer Institut für angewandte Strahltechnik, Bremen, FRG

1. Introduction

Hard TiN coatings are of increasing technical importance for wear
protection on wear loaded pieces like tools, machine elements etc.
/1/. One problem of producing TiN by conventional technics (sputtering or ion-plating, CVD etc) is the high thermal distortion of
the substrate (>500 °C). Laser-plasma deposition may solve this
problem /2/.

In this paper results are reported on producing TiN-layers by the
laser-plasma method under conditions of very low thermal load.
Parameters were changed in order to achieve hard and dense layers
with good adhesion to the substrate.

2. Experiments

2.1 Experimental set-up

A TEA-CO_2-laser has been used for the production of laser-plasma
deposited thin films. The experimental set-up of the deposition
chamber is shown in fig.1.

2.2 Experimental procedure

Sintered pellets of exact composition of TiN (purity: 99.5%) were
used as target material in order to achieve stoichiometric
deposition of thin TiN-layers on steel substrates (115CrV3). The
steel substrates were polished in order to smoothen the surface
and to remove contaminants.

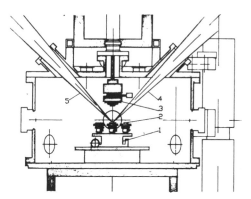

Fig.1: Laser-plasma coating device
1) target holder; 2) targets; 3) substrate holder
4) and 5) laser beams

In general the deposition parameters were hold constant as follows:

Geometric arrangement: Diameter of target and substrate: 25 mm
                        Distance target - substrate: 30 mm
Process conditions: Substrate temperature: 200 °C
                        Chamber pressure: 10^{-6} hPa
Laser parameters: Max. pulse energy: 2 J
                        Pulse repetition rate: 2 Hz
                        Pulse duration: 1 μs

Pulse energy and pulse length were detected from a partial deflecting mirror and kept constant. The deposition rate and the plasma parameters (density, degree of ionization, energy distribution of ions) were controlled in-situ by a combination of piezo-quartz detector and ion collector. The residual gas in the vacuum chamber was measured by a gas analyser.

In this evaluation we studied the influence of the energy density on the formation of thin TiN-layers. Energy density of the square shaped beam was varied by attenuators (focus position was hold below the surface). The substrate temperature was kept to a low value of 200 °C.

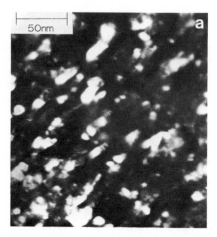
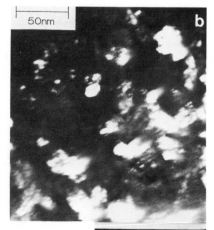
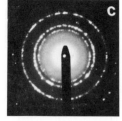

Fig.2: TEM-examination of TiN-coatings
       (dark field)
   a) deposition rate = 0.06 nm/pulse
   b) deposition rate = 0.04 nm/pulse
   c) diffraction pattern

3. Results

3.1 Deposition rate and structure

Deposition rate is very important for nucleation and growth. This parameter and the substrate temperature basically influence the formation and structure of laser-plasma deposits. In general, formation of polycrystalline, monocrystalline and amorphous structure is possible /3/.

Thin TiN-layers were produced under constant conditions and equal number of pulses (3600 pulses). By varying the energy density the deposition rate varied between 0.02 and 0.1 nm/pulse (temperature 200 °C). This means a mean deposition rate of >.2-1*10^5 nm/s during the time of evaporation! In this regime we always found a polycrystalline structure. This is in accordance to considerations that low temperatures and high deposition rates favour the formation of polycrystalline structures /3/. Fig.2 shows a TEM investigation of TiN-layers, formed under different deposition rates. The slightly lower mean grain size in fig.2a is due to higher deposition rates, which means a higher production of crystallisation centers in unit time. The lattice constant of 0.425 nm (see diffraction pattern, fig.2c) fits quite well to a cited value of 0.424 nm for the cubic-face centered structure of TiN /1/.

From the technical point of view -in particular, when producing high precision layer systems- it is important that the deposition rate is nearly constant over a wide range of energy densities. Fig.3 shows that this condition is fullfilled in the range of 20 to >60 J/cm². This is important in order to reproduce constant film thicknesses and could be explained by plasma formation, screening of the target surface and weakening the dependence of deposition rate on energy density /4/.

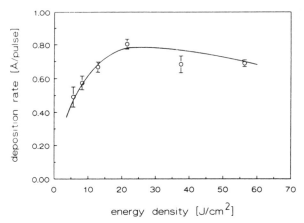

Fig.3: Dependence of deposition rate on energy density

3.2 Composition

Laser-plasma evaporation produces at sufficient power densities and short pulse durations heating rates of 10^{10} Ks^{-1}. This ensures practically the instantaneous reaching of highest evaporation temperature and the simultaneous evaporation of all elements in the target. Even when, due to the high temperature, dissociation of the target material occurs, all elements in the vapour are in the same stoichiometric ratio as in the target. This is a precondition for synthesis of stoichiometric films on the substrate surface /4/.

However, other parameters (laser, vacuum condition, reactive atmosphere etc.) should be fitted to the process. Glow discharge measurements of the composition profile of TiN-layers produced under conditions, see chapter 2.2, showed a lack of nitrogen which may be caused by incorporation of residual gases of the vacuum (for example <10 at% of oxygen and carbon in the analysis, fig.4). Titanium is very reactive to these elements. Recent experiments, reducing the residual gases to a partial pressure $<10^{-6}$ hPa and adding nitrogen as reactive gas in the 10^{-3} hPa regime reduced the contamination to less than 1 at%.

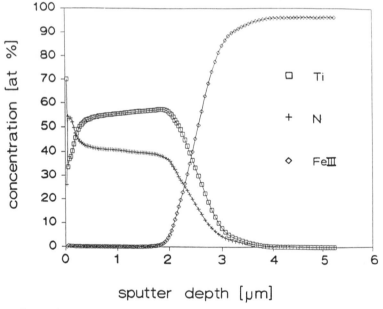

Fig.4: Glow discharge elemental analysis of TiN-layer

3.3 Mechanical properties and surface morphology

Under conditions as summarized in chapter 2.2 thin TiN-layers were produced in the range of 100 to 2000 nm. SEM-investigations revealed that the surface was smooth and dense even for thin layers. This is due to high deposition rates (in our experiments exceeding 10^5 nm/s) which are many orders of magnitude higher than those of conventional methods. By these high deposition rates the film is dense after depositing some monolayers /4/.

Fig.2 shows a fine micro-crystalline structure. The very small mean grain size of the structures (4 nm) indicates good mechanical properties of the coating (for example toughness). Though prepared under relatively low substrate temperatures the adhesion was good, which was prooven by qualitative scratch tests. Hardness was measured by a micro-identer. We found values between 1500 and 2000 HV.005. This is slightly lower than the optimized value of 2400 HV /1/ and might be explained by the fact, that there are some oxygen and carbon contaminants in the layer (see chapt. 3.2).

From a technical point of view a smooth surface is desired. Under bad laser conditions (pulse duration high and energy density low) surface smoothness could be affected by small droplets of molten particles (~1 μm, see fig.5) /4/. This -sometimes- is a problem of the laser-plasma deposition process but could be solved by adjusting the energy density. This is shown in Fig.5b with a minimum at 8 J/cm^2.

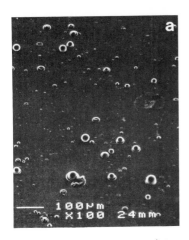
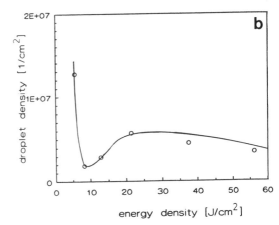

Fig.5: Droplets on TiN-coatings
   a) SEM-investigation
   b) dependence of droplet density on energy density

4. Conclusions

We used a TEA-CO_2-laser to produce hard, dense TiN-coatings with good adherence to steel substrates. Deposition temperature of the substrate was kept much lower (200 °C) than in many conventional processes. Under conditions of low temperature and high deposition rates always polycrystalline structure was found. Deposition rate was constant over a wide range of power densities. High energy densities caused a reduction of droplets on the surface with a minimum at 8 J/cm^2. Hardness was in the region of conventional TiN-deposits and should be improved by lowering residual contamination (mainly oxygen) in the layer.

5. References

[1] Pulker, H.K., Wear and Corrosion Resistant Coatings by CVD and PVD; Expert Verlag 1989
[2] Cheung, J.T, Sankur, H., Growth of thin films by laser-induced evaporation in "CRC Critical Reviews in Solid State and Materials Science", Dez. 1988, p.63
[3] Metev, S; Meteva, K.B., Nucleation and growth of laser-plasma deposited thin films, Applied Surface Science 43, 402(1989)
[4] Metev, S.M., Laser-plasma synthesis of thin film polycomponent materials in "Laser processing and diagnostics (II)",(Bäuerle, D.; Kompa, K.L.; Laude, L. edit.), Edition de Physique, 1986, p.143

Material Removal with High-Power-CO_2- and Nd:YAG-Lasers

M. Stürmer, Laser Zentrum Hannover e.V., FRG

Introduction

The removal of geometrically well-defined forms using advanced laser machining units is of increasing importance, due to considerable advantages in comparison with conventional cutting processes and electrical discharge machining (EDM). The efficiency of cutting machining processes is limited by material and cutting properties (e.g. hardness), and by the demanded shape of the workpiece. Cavity sinking by EDM also shows technological disadvantages, e.g. expensive electrode designing for every new workpiece geometry, wear, and the generation of pits and structural transformation, which both depend partially on the discharge current /1/. In comparison, material removal by laser radiation without tool wear is applicable even for electrically non-conductive materials which are hard to cut. The possibility of producing or refinishing NC-controlled forms and dies is given. Both metals and nonmetals (e.g. ceramics, quartz glass, graphite) and also synthetic materials can be machined. Nevertheless, well-known disadvantages of the process are variation in surface and removal quality, partial reagglomeration of the molten material and possible uneconomical removal efficiency. The results described in the following which answer these questions were given by investigations at the LZH, using pw/cw CO_2 and Nd:YAG lasers.

Process and technology of laser material removal

Figure 1 shows the principle of the process and one application example of laser shaped graphite. The critical part of the process is expelling the molten, vaporized or decomposed material which can be supported by processing gases, lead to the machining region. The most important parameters which have influence on the removal process are depicted in figure 2.

Of the different laser systems, generally CO_2-, and Nd:YAG-lasers are qualified for high removal rates in industrial applications. The amount of vaporized material can be increased by use of pw-Nd:YAG lasers with higher pulse peak power. At the same time, heat affection is reduced.

Q-switching of cw-sources increases this value to over 100 kW, but with lower avarage power. TEA- and excimer lasers are not very suitable for macro-removal of metals due to short irradiation times and shielding by plasmas.

For a long time, pulsed CO_2 lasers have been used for shaping rubber in roller production ("Flexodruck"). The computer-controlled keyed laser beam removes the rotating elastomere up to 10 mm maximum depth. Also for printing technique, grooves are formed on chromium-oxyde coated "Anilox" rollers.

The machining behaviour of rollers in sheet metal production can be improved by little cups on the material surface /2/.

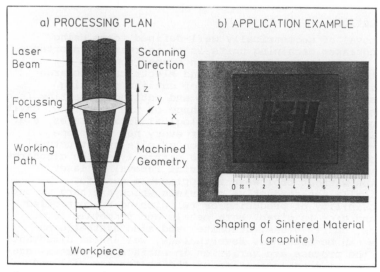

Fig.1: Processing concept and application example

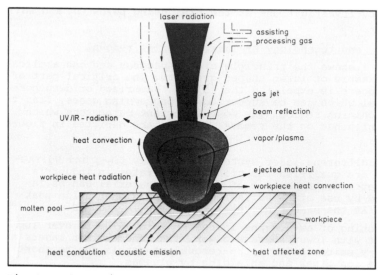

Fig.2: Interaction process between laser radiation, processing gas and material

The possibility of removing layers on steel by scanning the surface with the focussed CO_2 laser beam has been proved in earlier investigations /3,4,5/. In 1989, for the first time, a complete system for industrial shaping could be realized by adapting a pulsed 750 W CO_2 laser to a five axis milling unit ("Lasercav") /6/. Another concept for material removal contents two laser beams joined in the focal spot which seperate entire bulks from the workpiece without melting them completely. The control of the focal depth and the guarantee of complete expell of the material raise problems /7/.

Research results

The investigations described in the following were carried out using a RF-CO_2-laser source of 750 W output power and a flashlamp-pumped Nd:YAG laser unit of 300 W average beam power. Mild steel was chosen as sample material. The material was removed by unidirectional scanning of the workpiece surface and an additional gas flow.

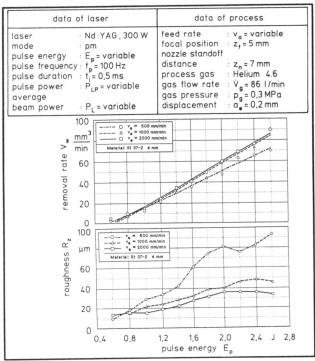

data of laser		data of process	
laser	: Nd:YAG, 300 W	feed rate	: v_a = variable
mode	: pm	focal position	: z_f = 5 mm
pulse energy	: E_p = variable	nozzle standoff	
pulse frequency	: f_p = 100 Hz	distance	: z_n = 7 mm
pulse duration	: t_i = 0,5 ms	process gas	: Helium 4.6
pulse power	: P_{LP} = variable	gas flow rate	: $\dot{V}_g$ = 86 l/min
average		gas pressure	: p_g = 0,3 MPa
beam power	: P_L = variable	displacement	: a_e = 0,2 mm

Fig.3: Results of material removal processing with solid state laser

Using CO_2 lasers, on steel, removal rates of more than 3000 mm^3/min at an average power of P_L=800 W could be obtained. The removal rates of about 800 mm^3/min attainable with roughing EDM are exceeded easily. Nevertheless, with use of oxygen, average power of over 400 W or feed below 8 m/min, a tendency towards uncontrolled material combustion can be observed. As general surface quality decreases with increasing removal rates, a compromise is to be found when determining the parameters. Therefore, average power was chosen to 400 W at the majority of the CO_2 experiments. In cw- or pw-processing without pulse peak stretching, the amount of vaporized material is low compared with the amount of molten metal.

Nd:YAG lasers render better beam coupling to the workpiece surface, showing less reflection. Figure 3 shows the results for pulsed Nd:YAG processing. As expected, removal rates raise with increasing pulse energy, but the removal quality depends on energy per track section. Therefore, a fall-off in surface quality can be noticed at pulse energy chosen too high, because the inner layers of the material are melted on.

The feeding rate related to these parameters, which is too low, cannot be increased in an unlimited way, otherwise, single pulse marks would become visible on the machined surface. For this reason, the high degree of overlap (>60%), which occurs with high pulse frequency, renders both high amount of material removal and high surface quality in avoiding a quilting seam. Thus it is is equal to the demand for a laser system with a pulse frequency as high as possible in order to reach high feeding rates.

So the machining with pulsed Nd:YAG lasers represents an economically and technologically interesting alternative to cavity sinking. Certainly, removed rates obtained by use of Helium (100mm^3/min) remained below the maximum values reached with CO_2 radiation, but advantages of the pulsed Nd:YAG laser exist concerning removal quality and surface roughness (R_z=10mm): The amount of vaporized material is higher, ensuring an effective expell of molten metal without using oxygen. As a conclusion, when using inert gases, considerably less heat affection of the marginal areas can be observed.

Figure 4 shows that in opposition to machining with reactive gases, pure surfaces without adherent oxide layers can be formed by using inert gases, but the attained removal rates are clearly reduced due to energy supply by laser only. Sharp and smooth crossings of surface elevations without erosion by heat are formed within the dimensions of the spot size. Another advantage of Nd:YAG radiation is the possibility of guiding the beam by light, uncooled mirrors. Additionally, very dynamic beam guidance systems can be realized within adequate construction expense.

<u>Outlook</u>

In comparison to cutting and roughing EDM, high removal rates on different materials can be achieved by laser machining. Nevertheless, the investigations brought out that machining below the theoretical maximum removal rate is necessary for the realization of smooth high-quality surfaces. The possibility of combining roughing laser machining at high beam power with following cavity sinking by EDM or fine finishing seems to be given.

When referring to the technique of manufacturing printing rollers as mentioned above, the laser shaping of metals using computers in connection with technological data bases has to be taken into account. Connecting geometry and material data, the laser beam could be controlled exactly as far as power and temporal keying are concerned. To avoid deviation of geometry from the demanded values, especially in depth, an additional surface and distance on-line control system (Figure 5) is favourable.

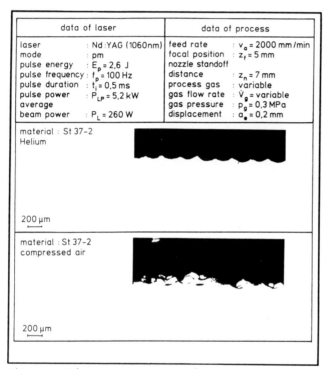

data of laser		data of process	
laser	: Nd:YAG (1060nm)	feed rate	: $v_a = 2000$ mm/min
mode	: pm	focal position	: $z_f = 5$ mm
pulse energy	: $E_p = 2,6$ J	nozzle standoff distance	: $z_n = 7$ mm
pulse frequency	: $f_p = 100$ Hz	process gas	: variable
pulse duration	: $t_i = 0,5$ ms	gas flow rate	: $\dot{V}_g$ = variable
pulse power	: $P_{LP} = 5,2$ kW	gas pressure	: $p_g = 0,3$ MPa
average beam power	: $P_L = 260$ W	displacement	: $a_e = 0,2$ mm

material : St 37-2
Helium

200 µm

material : St 37-2
compressed air

200 µm

Fig.4: Polish of cross section of workpieces after processing with Helium and compressed air as assisting gas

The influence of the processing gases, the material and the removal concept on the process remains of particular importance. Apart from pressure optimized nozzles (co- and off-axial outlets and Laval nozzles), at the LZH, suitable sight systems for process signals are beeing developed and investigated. Detection and analysation of IR and Plasma radiation as well as deviation probes and a guiding system for the focal depth rank with this topic. The adaption of laser power

576

controlling units, the equalization of the machined to the designed geometry and accompanying scaling of the parameters are the further aims of our research.

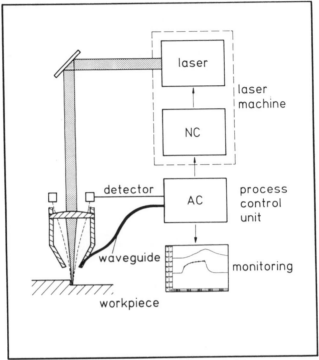

Fig. 5: Control and direction of the process

References

(1) Gehring, V., J. Hernandez Camacho: "Feinbearbeitung im Werkzeug- und Formenbau", VDI-Z Nr. 2/90, 51-55
(2) N.N.: "Lasertexturierung schafft hochwertige Blechoberflächen", LASER 5/90, 78-80
(3) H. Haferkamp, T. Vinke, K. Engel: "Abtragen mit CO_2-Laserstrahlung", Laser Magazin 6/88, 24-27
(4) H.K. Tönshoff, M. Stürmer: "Untersuchungen zum Formabtrag an Stahl mit Nd:YAG-Lasern", Laser und Optoelektronik 22(5) (1990) 59-63
(5) A. v.Krieken, J. Schaarsberg, H.J. Raterink: "Laser micro machining of material surfaces", SPIE Vol. 1022 Laser assistant processing, 1988, 34-37
(6) G. Eberl: "Laserbearbeitung für den Formenbau", LASER Nr.1 (1990) 24-27
(7) H.K. Tönshoff, F. v. Alvensleben: "Verfahren zur Qualitätskontrolle für den Materialabtrag mit dem CO_2-Laser", Laser und Optoelektronik 22 (2) (1990) 62-66

Influence of Process and Material Parameters on Ablation Phenomena and Mechanical
Properties of Ceramics and Composites for Excimer Laser Treatment

M. Geiger; N. Lutz; T. Rebhan; M. Goller,
Forschungsverbund Lasertechnologie Erlangen, Lehrstuhl für Fertigungstechnologie,
Universität Erlangen-Nürnberg, Germany

1. Introduction

A lot of work has been performed to study surface modification and micromachining
of ceramics and composites by excimer laser radiation [1,2,3,4]. The ablation process
has been characterized in terms of ablation thresholds and ablation rates. Few
experiments were made to measure parameters of the ablation phenomena [5] so that
the nature of excimer laser ablation processes is not fully understood at the moment.

Measurements of sample temperatures during ablation and SEM photographs of the
ablated structures indicate a thermal component of the ablation process [4].
Calculations based on a thermal ablation model gave a good agreement with measured
ablation rates [6].

In this paper measurements of the ablation phenomena are presented and their
influence on processing quality and mechanical properties is discussed. Blast wave and
plasma expansion have been studied using high speed flash photography, a gated high
speed camera system and optical emission spectroscopy. CO_2 laser heating was used
to investigate the influence of the substrate temperature on the excimer laser ablation
process. The mechanical properties of alumina bending samples after excimer laser
surface modification were measured with the 4 point bending test.

2. Experimental Setup

Surface modification and ablation was carried out with a XeCl excimer laser (2J, 50ns)
in a mask projection setup. The intensity distribution at the mask was measured with
a CCD-camera system [7]. The output signal of a fast photodiode illuminated by
reflected UV-radiation from the sample triggered the diagnostics. A high speed flash
(flash duration: 20 ns) illuminated a standard photographic camera to obtain time
resolved photographs of the blast wave, see fig. 1. A gated high speed camera system
(spectral range 180...820 nm) delivered photographs of the temporal and spatial
development of the ablation plume. The exposure time was kept constant at 10 ns.
An area of 5x5 mm^2 above the surface of the sample was imaged to the camera.

Spectral analysis of the plasma plume was performed with an optical multichannel
analyzing system (OMA) with a spectral resolution of 0,5 nm and a minimum exposure
time of 20 ms.

Surface modification by excimer laser radiation has been used to increase the
absorption of CO_2 and CO laser radiation in metals [8,9]. In contrary we developed a

dual wavelength technique to study the influence of substrate temperature on the ablation phenomena. A CO_2 laser (maximum power 20 W) heated the samples during excimer laser irradiation. Surface temperatures were measured with a pyrometer with a temperature range from 300 up to 1400°C.

Commercial alumina, zirconia, aluminum titanate, RRIM-PUR and carbon fibre reinforced epoxy resin samples were used for the experiments.

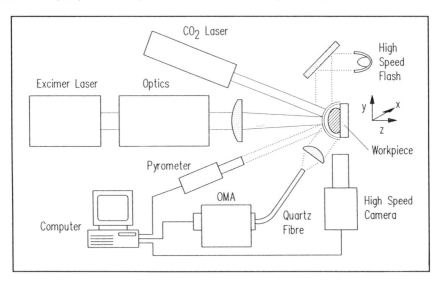

Fig. 1: Experimental setup for ablation diagnostics and dual wave length processing

A scan technique was used to irradiate one surface of alumina bending samples. The mechanical strength was then measured using the 4 point bending test.

3. Results

3.1 Time resolved measurements of the plasma plume

Fig. 2 shows photographs of the region above the irradiated surface of the sample that are obtained by high speed flash photography. The ablated material expands and collides with the air. As a result a supersonic blast wave is created near the surface as it was reported for the excimer laser ablation of polymers [10] and Y-Ba-Cu-O [11].

The velocity of the blast wave (dark line in the photographs) was calculated from the distance from the surface and the time interval between the beginning of the excimer laser pulse and the flash exposure time. Values of 10^3 m/s were obtained for alumina and RRIM-PUR.

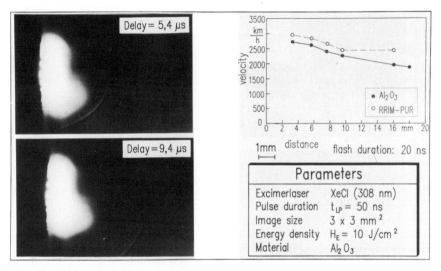

Fig. 2: High speed flash photographs of supersonic shock wave created by XeCl laser ablation of alumina in air

Plasma formation for pulsed infrared laser interaction with materials has been studied using different high-speed diagnostic techniques [12]. Time of flight measurements during excimer laser ablation of Y-Ba-Cu-O in an oxygen atmosphere have been performed to measure the range of the plasma plume [11]. Other experiments demonstrated a dependence of the colour and geometry of the plasma plume on material and process parameters [4].

The start phase is characterized by a fast three dimensional expansion with a hemispherical geometry, see fig. 3. A drastic decrease of the expansion in the z-axis was observed in our experiments typically 1 μs after the beginning of the laser pulse. The medium velocity of the luminous front of the plasma plume can be written as

$$\overline{v_{x,y,z}} = \frac{\Delta d_{x,y,z}}{\Delta t} \qquad \text{Eq. 1}$$

where $\Delta d_{x,y,z}$ are the distances of the luminous front between two measurements and Δt is the time interval. Larger plasma plumes and higher expansion velocities are observed at higher energy densities, see fig. 4. Different temporal and spatial developments of the plasma plume were obtained for other ceramics and composite materials [13]. There was also a strong dependence on the size of the irradiated area and surface geometry of the sample.

As a result plasma plume expansion has to be considered for excimer laser materials processing. Ablated material is deposited beside the irradiated area. A large range of the plasma plume may lead to deposition on the imaging lens. Lifetimes of 100 μs and more were measured for the plume giving limitations for the maximum excimer laser pulse frequency.

580

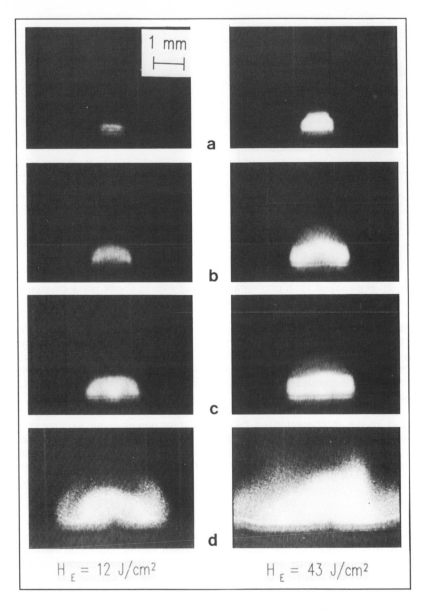

Fig. 3: High speed photographs of plasma plume for XeCl laser ablation
of alumina in air (image size 0,7x0,7 mm^2); (a) delay 200 ns,
(b) delay 300 ns, (c) delay 1 μs, (d) delay 10 μs

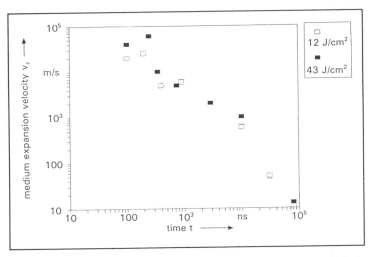

<u>Fig. 4:</u> Expansion velocities of plasma plume for XeCl laser ablation of alumina in air

3.2 Optical emission spectroscopy

Optical emission spectroscopy is a useful tool to determine the chemical nature of the ablated fragments and to identify newly formed compounds. It has been used to study excimer laser ablation of metals [14] and ceramics [15] . Line and band emission is dominant for a few microseconds after the laser pulse as it was measured for the excimer laser ablation of polyimide [16]. The dominance of atomic lines during the ablation of ceramics was interpreted as an evidence for thermal equilibrium of vapour near the surface [15].

In our experiments the obtained spectra were used as a fingerprint of the processed material. Spectra were recorded for the ablation of ceramics and composites in air. Line identifications were made using standard tabulations [17].

Fig. 5 shows a number of subsequent spectra obtained during the ablation of carbon fibre reinforced epoxy resin. During the first pulses a resin layer is removed. The prominent decay is identified to be a vibrational band of C_2 ($A^3\pi_g$ - $X^3\pi_u$; 468-564 nm). After exposing the carbon fibres a dramatic change of the spectra occurs. The intensities arise vastly and another vibrational band (CN: $B^2\Sigma$ - $X^2\Sigma$; 358-460 nm) appears.

Further investigations will be carried out to develop simplified diagnostic techniques based on optical emission spectroscopy for the excimer laser ablation of composites.

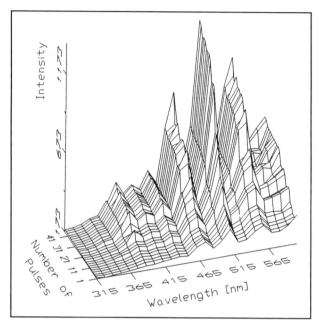

Fig. 5: Spectra recorded during XeCl excimer laser ablation of carbon fibre reinforced epoxy resin; energy density 4,7 J/cm²

3.3 Dual wavelength processing

For the first time the influence of the substrate temperature on excimer laser ablation thresholds and ablation rates of ceramics and composites has been studied. Heating of the sample in a furnace is a very gentle method to vary temperature. This method avoids thermal stresses because the whole sample is heated and heating and cooling rates are very low. Advantages of dual wavelength processing compared to furnace heating are flexibility, controllability and higher productivity.

Sample temperature was varied from room temperature to more than 1400°C. The ablation depths of zirconia for different CO_2 laser powers are shown in fig. 6 as a function of energy density. There is a pronounced increase of the ablation rate with substrate temperature and a decrease of the ablation threshold. This behaviour was expected from calculations [5]. Very similar results were obtained for aluminum titanate. Results that were obtained for other ceramics are reported in [13]. Dual wavelength processing also improves the quality of the obtained structures.

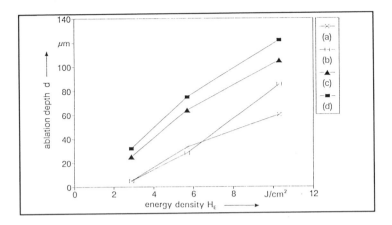

Fig. 6: Enhanced ablation rates for XeCl laser ablation of zirconia using dual wavelength processing, 1000 pulses; CO_2 laser power: (a) 0%, (b) 60%, (c) 80%, (d) 100%

3.4 Mechanical properties after excimer laser surface modification

Surface modification of ceramics has been investigated using energy densities below ablation threshold [4,18]. Melting and cooling of thin surface layers and local evaporation modify the surface topography. The results of the bending tests indicated only a moderate decrease of the mechanical strength although a pronounced change of surface topography due to excimer laser irradiation was observed, see fig. 7.

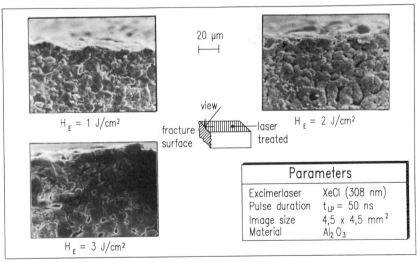

Fig. 7: Modified surfaces of alumina samples after bending test

4. Conclusion

Studies of the dynamics of XeCl laser produced plasma plume for ceramics and composites have been carried out using high speed photography and optical spectroscopy. A strong dependence of the range, velocity and spectral composition of the plasma plume on process and material parameters is observed. The results indicate that plasma plume expansion has to be considered to optimize processing quality. Dual wavelength processing incorporating a CO_2 laser for substrate heating has been used to study the temperature dependence of the ablation phenomena. A drastic increase of ablation rates and a decrease of ablation thresholds was measured. Further investigations will be carried out to simulate the ablation process and to improve the efficiency of excimer laser materials processing.

Bending tests of alumina probes irradiated with excimer laser radiation indicated only a moderate decrease of mechanical strength. Surface modification by excimer laser radiation might therefore be a suitable technique for finishing of ceramic parts.

Acknowledgements

The investigations presented are supported by the Bundesministerium für Forschung und Technologie (BMFT) under contract number 13N5629. The authors gratefully acknowledge support from Rofin Sinar Laser, Hoechst AG, Hoechst Ceram Tec and Linde AG.

References

[1] K.J. Schmatjko; G. Endres: Proc. Laser'87, Munich, 1987, Springer, 573

[2] M. Wehner; R. Poprawe; F.J. Trasser: Proc. SPIE, Vol. 1023, 1988, 179

[3] J. Arnold; F. Dausinger: Proc. 3rd ECLAT. Vol. 2, Erlangen, 1990. Coburg: Sprechsaal Publ. Group, 1990, 859

[4] N. Lutz; M. Geiger: Proc. 3rd ECLAT. Vol. 2, Erlangen, 1990. Coburg: Sprechsaal Publ. Group, 1990, 849

[5] Cl. Buerhop; R. Weißmann; N. Lutz: Appl. Surf. Science 54 (1992), 187

[6] H.K. Tönshoff; O. Gedrat: Proc. SPIE, Vol. 1377, 1990, 39

[7] M. Geiger; N. Lutz; S. Biermann: Proc. SPIE, Vol. 1503, 1991, to be published

[8] W. O'Neill; W.M. Steen: Proc. ICALEO'88, 1988, 90

[9] S. Sato; K. Takahashi; H. Saito; M. Sugimoto; T. Fujioka; S. Skono; O. Matsumoto; S. Beppu; K. Matsuda: Opto Elektronik Magazin 5(1989)6, 554

[10] R. Srinivasan; K.G. Casey; B. Braren: Chemtronics 4 (1989), 153

[11] P.E. Dyer; A. Issa; P.H. Key: Appl. Phys. Lett. 57 (1990)2, 186

[12] M. Hugenschmidt: Proc. SPIE, Vol. 1032, 1988, 44

[13] M. Geiger; N. Lutz; T. Rebhan; J. Hutfless: GCL'92, Heraklion, Crete, submitted

[14] R. Poprawe; M. Wehner; G. Brown; G. Herziger: Proc. SPIE 801, 1987, 191

[15] U. Sowada; P. Lokai; H.-J. Kahlert; D. Basting: Laser und Optoelektronik 21 (1989)3, 107

[16] G. Koren; J.T.C. Yeh: Appl. Phys. Lett. 44 (1984)12, 1112

[17] R.W.B. Pearse; A.G. Gaydon: The identification of molecular spectra. Chapman and Hall, London, 1976

[18] K.J. Schmatjko: Industrie-Anzeiger 99 (1989), 39

Contoured Material Removal Using cw-Q-switch Nd:YAG-Laser Radiation

B. Läßiger, M. Nießen, P. Ott, H.G. Treusch, E. Beyer; Fraunhofer-Institut für Lasertechnik, Aachen, FRG

1. Introduction

Industrial manufacturing processes for various materials by laser radiation become more important. Cw and modulated CO_2-laser as well as pulsed and cw-q-switched solid state laser radiation are used. Experimental results are reported in several publications (1,2,3). The advancement of the field will need a more fundamental insight into the processes involved, which is to be gained from modelling.

2. Modelling of the contoured material removal

In a first step of modelling the material removal process some simplifications like a spatial and temporal rectangular intensity distribution and a parallel beam propagation are made which are replaced step by step by realistic parameters used in the experiments. Up to now the model takes into account the feed rate, the spacing of the laser pulses, the intensity distribution in the beam cross section, the real beam propagation and the topography of the machined surface. The model determines the energy distributed on the treated part at the end of the scanning procedure in a projection to the initial surface.

In a first order approximation the depth of the machined cavity could be calculated by an energy balance. Because of the short pulse duration in the range of 100 ns and a peak power of 10 kW only vaporization as the main cause of material removal is taken into account, also for metals.

In a further step, the model accounts for the contributions of the pulse shape, the heat losses, the phase transitions, the melt dynamics and the laser-plume-interactions.

In a final stage the model should be able to determine the machining and laser parameters for a given contour in 3-dimensional space and to support possible methods of process control. Todays insight offers two approaches for process control:

- The actual geometry of the cavity is measured by appropriate sensors (4).
- Process relevant signals are used to determine the removed volume.

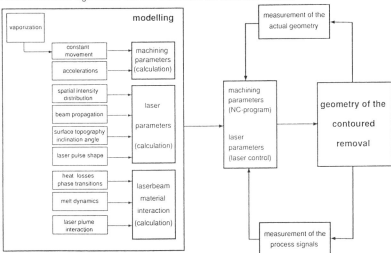

Fig. 1: Flow chart for modelling material removal and process control

In both cases the final geometry is met by adjusting laser and machining parameters. The different tasks of the model and process control are shown in figure 1 as a flow chart. The influence on the removed geometry of the parameters already included in the model is discussed in the following.

2.1. Machining parameters

For a contoured material removal the laser beam and the work piece are mutualy moved. For a pulsed laser system the lateral overlap of the single laser pulses dominantly influences the deposited energy distribution. The lateral overlap is a function of relative speed between the laser beam and the work piece, the pulse repetition rate, the beam diameter and the spacing (overlap) of adjacent tracks. Also for a top hat energy distribution in the beam cross section the influence of the overlap on the deposited energy distribution is significant especially during the acceleration and deceleration phase of the movement. Figure 2 shows the calculated energy deposition pattern on a plane surface taking an acceleration and deceleration ramp along a distance of 0.3 mm into account. In the feed direction the energy deposition is increased during the acceleration and deceleration by a factor of 2 compared to the value at constant speed. At the rim, when the material starts to interact with the laser beam, the

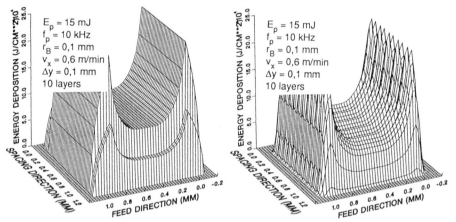

Fig. 2: Energy deposition pattern on a plane surface for a top hat energy distribution in the cross section of the laser beam; parallel beam is assumed.

Fig. 3: Energy deposition pattern on a plane surface for a super-Gaussian distribution in the cross section of the laser beam; parallel beam is assumed.

deposited energy drops to zero in the feed direction as well as in the spacing direction. The width of the area and the spacing of single contour lines is given by the beam radius. To reduce the influence of the acceleration on the energy distribution, the average output power of the laser has to be adapted to the actual speed by controlling the pulse repetition rate and the pulse energy, while the repetition rate of a cw-q-switched solid state laser could not be changed independently from the pulse energy.

2.2. Laser beam parameters

Replacing in the model the top hat energy distribution in the beam cross section by a so called super-Gaussian distribution representing a fit to the distribution of the multi mode beam (5), the influence of the spacing is increased. Figure 3 shows the distribution of deposited energy for the same spacing as in figure 2. The modulation depth of the energy distribution in spacing direction is increased. A reduction of the spacing by a factor of 2 (half beam radius)(see figure 4) leads to a smooth distribution of the deposited energy.

The influence of the beam propagation on the distribution of deposited energy is shown in figure 4. The beam focus is supposed at the former surface of the workpiece and the energy distribution is calculated with respect to this surface. The propagation of the super-Gauß-mode is taken to be that of a Gaussian (5) and the removed depth is assumed to be proportional to the deposited energy with a factor of 3×10^{-6} cm^3/J. Due to the increased beam diameter with removed depth the peak of the energy distribution is reduced and broadened. Also the energy distribution is smoothed due to the increased overlap in the

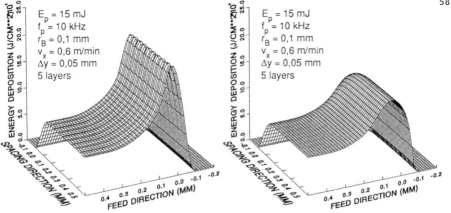

Fig. 4: Energy deposition pattern for a super-Gaussian energy distribution taking the depth of the machined cavity into account; left: parallel beam propagation; right: Gaussian beam propagation.

spacing direction because of the increased beam diameter.

2.3. Surface topography

The influence of the surface topography on the material removal process is determinated by the inclination angle of the laser beam with the surface plane. The enlargement of the laser spot, that entails a decrease of the energy density because of the inclined surface, and the Fresnel-absorption is considered. Both have countercurrent effects, because for metals the Fresnel-absorption is increased at inclination angles close to 90°.

3. Experimental results

Experimental investigations are performed with a 60 W cw-q-switch Nd:YAG-laser for stainless steel and aluminum. The cross sections of the removed volume in the feed direction is opposed in Figure 5 and 6 to the speed of the crossed linear tables used for handling the work piece. Different acceleration ramps can be realized by the CNC, but the maximum speed, which is choosen to 3 m/min for the results in figure 5, is not reached within the choosen distance of 5 mm for an acceleration time of 200 ms. The geometry of the removed volume shows a characteristic deepening on the side wall of the cavity in the feed direction as predicted by the model. The use of scanning optics like galvo mirrors will reduce the influence of the acceleration and deceleration on the removed geometry because of the smaller mass of those mirrors compared to linear tables.

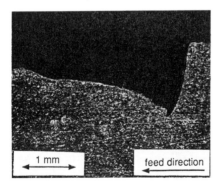

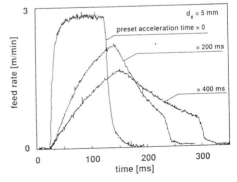

Fig. 5: Cross section of the machined cavity in feed direction, acceleration ramp duration 200ms (fig.6)

Fig. 6: Speed of the linear table at different acceleration ramp durations, travel distance 5mm

4. Process relevant signals

A laser material removal process needs a control mechanism to meet the specified tolerances of the requiered cavity contours. As mentioned above on line measurement of the actual depth of the cavity is one possibility to utilize a feed back to the process. Another possibility arises from the investigation of process relevant signals like the reflected laser radiation and the light emitted by the laser induced metal vapor plasma. A combination of both signals is an indicator for the removal rate in a single laser pulse. The short pulse duration of several 100 ns and pulse energy of a few mJ yield to a removal rate of µg per laser pulse, a sufficiently low rate to determine the actual removed volume with high resolution. Using on-line computer integration the geometry of the machined cavity can be controlled.

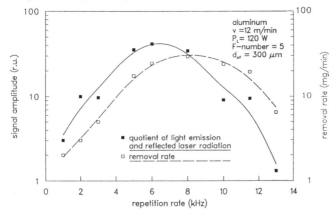

Fig. 7: Process relevant signals and removal rate processing aluminum as a function of pulse repetition rate

Figure 7 shows the removal rate in aluminum and the ratio of emitted light from the metal vapor plasma to the reflected laser light as a function of the pulse repetition rate. The characteristic performance of a cw-q-switched solid state laser system, which includes an increase of the average power with the pulse repetition rate up to 6 - 8 kHz and a significant decrease of peak power above 4 kHz, leads to a maximum of the removal rate and the amplitude of the process

relevant signals between 6 and 8 kHz. The agreement between removal rate and process relevant signals improves, if the removal rate is related to single laser pulses. The high travelling speeds above 10 m/min are achieved with a continuously rotating work piece and the process relevant signals are detected coaxially through the focussing optics.

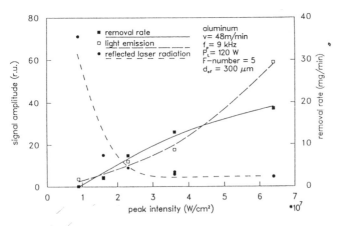

Fig. 8: Removal rate, emitted light and reflected laser radiation as a function of peak intensity

With an increasing depth of the produced cavity the beam propagation has to be taken into account. Assuming a constant focus position in space the change of the beam diameter on the actual surface leads to a change of the peak intensity and thus to a change in the removal rate. Figure 8 shows the amplitude of the reflected laser power and the emitted light of the vapour plasma as well as the concomitant removal rate as a function of peak intensity. The pulse repetition rate is kept constant and the peak intensity is varied by changing the beam diameter on the surface by moving the focus position. The pulse duration at a repetition rate of 9 kHz is 300 ns FWHM. Below a certain threshold intensity the removal rate is neglegible and the reflected laser radiation increases very strongly, above the threshold the removal rate increases linearly with intensity and so does the amplitude of the emitted light of the vapour plasma.

5. Conclusions

The first steps of modelling the process of material removal taking the spatial energy distribution and the propagation of the laser beam into account are showing encouraging results towards the possibility of a comprehensive process model. Cw-q-switched solid state lasers offer the advantage of a removal process dominated almost 100 % by vaporization of µg mass material, which in combination with well adapted process control may render geometries of 1 - 10 µm precission. Suitable control signals are the intensity of the plamsa light and the reflected laser radiation. The rate of this removal process is small compared to the laser caving process, which includes melt expulsion by a gas jet.

References

(1) B. Läßiger et al., Verfahrensentwicklung zum definierten Formabtrag mit Laserstrahlung
    Proceedings Laser 1991, München

(2) E. Meiners et al., Micro machining of hard materials using Nd:YAG laser
    Proceedings Laser 1991, München

(3) H.K. Tönshoff et al., Material Removal Processing on Steel with Nd:YAG-Lasers
    Laser und Optoelektronik 22(5), 1990

(4) E.Meiners et al., Approaches in Modelling of Evaporative   and Melt Removal Processes in Micro
    Machining; proc. ECLAT '92

(5):K.Du: priv.com.

Vapourization Cutting of Metal Sheets With Copper Vapour Lasers

H.W. Bergmann*, H. von Bergmann**, W. Klopper**, R. Kupfer***, J. Hofmann

* at present, **: Atomic Energy Corporation of South Africa Limited, PO Box 582, Pretoria, 0001, South Africa
*** now: FLUMESYS GmbH, Kropfersrichter Str. 6-8, D-8458 Sulzbach-Rosenberg
Forschungsverbund Lasertechnologie Erlangen (FLE), Universität Erlangen-Nürnberg, Lehrstuhl Werkstoffwissenschaften 2, Metalle, Martensstr. 5, D-8520 Erlangen

ABSTRACT

Laser beam cutting using pulsed CO_2- and Nd:YAG-laser systems is a well known and already established technique in industry.

For new applications, e.g. cutting of lead frame carriers, short pulsed lasers such as copper vapour lasers (CVL) show a potential in achieving precision cutting (narrow kerfs, minimized dross formation, small heat affected zones) due to the high pulsed power, high repetition rate and good focusability obtainable.

In this paper results are presented which demonstrate the influence of the energy density and the wavelength (511/578 nm) on the cutting speed and the quality of the cut samples for a number of high reflective materials such as copper, brass, iron based alloys and others. Influences of the resonator setup and the amplification of the laser light are taken into consideration.

1. INTRODUCTION

Most current discussion on laser beam cutting techniques concerns oxygen assisted burn cutting with the CO_2-laser /1-3/. The obtainable quality of the edge rounding and tip burning when processing steel sheets is acceptable to industry. For cutting Al and Ti plates or materials of large and medium thickness, melt cutting is used, with an inert gas for blowing out molten material /4/. In micromachining applications such as microelectronics and the jewelery industry, the CO_2-laser comes to its limits because of insufficient cut quality as well as thermal distortion. Here the Nd:YAG-laser can be more successful, if appropriate beam handling is applied (beam expansion, mode-locking) /5,6/. A third technique of laser cutting is vapourization cutting in which the material is melted, vapourized and ejected through plasma and shock wave formation /7/. In this way vapourization cutting is similar to drilling.

CO_2- and Nd:YAG-lasers have been used for drilling and cutting for a long period; more recently copper vapour lasers achieved more significancein these fields /8-15/. The laser is commercially available up to 100W average power. It is important that through a proper resonator setup a large portion of the beam has low divergence /16-18/. From a manufacturing viewpoint the advantages of vapourization cutting have to be compared to that of melt and burn cutting. There are advantages when thermal influence on the substrate must be avoided or when material could thermally decompose (e.g. plastics, Si_3N_4, biocompatible materials). Included in these are materials (Nb, Ti, U) which must be protected from atmospheric gases or those with high reflectivity and thermal conductivity. In precision mechanics parts have to be protected in finished component quality, therefore burning of tips and rounding of edges can not be tolerated.

The above mentioned reasons are adequate motivation for a thorough investigation of vapourization cutting with CVL's.

2. EXPERIMENTAL SETUP

The investigations were carried out using a master oscillator - power amplifier (MOPA) copper laser setup. The cavity of the oscillator was built as an unstable off-axis resonator where the hot spot was selected, amplified and monitored in a laser power chain, see Fig. 1.

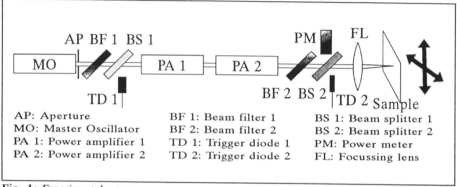

AP: Aperture   BF 1: Beam filter 1   BS 1: Beam splitter 1
MO: Master Oscillator   BF 2: Beam filter 2   BS 2: Beam splitter 2
PA 1: Power amplifier 1   TD 1: Trigger diode 1   PM: Power meter
PA 2: Power amplifier 2   TD 2: Trigger diode 2   FL: Focussing lens

Fig. 1: *Experimental setup*

Different materials, like copper and brass sheets, were treated in the experiments; the laser power was reduced by gray filters. Cuts were performed with both wavelengths (510/578nm) and with each of these two separately. Exact values are stated for each experiment. The cut limit (split the material with slight mechanical pressure) and the free cut limit were obtained. The oxidation behaviour and the cut quality were investigated.

3. RESULTS

In Fig. 2a) a typical curve of the linear correlation between cutting speed and laser power is given for different brass sheets. The cut limit and the free cut limit show parallel curves. Fig. 2b) demonstrates the influence of the sheet thickness on the cutting speed resulting in the well known hyperbolic dependancy either fo copper or brass material and comparable laser output powers. The dependancy of the sheet thickness on the cutting speed is already known. Fig. 3a) illustrates that the drilling speed in metal plates of larger thickness varies for different focal lengths. For vapourization cutting a correlation between the drilling time and the cutting time exists; these results reflect on the efficiency of the process. The focal plane position has to be choosen very exactly, as can be seen in Fig. 3b). Best values were obtained when the focal plane position was placed in the upper part of the sheet.

The synchronisation of the pulses in the laser chain allowed a tayloring as shown in Fig. 4a). The first part of the pulse has a large divergence as the second part is reduced in divergence. Performing cuts using only the yellow line lead to higher cutting speeds (when the same output power is used). This can be explained by the intensity ratio of 2:1 (green/yellow) where the green line (and a large part of the high divergent pulse) will be amplified more than the yellow line.

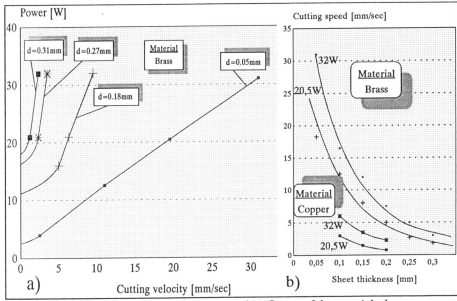

Fig. 2: a) Cutting limits of different brass sheets; b) influence of the material, the output power and the sheet thickness on the cutting speed

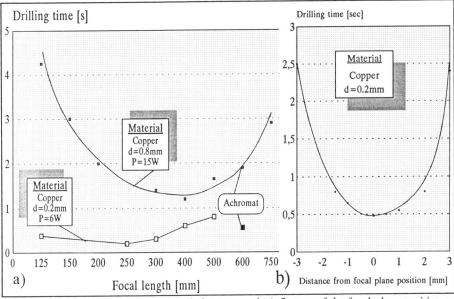

Fig. 3: a) Drilling time vs. focal length for copper; b) influence of the focal plane position on the drilling time

Materials

Metal sheets with different thickness of up to 0.35 mm could easily be cut showing no detectable heat affected zone. Contamination products and slags sticked at the cut edge, but they could easily be removed. Plunge-cutting indicated no influence on the geometry of the cut kerf. By crossing two cut kerfs, cut tips with different angular positions were received. The distortion of the cut angle from the programmed contour at the tip end was smaller (less than 5 μm) than the accuracy of the translation stage. The geometrical distortion of the cut kerf was minimal when using the MOPA setup; using a single device, derivations of more than 15° could be observed, as shown elsewhere /19/.

In the following table a list of investigated materials and a short qualitative description of the process is given. Additionaly, the required pulse tayloring to process the different material groups is added.

Table 1: *Qualitative description of the process and required pulse tayloring*

Metals and alloys	
Mild steel Stainless steel Duplex steel	Due to low thermal conductivity a large melt volume is generated = = > low intensity of the first peak, higher intensity of the second peak
Copper Brass	Cu: high thermal conductivity, small melt volume = = > higher int. first peak, lower int. for the second peak Zn lead to high vapor pressure = = > balanced double peak
Ceramics	
Al_2O_3	hot spot with low divergence = = > cutting, smooth cut edges
Plastics	
thermoplastics	no striations, no droplet formation, minimal geometric derivation, cutting of metal fibre reinforced plastics easily possible

4. RELIABILITY OF THE LASER TOOL

Regarding the above mentioned results, the question may arise whether the reliability of the laser as a processing tool is sufficient enough for harsh industrial uses. The following statements will focus on some problems concerning these lasers. Reffering to this some suggestions to overcome these inadequacies are submitted.

Power supply and electrical setup

Using the excitation power for both, heating of the plasma tube and generating the population inversion of the laser medium might not be the best solution. Only small fluctuations in the main power circuit will lead to a temperature shifting in the laser tube and as a consequence, the green/yellow-ratio will shift also. Possible new concepts can be a line regulated and stabilized power supply for excitation as well as a separate source for heating.

The quality of the copper lumps inside the insulated laser tube has a great influence on the beam quality especially after a regular refill service. Instabilities in the gain medium cause fluctuations of the LC-circle and hence, the life-time of the thyratron is reduced. To replace this high-power switch, an all solid state switching and/or pulse compression can be introduced.

The insulation material itself and the regularity of the packing around the hot plasma tube has to be regarded as a sensitive factor in achieving a uniform metal vapour distribution along the plasma tube. Although a top hat intensity profile across the tube diameter can be achieved in the near field, the far field intensity show some distortion of the even distribution.

Cooling system

The cooling system has to keep the heat balance at a certain level. Fluctuations of the cooling water temperature in the range of 15-35° showed little influence on the processing results although the power dropped by about 10%. Comparing this with others, one has to mention that a increasing temperature for all other laser types will lead to serious problems either in the achievable output power or the beam quality, which are caused by thermal depopulation or thermal lensing effects.

Maintainance

During an operation cycle of 5h x 5d in 5 weeks (equal to 3×10^9 pulses), no maintance was necessary after an initial setup. The power shifting along one working day resulted as following: after an slight increasing, the laser output power dropped down after some hours of operation. The amplitude was within +/- 2% of maximum mean output power. As a consequence, the pointing stability (free running beam for 27m) was within the range given in Fig. 4b).

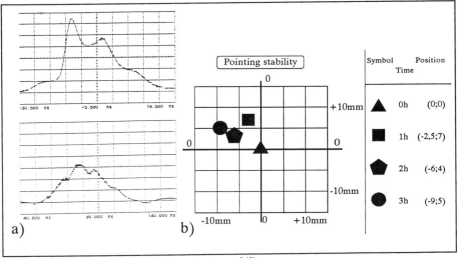

Fig. 4: *a) Taylored pulse shapes; b) pointing stability*

596

5. CONCLUSIONS

The above mentioned results demonstrate the potential of copper lasers in applications for vapourization cutting tasks. Narrow cut kerfs, small heat affected zones and a very precise geometry of the cut kerf are obtainable when using a MOPA laser system. This setup generates an excellent beam quality with easily adjustable resonators which hardly could be received with others lasers. The possibility of a pulse tayloring offers a wide range of flexibility for material processing.

6. REFERENCES

/1/ R. Nuss: *Untersuchungen zur Bearbeitungsqualität im Fertigungssystem Laserstrahlschneiden.* München: Carl Hanser Verlag (1989)

/2/ D. Petring, P. Abels, E. Beyer, G. Herziger: *Schneiden von metallischen Werkstoffen mit CO_2-Hochleistungslasern.* F&M, **96** (1988) 9

/3/ W.M. Duley: *CO_2-Lasers, Effects and Applications.* Academic Press, New York (1976)

/4/ G. Sepold, R. Rothe: *Laser beam cutting of thick steel.* L.I.A. ICALEO, Vol.38 (1983)

/5/ H. Haferkamp, A. Homburg: *Thermisches Trennen mit Nd:YAG-Hochleistungsfestkörperlasern.* Proc. ECLAT'90. Coburg, Sprechsaal (1990)

/6/ P. Schäfer: *Metalle gepulst flexibel und präzis schneiden.* Laser-Praxis **6** (1989)

/7/ E. Beyer, O. Märten, K. Behler, J.M. Weick: *Schneiden mit Laserstrahlung.* Laser und Optoelektronik, **3** (1985)

/8/ B. Haba, B.W. Hussey, A. Gupta, R.J. Baseman: *Copper Vapour Laser Used in Etching and Deposition.* Mater. Res. Soc. Symp. Proc., 158 (1990)

/9/ M. Nichonchuk: *Copper Vapour Laser Precision Processing.* Gas and Metal Vapour Lasers and Applications. Proc. SPIE 1412 (1991)

/10/ M. Nichonchuk, J. Polykov: *Efficiency of vaporization cutting by copper vapor lasers.* Proc. SPIE 1628 (1992)

/11/ H.W. Bergmann, R. Kupfer, H. Lassner: *Further aspects of materials processing with Copper Vapor Lasers.* Proc. ECLAT '90, Coburg: Sprechsaal (1990)

/12/ R. Kupfer, H.W. Bergmann: *Materials processing with Copper Vapour Lasers.* OPTO ELEKTRONIK MAGAZIN 6 (1990) 1

/13/ R. Kupfer, H.W. Bergmann, M. Lingenauer, H. Lassner: *Materialbearbeitung mit Kupferdampflasern.* Laser und Optoelektronik in der Technik. Berlin: Springer (1992)

/14/ R. Kupfer, H.W. Bergmann, M. Lingenauer: *Materials influence on cutting and drilling of metals using copper vapour lasers.* Lasers in Microelectronic Manufacturing. SPIE Proc. 1598 (1992)

/15/ H.W. Bergmann, R. Kupfer, M. Lingenauer, G.A. Naylor, A.J. Kearsley: *Micromachining of metals with copper vapor lasers.* Proc. CLEO'92, Vol. 12 (1992)

/16/ N.N.: *Cavity Design for Metal Vapour Lasers.* Technical Note No.1, Oxford Lasers Ltd., Oxford (1988)

/17/ R. Pini: *High efficiency diffraction limited operation of a copper vapour laser.* In: Optics Communications. **81** (1991)

/18/ M. Nichonchuk, I. Polyakov: *Focal plane intensity distribution of copper vapour lasers with different unstable resonators.* Gas and Metal Vapour Lasers and Applications. Proc. SPIE 1412 (1991)

/19/ R. Kupfer: *Laserstrahlbohren mit Kupferdampflasern.* Doctoral Thesis, Erlangen, 1992, in print

Process of generating three-dimensional microstructures with excimer lasers

H.K. Tönshoff, Laser Zentrum Hannover e.V., Hannover, FRG
J. Mommsen, Laser Zentrum Hannover e.V., Hannover, FRG

Introduction

Three dimensional material removal with lasers is a new a field of laser material processing, and at present, CO_2- and Nd:YAG-lasers are mostly used for the well known laser caving (1-2). This material removal, especially in the mould making must be titled as makro material removal (3). The excimer laser, with a short wavelength and low penetation depth is a good tool for micro material removal of non-metals. But, as for the high power lasers, the controlling of the removal depth is the main problem in generating three-dimensional microstructures with excimer lasers (4).

Process of 3D-material removal with excimer lasers

Normally, material removal with excimer lasers is restricted to defined geometries, because the beam can only be shaped by mechanical masks. During the removal process, it is complicated, or mostly impossible, to change the masks without decreasing the removal quality. The removal depth is controlled by the number of pulses given on the irradiated surface. Mostly, the excimer laser is used for applications like drilling, cutting and surface treatment where no changing of the mask during the removal process is necessary. Simple structures, e.g. cuts, are produced by putting together or overlapping single removals. For this, nessessary NC-programms normaly include only simple moving commands for the work piece handling and a laser on/off command. The pulse frequency will not be changed during the process. In this case, a definded coupling of the laser trigger, pulse frequency and the work piece handling is not practiced. For producing three-dimensional structures under these presuppositions very intensive work must be done.

First, the geometry which must be produced has to be defined in a point scan for the x/y-plane, and then after addition of the information of the depth, to be taken into the third dimension (z-plane). For excimer lasers, this step is done by fixing a number of pulses for a defined position. The necessary number of pulses is a function based on the characteristic removal rate of the material. Scanned material removal with only one mask geometry in shown in fig. 1.

For producing this structure, about 70.000 pulses and 1800 NC-data blocks were needed. Using a pulse frequency of 50 Hz, the time for the removal amounted to 35 minutes.

598

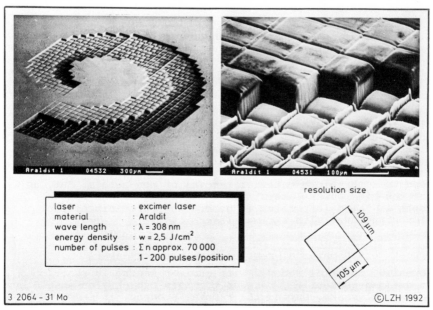

Fig. 1: CNC-controlled 3-D micro material removal

The time for treatment shows that this process is very slow
and will need even more time for more complicate structures,
larger removal areas or higher removal depths.

Therefore, a special system has been developed which allows
variable mask cross-sections during the removal process. Using
large mask cross-sections improves the exploitation of the
beam, and decreases the total time for material removal.
Nevertheless, because of the possibility of using small masks
too, the resolution accuracy should not be reduced.

Figure 2 shows, that the removal geometry has to to sliced
into separate layers. The chosen layer thickness is once more
a function of the available removal rate of the material.
Removal areas which are defined in every layer are filled with
large removal cross sections, and then with smaller cross
sections. The smallest is defined by the grid width of the
removal scan, which has been fixed by the surface measurement.
Because of the complex calculation operations, it is nesessary
to establish some mask cross sections before running the
simulation program. Multiple of the smallest mask cross
section have been exposed as usefull. After this, the mask
position data of every layer are compared and summed up. This
means that if identical removal cross sections are detected on
the same position in layer 5,10,11 and 22, they will be
removed by four pulses, one after the other, in layer 5.

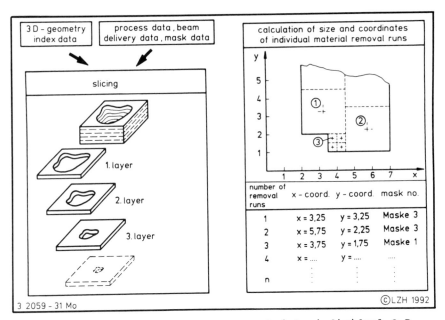

Fig. 2: Slicing the 3-D volume removal into individual 2-D geometries

For some materials and deeper structures with more than 40 layers it is better to remove not more than 20 layers with the described optimization. The reason is the conical shape of the material removal, which deviates more and more from the index-cross section after several pulses. Because of the difference between index- and actual-material removal, some bridges can be detected on the treated material. Nevertheless, with this optimisation the total time for the removal process can be reduced because less delay and positioning times are needed. Depending on the workpiece handling, this time can amount to 30 % of the total time. Therefore, ramping and delay times have to be shortened, not only to guarantee high speed, but also position accuracy.

The possibility of using various mask cross sections increases the calculation time up to the construction of the NC-program indeed, but diminishes decisively the number of pulses. Fig. 3 shows the different qualities for single and multiple overlap removals, in comparison.

In figure 3b, the use of different mask cross sections is simulated by using the shown removal scheme. The valleys at the bottom of the removal, which can be seen in fig. 3a, were now dissapeared. The low edge steepness is caused by this special removal scheme, and can be you greatly improved.

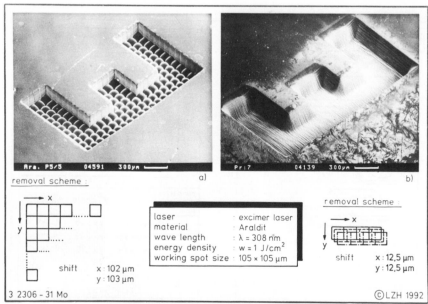

Fig. 3: Removal results from single and multiple overlapping of the single material removal runs

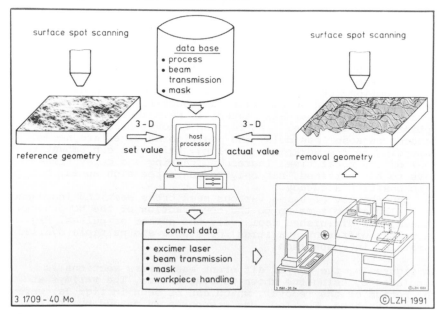

Fig. 4: Control system for the 3-D microstructuring of technical surfaces with excimer lasers

The generation of three-dimensional structures, like embossed leather with excimer lasers is based on the combination of a well-known removal process of special materials combined with a complex system technology.

First, it is nessesary to generate the index three-dimensional data. This will be done by using an optical surface measurement system (5). This is followed by breaking down the 3-D removal data into 2-D geometries. These single geometries are compared with the process, beam guidance, and mask data in a databank . Figure 4 shows that all the control data for the laser, the mask, the beam guidance system and workpiece handling system are used in a processing station to obtain defined 3-D material removal geometries. , Figure 5 shows a structure which was created as a result of the described automatic process on a leather surface.

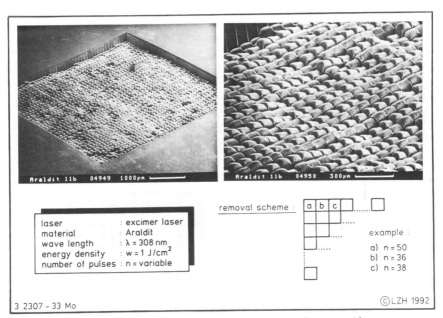

Fig. 5: Reproduction of a complex surface in Araldit

The Material removal was realized with only one size of mask, so that the valleys between the single removals have to be accepted. The rise ground, existing of three pixels, is based on a fault in the optical measurement system which is taken over without correction of the data of the index geometry.

Conclusion

To improve and optimize this process, very intensive work is being done at the moment. An economic use of this process to structure large areas, e.g. embossing rollers, might be possible in the not-too-distant future.

References

(1) H.K. Tönshoff, M. Stürmer: Untersuchungen zum Formabtrag an Stahl mit Nd:YAG-Lasern, Laser und Optoelektronik 22(5) (1990) 59-63
(2) H.K. Tönshoff, F. v. Alvensleben: Verfahren zur Qualitätskontrolle für den Materialabtrag mit dem CO_2-Laser, Laser und Optoelektronik 22 (2) (1990) 62-66
(3) G. Eberl: Laserbearbeitung für den Formenbau, LASER Nr.1 (1990) 24-27
(4) O. Gedrat: Grundlagen der Wechselwirkung von Excimer-Laserstrahlung mit Materie, Informationsveranstaltung Fa. Polytec Waldbronn (Mai 1991)
(5) L. Overmeyer, K. Dickmann: Dynamischer Autofokussensor zur dreidimensionalen Mikrostrukturerfassung, Technisches Messen TM (Dezember 1991)

Marking of Silicate Glasses with Excimer Laser Radiation
- Influence of Glass Properties and Laser Parameters -

Cl. Buerhop[1], N. Lutz[2], R. Weißmann[1], M. Geiger[2]

1) Institut für Werkstoffwissenschaften 3 (Glas und Keramik),
   Universität Erlangen-Nürnberg, Erlangen, Germany
2) Lehrstuhl für Fertigungstechnologie,
   Universität Erlangen-Nürnberg, Erlangen, Germany

1 Introduction

Pulsed high power lasers are favourable candidates for micromachining, structuring, and marking of brittle materials [1]. The advantage of excimer laser marking of glass is the extreme short interaction time, resulting in minimal damage to the surrounding unexposed material, and the small laser wavelength, allowing patterns to be structured on a microscopic scale.

The unique properties of pulsed UV-radiation can be used to remove minute amounts of material with high precision, thus obtaining easily visible and abrasion resistant marks in a few seconds or less. Compared to conventional etching techniques and sand abrasion, laser marking is a fast and precise alternative. The material is ablated very cleanly, leaving well defined sharp and a modified surface topography, which diffusely reflects light. As shown in fig. 1, these two effects combine to make the mark clearly visible.

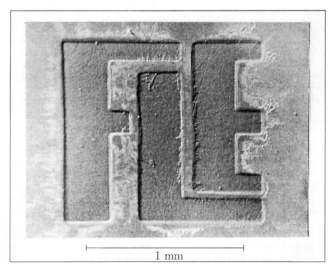

Figure 1: Floatglass (bath-side), marked with Excimer-Laser radiation using the mask projection technique: $H_E = 12$ J/cm^2, $f = 5$ Hz, 3 pulses

2   Experimental procedure

Commercial silicate glasses are irradiated using a pulsed high power XeCl-Excimer-Laser, Siemens XP2020. As presented in table 1, this laser is characterized by a large beam cross section, a uniform intensity distribution, and a nanosecond pulse duration. Using the mask projection technique, the desired structure is copied on the glass without any contact.

Laser-System	XeCl-Excimer-Laser
Wavelength	308 nm
Pulse duration	50 ns
Pulse energy	2 J
Max. repetition rate	20 Hz
Beam cross section	45 x 55 mm^2
Max. average power	40 W

Table 1: Data of the Laser-system

As presented in table 2, the investigated glasses are characterized by several optical and thermal properties. The absorption depth β^{-1} at 308 nm is determined by transmission measurements and n_D is the refractive index at $\lambda=587$nm. The transformation temperature T_g and the thermal expansion coefficient α are measured by dilatometry.

Glass Type	fused silica	borosilicate glass	soda-lime-silicate-glass		lead glass
Brand Name		Duran	floatglass	Athermal	crystal glass
Fe$_2$O$_3$ [wt.-%]			0.1	5	0.02
n_D	1.4589	1.473	1.52	1.54	1.55
β^{-1} (308 nm) [μm]	37000	4400	470	96	1700
T_g [°C]	1100	568	530	503	465
α [$10^{-6} \cdot$ K^{-1}]	0.5	3.2	9.35	8.7	9.8

Table 2: Glass properties

The modified surface topography is studied using light microscopy and scanning electron microscopy. The depth of the ablated area is measured by profilometry and light microscopy.

3   Results and Discussion

The interaction of XeCl-Excimer-laser-radiation with glass is strongly determined by the glass' optical properties. Most glasses are fairly transparent at 308 nm, so, at low energy densities no material damage occurs. At energy densities above a certain threshold, very thin layers of material are removed with each pulse. This ablation effect is used for marking and structuring surfaces. But different glasses do not show always the same process quality. Even opposite sides of the same glass can exhibit different behaviour, i.e. as seen with the commercially most important floatglass, the so called window glass. In this particular case, the atmosphere (air) side reacts completely different than the bath-side (containing tin-oxide according to the manufacturing process) [2]. The tin oxide at the glass surface increases n_D, T_g, hardness, and the chemical resistivity [3].

Characteristic parameters for Excimer Laser treatment of glass are: the ablation rate ϕ, as well as the ablation threshold $H_{E,thr}$ and the number of start-pulses for ablation N_0. $H_{E,thr}$, and N_0 are important because a certain amount of absorbed energy is necessary before the ablation process begins. The ablation rate is defined by the ratio of ablation depth d versus $(N-N_0)$, with N the total number of applied pulses. Therefore, the ablation depth $d(H_E,N)$ is a function of the ablation rate, number of applied pulses and the energy density H_E.

$$d(H_E, N) = 0, \quad \text{for} \quad N \leq N_0(H_E)$$
$$= \phi(H_E) \cdot (N - N_0), \quad \text{for} \quad N \geq N_0(H_E)$$

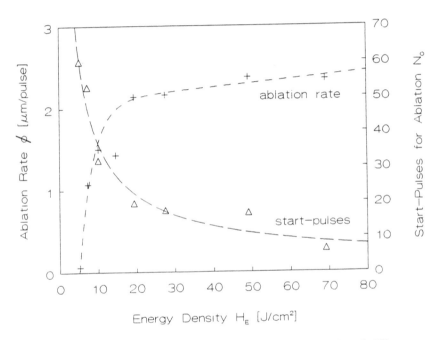

Figure 2: Energy dependent ablation parameters for floatglass, f=5Hz

As shown in fig. 2, the reduction in the number of start-pulses with increasing energy density correlates to an increase in absorption, i.e. solarization which is due to an increase in absorption coefficient $\alpha = f(N \cdot H_E)$. This effect has been measured by transmission spectroscopy on lead crystal glass. Some glasses even show visible solarization effects. Floatglass shows a light brown-grey colouration right after exposure. Lead glasses, containing arsenic oxide as a refining agent, turn yellow after irradiation [4]. The colour vanishes only after a moderate heat treatment.

As shown in fig. 3, two characteristic ablation phenomena can be distinguished for glass. Glasses with high absorption coefficients show low ablation rates and only a slight increase with increasing energy density. On the other hand, the ablation rates of glasses like fused silica and borosilicate glass decrease with increasing energy density.

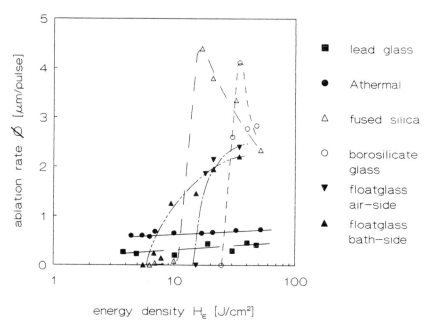

Figure 3: Ablation rates of various glasses

Just as the ablation rates of glass differ, so does the surface morphology after irradiation [5], [6]. Lead glass and Athermal glass show a smooth surface and well defined edges. The air side of float glass shows the uneven surface typical of a melted and quickly resolidified glass. The bath side is well structured up until the tin-enriched surface layer is removed and the base glass is reached. Fused silica and borosilicate glass surface morphologies are very rough and the edges contain numerous break-outs. In addition, local surface damage is observed at the rear side [7]. This behaviour makes those complicated to produce with precise and homogeneous structures.

Common to the more transmitting glasses is a high ablation rate and a rough surface. The glasses with high refractive indices exhibit low ablation thresholds and low ablation rates. This phenomenon may be due to non-linear optical effects, such as two-photon excitation and self-focussing, and thus determined by the non-linear optical properties.

As a result of the above mentioned differences, selected glasses are "predestinated" for Excimer-laser micromachining. Precise structures down to several micrometers can be achieved. Fig. 4 shows a line pattern with a width of 15 μm in floatglass.

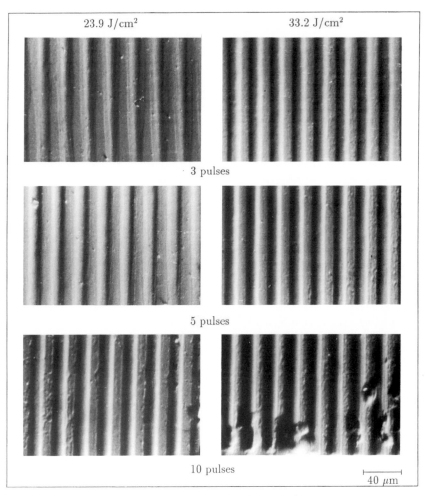

Figure 4: Structuring of floatglass (bath-side): f = 5 Hz

By applying only a few pulses, lines of a micrometer scale can be produced. Additional pulses deepen the grooves and sharpen the edges. If the number of applied pulses is too high, the fine lines are destroyes by melting, rounding of the edges, and break-outs. Depending on the energy density, the number of pulses has to be carefully adjusted to achieve the best results. For example, with a high energy density, and low number of pulses, only a small interval for the pulse number exists for reasonable results.

4   Conclusions

Pulsed Excimer-Lasers are adequate for marking and structuring glass fast and precisely. In general, the interaction of UV-radiation with glass is strongly determined by the optical properties of the glass.

After applying only a few pulses, easily visible marks of a millimeter size, as well as line structures on a micrometer scale are attainable. If the energy density and the number of applied pulses are not precisely adjusted, the quality of the pattern may be negatively influenced. This occurs because the edges become rounded and the contrast between the irradiated area and the unexposed surface is lost. Thus, parameters must be optimized to abtain well defined patterns with sharp edges and smooth surface morphology.

References

[1] T. Watanabe and Y. Yoshida.
Numerical Calculation of shaped machined with pulsed YAG Laser.
J. of Mechanical Working Technology, 17 (1988) aug., 297-306.

[2] C. Buerhop and R. Weißmann.
Precise Marking of Silicate Glasses by Excimer Laser and TEA-CO_2-Laser Irradiation.
XVI Int. Congress on Glass in Madrid, (1992), to be published.

[3] W.E. Baitinger, P.W. French, and E.L. Swarts.
Characterization of tin in the bottom surface of float glass by ellipsometry and xps.
J. of Non-Cryst. Sol., 38&39 (1980), 749-754.

[4] K. Bermuth, A. Lenhart, H.A. Schaeffer, and K. Blank.
Solarisationserscheinungen an Cer- und Arsenhaltigen Kalk-Natronsilicatgläsern.
Glastechn. Ber., 58 (1985) 3, 52-58.

[5] C. Buerhop, B. Blumenthal, R. Weißmann, N. Lutz, and S. Biermann.
Glass Surface Treatmant with Excimer and CO_2 − Lasers.
Appl. Surf. Sc., 46 (1990), 430-434.

[6] C. Buerhop, R. Weißmann, and N. Lutz.
Ablation of Silicate Glasses by Laser Irradiation: Modelling and Experimental Results.
Appl. Surf. Sc., 54 (1992), 187-192.

[7] J. Ihlemann.
Excimerlaser Ablation of Fused Silica.
Appl. Surf. Sc., 54 (1992), 193-200.

High Resolution Excimer Laser Based Micromachining

B. Burghardt, H.-J. Kahlert*, D. Basting, Lambda Physik,

Göttingen, FRG

*MicroLas Lasersystem GmbH, Jena, FRG

Abstract

Excimer lasers are pulsed UV-lasers emitting at wavelengths between 351 and 193 nm. Micromachining applications of these lasers have already become an important role in the semiconductor industry.

The paper will report on recent investigations of large field polymer ablation experiments. Optical beam delivery system concepts for homogeneous 5 x 5 mm^2 field size micromachining will be presented. Additionally sample processing results and micromachining quality will be discussed.

Introduction

Excimer lasers have been applied in research laboratories since 1977. About 10 years later, they were introduced into industrial processing and manufacturing. Due to the unique properties of pulsed ultraviolet laser radiation excimer lasers are used in a wide range of micromachining and microprocessing of plastics, ceramics and metals.

Depending on the beam parameters the following processes can be induced on material surfaces

- chemical surface alteration
- heating
- melting and solidification
- vaporization

while principally the thermal processes can also be induced by other lasers, chemical surface alteration is something completely different and is used for ablation and marking of polymers like polyimide.

Due to the high temperature stability of polyimide and its high electric break-down resistance this polymer is widely used in the electronic industry as an isolating material.

The high intensity of the excimer laser radiation results in a high concentration of small molecular fragments which do not stick together but ablate from the plastic surface. This happens on such a short time scale that no heat is transferred to the remaining solid polymer. The result is a superior quality of excimer laser based micromachining of polyimide.

The photochemical mechanism of this process for example has been investigated by Srinivasan and co-workers (1).

The Beam Delivery System

For controlled high resolution excimer laser ablation processing a beam delivery system has to be designed to match the required process parameters.

For micromachining of polyimide the excimer laser wavelength of 308 nm or 248 nm is used with energy densities of 0.1 to 2.0 J/cm^2. But the results and quality of the process not only depends on the energy density, but also on the spatial and angular energy distribution and the tilt angle of the principal rays in the work plane.

A typical optical set-up for laser ablation is shown in fig. 1. The user field of the excimer laser beam is transferred by a relay optics into a beam homogenizer. The homogenizer scrambles the incoming laser light and creates a top head beam profile on a mask.

The mask pattern is then imaged by a transfer lens down into the work-plane. A monitor system is used to control the laser beam and imaging parameters.

For processing a larger field size, a telecentric set-up provides perpendicular principal rays to enter the image plane which results in parallel microstructures within the processing area and depth of focus (DoF). The spatial and angular energy distribution and the light cone α^* in the image plane control the wall angle and the 3D-symmetry of the structures.

Results

With a beam delivery system as described polyimide samples have been processed.

The REM-photo in fig. 2 shows the high quality of the microstructures, when etched by excimer laser light. A positive wall angle of the holes can be seen; this means, the exit hole is smaller in size than the entrance hole.

To investigate the dependence of the wall angle on the energy density and the optics parameters, a telecentric imaging lens was used with a focal length of 25 mm and a numerical aperture of 0.16. The depth of focus was ± 10 μm. The polyimide samples are positioned with the top surface in the image plane and exposed at different energy densities between 1.4 and 22 J/cm^2.

With these parameters, a negative wall angle is achieved at an energy density above 2 J/cm^2. For lower energy densities the wall angle becomes positive (see fig. 3).

For comparision the resulting wall angle of an imaging system with a NA of 0.01 is given, too. The wall angle follows the light cone angle α^* but is in general lower. A minimum light cone angle is necessary to create a negative wall angle. This value depends on the material ablation rate and the energy density.

Different laser beam divergence for horizontal and vertical direction has an impact on the ablation microstructures.

The angular energy distribution becomes unsymmetric and creates unsymmetric microstructures within and outside the depth of focus. Circular holes at the entrance side become elliptical at the exit hole as to be seen in fig. 4. The beam divergence has to be matched either by a mode aperture, which results in medium to high energy losses, or by a homogenizing illumination optics, which gives symmetric microstructures with high optics efficiency.

Tilting the direction of the main rays in the image plane against the polyimide surface results in a tilted axis of the ablated structures. The axis follows the direction of the main ray as given in table 1.

Summary

Polyimide can be precisely processed by excimer laser ablation. Microstructures with structure size of 10 to 50 μm with a resolution of less than 10 μm can be achieved.

Resolution, wall angle and 3D-symmetry of the microstructures are controlled by the laser parameters and by the design of the optical beam delivery system.

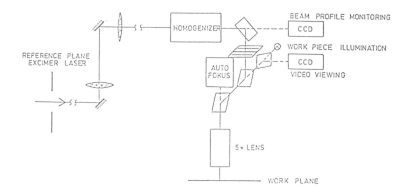

Fig. 1: beam delivery system.

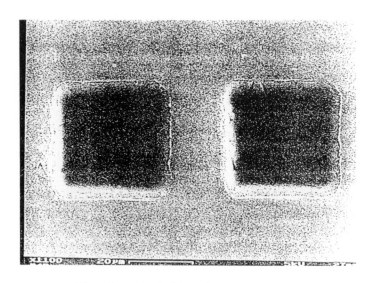

Fig. 2: polyimide microstructure
(248 nm, 800 mJ/cm^2).

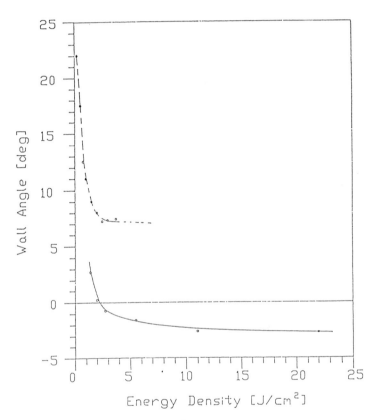

Fig. 3: wall angle versus energy
density for different NA optics.

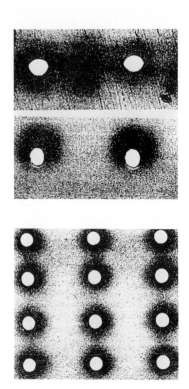

Fig. 4: unsymmetric and
symmetric angular illumination
results in Kapton ablation.

α	$0°$	$3°$	$6°$	$9°$	$12°$	$15°$
β	$(0 \pm 1)°$	$(3 \pm 1)°$	$(7 \pm 1)°$	$(10 \pm 1)°$	$(14 \pm 1)°$	$(16 \pm 1)°$

Table 1: hole telecentricity follows principal ray tilt.

Microstructuring of glass with excimer lasers

B. Wolff-Rottke, H. Schmidt, J. Ihlemann, Laser-Laboratorium Göttingen, Göttingen, Germany

Introduction

The mechanical machining of glass is often rather difficult, especially if small components or parts with complex geometry are required. Therefore laser processing of glass surfaces such as polishing, and dry etching of microstructures may be a promising alternative as has been demonstrated already for several other materials like polymers (1,2,3), ceramics (4,5), and inorganic crystals (6,7). Borosilicate glass and heavy flint glass transmit radiation in the wavelength range 0.4 to 1.4 μm to a high degree (figure 1). They are used for various optical components (8). On the other hand uv-radiation is strongly absorbed, so that laser radiation with short wavelength is efficiently coupled into the substrate (9). The high power and the short wavelength (193nm, 248nm) of the excimer laser thus make this laser type well suited for precise micro machining of glass components.
In this paper we investigate laser ablation of several glasses with excimer laser radiation.

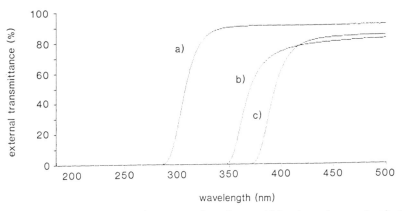

Fig. 1: External transmittance for borosilicate glass (a) BK7) and heavy flint glass (b) LaSF N9, c) SF11) of 5mm thickness.

Experiment

The materials used in the experiments were borosilicate glass (BK7) and heavy flint glass (SF11, LaSF N9) from Schott. The plates were 5mm thick and polished on the back and front side (arithmetic average roughness $R_a \leq 15$nm).
Ablation experiments were performed by mask imaging using a Lambda Physik laser EMG 301 MSC at 248nm or 193nm. The

experimental setup is given in fig. 2. Pulse durations are about 20-30 nsec FWHM, and pulse energies up to 1J. The fluence is adjusted by tilting a dielectrically coated fused silica plate placed directly in front of the excimer laser. The beam is directed onto the mask either unfocused or slightly focused by a 500 mm focal length fused silica lens. The mask is imaged onto the sample surface using a fused silica lens (f=100mm) or a reflecting objective (Ealing x15/0.28) with a dielectrical coating for 193nm or 248nm. All experiments were performed at air atmosphere. On-line control of the ablation process is possible with a CCD-camera, which views the sample surface through the reflecting objective. A HeNe laser, coupled into the excimer laser beam path in front of the mask, generates a visible image of the mask on the sample surface, which allows a precise adjustment of the sample. The sample is mounted on a three axis translation stage driven by computer controlled stepper motors with $0.5\mu m$ resolution. The same AT-computer also controls the laser trigger and the beam attenuation.

Ablation depth and surface roughness were measured by a Dektak 3030 Auto II stylus profilometer and by optical microscopes. The morphology of the ablation holes was investigated by a scanning electron microscope (Zeiss DSM 962).

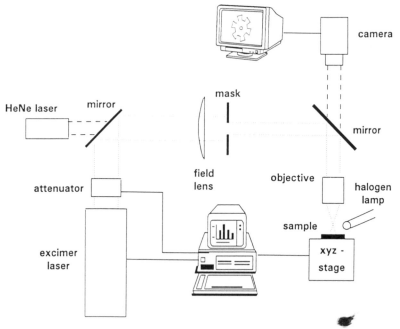

Fig.2: Experimental setup for excimer laser micro machining

Results

Excimer laser radiation at 193nm and 248nm is strongly absorbed by borosilicate glass and the heavy flint glasses (figure 1).

Laser ablation starts, if the fluence on the sample surface is higher than a certain minimum value (threshold fluence), which depends on laser wavelength and the material. The quality of the generated surfaces and edges is different for both wavelengths and depends on the type of glass.

Material removal
248nm
The ablation rate (ablation depth per pulse) for the heavy flint glasses with 248nm laser radiation is significantly lower than the ablation rate for BK7 (fig. 3a). Laser drilled holes in the materials SF11 and LaSF N9 have smooth walls and the bottom of the holes is only slightly wavy (fig. 4a,b). This wavy structure is irregular and most pronounced at high fluences. Therefore an interaction between the sample surface and the ablation plume may cause this structure. The debris is granular with grain sizes below 0.3 μm. The laser machining of BK7 at medium fluences (around 6J/cm^2) shows two phases which have already been observed for laser ablation of fused silica (10). Up to ten pulses there is nearly no ablation (<10nm/pulse), but the irradiated area can be cleary distinguish from the unirradiated surrounding material. The generated surface is still smooth. With an increasing number of pulses the ablation rate strongly increases, reaching the values shown in figure 3a, and the ablated holes develop rough walls and bottoms (fig. 4c). This surface modification during the first laser pulses may be comparable to the incubation of polymers (11). For all three types of glass the surrounding material seems to be nearly unaffected by the laser ablation process.

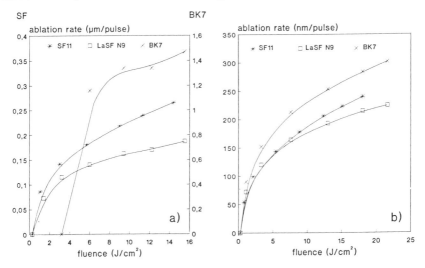

Fig. 3: Ablation rate depending on the fluence for three types of glass a) 248nm, 320μm spot diameter, b) 193nm, 200μm spot diameter.

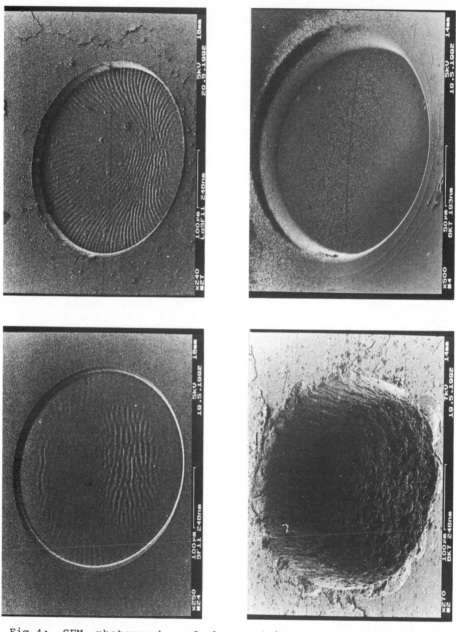

Fig.4: SEM photographs of laser drilled holes in different glasses: a) SF11, 248nm, 14.4 J/cm^2, 20 pulses; b) LaSF N9, 248nm, 6 J/cm^2, 100 pulses; c) BK7, 248nm, 9.4 J/cm^2, 100 pulses; d) BK7, 193nm, 21.6 J/cm^2, 100 pulses.

193nm

The ablation rate of BK7 at 193nm is much lower than at 248nm and only slightly higher than the ablation rate for the heavy flint glasses, which do not change significantly (fig. 3b). The quality of the laser drilled holes is nearly the same for all three glasses at this wavelength. The generated walls and surfaces are very smooth. Wavy structures on the bottom of the holes do not appear (fig. 4d). The edges are sharp and show only very tiny cracks (≤50nm wide, 10μm long). There is less debris around the ablated areas.

Surface treatment

As already mentioned above the surface of the three types of glass show a very smooth structure after laser ablation with 193nm radiation (4d). The smoothing of the surface by laser ablation was investigated for borosilicate glass at 193nm. A sample with a rough surface ($R_a \approx 100nm$) was ablated with a fluence of 3 J/cm². After 100 laser pulses a surface roughness of $R_a \leq 15nm$ was achieved.

Instead of surface smoothing sometimes a definite roughness of a small surface area is desired. This can be generated by using small nets as a mask. Figure 5 shows the surface structure after ablation. The original roughness of the sample surface was $R_a \leq 10nm$ and the generated structure corresponds to a roughness of $R_a \approx 100nm$. The value of the roughness can be varied by choosing

nets with different net sizes and different pulse numbers.

Both machining processes, roughening and smoothing, allow for an individual surface modification on a very small scale.

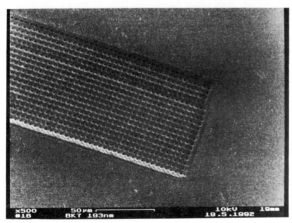

Fig. 5: SEM photograph showing definite roughening of a BK7 surface (Ra≈100nm) by laser ablation.

Micro machining

The production of micro structures by means of photo ablation using the method of mask imaging is demonstrated for the heavy

620

flint glass SF 11. Figure 6 shows a SEM photograph of a gear cut into the glass by laser ablation at 248nm. The mask structure is reproduced in a very precise way. The edges obtained are sharp and the walls are smooth. Using the method of mask imaging small and complex structures can be produced in a very fast manner.

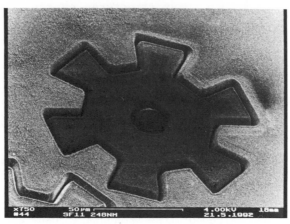

Fig. 6: SEM photograph of a gear cut into SF11 by laser ablation at 248nm.

Conclusions

Laser ablation with excimer lasers is a well suited method for machining borosilicate and heavy flint glasses. Especially at the wavelength 193nm the generation of precise structures smaller than 10μm is possible. Besides cutting and drilling also a definite roughening and smoothing of the surfaces can be achieved by laser ablation.

References

(1) R. Srinivasan, B. Braren, Chem. Rev. 89 (1989) 1303
(2) J.H. Brannon, J. Vac. Sci. Technol. B 7 (1989) 1064
(3) S. Lazare, V. Granier, Laser Chem. 10 (1989) 25
(4) U. Sowada, P. Lokai, H.-J. Kahlert, D. Basting, Laser und Optoelektronik 21 (1989) 107
(5) M. Geiger, N. Lutz, S. Biermann, Proc. SPIE Vol. 1503 (1991) 238
(6) W. Zapka, A.C. Tam, J.L. Brand, D.C. Cheng, P.Simon, Bull. Am. Phys. Soc. 34 (1989) 1687
(7) M. Eyett, D. Bäuerle, W. Wersing, H. Thomann, J. Appl. Phys. 62 (1987) 1511
(8) Optics Guide 5, Melles Griot
(9) C. Buerhop, B. Blumenthal, R. Weissmann, N. Lutz, S. Biermann, Appl. Surf. Sci.46 (1990) 430
(10) J. Ihlemann, B. Wolff, P. Simon, Appl. Phys. A 54 (1992) 363
(11) S. Küper, M. Stuke, Appl. Phys. A 49 (1989) 211

Laser Texturing R-Ba-Cu-O Superconductors

D. Dubé, B. Arsenault, C. Gélinas and P. Lambert
Industrial Materials Institute, CNRC, Boucherville, Québec, Canada

Introduction

The improvement of critical current densities in bulk HTSC is obtained through the orientation (texturing) of superconducting grains, and melt-processing in high thermal gradient is presently the best route to alignment of superconducting grains in bulk.

The high thermal gradients generated in the melt pool give a potential advantage to the laser heat source over other melt-processing methods. Laser melt-processing has been used in the past for texturing $YBa_2Cu_3O_{7-x}$ (123) grains in rods and fibers [1,2] with growing rates of the order of a few microns per second. Laser processing at higher scanning velocities (2.5 to 25 cm/s) has also been used to restore the super-conductivity in plasma sprayed YBaCuO coatings [3,4] and for melting-and-quenching $YBa_2Cu_3O_x$ in order to prepare precursors for subsequent texturing operations [5,6]. Very little on microstructural characterization has been published.

An effort is currently underway to achieve texturing of 123 grains in ErBaCuO-metal composites using the laser beam [7]. Composites offer a better structural resistance than YBaCuO ceramics and are also required for many electrotechnical applications. In this paper, we present the effect of various processing parameters on the microstructural aspect of laser melt-processed coatings.

Experimental

Coating preparation. The superconductive $ErBa_2Cu_3O_{7-x}$ powder (size ranging between 1 and 10 µm) was obtained commercially from SSC Inc. (Seattle, WA, USA). Coatings were prepared by spraying a slurry of 123 particles in trichloroethane onto thin nickel substrates using nitrogen as atomizing gas. After complete drying, coatings were consolidated by cold pressing. The thickness of coatings prepared by this technique ranged between 50 and 150 µm with uniform thickness. The resulting coatings (10 cm long x 0.8 cm wide) showed 75% of theoretical density (Fig. 1).

Laser set up. Laser processing experiments were carried out using a 200 W CO_2 cw laser (MPB LVPS-200) in the TEM_{oo} mode. As sketched in Fig. 2, the laser beam was focalized through a Zn-Se cylindrical lens (f=62.5 mm) and projected at normal incidence on the surface of the samples. The shape of the incident laser beam was nearly rectangular. A continuous flow of air (10 L/min) was used during laser process-ing to protect lenses. Some experiments were performed at the focal distance, i.e., 62.5 mm (line 170 µm wide x 8 mm long) or out of focus, i.e., at a working distance of 75 mm (1.5 mm wide x 8 mm long) in order to lower the thermal gradient roughly estimated as about 10^5 K/cm in the melt pool under focalized conditions. The linear beam was oriented transversely on the surface of the coating. Samples were either held at room temperature or preheated for 30 min on a hot stage. The melt pool tem-perature was monitored using optical pyrometry and a double wavelength configu-ration. The incident laser power was computer controlled, helping to maintain a constant temperature during processing.

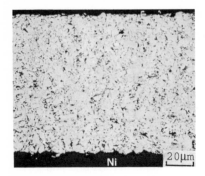

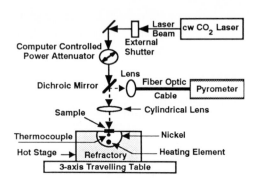

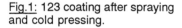

Fig.1: 123 coating after spraying and cold pressing.

Fig. 2: Schematic set up of laser processing unit and real-time temperature monitoring system.

Parameters for dynamic laser processing (texturing). In order to texture coatings, many experimental parameters were varied during dynamic laser processing of coatings: incident laser power density, interaction time (function of beam width and relative scanning velocities) and substrate temperature. The relative displacement of the beam over the surface of the coating was obtained using a 3-axis Klinger table (Fig. 2). Coatings were laser processed using a single scan at relative velocities ranging between 3 and 1600 µm/s. Incident powers ranged between 15 and 40 W in unfocused conditions and between 15 and 100 W with focalized beam. Power densities were varied by changing the incident power and the working distance.

Parameters for stationary laser processing. Static heat treatments were carried out in order to study the effect of incident power and dwell time on the microstructural aspect of the coatings. These experiments provided additional insights for understanding the transformations occurring during dynamic laser processing. They consisted simply in statically heating the coatings for fixed periods of time and in examining the microstructure of coatings. Coatings were heated with the laser for durations varying between 5 and 270 s. Incident laser power ranged between 15 and 40 W.

Characterization. The microstructure of coatings was characterized by Scanning Electron Microscopy (SEM) and the chemical composition of phases was assessed using Energy Dispersive Spectroscopy (EDX). Note that the erbium-based HTSC oxide alloy was used in this study because of the very high contrast between 211 and 123 grains in the Backscattered Electron Image mode (BEI).

Results and discussion

Coatings not preheated. Without preheating, the laser beam had to be focused to supply the high densities of energy (1 kW/cm^2) necessary for melting the 123 coatings. As shown in Fig. 3, after laser processing under these conditions, Er_2O_3, Er_2BaCuO_5 (211), $BaCu_2O_2$ and $BaCuO_2$ phases formed in the coatings without any crystallized 123 grains. Some of the low melting point Ba-Cu-O phases infiltrated the unmelted 123 grains of the coating depleting the molten layer in barium and copper. The presence of Er_2O_3 grains and long 211 grains indicate that the molten zone reached extreme temperatures and that solidification started from the upper peritectic event.

The Er_2O_3, 211 and Ba-Cu-O phases found in these microstructures could allow the formation of 123 grains upon post treatments but at the expense of prolonged diffusion periods. In attempts to reduce the temperature of the melt, the incident power of the laser beam was decreased and the scanning speed increased but these changes resulted only in a thinner melted layer and they yield the same undesirable phases.

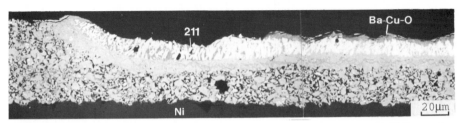

Fig. 3: Cross section of a laser processed 123 coating. (Focused beam; 25W; 7µm/s; no preheating).

Static heating tests at these high energy densities also gave strong evidence of thermocapillarity and demixing. Fig. 4 shows a top view of a laser trace produced by heating at 25 W for 15 s. Molten materials migrated toward cold zones on each side of the laser line and crystallized in the form of rims. Fig. 5 illustrates the effect of heating time on the shape of the laser trace and the growth of the rims. Demixing of Ba-Cu-O liquid phases and erbium-rich phases occurs very quickly upon melting (Fig. 5a and 5b) and after only 20 s of stationary heating the ridges became visible (Fig. 5c). In texturing conditions, this material transfer could significantly affect the melt pool composition and modify the crystallization process.

Fig. 4: Top view of the laser trace. (Focused and stationary laser beam; 25W; 15s; no preheating).

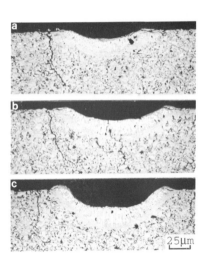

Fig. 5: Cross section of coatings heated with a focused beam for (a) 5s, (b) 10s and (c) 20s. (25W; no preheating).

Preheated coatings. With preheated coatings, less energy was required for melting. The energy densities were reduced by one order of magnitude (0,1 kW/cm²). The Er_2O_3 disappeared and 123 and 211 phases formed in the textured coatings indicating that the crystallization occurred in a lower temperature range. It was noted, as expected, that preheating at temperatures above about 900°C inhibited the formation of a ridge in front of the melt pool since the migrating Ba-Cu-O rich liquid phase solidifies below that temperature. On the other hand, preheating at temperatures above 980°C did not allow formation of many 123 grains in the coating, leaving numerous undesirable phases such as CuO and $BaCuO_2$. Thus, with unfocused beams, the preheating temperature was set at 950°C.

The textured microstructure is strongly dependent on the incident power and scanning velocity, two closely related parameters. With incident energy between 15 and 40 W, texturing at low speeds (below 3 μm/s) generally produced many 211 grains, many voids and undesirable reactions with the nickel substrate. The increased proportion of 211 phases indicates that the Ba-Cu-O rich liquid phase was lost during texturing. Texturing at high speeds (over 30 μm/s) usually produced non-textured structures. Fig.6a and b show the effect on the microstructure of the incident power (15 and 20 W) at a speed of 30 μm/s. After texturing at 15 W, a porous top layer with 211 grains appears in the coating (Fig.6a). Some of the Ba-Cu-O rich liquid phase infiltrates the unmelted portion of the coating. Increasing the incident power to 20 W increases the molten layer thickness and allows the formation of 123 grains in the center of the coating (Fig.6b). No 123 grains are detected in the upper part of the coating. Increasing the incident speed to 1.6 mm/s reduces the thickness of the molten layer and the infiltration of liquid in the unmelted subjacent zone (Fig.6c). Very fine 123 grains are present in the lower zone of that molten layer.

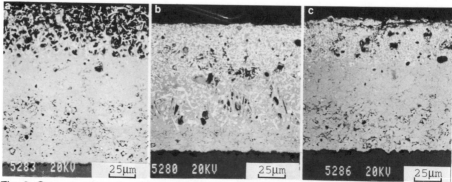

Fig. 6: Cross section of preheated coatings textured (a) with 15 W at 30 μm/s, (b) with 20 W and 30 μm/s, and (c) 20 W and 1600 μm/s. (Beam out of focus).

In order to increase the proportion of 123 in textured samples preheated at 950°C, an incident power of 25 W and a texturing speed of 7 μm/s were used. Fig. 7 shows the microstructure obtained with these parameters. The proportion of textured 123 grains is very high and concentrated in the middle of the coating with an orientation mostly colinear with the scanning direction. However, we observed that the orientation of grains is frequently disrupted in the textured coating. Pores in the melt and NiCuO-rich reaction products at the nickel-coating interface may influence the nucleation process and explain this disruption. The top of the coating is porous and rich in 211 grains dispersed in a Ba-Cu-O matrix. Fig. 8 shows the measured temperature in the melt pool and power densities recorded during laser processing. A typical temperature variation of about ±5°C in the melt pool could be responsible for the irregular texture orientation. Efforts are underway to reduce these fluctuations.

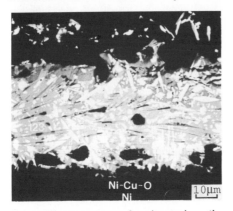

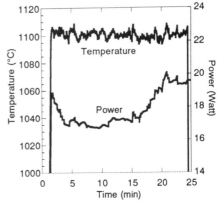

Fig. 7: Microstructure of preheated coatings textured at 7 μm/s with 25 W (out of focus).

Fig. 8: Incident laser beam power and melt pool temperature during texturing.

Using thicker coatings with these parameters, it was not possible to texture the 123 grains. The thickness of the coating is therefore a critical parameter for obtaining a regular textured structure since, for given heating conditions, the temperature of the melt pool varies with the thickness of the coating. The microstructure of stationary heated coatings reveals that the shape of the melt pool is affected by heating conditions. For instance, with longer heating time heat diffuses in the nickel substrate and reverses the melt pool profile (Fig. 9a and b). We also observed that reducing the thickness of statically heated coatings (or increasing the incident power density) results in the same trend. During dynamic processing, the solidification front could be affected accordingly and therefore influence the grain growth direction. For example, the shape of the melt pool corresponding to the texturing conditions in Fig.7 is shown in Fig. 10 a and b. The solidification front (Fig.10a) is rounded with grains growing in the texturing direction whereas the melting front (Fig.10b) is oriented downward similar to that of Fig. 9b.

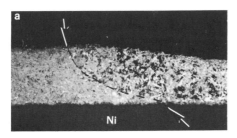

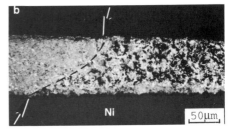

Fig. 9: Laser melt pool shape at melt-solid interface. Preheated coatings processed with a stationary and unfocused laser beam using 25 W for (a) 45s and (b) 270s.

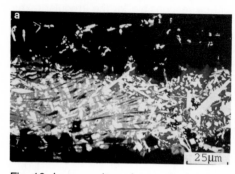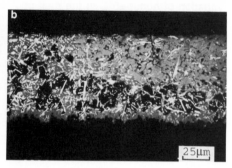

Fig. 10: Laser melt-pool quenched at the end of texturing showing (a) the solidification front (b) the melting front. (Unfocused beam; 25W; 7μm/s; preheated coatings).

Conclusion

These experiments showed that laser may be used for texturing 123 coatings with grains colinear with the scanning direction. The influence of laser heating parameters, preheating temperature and coating thickness was examined and adequate experimental parameters are proposed for texturing 123 coatings. The development of clean and textured coatings will require the development of more inert substrates. Additional work required for understanding the crystallization kinetics and for developing the new substrates is presently underway.

Acknowledgements

The authors are grateful to B. Harvey, M. Thibodeau and P. E. Mongeon (IMI-NRC) and to W. Jamroz, J. Tremblay and R. Bilodeau (MPB technologies Inc.) for their technical assistance during experimentation. The project was financially supported by the Canadian Space Agency.

References

(1) X.P. Jiang, J.G. Huang, Y. Yu, M. Jiang, G.W. Qiao, Y.L. Ge, Z.Q. Hu, C.X. Shi, Y.H. Zhao, Y.J. Wang, G.Z. Xu and Y.E. Zhou, Supercond. Sci. Technol. 1 (1988) 102.
(2) M.J. Cima, M.C. Flemings, A.M. Figueredo, M. Nakade, H. Ishii, H.D. Brody and J.S. Haggerty, submitted recently to J. Appl. Phys.
(3) U. Varshney, R.J. Churchill, H.P. Groger and A.I. Kingon, J. Supercond. 2 (1989) 293.
(4) H. Miyazawa, K. Hotta, S. Watanabe, S. Miyake, H. Hirose and M. Murakawa, Supercond. Sci. Technol. 4 (1991) 491.
(5) S. Nagaya, M. Miyagima, I. Hirabayashi, Y. Shiohara and S. Tanaka, IEEE Trans. on Magnetics 27 (1991) 1487.
(6) B. Arsenault, B. Champagne, D. Dubé and P. Lambert, Paper presented at the National Thermal Spray Conf., Long Beach, CA, May 4-10 1990.
(7) B. Arsenault, D. Dubé and C. Gélinas, Proc. of the 2nd Laser Advanced Materials Processing Conf. (LAMP-92) Nagaoka, Japan (1992).

VII.
Residual Stresses

Residual Stresses and Microstructures in the Surface Layers of Different Laser Treated Steels

K.-D. Schwager, B. Scholtes, B.L. Mordike[*] and E. Macherauch
Institut für Werkstoffkunde I, Karlsruhe, FRG
[*]Institut für Werkstoffkunde und Werkstofftechnik, Clausthal-Zellerfeld, FRG

Materials and experimental details

The materials investigated were plain carbon steels of the German grades Ck 22 (0.24% C, 0.20% Si, 0.52% Mn, 0.011% P, 0.010% S, balance Fe), Ck 45 (0.47% C, 0.19% Si, 0.65% Mn, 0.013% P, 0.009% S, balance Fe) and C 75 (0.76% C, 0.28% Si, 0.64% Mn, 0.015% P, 0.006% S, balance Fe, all specifications in wt-%). For basic experiments specimens with a cross-section area of 50x50 mm^2 and a thickness of 10 mm were manufactured from the raw material. Furthermore for bending fatigue tests specimens from Ck 45 with a length of 150 mm and a rectangular cross-section of 9.6x17 mm^2 in the gauge length were produced. In both cases the surfaces provided for laser treatments were ground. All specimens were normalized after machining to achieve comparable starting conditions. Afterwards they were graphite coated to optimize absorption during the laser process. Laser treatments were performed using a 5.5 kW CO_2-laser (model C-76, Heraeus) with the following processing parameters:

specimens	flat specimens	fatigue specimens
material	Ck 22, Ck 45, C 75	Ck 45
power	1.4 kW	2.9 kW
feed rate	250 mm/min	270 – 300 mm/min
width of the track	8 mm	15 mm

Residual stresses were determined by X-ray diffraction with Cr Kα-radiation, employing the $\sin^2\psi$-method (1). For each stress evaluation 9 lattice strain measurements in the range of $-60° \leq \psi \leq +60°$ were performed on {211}-planes of near surface grains of ferrite and/or martensite. For the stress calculations Young's modulus E = 210000 MPa and Poisson's ratio $v = 0.28$ were used. The indicated integral breadths are averages of the interference lines at $\psi = 0°$ and $\psi = \pm 15°$. Alternating bending tests were carried out using a Flato-type bending fatigue device (Fa. Schenck) with an approximately constant loading amplitude with zero mean stress. The loading frequency was 25 Hz.

Experimental results

In Fig. 1 the distributions of the surface longitudinal and transverse residual stresses and of the integral breadths as a function of the distance from the centre of the laser tracks are shown for Ck 22, Ck 45 and C 75. The longitudinal direction is always parallel to the processing direction. Although identical parameters were used for the laser treatments, a comparison of the results reveals characteristic differences between the measured materials. For Ck 22 in both directions compressive residual stresses are measured with maximum values in the centre of the

630

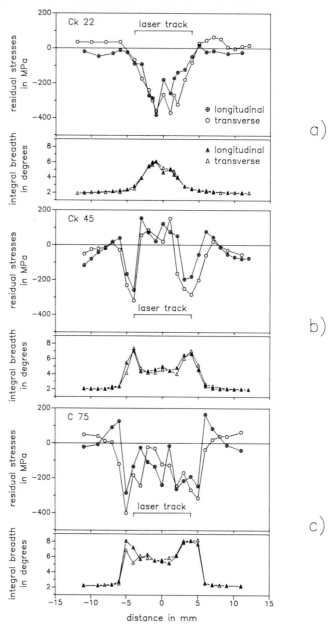

Fig. 1: Distributions of surface longitudinal and transverse residual stresses and of integral breadths as a function of the distance from the centre of the laser tracks a) Ck 22; b) Ck 45 and c) C 75.

laser track. At the edges of the track, only small tensile stresses occur in transverse and compressive stresses in longitudinal direction. Ck 45 exhibits a pronounced maximum of compressive residual stresses at the edges of the laser track. Within the hardened zone, the stresses rapidly change their signs and tensile residual stresses with locally fluctuating absolute values exist. In the case of C 75, beyond the edges of the laser track, maximum tensile residual stresses are observed in longitudinal direction. These stresses change with a sharp stress gradient to maximum compressive residual stresses near the edges of the laser track. The hardened zone exhibits compressive residual stresses with oscillating absolute values. In all cases, the integral breadths increase rapidly from the values of the normalized starting condition to those of the hardened laser track. It is interesting to note that both for Ck 45 and for C 75 maximum integral breadths are measured near the track edges and relative minima at the centre of the track. Contrarily, Ck 22 exhibits maximum values of the integral breadths at the centre line of the track.

Micrographs of the near surface areas of the laser tracks showed in all cases small ledeburitic surface layers containing considerable amounts of retained austenite. This is obviously a consequence of the graphite coating applied for improving the absorption conditions of the laserlight. Below these layers martensitic structures were found. In the case of Ck 22 e.g., a mixture of martensitic and ferritic grains appeared. For C 75, small cracks normal to the specimen surface were detected in the martensite below the ledeburitic layer. The depth-distributions of the microhardness HV 0.2 measured in the centre of the laser tracks are shown in Fig. 2. With the exception of the values measured directly near the surface, the mean hardness values below the surface increase with increasing carbon content of the steels investigated. In addition, the thickness of the hardened layer is also enlarged. Ck 45 as well as Ck 22 show an absolute or a relative hardness maximum, respectively, at certain distances below the surface. Measurements accross the laser tracks clearly show that hardness distributions within the tracks are the more inhomogeneous the smaller the carbon content is. This is due to the fact that a decreasing carbon content leads to an increasing fraction of ferrite grains in the

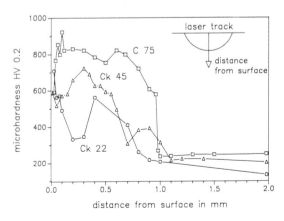

Fig. 2: Microhardness vs. distance from the surface for the steels investigated.

normalized ferritic-pearlitic matrix, thus producing a more inhomogeneous austenitizing during the rapid heating process.

Fig. 3 shows how the residual stress fields of individual laser tracks are changed if interactions with neighbouring tracks occur. The results were obtained from laser

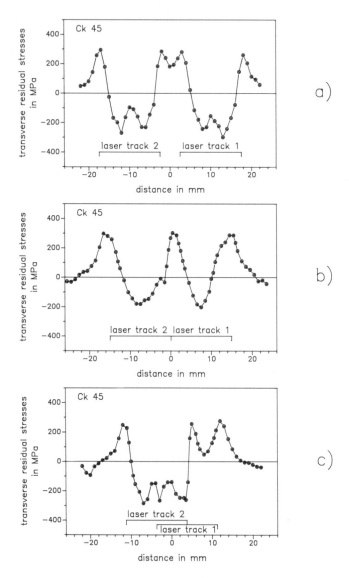

Fig. 3: Distributions of transverse residual stresses perpendicular to two tracks with different distances to a neighbouring track.

treated Ck 45 fatigue specimens. From Fig. 3a, one can see that maximum tensile residual stresses developed at the edges of two individual laser tracks having a distance of 20 mm between their centre lines. Within the hardened zone, compressive residual stresses are measured with a relative minimum of the absolute values in the centre of each track. Remarkable differences in the resulting residual stress distributions around the individual tracks occur in comparison with the residual stress distributions of the single track shown in Fig. 1b. These differences are due to the different laser parameters used. If the two laser tracks are positioned in such a way that they come in contact at their edges transverse residual stresses develop, as can be seen from Fig. 3b, which are very similar to those shown in Fig. 3a. Maximum tensile residual stresses of nearly the same amount occur at the outer as well as at the contacting edges and compressive residual stresses near the centre lines of the individual tracks. The consequences of partly overlapping laser tracks on the transverse residual stress distributions are shown in Fig. 3c. The 50%-overlapping laser track exhibits almost the same residual stresses as the single tracks shown in Fig. 3a. Consequently, the distribution of the transverse residual stresses in Fig. 3c can be understood as the superposition of a single track on the remaining residual stresses of the laser track 1 at the right-hand side of the specimen.

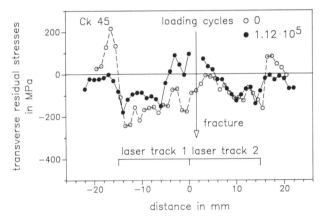

Fig. 4: Distribution of transverse residual stresses before and after fatigue testing of specimens with two adjacent laser tracks. The specimen fractured after $1.12 \cdot 10^5$ loading cycles.

In Fig. 4, the consequences of cyclic bending loading on the laser induced surface residual stress distributions are exemplarily shown for a Ck 45 specimen with two adjacent laser tracks. The tracks were placed in the gauge length of the bending specimen perpendicular to the direction of the bending stress applied. After the laser treatment a quite different residual stress distribution is found than in the case shown in Fig. 3b. In particular the absence of the tensile residual stresses at and near the contacting edges of the tracks is striking. Obviously not only the slight changes of the laser conditions, but also perhaps a slight superposition of the adjacent tracks in the envisaged contact area, change the resulting residual

634

stress distribution markedly. Cyclic loading of the laser treated specimen with a nominal bending stress amplitude of 340 MPa yielded $1.12 \cdot 10^5$ cycles to fracture. The macrocrack developed near the envisaged contacting edge of the tracks. This position is indicated in Fig. 4 by an arrow. The residual stress distribution after fracture of the specimen is shown in Fig. 4 by full dots. As can be seen, considerable differences occur compared with the state after laser treating. In particular, the tensile residual stress peaks near the outer edges of the laser tracks were released and tensile residual stresses were built up near both sides of the fracture plane.

Discussion

The results presented confirm the observations already made in (2) that residual stress distributions resulting from laser tracks are comparable with welding residual stresses existing in dummy welds and that the same sources of residual stresses, i.e. shrinking, quenching and transformation processes, are active. Consequently, the final laser-induced residual stress distributions are due to a complex interaction between these processes, which are in turn considerably influenced by the laser parameters chosen. However, the extremely high heating and cooling rates cause non-equilibrium microstructures. A detailed prediction of the sequences of the local transformation processes is rather complicated. It is interesting to note that interactions between the graphite coating and the near surface material may occur, which produce characteristic microstructures and influence the resulting residual stresses. The results shown in Fig. 3 clearly indicate that in the case of overlapping laser tracks, the final residual stress distribution is the result of a simple side-by-side arrangement of the residual stresses of single laser tracks, whereby the last track wipes out the already existing residual stress fields of precedent tracks. In (3) it was shown that the residual stress state existing after laser hardening considerably influences the lifetime of bending fatigue specimens. The results presented in Fig. 4, which are part of a systematic series of investigations, indicate that the complex residual stress state of contacting or overlapping laser tracks may be rearranged during a fatigue test. Consequently, to predict the effects of residual stresses, their stability during fatigue loading must be known.

The financial support of the work by the German Research Association (Deutsche Forschungsgemeinschaft) is gratefully acknowledged.

References

(1)   E. Macherauch, P. Müller: Das $\sin^2 \psi$-Verfahren der röntgenographischen Spannungsmessung, Z. f. angew. Physik 13 (1961), pp. 305-312.
(2)   K.-D. Schwager, B. Scholtes, B.L. Mordike, E. Macherauch: Residual Stress States in Laser-Treated Plain Carbon Steels, Proceedings International Conference on Residual Stresses ICRS3, Tokushima, 1991, pp. 858-863.
(3)   B. Scholtes, B.L. Mordike: Residual Stress States of Laser-Surface-Hardened Steel Specimens, In: Residual Stresses in Science and Technology, DGM (1987) Vol. 2, pp. 655-662.

Prediction of thermal, phase transformation and stress evolutions during laser hardening of steel pieces

M. BOUFOUSSI, S. DENIS, J. Ch. CHEVRIER, A. SIMON
L.S.G.2.M. URA 159 CNRS, Ecole des Mines de Nancy (FRANCE)
A. BIGNONNET - PEUGEOT S.A. BIEVRES (FRANCE)
J. MERLIN - INSA - GEMPPM - CALFETMAT LYON (FRANCE)

Introduction

Laser surface hardening is one of the promising application of lasers especially for localized treatment of complex geometric shapes. From the practice, it is known that the choice of the operating parameters (power, velocity and energy distribution of the beam) affect significantly the properties of the treated piece. The numerical simulation of the treatment may bring a help for a better control of the treatment process. In this paper, we present shortly the calculation model. Then the model is applied to laser surface hardening of plates. We shall analyse the evolution of microstructures and internal stresses during the treatment. The calculated distributions of hardness and residual stresses are compared to experimental ones.

Calculation model

For the prediction of residual stresses after a surface hardening treatment it is necessary to calculate the temperature fields, the phase transformation and the stress and strain evolutions during the treatment by considering the couplings between these evolutions. The phase transformation calculation model developed at Ecole des Mines de Nancy (1,2) has been put together with the finite element code SYSWELD (3). These models have been described elsewhere and are just summarized in the following.

– Thermal and phase transformation model

The thermal calculation (3) is based on the solution of the equation governing the heat conduction problem by finite elements. The coupling with phase transformations is taken into account through an internal heat source term which is related to the latent heat and to the rates of transformations. The thermophysical properties (specific heat, conductivty, density) depend on temperature and on the volume fractions of the different phases (linear mixture rules). The metallurgical model (1,2) is based on a principle of additivity that allows to calculate the anisothermal kinetics of phase transformation from isothermal ones on heating and on cooling. The isothermal kinetics are modelled by the Johnson-Mehl-Avrami's law. On heating, for an original ferrite-pearlite microstructure, the austenitization kinetics is described in two steps : firstly the pearlite dissolution, then the transformation of ferrite. The inhomogeneity of carbon content in austenite and the austenite grain growth are modelled. On cooling, for diffusion controlled transformations, the incubation period is calculated (by Scheil's method) before the growth period. The progress of the martensitic transformation is described by the Koistinen-Marburger's law. The effect of the inhomogeneity and grain size of austenite is taken into account on the kinetics of phase transformations during cooling. A hardness calculation is associated with this phase transformation calculation.

- **Mechanical model** (3)

The mechanical model considers a thermoelastoplastic behaviour law of the material taking into account the transformation strain (volume change and transformation plasticity). The elastic properties (Young's modulus, Poisson's ratio) are temperature dependent, the plastic properties (yield stress, hardening constant) are dependent on temperature and on the volume fractions of the different phases through mixture laws. The model takes into account possible "recovery" of strain hardening during transformation (the new phase formed can forget partially or totally the previous hardening) (4). The transformation plasticity deformation is taken into account on cooling. It is written in the

form $\qquad \dot{\varepsilon}_{ij}^{tp} = \frac{3}{2} Kf'(z) \dot{z} s_{ij}$ with $f(z) = z(1-\ln z)$

for martensitic transformation (z volume fraction of new phase formed, s_{ij} stress deviatior, K material parameter taken from experiments). The effect of stresses on the kinetics of phase transformation (5) is not taken into account.

Application to the laser hardening treatment

- **Laser treatment of plates**

A CO_2 laser of 4 kW continuous power has been used for the treatment of plates with dimensions 100 x 50 x 20 mm. The plate moves beneath the laser beam in order to obtain a single hardened track. The energy distribution resulting from the transformation of the initial beam through an optical device (6) has a rectangular shape.

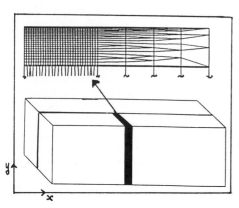

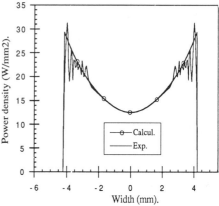

Fig. 1 : Finite element mesh

Fig. 2 : Energy distribution of the laser spot

Before the treatment, an absorbing coating is realized on the plates by an oxydation in a melted salt bath. The process parameters and the size of the laser spot are given in table 1. The plates are made of an hypoeutectoïd plain carbon steel (0.42 % carbon) with a ferrite-pearlite initial microstructure.

- **Assumptions and input data for the calculation**

In this work we have carried out two dimensional calculations. For

Treatment	Total power (W)	Beam velocity mm/s	Spot dimensions (mm)	
			$//\vec{v}$	$\perp\vec{v}$
1	960	4.5	6	8.4
2	960	6.5	6.2	8.4

Table 1 : Process parameters

symmetry reasons, only one half of the transversal section of the plate needs to be considered for the finite element mesh (fig. 1).
- For the thermal calculation we use the measured energy distribution considering only the distribution (smoothed by a polynomial) in the direction perpendicular to the beam displacement (fig. 2). (It is supposed constant along the displacement). The absorption has been measured versus temperature (7). The thermophysical properties and enthalpies of phase transformations as well as the data for the metallurgical model (IT heating diagram, IT cooling diagram...) have been determined in a previous study (2).
- The mechanical calculation is performed under generalized plane strain conditions. We assume kinematic hardening of the different phases without memory of previous hardening when transformation occurs. The plastic properties for the different phases as functions of temperature have been obtained from different studies carried out at LSG2M (8)... Linear elastic-linear strain hardening curves are fitted to experimental stress-strain curves up to 1 % plastic strain. The data are taken at a deformation rate of the material close to the ones it has during the treatment (here about $10^{-2}s^{-1}$). As the model considers only two different hardening constants, one for austenite and one for the other phases, we have chosen to introduce the data corresponding to austenite and to the initial microstructure.

- Results
The simulation is performed in transient conditions and in two steps firstly the thermal and metallurgical calculation and secondly the mechanical calculation.
• Temperature and microstructure evolutions
Fig. 3 shows the calculated temperature evolutions in the middle of the laser track at different depths for treatment 1. The maximum surface temperature is 1075°C.
The kinetics of austenitization are also calculated. At the end of heating, austenite is homogeneous in a superficial area and leads to homogeneous martensite on cooling as shown on the calculated distribution of microstructures at the end of cooling (fig. 4). In zones where austenite has not reached homogeneity, the microstructure presents high carbon martensite and low carbon martensite. For treatment 2, the final distribution of microstructures is similar but in that case, no homogeneous martensite appears. Indeed, the heating conditions, particularly the maximum temperatures reached, are lower (998°C at the surface) and don't allow the homogeneization of the austenite. It is interesting to mention that these microstructure distributions are close to the ones obtained for an induction hardening treatment (2), solely the heat affected zone is much less deep in the case

of laser treatment. The calculated hardness profiles are presented on fig. 5 for both treatments. Measurements are also reported. It appears that the calculation describes correctly the evolution of hardness and the depth of the hardened zone. Other comparisons between experimental and calculated results have shown that the whole geometry of the hardened zone is well described by the calculation. It can be noted that 3D stationary simulations have lead to similar results.

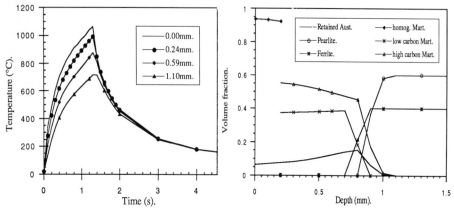

<u>Fig. 3</u> : Temperature evolutions at different depths. Treatment 1

<u>Fig. 4</u> : Final distribution of microstructures. Treatment 1

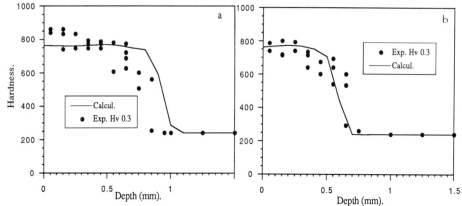

<u>Fig. 5</u> : Hardness profiles for treatment 1 (a) and 2 (b)

- <u>Internal stresses evolutions</u>

We present here only the results concerning the transverse stress. Indeed 2D calculations lead to reliable results only for the transverse stress (9). Fig. 6 shows the evolution with time of the transverse stress at the surface of the track for both treatments. Fig. 7 gives the evolution of the cumulated effective plastic strain ε^p and transformation plasticity strain ε^{pt} for treatment 1 only (for clearness reasons).

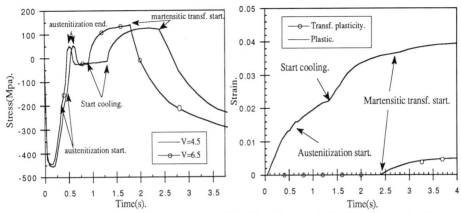

Fig. 6 : Evolution of transverse stress with time at the surface

Fig. 7 : Cumulated effective strains versus time at the surface

Let us consider firstly treatment 1. At the beginning of heating until the austenitization starts the stresses are purely thermal in origin. The high thermal gradients generate a compressive stress (maximum value of about 450 MPa) associated with plastic strain. Then austenitization occurs and leads to a small tensile stress (50 MPa). During the heating of austenite the stress remains at a low level due to the low mechanical properties of austenite and plastification goes on. The beginning of cooling is characterized by an increase of the tensile stresses and plastic strain. The martensitic transformation (associated with volume change and transformation plasticity) leads to a significant stress relaxation. For treatment 2, the stress evolution is very similar. Solely, a shift in the time scale is observed according to the temperature evolutions. It is noticeable that the maximum stress levels (in compression and in tension) are close to the ones calculated for treatment 1. Moreover, the evolutions of ε^p and

ε^{pt} are close for both treatments. On fig. 8 we have reported the calculated residual stress profiles for both treatments. They are compared to measurements performed by X-ray diffraction (10). The calculated profiles are similar for both treatments. An approximately constant compressive stress (about 340 MPa) appears in a superficial area. The stress becomes tensile with sharp gradients until the transition between the hardened zone and the base metal is reached. The comparison with the experimental results shows a correct agreement.

At this stage, it is necessary to study the influence of the metallurgical and mechanical behaviour of the material on the evolutions of internal stresses and strains and on the residual stress profiles in order to point out the significant parameters. For example a recent result has shown that the hardening behaviour of the base metal has a significant effect on the level of the residual tensile stress at the transition between the hardened layer and the base metal. We think also that the martensitic transformation plasticity range has a great effect on the level of

the compressive stress in the superficial area as we have shown it
for an induction hardening treatment (11).

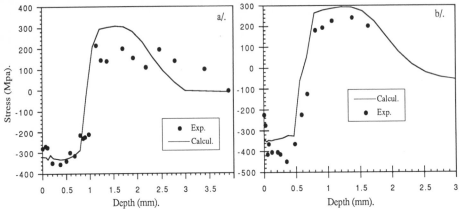

Fig. 8 : Calculated and experimental residual stress profiles
for treatment 1 (a) and 2 (b)

Conclusion

The model that we have presented allows to calculate thermal,
metallurgical and stress evolutions during laser hardening of
steels. The concepts that are taken into account and the set of
input data that has been established allow to describe in a
satisfactory way the geometry of the hardened zone, the hardness
distributions and the residual stress distribution for well
controlled laser treatment conditions. One further step of this
study is to get a better knowledge of how the process parameters
(power density distribution and beam velocity) may affect the
stress and strain evolutions and the residual stresses. The final
step might be to define an optimal laser treatment in connection
with the studies on fatigue properties of the treated workpieces.

References

(1) F. FERNANDES, S. DENIS, A. SIMON : Mem. Etud. Scient. Revue
Metall. Juillet (1986), 355.
(2) S. DENIS, D. FARIAS, A. SIMON : ISIJ International Vol. 32,
n°3 (1992) 316.
(3) SYSWELD, User's manual, FRAMASOFT + CSI.
(4) J.B. LEBLOND : Int. Journal of Plasticity, Vol. 5 (1989) 573.
(5) S. DENIS, S. SJÖSTRÖM, A. SIMON : Metall. Trans., Vol. 18A
(1987) 1203.
(6) C. RENARD, Jun Chand Li, T. MANDERSCHEID, J. MERLIN : Mem.
Etud. Scient. Revue Metall., Juin (1991) 341.
(7) V. GRANIER, D. KECHEMAIR, R. FABBRO : Rap. inter. ETCA (1990).
(8) Ch. LIEBAUT, E. GAUTIER, A. SIMON : Mem. Etud. Scient. Revue
Metall. 85 (1988) 571.
(9) J.M. BERGHEAU, D. PONT, J.B. LEBLOND : Proc. IUTAM SYMPOSIUM
"Mechanical Effects of Welding", Edit. L. KARLSSON, L.E. Lindgren,
M. Jonsson, Springen Verlag (1992) 85.
(10) J.L. LEBRUN : Rapport interne E.N.S.A.M. (1992).
(11) M. ZANDONA, A. MEY, S. DENIS, A. SIMON : Mem. Scient. Revue
Metall (à paraître).

VIII.
Beam-Material-
Interaction

Laser Target Interaction: a Comparison between CO and CO₂ Lasers

M. Stöhr, E. Zeyfang, DLR, Institut für Technische Physik, Stuttgart, Germany.

Introduction

The CO laser with its wavelength around 5 μm (multiline, 4.8 to 5.8 μm) is lying between the CO_2 laser (10.6 μm) and the Nd:YAG laser (1.06 μm). Due to its specific properties, the CO laser is expected to combine advantages of both other laser types: more efficient material processing and easier fiberoptic beam delivery compared to the CO_2 laser, scalability to high average beam power and good beam quality compared to the Nd:YAG laser. In spite of these expectations, CO lasers up to now still exist only as laboratory prototypes, maybe caused by the necessity to cool the laser gas, which results in a more complex construction of the laser.

In order to investigate possible advantages of the CO laser and to confirm results which have been reported since several years (1,2), comparative experiments with CO and CO_2 lasers are carried out at DLR Stuttgart. Results on cutting and welding experiments are reported in this paper.

Part of this work was done in the frame of the EUREKA EU 113 project (3), part of it was supported by the Commision of European Communities (CEC DG XII/D2, Contract Nr. FI 2D.0028.C)

Characteristic Laser Data

The CO laser used in these experiments has been developed at the DLR Institut für Technische Physik, Stuttgart. The laser gas is excited by a RF-discharge and cooled by means of a gasdynamic expansion in a supersonic nozzle (4, 5). The closed cycle system enables operation times of several hours depending on the output power. Due to a folded resonator design the output beam is linearly polarized. Inserting apertures inside the resonator near the output mirror, the beam diameter can be varied between 6 mm and 18 mm according to Fresnel Numbers of 1.4 and 13 resp. This allows to adapt the quality and the focussability of the beam to the specific experiment. On the other hand, the maximum beam power varies from 100 W at 6 mm up to 850 W at 18 mm beam diameter.

In order to get comparable beam characteristics of CO and CO_2 lasers at beam - workpiece interaction experiments, detailed profile measurements of the CO laser beam have been carried out. Table 1 shows a summary of the results obtained using mirrors and lenses of different focal lengths at different beam diameters. The beam quality is characterized by the factor K, which is 1 for a Gaussian beam profile. Subtracting the calculated spherical aberration of the optical system from the measured focal spot diameter yields in a corrected factor K_{CORR} which is a measure of the beam quality itself. In the case of the multiline CO laser beam the chromatic aberration is still included, which is essential only at lenses of short focal length. As a good compromise referring to beam power and beam quality the diameter of 12 mm was chosen to carry out the experiments reported in this paper.

644

Optics	P_{laser}	R_B (mm)	R_F (μm)	I_{MAX} (W/cm^2)	I_{MIT} (W/cm^2)	K	Sph.Aberr. (μm)	K_{CORR}
mirror f=150mm	100W TEM$_{OO}$	3.3 ±5%	94	8.5×10^5 ±5%	3.5×10^5 ±3%	0.78		
lens f=127mm			79	1.2×10^6 ±5%	5.0×10^5 ±3%	0.79	0.6	0.80
lens f=63.5mm			55	2.8×10^6 ±5%	1.0×10^6 ±3%	0.56	2.5	0.59
lens f=63.5mm	350W medium order	6 ±5%	129	1.6×10^6 ±5%	6.8×10^5 ±3%	0.13	14	0.15
lens (silicon) f=50mm			90	2.8×10^6 ±5%	1.4×10^6 ±3%	0.15	3.5	0.16
mirror f=150mm	300W high order	7.75 ±5%	290	2.0×10^5 ±5%	1.5×10^5 ±3%	0.11		
lens f=63.5mm			140	1.0×10^6 ±5%	6.5×10^5 ±3%	0.095	28	0.12

Table 1: CO laser beam characteristic data R_B: beam diameter, R_F: focus diameter, I_{MAX}: max. focal intensity, I_{MIT}: medium focal intensity, Sph. Aberr.: calculated spher. aberration, K_{CORR}: beam quality factor K corrected for spher. aberration, $K = (\lambda f/\pi)/R_F R_B$

In order to get similar spot sizes and power distributions at the surface of the workpiece, the beam of a commercial 1-kW axial flow CO_2 laser was modified by means of an aperture, too.

Cutting Investigations

It should be mentioned that in the case of cutting and welding materials with lasers we have two overlapping effects (3):
- there are different absorption behaviours, depending on the wavelength of the laser radiation (6,7).
- the diameter of the focal spot scales with the wavelength and hence the area with λ^2; this is valid only for equivalent f-numbers of the optics and equivalent radiation characteristic factors of the different laser beams.

In our cutting experiments we concentrated on the first topic and both laser beams were modified to achieve equal focal spot diameters as described in the previous chapter (CO laser: D = 348 μm; CO_2 laser: D = 336 μm). A mirror optic system of f = 150 mm was used, and the experimental conditions were equal for both the CO and CO_2 laser experiments.

"Cutting" in these experiments is defined as just a complete separation of the sample without regarding dross-formation or cut quality.

Cutting Results

Cutting experiments of mild steel were performed with oxygen as assist gas (Fig. 1). Furthermore there were done experiments with nitrogen or argon as assist

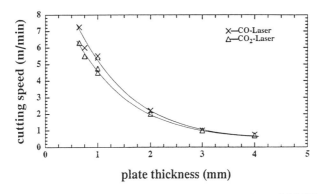

<u>Fig. 1</u>: Oxygen assisted cutting of mild steel. P_L = 250 W.

gas (Fig. 2). The plots show the expected hyperbolic dependence of the cutting speed on the material thickness.

The oxygen assisted cutting of mild steel shows a slightly improved cutting efficiency using the CO laser in comparison to the CO_2 laser. With increasing material thickness the cutting speeds become more and more equal. To understand this behaviour it is necessary to look at two different points. A simple calculation of the heat conduction shows that this is the dominat term of energy loss at cutting speeds below 2 m/min. The effect of a chemical reaction between the sample material and the assist oxygen is not pronounced when using thin samples (d ≤ 2 mm) (8). Hence the cutting behaviour of samples thicker than 2 mm is predominantly controlled by the chemical oxydation energy and the heat losses. Both effects are independent on the wavelength of the laser beam and therefore the cutting speeds are equal for both lasers.

The cutting behaviour of thin material (d < 2 mm) is dominated by the different absorption characteristics of CO and CO_2 laser radiation, which is also shown by the fact that the cutting speed in this case varies proportional with the laser power (3).

The nitrogen and argon assisted cutting experiments show an improved cutting efficiency for sample thicknesses below 1 mm when using CO laser radiation compared to CO_2 laser radiation (Fig. 2). The theoretical treatment is more easy in this case of non-reactive gas cutting. Table 2 shows the different power losses while cutting. The heat conduction loss is calculated by means of an approximation formula derived by Petring (9). The results show, that the main power loss is caused by heat conduction, which therefore dominates the cutting process for the low powers applied in our experiments. Only at a sample thickness of 0.5 mm the influence of the heat conduction is reduced. In this case the angle of the cutting front is about 55° (10) and the absorption coefficients for this angle are comparable to the results of Stern (7).

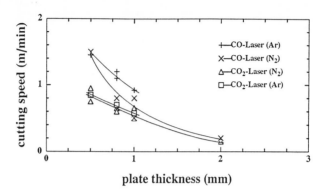

<u>Fig. 2:</u> Inertgas assisted cutting of stainless steel. P_L = 250 W.

	Material thickness [mm]	Kerf width [mm]	V [m/s]	P_S [W]	P_W [W]	P_H [W]	A_{exp}
CO_2-Laser	5.0	0.28	0.0125	3.5	13.4	12.0	0.12
	0.8	0.26	0.0108	4.8	18.6	16.0	0.16
	1.0	0.26	0.0092	5.2	22.0	17.8	0.18
	1.0	0.26	0.0108	6.0	21.5	20.6	0.19
	2.0	0.26	0.0025	2.8	36.6	9.6	0.19
CO-Laser	0.5	0.22	0.0250	5.3	10.0	17.9	0.17
	0.8	0.25	0.0180	8.3	16.5	27.5	0.2
	1.0	0.24	0.0130	6.6	21.0	22.8	0.2
	2.0	0.24	0.0033	3.3	35.0	11.3	0.2

$$A \cdot P_L = P_H + P_S + P_W$$

A: Absorption coefficient
P_L: Laser power

P_H: Heating power
P_S: Melting power
P_W: Heat losses (conduction)

<u>Table 2:</u> Power balance of the cutting process, stainless steel, P_L = 250 W.

The experiments generally show an increased cutting speed for the CO laser compared to the CO_2 laser. This effect might be seen more clearly at increased laser powers.

<u>Welding Investigations</u>

The welding experiments have been carried out using transmissive optics of short focal length, because higher power densities are required compared to cutting. The beam modifications resulted in nearly the same focal diameter D (CO: f = 50 mmSi, D = 106 μm; CO_2: f = 63,5 mm ZnSe, D = 97 μm), but the power density differs by a factor of 1.4 (CO: 2.2·10⁶ W/cm² max., CO_2: 3.2·10⁶ W/cm² max.).

Experiments were performed with argon and helium as shielding gas resp. The experimental conditons were equal for both lasers. When welding with the CO_2 laser, a bright blue plasma (which is well known) appears near the surface. Using

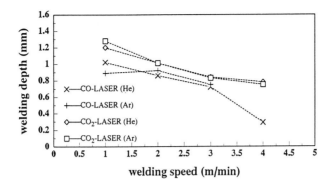

<u>Fig. 3</u>: Welding depth as a function of welding speed. P_L = 400 W.

the CO laser, this plasma can not be observed. This effect is expected from theoretical considerations depending on the shorter wavelenght of the CO laser.

<u>Welding Results</u>

Fig. 3 shows the depth of the welding seam for different welding speeds. There is no significant difference between experiments using helium or argon. The drop of the welding depth at 4 m/min for the CO laser is caused by a breakdown of the keyhole, thus only conduction welding is possible. The drop at low speed CO laser welding when using argon cannot be explained up to now. Both experiments show a slightly higher welding depth for the CO_2 laser. This corresponds to the higher power density at the target surface. More information yields the analysis of the cross-sectional area of the molten bath pool. Neglecting heat losses this area is proportional to the line energy (laser power/welding speed) and gives information of the efficiency of the energy transfer to the material. The results are shown in

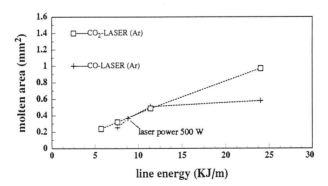

<u>Fig. 4:</u> Cross-sectional area of molten bath pool. Shielding gas Ar. P_L = 400 W.

Fig. 4 and Fig. 5. There is no difference between the CO and the CO_2 laser for the medium line energies. The drop at 5.5 kJ/m (CO laser) corresponds to the break down of the keyhole as mentioned already.

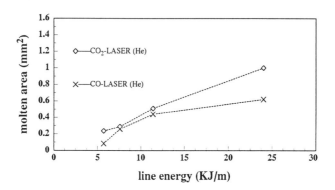

Fig. 5: Cross-sectional area of molten bath pool. Shielding gas He. P_L = 400 W.

At low welding speeds (high line energies) the energy transferred to the material drops for both CO-welds. Possible explanations could either be a better energy coupling of the CO_2 laser radiation (via the plasma?) or a worse energy coupling of the CO laser radiation.

References

(1)   S. Sato et al.: J. Appl. Physics, Vol. 58, No. 11 (1985) 3991.
(2)   S. Shono et al.: Proc. ICALEO 69 (1989) 130.
(3)   F. Maisenhälder et al.: EUREKA Project EU 113, Final Report - Industrial CO Laser Evaluation - (1990).
(4)   H.v. Bülow, E. Zeyfang: Proc. 8th GCL, SPIE 1397 (1990) 499.
(5)   E. Zeyfang, H.v. Bülow, M. Stöhr: Proc. 8th GCL, SPIE 1397, (1990) 449.
(6)   F.Dausinger, T. Rudlaff, J. Shen: Laser und Optoelektronik 23, 2 (1991) 43.
(7)   G. Stern: Proc. ECLAT (1990) 25.
(8)   M. Lepore et al.: Proc. ICALEO 69 (1989), 130.
(9)   D. Petring et al.: Feinwerktechn. u. Meßtechn. 96, 9 (1988) 364.
(10)   F. Dausinger: Proc. ECLAT (1990) 1.

Characterization of excimer laser treatment of metals using short time diagnostics; influence on possible industrial applications

K. Schutte, R.Queitsch, H.W. Bergmann

Forschungsverbund Lasertechnologie Erlangen, Friedrich-Alexander-Universität, Lehrstuhl Werkstoffwissenschaften 2 (Metalle), Erlangen, FRG

Abstract

Materials processing of metals using excimer lasers is limited by the formation of a metal vapour plasma. The condensed material and metallic oxide layers beside the treated areas are critical points for large area treatments. The possibility of suppressing those contaminations and optimizing the process for typical industrial applications will be discussed.
Experiments using a short time, high speed diagnostic were carried out to determine the influence of the plasma wave front propagation on the processing results of the laser treated areas. Changing the laser parameters (energy density, number of pulses, etc.), the substrate material and the surrounding atmosphere leads to a different behaviour of plasma formation and processing quality.

Introduction

Nowadays excimer laser systems have become a reliable tool for the industry. Many processes like desoxidation, cleaning, ablation etc. are examined by several authors and are reported in literature (1), (2), (3), see figure 1. But there are only few industrial applications (4) using the excimer laser technology integrated in manufacturing lines. The main processes used in industry are marking (by roughening the surface) and microstructuring (by ablation of material) of glass and ceramic materials.

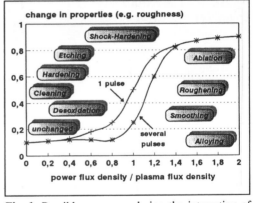

Fig. 1 Possible processes during the interaction of UV-light with metallic surfaces (see ref. (1))

Especially for large area surface modifications of metals two facts have to be considered. First the quality of the treated area depends significantly on the applied laser beam intensity profile (e.g. profile homogeneity, pulse shape, variation in energy density) and the laser optics, see figure 2. Secondly illuminating a surface step by step will often lead to a pattern of treated zones on the sample, see fig. 3. A typical effect by scanning a samples surface is the generation of condensed material and oxide layers beside and between the laser spots, see figure 2. This results in a reduction of the surface quality (roughness, adhesion, wear resistance, reflection) and can not be accepted by an industrial user. Aim of the present investigation was to determine the influence of the plasma propagation during the excimer laser surface treatment of large surfaces on the surface quality.

Fig. 2 Macrograph of a typical surface pattern on a excimer laser treated sample

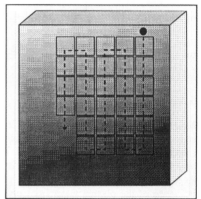

Fig. 3 Schematic drawing of a surface treatment by scanning the work piece

Experimental Setup

The experimental investigations were carried out using a high power excimer laser (Siemens-type Excimer Laser XP2020) and a high speed diagnostic system. The characteristic data of the applied beam source are summarized in the table below.

Laseractive Medium	Wave Length	Energy per Pulse	Repetition Rate	Pulse Duration	Cross Section
(XeCl)*	308nm	2J	20Hz	FWHM 55ns	45x55mm^2

Figure 4 exhibits a schematic drawing of the arrangement for the time resolved measurements of the laser plasma formation and propagation. The short time camera (DICAM, PCO computer optics) consists of a very fast gated photo cathode, an integrated micro channel plate and an imaging intensifier including CCD-sensor (5),(6). The image information can be stored by a fast frame grabber in the imaging system memory board of a personal computer. Applying the rising current signal of the laser as a trigger signal to the control unit it is possible to use this camera in a framing

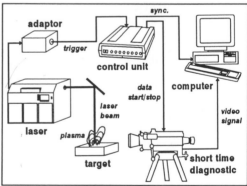

Fig. 4 Schematic drawing of the experimental setup

mode. An integrated delay generator allows to vary the delay between trigger pulse and start of exposure from 1ns to 1ms. Exposure times down to 5ns can be realized with this system.

Experimental Results

Plasma Formation and Propagation
The authors examined the formation of a gas/metal vapour plasma using the high speed diagnostics mentioned above.

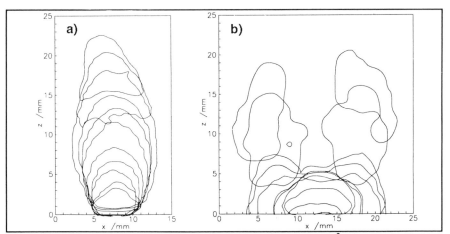

Fig. 5 Front Lines of plasma propagation for copper. **a)** vacuum, 10^{-2} mbar **b)** air, 1bar Each line traces the expanding plasma front with the highest intensity detected.

Changing both the laser parameters (energy density, number of pulses) and the processing parameters (physical properties, surface roughness and conditions, pressure of working gas) the propagation of the particles ablated by the excimer laser illumination can be investigated.

50ns after the beginning of the optical pulse an irradiation from the plasma can be detected. It is possible to study significant changes 300ns after the end of the laser pulse. In this time regime a very fast plasma broadening to the edges of the illuminated area, parallel to the targets surface, can be observed. Several μs later this motion turns into a slower

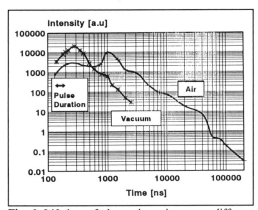

Fig. 6 Lifetime of plasma intensity at two different pressures (in air 1bar and vacuum 10^{-2}mbar)

propagation of the metal vapour directed to the original laser beam direction. Figure 5 shows the front lines of the plasma propagation in air a) and for vacuum b). Due to the reduced gas particles density the absorption length for the plasma ions and electrons in vacuum is large compared to those at normal pressure. No broadening of the plasma to the edges of the laser treated zones occurs in vacuum. The different lifetimes of the plasma corresponding to the working pressure are shown in figure 6. The maximum plasma intensity in vacuum is detected 50 - 100ns after the end of the optical pulse. Irradiating the sample in air causes a second maximum of the plasma intensity. The quasi-adiabatic expansion of the plasma wave front then leads to a slight decrease of intensity for both investigated pressures.

Influence of Laser Parameters

In the first chapter the possible processes during the interaction of intensive UV-light and metallic substrates are reported, see figure 1. Using the high speed diagnostic system it is now possible to verify this phenomenological description of the interaction mechanism. If the pulse length is kept constant and the energy density is varied different power flux densities can be applied on a copper sample with equal surface conditions.

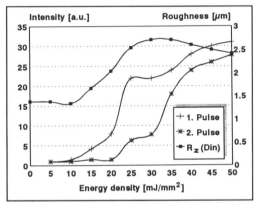

Figure 7 traces the relative plasma intensity for the first and second laser pulse and additionally the change in roughness with increasing energy density. The experimental results attained with this

Fig. 7 Plasma intensity and roughness as function of energy density for the first and second laser pulse

diagnostic method fit in the phenomenological model very well, compare figure 1 and figure 7.

Influence of Initial Surface Roughness

The influence of the initial surface conditions on the plasma formation was investigated by excimer laser treatment of copper samples of different roughness and grinding direction.

A higher roughness is strongly correlated with an increased active surface of the treated areas. This effect leads to an increased generation of metallic oxide layers and to an improved absorption behaviour of the laser light itself by multiple reflections. Figure 8 exhibits the plasma intensity as a function of the surface roughness. The maximum intensity can be detected if the initial surface roughness fits the chosen wave length of about 0,3 μm.

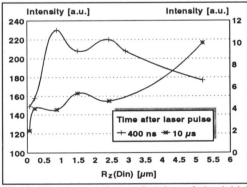

Fig. 8 Plasma intensity as function of the initial surface roughness of the treated samples

It was found that the plasma propagation and distribution is significantly influenced by the grinding direction to the applied beam and the beam geometry (rectangular or circular) itself. Perpendicular to the grinding lines on the samples surface the plasma propagation is forced into the channel between two grinding lines which leads to a high pressure on the walls. The velocity of the parallel expanding plasma is very fast compared to the perpendicular vector of this movement. The rectangular laser pulse geometry, which is commonly used for excimer laser applications, will lead to an unisotropic distribution of the pressure wave by plasma formation and has to be considered for large area surface treatments of metals.

Influence of Materials Properties

The authors investigated several different metallic materials to detect the influence of the materials properties on the lifetime of the observed plasma and its intensity.

In figure 9 on the right side the experimental results for Titanium, Aluminium, Copper, Brass (CuZn37) and Steel (St37) using different delays after the laser pulse are summarized. Up to a delay of about 2 - 4µs no significant differences between the materials can be observed.

The plasma intensity does not depend on the materials choosen but the lifetime of the plasma does. Therefore the authors

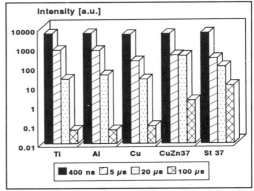

Fig. 9 Plasma intensity for different materials using four typical delays after the optical pulse

will suggest that the plasma is not only a pure metal vapour plasma but a metal vapour induced gas plasma. The formation of this special type of plasma can be possibly explained by ionization processes caused by the laser beam in the processing atmosphere itself and additionally by ionization of the working gas molecules by impact of metal ions from the ablated volume of the treated surface.

Conclusions and Outlook

The work presented in this paper outlines the effect of plasma formation and propagation on the processing quality for large area surface treatments using an high power excimer laser. The phenomenological description of the interaction of intensive UV-radiation with metallic surfaces could be proved by the experiments and the characteristic influence of both the laser parameters and the processing parameters on the plasma formation is shown.

In future there have to be more detailed studies of the phenomena, discussed by the authors, using spectrometric and time resolved methods. This may result in a further optimization of the processing parameters (type and pressure of working gas) for each of the metallic substrates, which are of great interest for industrial applications, and will lead to an improved control of the process.

References

(1)   H.W. Bergmann, E. Schubert, K. Schutte
       Surface modifications using excimer lasers; Fundamentals and applications
       Journal de Physique IV, Colloque C7, Vol.1 (1991), p. C7-7
(2)   H.W. Bergmann, K. Schutte, R. Queitsch
       Surface Treatments of Metals Using Excimer Lasers; Possible Applications for the
       Automotive Industry
       Proc. 25th ISATA '92, Conf. Laser Applications in the Automotive Industries (1992),
       p. 365
(3)   H.K. Tönshoff, O. Gedrat
       Removal process of ceramic materials with excimer laser radiation
       Proc. SPIE, Vol. 1132 (1989), p. 104
(4)   F. Bachmann
       Application of excimer lasers in an industrial fabrication line
       Proc. ECLAT '90, H.W. Bergmann and R. Kupfer eds., Sprechsaal
       Publishing Group Coburg (1990), Vol.2, p. 825
(5)   W. Turba
       Eine Video-Meßkamera erfaßt ultraschnelle Vorgänge
       LASER MAGAZIN, No.4 (1990), p. 6
(6)   W. Turba
       Bildverstärkte CCD-Meßkameras
       LASER UND OPTOELEKTRONIK MAGAZIN, Vol.22, No. 1 (1990), p. 74

Two-Dimensional Model for Materials Removal for Laser Drilling

A. Kar, J. Mazumder, Center for Laser-Aided Materials Processing, University of Illinois at Urbana-Champaign, Urbana, IL 61801, USA
T. Rockstroh , General Electric, Aircraft Engine Business Group, Evandale, OH 45215, USA

Introduction

Laser provides a contactless and an inertialess tool for carrying out drilling, which means that laser can be used for drilling fragile materials such as ceramics as well as very hard materials without being concerned about the tool wear. Also, by using laser, holes can be drilled at a very high speed, very small hole diameter and a high aspect (depth to width) ratio can be achieved, and components that have complex geometries can be drilled easily. In spite of these advantages of laser drilling, very little information on theoretical and experimental studies on laser drilling can be found in the literature. However, laser drilling can be classified as a process of material damage in a controlled manner, and in light of this, a brief review of various studies on laser-materials interaction can be found in (1). Numerous mathematical models for laser material processing can be found in (2-4). The physics of materials removal during laser irradiation and models for laser drilling have been discussed in (5-14).

The purpose of this paper to study the effects of multiple reflections and the shear stress-induced liquid metal flow caused by the assist gas on three important cavity parameters such as the cavity depth, recast layer thickness, and cavity tapering. The effects of polarization on absorptivity, and the propagation of laser beam through the liquid metal film are considered by using an effective absorptivity. In order to determine the maximum drilling rate, the absorbed laser energy is considered to only melt and vaporize the material.

Mathematical Model

The mathematical model of this study utilizes the conservation of energy, that is the Stefan condition at the solid-liquid and the liquid-vapor interfaces, in particular, at the points C, D, E, and G (Fig. 1). The details of the model and the solution procedure can be found in (1, 14). Figure 1 represents the model geometry of an axisymmetrically laser-induced cavity considered in this study. The laser beam propagates in the x-direction when the work-piece is irradiated with a laser beam, and the distance of any ray of the beam is measured along the y direction from the x-axis. The origin of the x,y coordinates system is located at the top surface of the work-piece, that is at $z = h$ where h is the thickness of the work-piece. The height and the radial distance of any point of the work-piece are measured along the z and r directions, respectively. The origin of the coordinate system (r,z) is at the bottom of the work-piece such that the work-piece is axisymmetric around the z-axis. In Fig. 1, the curves CEGH and DFG'H represent the liquid-vapor and the solid-liquid interfaces, respectively, and the curves CE and DF are taken to be parabola while EG, GH, and FH are considered to be straight lines in order to carry out interpolation for mathematically representing the interfaces. The governing equations and solutions for the location of the points D, C, E, and G can be found in (14).

Results and Discussion

Results for laser irradiation of In-718 with a spatially Gaussian and temporally triangular pulse laser are given in this paper. The duty ratio, that is the ratio of the time taken by each pulse to attain its peak energy to the pulse-on time, is taken to be 0.21 in this study. While Figs. 2-7 are based on 85% absorptivity, Fig. 8 examines the effect of absorptivity on the cavity depth. In most of the figures,

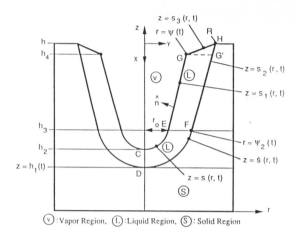

<u>Fig. 1.</u>   Geometric model of the crater formed during laser irradiation. (V): Vapor region, (L): Liquid region, (S): Solid region.

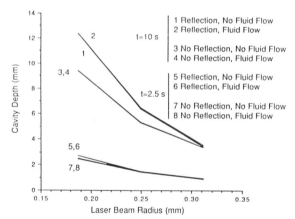

<u>Fig. 2.</u>   Effects of liquid metal flow on the cavity depth for laser power 250 W and laser pulse-on time 2.5 msec.

sharp corners are present because the points representing the numerical results are joined by straight lines instead of smoothing out the results.

Figures 2 and 3 show the effects of various laser parameters on the cavity depth and the recast layer thickness. The results for cases 1 and 2, 3 and 4, 5 and 6, and 7 and 8 are so close to each other that they almost coincide with each other suggesting that the liquid metal flow caused by the shear stress is not large enough to significantly affect the cavity parameters. This is so, because after the irradiation is carried out for a while, a very thin (about a few tens to several hundreds of micron thick) liquid metal film is formed at the solid surface of the cavity. Since the velocity of liquid is zero at the solid surface,

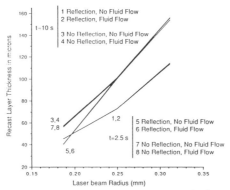

Fig. 3. Effects of liquid metal flow on the recast layer thickness for laser power 250 W and laser pulse-on time 1.7 msec.

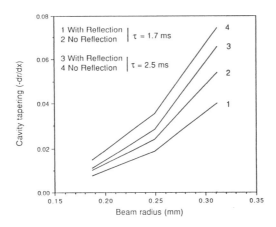

Fig. 4. Variation of cavity tapering with beam radius, at t = 10 s for laser power 250 W, and no liquid metal flow.

the velocity profile of such a thin liquid metal film would be almost zero, and consequently, the liquid metal removal rate would be negligible. However, in the initial stage of irradiation, the cavity is mainly filled with liquid metal and the rate of liquid metal removal by the assist gas is relatively high.

Figure 4 shows the results for cavity tapering, which is obtained for various laser parameters at the irradiation time, t = 10 s, and is determined at the middle of the points C and E, that is at $z = (h_2+h_3)/2$ by using the equation of the parabola CE. Mathematically, it is calculated by using the expression, $-dr/dx = r_0/\{\sqrt{2}\,(D_2-D_3)\}$, where r_0 is the radius of the point E, and D_2 and D_3 are the depths of the points C and E, respectively, that is, $D_2 = h-h_2$ and $D_3 = h-h_3$. It should be noted that the cylindricity of the cavity depends on the cavity tapering in such a way that the smaller the cavity tapering, the more vertical the side wall of the cavity would be.

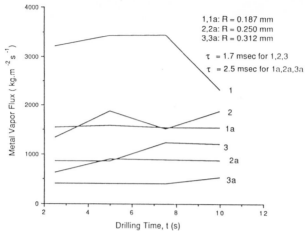

Fig. 5. Metal vapor flux at the bottom of the cavity when laser power is 250 W and multiple reflections are considered.

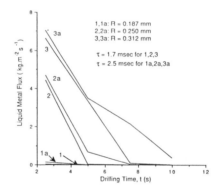

Fig. 6. Liquid metal flux at the bottom of the cavity when laser power is 250 W and multiple reflections are considered.

Figures 5 and 6 represent the mass of the material removed per unit area per unit time at the bottom of the cavity, that is at the point C (Fig. 1). In these figures, the results represented by the curves 1 and 1a, 2 and 2a, and 3 and 3a are for laser beam radii 0.187, 0.25, and 0.312 mm, respectively. Also, the curves 1, 2, and 3; and 1a, 2a, and 3a correspond, respectively, to the results obtained with lasers of pulse-on times 1.7 and 2.5 msec. Figures 5 and 6 show the metal vapor and the liquid metal fluxes, respectively, for laser power 250 W with the beam undergoing multiple reflections inside the cavity. It can be seen from these figures that the metal vapor flux is several orders of magnitude larger than the liquid metal flux. Also, the liquid metal flux drops to almost zero within a very short period of time.

Figure 7, which is based on the numerical results up to 10 s after the irradiation is initiated, shows how the cavity depth varies with the gross laser intensity, I_g, which is defined as $pt/(pR^2t)$, where p is the

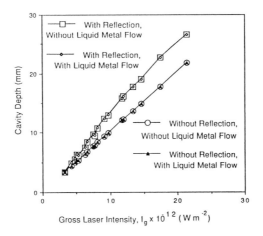

Variation of the cavity depth with the gross laser intensity.

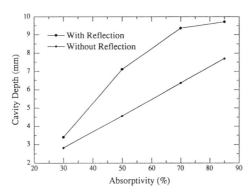

Fig. 8. Variation of the cavity depth with absorptivity at drilling time, t=10 s in the absence of liquid metal flow, when the laser power, beam radius, pulse-on time, and the number of pulses per second are 250 W, 0.25 mm, 1.7 ms, and 4 respectively.

laser power, t is the irradiation time, R is the beam radius, and t is the laser pulse-on time. The upper curve represents the results obtained by considering multiple reflections of the laser beam in the presence as well as in the absence of the liquid metal flow, and the lower curve is for the case of cavity formation without any reflections. The cavity depth obtained by considering no reflections of the laser beam varies linearly with the gross laser intensity, whereas this variation becomes nonlinear when multiple reflections are taken into account in the mathematical model. Finally, Fig. 8, which is plotted by determinig the cavity depth for 30%, 50%, 70%, and 85% absorptivities, shows the variation of the cavity depth with absorptivity.

Conclusions

A simple two-dimensional axisymmetric model is developed for laser-induced materials removal by taking multiple reflections of the laser beam inside the cavity, and shear stress-induced liquid metal flow into account. The conclusions of this study can be summarized as follows:
(a) The laser-induced cavity depth increases as the laser intensity increases. Multiple reflections lead to the formation of deeper cavities than when multiple reflections are absent.
(b) The recast layer thickness decreases as the laser intensity increases. Multiple reflections generate thinner recast layer than when reflections are not taken into account.
(c) The cylindricity of the laser-induced cavity increases as the laser intensity increases. The cavity is more cylindrical in the presence of multiple reflections than in the absence of any reflections.
(d) The effects of the liquid metal flow, which is caused by the shear stress at the liquid-vapor interface, on the cavity parameters such as the cavity depth and the recast layer thickness are found to be insignificant in this study. The rate of material removal due to vaporization is found to be several orders of magnitude higher than that due to the liquid metal flow.
(e) The cavity depth varies linearly with the gross laser intensity when reflections of the laser beam inside the laser-induced cavity are not taken into account. However, this variation is found to be nonlinear when multiple reflections are considered in the mathematical model.

Acknowledgment

This work was performed under a contract (Purchase Order No. 200-14-14U11788) from the General Electric Aircraft Engines Business Group. Continued encouragement from Dr. P. J. E. Monson and his helpful suggestions are appreciated.

References

(1) A. Kar and J. Mazumder: J. Appl. Phys. 68 (1990) 3884.
(2) W. W. Duley: "Laser processing and analysis of materials", Plenum Press (1983).
(3) M. Bass: "Laser material processing", North-Holland (1983).
(4) J. Mazumder: "Laser Surface Treatment of Metals", Martinus Nijhoff (1986) 185.
(5) C. L. Chan and J. Mazumder: J. Appl. Phys. 62(1987) 4579 .
(6) E. Armon, Y. Zvirin, G. Laufer, and A. Solan: J. Appl. Phys. 65 (1989) 4995.
(7) E. Armon, M. Hill, I. J. Spalding, and Y. Zvirin: J. Appl. Phys. 65 (1989) 5003.
(8) J. F. Ready: J. Appl. Phys. 36 (1965) 462.
(9) M. V. Allman: J. Appl. Phys. 47 (1976) 5460.
(10) J. G. Andrews and D. R. Atthey:J. Inst. Math. Appl. 15 (1975) 59.
(11) M. G. Jones, G. Georgalas, and A. Brutus: "Laser in Materials Processing", American Society of Metals (1983) 159.
(12) M. K. Chun and K. Rose: J. Appl. Phys. 41 (1970) 614.
(13) G. Kinsman and W. W. Duley: Appl. Phys. Lett. 56 (1990) 996.
(14) A. Kar, T. Rockstroh, and J. Mazumder: J. Appl. Phys. 71, 15 March (1990), in press.

Laser Induced Thermal Shock Cracks in High Temperature Materials inside a Scanning Electron Microscope

S. Menzel*, K. Wetzig*, J. Linke**
* Institut für Festkörper- und Werkstofforschung Dresden e. V.,
  Dresden, Bundesrepublik Deutschland
** Forschungszentrum Jülich GmbH,
  Jülich, Bundesrepublik Deutschland

1. Introduction

Carbon based high temperature materials are mostly used for the protection of highly loaded areas of the plasma facing side of the first wall of nuclear reactors (1). Both fine grain graphite and carbon fibre reinforced carbons (CFC materials) possess an excellent thermal shock behaviour. For materials characterization and data basing the fine grain graphite EK 98 (Ringsdorff-Werke, Germany) and the CFC-materials FMI were tested in laser exposure tests inside a laser coupled scanning electron microscope (LASEM) (2). The schematic section of LASEM and experimental parameters are discussed in a further article of these proceedings (3).

Here a free running Nd-YAG-laser with maximum pulse energy 12 J and a pulse duration time 10 ms was used. The laser beam was focussed to a spot area of 0,1 - 30 mm^2. For detection of cumulative effects the pulse number could be varied. Characteristic parameters of tested materials, published by Koizlik et al.(4), are given in Tab. 1.

Surfaces of radiated materials are observed inside a SEM in order to study thermal shock crack processes. After ion beam slope cutting (IBC) (5) a vertical analysis of crack formation was possible. An image processing system was used to analyze characteristic crack network parameters.

Tab. 1: Characteristic parameters of tested materials

material	reasons for use	limitative factors
fine grain graphite	- low Z value	- thermal shock behaviour
	- knowledge from HT reactor	- thermal conductivity λ
		- strength value σ
carbon fibre reinforced carbon (CFC)	- excellent thermal shock behaviour	- chemical sputtering
	- high λ value	- radiation enhanced sublimation
	- high σ value	

2. Results and Discussion

2.1. Thermal shock crack formation

The material EK 98 shows an isotropic thermal conductivity. By laser irradiation mostly matrix material and also material from pores which was incorporated by preparation process is vaporized resulting in an increased porosity of the spot area (see Fig.1).

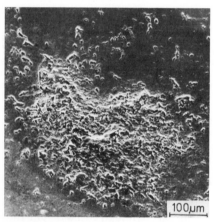

Figure 1:
Increased porosity at graphite after laser surface irradiation; $j = 0,3$ J mm^{-2}

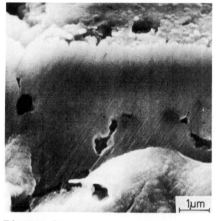

Figure 2:
Ion beam slope cutting (5) at an irradiated graphite EK 98 surface; $j = 0,3$ J mm^{-2}

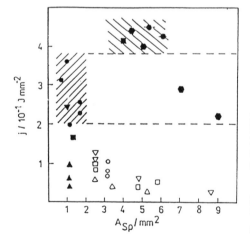

Figure 3:
Damage states for graphite EK 98 surface, irradiated by different laser beam energies and focussing conditions
(o, ⬡ full energy;
▽, □, △ decreased energy by dimming)
full points: crack formation
///// : lamellar structure
\\\\\ : macroscopic cracks

The cross section of irradiated materials surface after IBC shows also high porosity and deep thermal shock cracks (see Fig. 2). Damaged states of the material EK 98 are visible in Fig. 3. Thermal shock cracks exist only above a critical value of laser pulse energy density and beneath a critical spot area value. Crack formation begins with a microscopic crack in material, followed by crack growth in length and width (Fig.4). At an energy density value of about $2 \cdot 10^{-1}$ J mm^{-2} lamellar crack structures in graphite are visible (see Fig. 5). At high energy density values ($j = 3,8 \cdot 10^{-1}$ J mm^{-2}) macroscopic crack structures with radial and tangential orientation can be evaluated by image processing. It can be noted that typical dimensions as length, distance and orientation of cracks depend on the angle of incidence of the laser beam.

Damaged states of carbon fibre reinforced carbon (FMI) are essentially determined by multidirectional reinforced fibres (diameter - 10 μm). It results in anisotropic material and consequently in anisotropic damage characteristics. Irradiation was realised parallel and perpendicular to fibres orientation. Both crack formation and crack growth begin in binder material (see Fig. 6a). At high energy densities ($j > 0,4$ J mm^{-2}) or

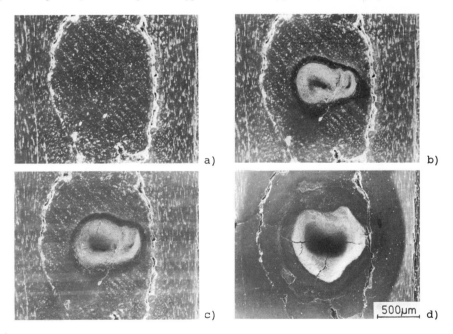

a)   b)   c)   500µm d)

Figure 4:
Damages at polished CFC material FMI after laser irradiation
a) unradiated; b) 1 pulse; c) 2 pulses; d) 100 pulses;
$j = 0,75$ J mm^{-2}

664

Figure 5:
Lamellar crack structure
at irradiated graphite
EK 98 surface
$j = 0,3$ J mm^{-2}

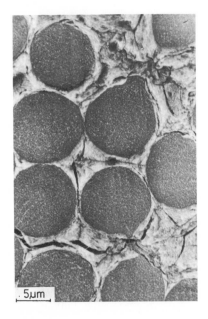

a)

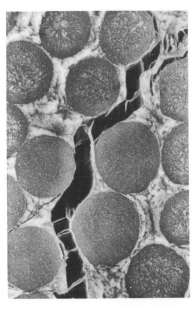

b)

Figure 6:
Thermal cracks in CFC material FMI
a) formation and growth in binder phase ($j=0,4$ J mm^{-2}, 1 pulse)
b) macroscopic crack at binder-fibre-boundary
   ($j=0,75$ J mm^{-2}, 10 pulses)

multiple irradiation macrosopic cracks are mostly located at
binder-fibre-boundaries (see Fig. 6b). At high beam energy
densities often cracks with a radial structure in fibre cross
section are visible (see Fig. 7). By this way high tangential
stresses in fibres may be compensated.

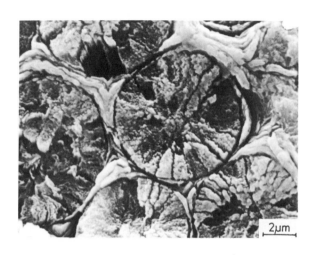

Figure 7:
Radial crack forma-
tion in fibres of CFC
material FMI after
10 laser pulses;
j = 0,75 J mm^{-2}

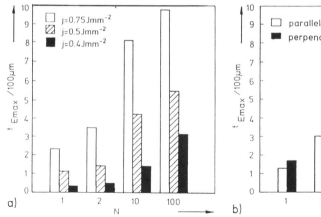

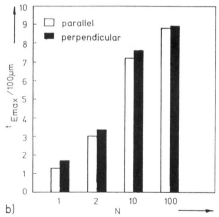

Figure 8:
Maximum depth t_{Emax} of erosion craters at EK 98 (a) - and FMI (b)
- surface as a function of laser beam energy density j, pulse
number N and materials orientation

2.2. Erosion

From irradiated surfaces material is evaporated and erosion craters are observed. At the isotropic material EK 98 the crater profile corresponds with the incident beam energy density profile, but at CFC material FMI the crater profile is also influenced by the anisotropic thermal conductivity.

Quantitative erosion depth t_{Emax} as a function of pulse number N is shown in Fig. 8. The erosion depth for $j = 0,5$ J mm^{-2} amounts to about 100 μm/pulse. It should be noted that a saturation in erosion is obtained which was also detected by Linke et al (6). This process is not quite understood, but effects like redeposition, change in reflectivity and increase of the effective spot area after irradiation may cause this behaviour.

The binders erosion is higher than that of the fibres but at high energy densities a more and more homogeneous erosion takes place ("smooth effect"). Irradiation perpendicular to the fibre orientation generally results in deeper erosion craters because of the lower thermal conductivity across the fibre.

References:
(1) H. Bolt et al.: Bericht der KFA Jülich, Jül-2086 (1986)
(2) K. Wetzig et al.: Scanning 9 (1987) 99.
(3) K. Wetzig et al.: published in this volume
(4) K. Koizlik et al.: Proc. 16th Symposium on Fusion Technology London, Sept. 1990.
(5) W. Hauffe et al.: Prakt. Metall. 25 (1988) 517.
(6) J. Linke et al.: Proc. 10. PSI-Conf., Monterey, Apr. 1992 (in press)

Effect of Ambient Pressure and Shielding Gas on Penetration Depth and CO_2 Laser-induced Plasma Behavior

M.Kabasawa, M.Ono, K.Nakada and S.Kosuge

Steel Research Center, NKK Corporation, Kawasaki-ku, Kawasaki 210, Japan

1. Introduction

Laser welding is performed by focusing a high power laser beam onto a material surface. This focused and highly collimated heat source evaporates some material to be welded and creates a keyhole filled with metal vapor through the thickness of the workpiece, where molton metal surrounds the keyhole. Vaporized materials, together with gases used to shield the weld pool, are heated up to a extent of excitation state with partial ionization, so-called a plasma forms not only in the keyhole, but also above the workpiece. For a given material to be welded, the quantity of plasma is dependent upon the kind of shielding gas as well as the laser beam radiation conditions such as output power, focused spop size and welding speed.

It was reported that a strong plasma was generated above the workpiece when welding at lower speed below 50 cm/min using Ar shielding gas (1). The plasma formed above the workpiece tends to absorb considerably the laser beam entering to the workpiece and reradiates it in all directions, thereby resulting in the reduction in the penetration capability of the beam (2)(3). That is why the penetration depth achieved in laser welding is much smaller as compared with that in vacuum electron beam welding particularly at low welding speed.

A high velocity herium jet is usually directed to weld surface in order to blow away plasma (4). In addition, the other techniques have been developed to suppress the plasma formation, including a extracting process which evacuate plasma using a nozzle connected with a vacuum pump and a laser beam pulsation process with a high frequency so that the duration of each pulse is shorter than the time necessary to generate plasma. However, it is difficult to completely suppress the plasma formation above the workpiece and impossible to evacuate plasma generated in the keyhole with these techniques.

In this paper, in order to solve such a problem, laser welding at low ambient pressure was conducted using a vacuum chamber capable of varying the ambient pressure from 10^{-2} to 760Torr. It is expected that this process would be significantly effective to evacuate plasma in the keyhole as well as above the workpiece. Photoluminescence spectroscopy of laser-induced plasma was examined in the range of 190-400 nm.

2. Experimental procedure

12mm-thick austenitic stainless steel (AISI 304) and pure iron was used as listed in Table 1. Welding was carried out in a vacuum chamber. Vacuum chamber consists of a vacuum pump system, a work table, a ZnSe window and a beam-guided nozzle with a device developed for protecting a ZnSe window from fume and sputter. The pressure inside the vacuum chamber can be varied from 10^{-2} to 760 Torr by controlling the flow rate of shielding gas supplied from the top of beam-guided nozzle.

Bead-on-plate melt runs were conducted to examine the effect of the ambient pressure on the penetration behavior and the plasma formation with 4kW CO_2 laser. The laser is a transverse flow machine using an unstable optical resonator of which beam has an annular configuration. The welding trials were performed under the laser radiation conditions as listed in Table 2.

Laser beam was brought to a focus just below the top surface of workpiece. After welding, transverse sections of the welds were taken to examine the weld shape and the weld defect.

Photoluminescence spectroscopy of laser-induced plasma was examined in the range of wavelength 190-400nm by a 0.25 meter focal length spectrometer. An optical multichannel analyzer was used for recording the spectra.

Table1 Chemical composition of base metal(wt%)

	C	Si	Mn	Ni	Cr	Fe
AISI304	0.05	0.48	1.20	8.76	18.4	bal.
Pure Fe	–	–	–	–	–	99.9

Table2 Laser radiation conditions

Laser power	4.0kW
Focal point	0 mm(Focal length 508mm)
Welding speed	0.25~5.0m/min
Ambient gas	He, N_2, Ar
Ambient pressure	0.1~760Torr

3. Results and dicussions

3.1 Effect of ambient pressure on penetration

Test results obtained were shown in Fig.1 and Fig.2 together with selected macrosections in Fig.3 when welding in He shielding gas. It was found that the penetration depth was significantly affected by the ambient pressure particularly at a welding speed of less than 1m/min, and it increased abruptly by reducing the pressure down to the level of less than 100Torr. The penetration depth at the pressure below 100Torr reached more than 12mm at a welding speed of 25cm/min, which was around twice as deep as that in an atmospheric pressure.

The results when using N_2 or Ar shielding gas are given in Fig.4 and Fig.5 together with representative macrosections in Fig.3. The penetration depth reached 11.5mm at Ar gas pressure of 0.1Torr and a welding speed of 25cm/min, while only a shallow penetration of approximately 0.5mm was obtained under 760Torr atmospheric pressure as shown in Fig.3. The critical ambient pressure corresponding to level-off were 10Torr in N_2 gas and 0.1Torr in Ar gas, while 100Torr in He gas.

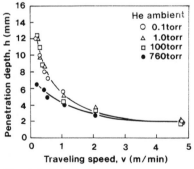

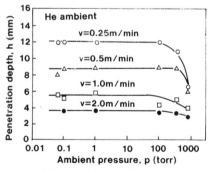

Fig.1 Relationship between penetration depth,welding speed and ambient pressure.

Fig.2 Effect of He ambient pressure on penetration depth and welding speed

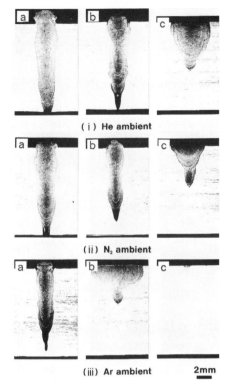

(i) He ambient

(ii) N₂ ambient

(iii) Ar ambient   2mm

Fig.3 Macrosections. Laser radiation
condition; P=4.0kW, v=0.25m/min
(a)p=0.1Torr,(b)p=100Torr,(c)p=760Torr

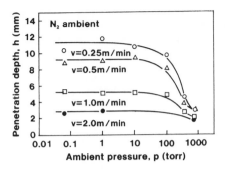

Fig.4 Effect of N_2 ambient pressure
on penetration depth and welding
speed

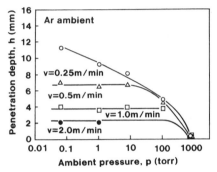

Fig.5 Effect of Ar ambient pressure
on penetration depth and welding
speed

This reduction in ambient pressure had another benefit that porosity in
weld metal remarkably decreased due to the improvement of weld shape related
to the solidification stability.   It is mainly because the promotion of metal
vaporization by reducing the ambient pressure has a remarkable effect on
preventing the molton walls around the keyhole from collapsing, resulting in
the improvement of the keyhole stability.   At higher welding speeds of more
than 2m/min, the effect of pressure on the penetration depth was found very
little due to a small quantity of plasma formation.

3.2 Laser-induced plasma

The laser-induced plasmas forming at various pressures of 0.1 - 760Torr
in He, N_2 and Ar gases are shown in Fig.6.   As is evident from this figure,
the visual size of plasma was Ar ≫ N_2 > He gas.   The plasma generation was
dependent mainly upon the physical properties of gas such as heat conductivity
and ionization potential.   Higher heat conductivity and ionization potential
gave rise to smaller plasma.

As a result, He gas which possesses the highest ionization potential and
heat conductivity among these gases was the most suitable for laser welding.
The size of plasma decreased as the ambient pressure decreased.   The plasma

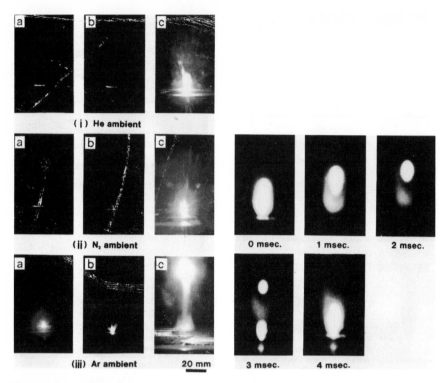

Fig.6 Plasma behavior induced by laser radiation. Laser radiation condition; P=4.0kW, v=0.25m/min, (a) p=0.1Torr, (b) 100Torr, (c) 760Torr

Fig.7 Film frames of the development of the laser-induced plasma

in He and N_2 gas completely disappeared at the pressures of less than 100Torr, resulting in a significant increase in penetration to virtually the same depth as that achieved in electron beam welding. However, a small amount of plasma was still observed in Ar gas at the pressure of 100Torr.

As can be seen from Fig.7, the cyclic laser-induced plasma formation behavior, including plasma generation, disappearance and regeneration, was obviously observed under atmospheric pressure of Ar gas. The first visible plasma in the frame of t=0msec initiated close to the surface of workpiece, and then climbed up to 40mm above the workpiece at a rising speed of approximately 20m/sec, corresponding to the frame of t=2msec. When the laser plasma was floating above the workpiece, the floating plasma blocked the laser beam entering to the workpiece, as shown in the frame of t=3 msec. As the laser plasma rose, the energy in laser beam decreased because the plasma was irradiated by far defocused beam with lower power density at its position. Thus, finally the floating plasma completely disappeared. Then quantity of laser beam possible to enter to the workpiece remarkably increased due to very little laser power loss, resulting in regeneration of laser plasma close to the surface of workpiece.

This plasma formation phenomenon was repeated periodically at an interval of 3 msec. Such a sequence was not observed when welding in He or N_2 shielding gas where the laser plasma formed only close to the surface of the workpiece.

3.3 Spectroscopic Analysis of Laser-induced plasma

From the results of the spectroscopic analysis of laser-induced plasma, its formation mechanism could be considered as follows. Initially, metal vapor is generated in the keyhole and a large amount of vapor escape from the top of keyhole. Then, thermal electrons would be generated from metal vapor. The metal vapor is considered to have a greater probability for electron criation during laser welding due to its lower ionization potential as compared with ionization potential of gas element.

When welding pure iron, some Fe vapor was found to be ionized as shown in Fig.8. The intensity of Fe ionization was dependent strongly upon the kind of shielding gas. Since the order of ionization intensity was Ar $\gg N_2 >$ He, He gas was the most suitable to suppress the Fe ionization because its very high thermal conductivity could prevent the metal vapor from ionizing. These results supported the above assumption that electrons would be emitted from the metal vapor plasma consisting of electrons, ions and neutral atoms.

Subsequently, electrons collide with the gaseous molecules, which leads to heat them up to its ionization temperature. Although CO_2 laser light with 10.6µm wavelength are all transparent for these gases in the ground state, it can be absorbed by such gases in the state of excitation with partial ionization, resulting in an enhancement of gas ionization.
The ionization intensity of Ar gas was found to be stronger as shown in Fig.8, while the ionization of He gas was observed very little.

The plasma temperature estimated by a line intensity ratio method was about 13000K irrespectively of the kind of gas element as shown in Fig.9. The electron density of laser-induced plasma was estimated by Saha equation was about $10^{17} cm^{-3}$.

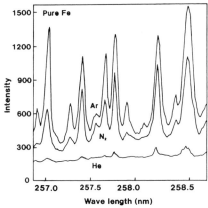

Fig.8 Spectrum analysis of plasma induced by laser radiation onto pure Fe substrate in He, N_2, Ar ambient

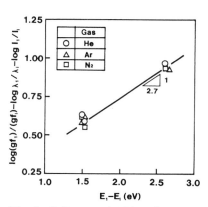

Fig.9 Boltzman plotting for measuring plasma temperature

It was concluded that He gas was the most beneficial to decrease the amount of laser plasma which consisted of the metal vapor and shielding gas in the excitation state with partial ionization. In addition, it would be assumed that the reduction in ambient pressure is very effective to completely suppress the plasma formation due to the elimination of plasma media vapour and shielding gas.

4. Conclusion

The effect of ambient pressure on behavior and weld defects for AISI 304 stainless steel and pure iron in CO_2 laser welding with 4kW output power. In addition, the laser plasma formed just above the workpiece was examined by spectroscopic analysis.

1) The penetration depth significantly increased by reducing the ambient pressure. At the same time, the quantity of the laser-induced plasma above the workpiece also decreased as the ambient pressure decreased.
   The plasma in He, N_2 and Ar ambient completely disappeared under the pressure of 100, 10 and 0.1Torr, respectively, resulting in the significant increase in the penetration depth.
   The penetration depth reached more than 12mm under the pressure of less than 100Torr in He ambient at a welding speed of 25cm/min, which was around twice as deep as that under the atmospheric pressure.
   It was mainly because the formation of plasma which blocked the laser beam entering to the workpiece was greatly suppressed under the low pressure.
2) The plasma generation was dependent upon the physical properties of gas element as well as the ambient pressure.
   The higher the ionization voltage and heat conductivity of the gas element was, the larger the plasma generated.
3) Laser-induced plasma consisted of rich emission lines of excited and ionized metallic and gas element. The intensity of atomic emission lines from the plasma was dependent upon the physical properties of the gas element such as ionization potential and heat conductivity.
   The plasma temperature estimated by a line intensity ratio method was about 13000K irrespectively of the kind of gas elements. The electron density of laser-induced plasma was about $10^{17}cm^{-3}$.

References

(1) Y.Arata,N.Abe et al: Fundamental Phenomenon during Vacuum Laser Welding, EBW-381-86, 48th committee of EBW, The Japan Welding Society.
(2) G.K.Lewis and R.D.Dixon: Plasma Monitoring of Laser Beam Welds, Welding Research Supplement, February, 1985.
(3) R.D.Dixon and G.K.Lewis: Electron Emission and Plasma Formation during Laser Beam Welding, Welding Research Supplement, March, 1985.
(4) M.N.Watson:Laser Welding of a 12.5mm thick Niobium Microalloyed Structural Steel at Powers of up to 9kW, TWI Report, July, 1984.

Materials processing using a combination of a TEA- and a cw-CO$_2$-laser

R. Jaschek, R. Taube, K. Schutte, A. Lang, H.W. Bergmann

Friedrich-Alexander-Universität Erlangen-Nürnberg, Institut für Werkstoffwissenschaften 2, Metalle, Martensstr. 5, 8520 Erlangen, FRG

1. Abstract

The radiation from a CO$_2$-laser is highly reflected by metallic surfaces. Additional coatings of graphite or ceramic suspensions increase the absorption during laser beam transformation hardening and hence the process efficiency. High absorption is also observed, when an increased energy density creates a plasma and a key-hole is formed during laser beam welding. However melting of the surfaces is not tolerable during laser beam transformation hardening. Therefore the following process is proposed. A plasma is generated using a TEA-CO$_2$-laser with a pulse peak power of 3,6 MW, a pulse width of 50 ns and a mean power of 18 W. This plasma is used to increase the absorption of a cw-CO$_2$-laser during laser beam transformation hardening without any harmful melting of the surface. An increased case depth could be achieved during laser beam transformation hardening of heat treatable steel (DIN C 45 and DIN C 60) without pre-oxidizing the metal surface or using additional coatings. No further oxidation of the surface could be observed after the laser beam transformation hardening process. The beam of the cw-CO$_2$-laser radiating at a wavelength of 10,6 μm and that of the TEA-CO$_2$-laser (9,6 μm) are made coincident by use of a copper mirror with a diffraction grating. The described system allows beam shaping and beam focusing with the same optics. Characteristics of the beam-solid interactions are detected by using spectroscopic and short time diagnostic methods. The short time photography gives information about the formation, the direction and the lifetime of the plasma initiated from the TEA-CO$_2$-laser. Running both lasers simultaneously the changes in plasma are determined.

2. Introduction

The process efficiency during laser beam transformation hardening is very low due to the poor absorption of metallic surfaces for the wavelength of the CO$_2$-laser. In the past many physical methods have been developed to increase the absorption :

a) An increased work piece temperature raises the electrical resistance and therefore the absorption [1]. The maximum temperature for each individual steel is limited by the martensitic start temperature.

b) The absorption can also be increased using dielectric coatings (e.g. an oxide layer) or non conductive coatings of isolated particles (e.g. graphite or ceramic suspension) [2,3]. Two more steps are needed in this process, because the work pieces have to be coated and the coating has to be removed after the laser treatment. Beside the case depth is strongly dependent on the thickness of the coating.

c) A polarized laser beam at the Brewster angle enlarges also the absorption [4,5,6]. The distortion of the laser beam may cause some trouble during laser beam hardening of work pieces with a geometrically complex shape.

d) A higher absorption is observed, when an increased energy density creates a plasma and a key-hole is formed, e.g. during the laser beam welding process [7,8,9]. To achieve a higher absorption of laser radiation during surface treatment of metals, the formation of a plasma would also be desirable in this case. Caused by the high intensities necessary, this effect can not be used for surface treatments of materials in the solid state, e.g. laser beam transformation hardening process because excessive melting of the surfaces is not tolerable.

3. Experimental setup

The investigations were carried out using a commercial TEA-CO$_2$-laser. This laser is of importance in many technological and scientific applications such as spectroscopy, drilling and cutting, as well as marking of glass and other materials [10]. Table 1 summarizes the characteristic parameters of this TEA-CO$_2$-laser.

Table 1: Specifications of the TEA-CO$_2$-laser

λ	pulse energy	P	P$_{max}$	repetition rate	pulse duration	efficiency
10,6(9,6) μm	180 mJ	18 W	3,6 MW	100 Hz	50 ns (FWHM)	~7%

The beam of the cw-CO$_2$-laser radiating a wavelength of 10,6 μm and that of the TEA-CO$_2$-laser (9,6 μm) can be made coincident by use of a copper mirror with a diffraction grating. The described system, which is shown in Fig. 1, allows beam shaping and beam focusing by the same optics.

The investigated metals for laser beam transformation hardening are the heat treatable steel DIN C 45 and DIN C 60, which are grounded and cleaned with acetone prior to each experiment. Nitrogen is used as a shielding gas to avoid oxidation of the steel surface. A plasma is generated on the surface of the work piece using the TEA-CO$_2$-laser. This plasma is used to increase the absorption of a cw-CO$_2$-laser during laser beam transformation hardening [11].

Characteristics of the beam-solid interactions are detected by using spectroscopic and short time diagnostic methods. The short time photography gives information

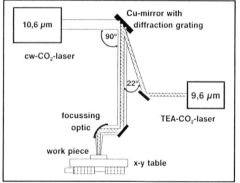

Figure 1: Combination of a TEA- with a cw-CO$_2$-laser

about the formation, direction and lifetime of the plasma, which is induced from the TEA-CO$_2$-laser on the surface of a DIN C 45 steel at room temperature. Running both lasers simultaneously changes in plasma development are determined. The experimental set up is shown as a block diagram in Fig. 2. A DiCAM camera from the company PCO Computer Optics is used for short time photography. DiCAM is a combination of a high resolution solid-state two-dimensional CCD sensor and an ultra-fast gatable image intensifier which yields an optical measuring and imaging system with a pixel number of 756 x 580.

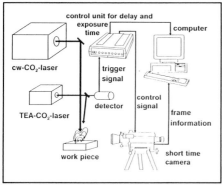

Figure 2: Block diagram of the experimental setup

This system covers a wide range of exposure times from 5 ns to 1 ms and incoming intensities. Fast and transient events can be detected with a spectral resolution between 200 nm and 810 nm. During these experiments the exposure time was 10 ns. The delay between trigger signal from a pyroelectric detector and control signal for the camera ranges between 1 ns and 1 ms. A mirror specially coated transmits 20 % of the radiation from the TEA-CO_2-laser to the detector. The other 80% of the radiation are reflected onto the surface of the work piece. In addition the spectra of the plasma radiation generated by the TEA-CO_2-laser on different materials like copper and DIN C 45 with various shielding gases like argon and nitrogen are recorded with an optical spectrometer between the wavelengths from 300 nm to 700 nm.

4. Experimental results

4.1 Laser beam transformation hardening

At room temperature an increased case depth could be achieved during laser beam transformation hardening of DIN C 45 (Fig. 3) and DIN C 60 (Fig. 4) without oxidizing the metal surface or using additional coatings.

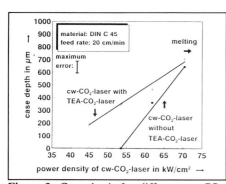

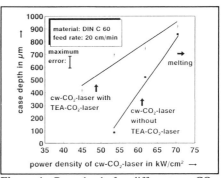

Figure 3: Case depth for different cw-CO_2-laser power intensities

Figure 4: Case depth for different cw-CO_2-laser power intensities

For DIN C 45 a higher absorption is observed between 45 kW/cm^2 and 70 kW/cm^2 when both laser beams are directed to the same spot on the surface. When the cw-CO_2-laser beam intensity is higher than 70 kW/cm^2, undesired melting of the metal surface is seen leading to different conditions of heat conduction within the material. Even at higher intensities of the cw-CO_2-laser beam a plasma generated from the TEA-CO_2-laser could not be observed anymore. The same effects are also observed for the DIN C 60 with an even increased case depth.

4.2 Short time photography

Fig. 5 shows the characteristic parameter used for both lasers during the experiments with short time photography. The beam of the TEA-CO$_2$-laser is directed at an angle of 30° to the surface of the work piece focusing on the target surface. The scale in Fig. 5 is the same as in Fig. 6 and Fig. 7 which show the different framing images obtained with the DiCAM camera for different delays. The closed lines display the same radiation intensities. The increase of intensity between each line is scaled in equal steps with the highest intensity in the

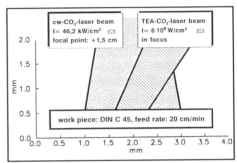

Figure 5: Characteristic parameters

middle of the plasma. Plasma formation is detected 50 ns after each optical laser pulse in a horizontal and vertical direction. During the beginning of the plasma development the vertical and horizontal velocities have a value of $3,8\cdot10^3$ m/s and $1,4\cdot10^3$ m/s, respectively. The velocity of the plasma formation will decrease after 100 ns and after 200 ns the velocity of the vertical plasma spreading is even zero. The existence of a LSD wave is expected due to the supersonic velocity of the plasma spreading.

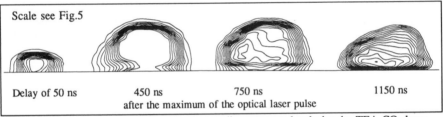

Figure 6: Framing images of the plasma spreading generated only by the TEA-CO$_2$-laser

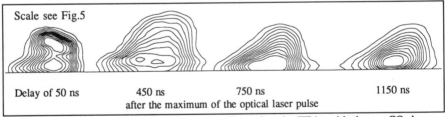

Figure 7: Framing images of the plasma spreading using the TEA- with the cw-CO$_2$-laser

The plasma stays in close proximity to the surface thus providing a good thermal coupling between the plasma and the target surface. Even for the TEA-CO$_2$-laser pulse intensity of $6\cdot10^8$ W/cm^2 and a FWHM of 50 ns the plasma does not detach from the target. The area of the TEA-CO$_2$-laser beam is small in comparison with the area covered by the plasma, which is then almost identical with the area of the CO_2-laser beam. Fig. 8 exhibits the average intensity of the plasma radiation as a function of time. The plasma spreading starts right with the steep increase of the optical laser pulse. The plasma generated by the TEA-CO_2-laser has first a lower average intensity than the plasma induced by the combined lasers. This effect changes after 250 ns, when the average intensity of the plasma using both lasers is then lower than the intensity of the plasma using the TEA-CO_2-laser itself. These values stay in good comparison with the framing images in Fig. 6 and Fig. 7. The maximum of the average inten-

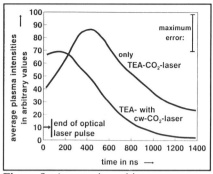

Figure 8: Average intensities

sity of the plasma radiation is at 450 ns for the TEA-CO_2-laser, using both lasers the maximum value is already observed at 200 ns. The duration of the plasma propagation is much longer than the duration of the optical laser pulse.

4.3 Plasma spectroscopy

Fig. 9 and Fig. 10 show the spectra obtained from the plasma radiation for different materials and shielding gases. No changes in the spectra are observed when the TEA-CO_2-laser creates a plasma on the surface of various metals like DIN C 45 or copper. A contrast in the spectra is only dedected when different shielding gases are used on the same material. The plasma with argon as shielding gas has a higher intensity and a typical shape with the line for the ionized argon atoms. These results yield the conclusion that mainly ionized atoms from the shielding gas build up the plasma. With the additional cw-CO_2-laser a change of the spectra could not be confirmed.

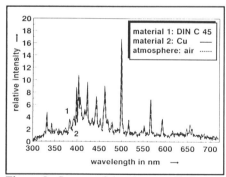

Figure 9: Spectra of the plasma radiation

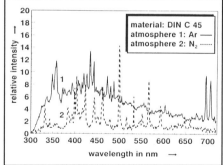

Figure 10: Spectra of the plasma radiation

5. Discussion and Conclusions

When the intensity of a laser beam exceed a critical value a plasma is induced on the surface of the work piece. The optical properties of the material and the absorption change depending on the electron density and the gas density of the plasma [4,12]. A plasma is generated by a TEA-CO_2-laser with a laser intensity of $6 \cdot 10^8$ W/cm^2. This effect is used to increase significantly the absorption of a cw-CO_2-laser during laser beam transformation hardening using nitrogen as shielding gas. An increased case depth could be achieved during hardening of heat treatable steel (DIN C 45 and DIN C 60) without oxidizing the metal surface or using additional coatings. This result shows that this plasma could be partly transparent for the radiation of the cw-CO_2-laser. In literature, three mechanisms can be found to provide an additional heat flux from a near surface plasma to a solid target described as the so called "enhanced thermal coupling" effect: **a)** The temperature of a near surface plasma is typically equal to 1 - 3 eV. Energy from the plasma can be transfered to the solid material by normal electron heat conduction due to the high temperature gradient between the plasma and the target [13]. **b)** The thermal plasma radiation in the short-wavelength region is efficiently absorbed by the metal surface [13,14]. **c)** The pressure at the target surface is strongly enhanced by the recoil of the expanding plasma [7] and can reach values in the range between 100 - 1000 bar [12]. Due to the effects mentioned above the content of thermal energy within a thin surface layer may increase significantly for a short time. Then it seems possible that the number of defects within the crystal structure is also increased, because the energy needed to produce lattice defects like radiation induced vacancies are only in the range of 1 eV [15]. These effects could be a possible explanation for the better absorption of the radiation from the additional cw-CO_2-laser and will be further investigated.

6. References

[1]   R. Hummel: "Optische Eigenschaften von Metallen", Springer-Verlag, Berlin, 1971
[2]   A. Zwick, A. Gasser, E.W. Kreutz, K. Wissenbach: "Surface remelting of cast iron camshafts by CO_2 laser radiation" in *Proc. of ECLAT'90*, edt. by H.W. Bergmann, R. Kupfer, Erlangen, 1990, p.389-398
[3]   J. Bach, R. Damaschek, E. Geissler, H.W. Bergmann: "Laser transformation hardening of different steels" in *Proc. of ECLAT'90*, edt. by H.W. Bergmann, R. Kupfer, Erlangen, 1990, p.265-282
[4]   E. Beyer, K. Wissenbach, G. Herziger: "Werkstoffbearbeitung mit Laserstrahlung - Teil 4: Absorption von CO_2-Laserstrahlung" in *Feinwerk & Messtechnik*, 92, 3, 1984, p.141-143
[5]   R. Nuss, S. Biermann: "Auswirkung der Polarisationsrichtung beim Laserstrahlschneiden" in *Laser und Optoelektronik*, 4, 1987, p.389-392
[6]   F. Dausinger, T. Rudlaff: "Steigerung der Effizienz des Laserstrahlhärtens" in *Proc. of ECLAT'88*, DVS, Düsseldorf, 1988, p.88-91
[7]   G. Herziger: "The influence of Laser-Induced Plasma on Laser Materials Processing" in *The Industrial Laser Annual Handbook*, edt. by D. Belforte, Pennwell Books, Tulsa, Oklahoma, 1986, p.108-115
[8]   M. Schellhorn: "Transientes Absorptionsverhalten von Metallen in der Startphase des Laserschweißprozesses" in *Opto Elektronik Magazin*, 4, 2, 1988, p.156-159
[9]   M. Beck, F. Dausinger: "Modelling of Laser Deep Welding Process", in *European Laser Workshop on Mathematical Simulation, Lissabon 1990*, edt. by H.W. Bergmann, Sprechsaal Verlag, Coburg, 1990, p.201-216
[10] H. Albrecht, J. Lademann, K. Seliger, R. Jaschek, H.W. Bergmann, et.al.: "A Mini-TEA-CO_2-Laser, Development and Application" in *Opto Elektronik Magazin*, 7, 2, 1991, p.118-125
[11] R. Jaschek, H.W. Bergmann, et.al.: "Experiences in materials processing using TEA-CO_2-Lasers" in *Proc. of 10th. Int. Congress Laser 1991*, Springer-Verlag, in press
[12] M. Hugenschmidt: "Repetierend gepulste thermomechanische Laserenergieübertragung auf Werkstoffe" in *Opto Elektronik Magazin*, 5, 7/8, 1989, p.615-623
[13] M. v. Allmen: "Laser-Beam Interactions with Materials", Springer Verlag, Berlin, 1987
[14] A.N. Pirri: "Plasma Energy Transfer to Metal Surfaces Irradiated by Pulsed Lasers" in *AIAA Journal*, 12, 16, 1978, p.1296-1304
[15] G. Schulze: "Metallphysik", Akademie-Verlag, Berlin, 1974

IX.
Modelling and
Simulation

Modeling CO_2-Laser Welding of Metals

G. Simon, U. Gratzke, J. Kroos, B. Specht and M. Vicanek
Institut für Theoretische Physik, Technische Universität, Braunschweig, FRG

1 Introduction

The use of high-power laser beams plays a central role in industrial applications of automated material processing such as cutting, drilling or welding of metals and ceramics. For an optimum performance, a detailed knowledge and understanding of the underlying physical processes is necessary.

In this communication we concentrate on laser welding of metals and present mathematical models recently developed to describe various relevant aspects of the welding process. In particular, we present results in connection with the laser-induced plasma and its influence on absorption, the static and dynamic properties of the keyhole, the temperature distribution under cw and pulsed operation, and the capillary instabilities which may lead to humping. It turns out that the entire process is quite complex and cannot easily be decomposed into isolated parts with the intention to study each of them separately, regardless of the others. The partitioning according to the sections in this paper thus merely reflects the logical frame of presentation; in reality, the various aspects discussed in each section are strongly interlinked and each model uses as input results from the others in a self-consistent way.

2 Absorption and laser-induced plasma

It is generally accepted [1] that in deep penetration laser welding of metals, a keyhole forms as a result of the high energy density in the focus. The workpiece is heated up, resulting in a molten region around the beam. Some of the workpiece material will evaporate, and will be exposed to the intense radiaton. Initially, there will be a small number of free electrons which, however, may be accelerated by the laser field to produce secondary electrons via ionizing collisions with metal vapor atoms. Thus, a laser-induced plasma is formed within the keyhole. In general, the plasma is not strictly confined to the keyhole region; part of it is visible as a plume above the surface of the workpiece. However, as the plume appears to disturb the process, it is usually removed by applying an assist gas jet.

The laser induced plasma, in its turn, has a feed-back effect on the absorption properties and hence on the evaporation rate which provides new material to maintain the plasma. The resulting influence on the energy transfer from the laser to the workpiece is still not fully understood. On one hand, there exist claims [2] that a carefully adjusted plasma will enhance energy transfer. On the other hand, even without a plasma, a substantial part of the radiation may be absorbed directly at the surface, bearing in mind the high absorptivity at extreme angles of incidence and also the possibility of multiple reflections within the keyhole [3]. Thus, a plasma may be unneccesary or even unwanted since it may act as a shield to the radiation and so inhibit energy transfer [4]. The issue can only be resolved on the basis of firm knowledge of the plasma parameters that govern the absorption and transport of energy, e.g., electron density and temperature, and ionized and neutral species fluxes. Here, plasma diagnostics experiments [5] can provide valuable information.

On the theoretical side, a self-consistent model of the detailed process is highly desirable. The picture which is emerging [6] contains the following ingredients: (1) Electrons in the plasma absorb the laser energy via *inverse Bremsstrahlung*. (2) These hot electrons transfer energy to the neutrals in ionizing collisions. (3) The ions are transported from their point of origin to the surface of the workpiece, the keyhole wall. (4) At the wall, the ions release their kinetic energy and, due to recombination, their ionization energy. A quantitative model has been set up [7] which includes the above steps on the basis of a Boltzmann equation balancing the ion production and transport within the keyhole volume (pre-sheath), and a Vlasov equation describing the coupling between the plasma and the keyhole wall in a thin layer (sheath) [8, 9]. Input quantities are the laser power P_{Laser}, the workpiece thickness d, the keyhole radius as well as several material parameters. Output quantities are the absorbed laser power P_{abs}, the electron temperature T_e and density n_e, the degree of ionization $n_e/(n_e + n_n)$ as well as the ion drift velocity v_i.

In figure 1 we plot the electron density and temperature as a function of the absorbed laser power per workpiece thickness. Typical densities are of the order of 10^{23} electrons per m^3, and temperatures are around 10 000 K. Figure 2 shows the absorbed laser power P_{abs} versus the supplied laser power P_{Laser}. There is a threshold value below which no plasma is formed. For laser powers exceeding this threshold, there are formally two solutions; however, only the upper branch (solid line) is a stable solution, whereas the lower branch (dashed) is unstable and will thus not be found in reality. The dotted line is a reference denoting 100% absorption. From figure 2 it is clear that the laser-induced plasma may lead to efficient absorption and thus notably enhance the energy transfer to the workpiece.

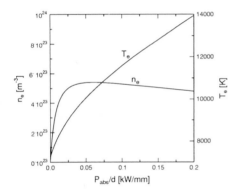

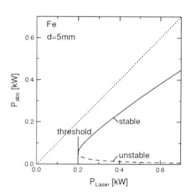

Fig. 1: Electron density n_e and temperature T_e in a keyhole plasma as a function of absorbed laser power per workpiece thickness.

Fig. 2: Absorbed versus supplied laser power in a keyhole plasma.

3 Laser power threshold and keyhole formation

In the previous section we have tacitly assumed that a keyhole exists which confines the laser-induced plasma. In this section we study the conditions under which such a keyhole may form and remain stable [10]. The force acting to create a cavity under the laser beam is generated by the recoil pressure p_{abl} of the ablating surface particles. This force must be balanced by a counteracting surface tension term p_γ of the molten metal. There are additional forces resulting from the convective motion of the melt and hydrostatic pressure, but these are of minor importance [10] and may be neglected in the first step.

A complication lies in the calculation of the ablation pressure, since it involves the ablation rate which, in turn, is strongly dependent on the surface temperature. The pressure balance has thus to be augmented by the energy balance in order to determine the surface temperature. Here, the absorbed laser energy-flux is balanced by losses due to heat conduction into the workpiece and due to the energy carried away by the ablating particles. In our model we assume full absorption of the fraction of laser power which actually enters the keyhole, without specifying in detail the absorption mechanism. The ablation of metal particles is modeled as a non-equilibrium evaporation process.

In combining the balance of forces on the one hand with the balance of energy flow on the other, we thus arrive at a self-consistent model for the calculation of the average surface temperature, the keyhole radius and the keyhole pressure. The general behavior of the two competing pressures p_γ and p_{abl} is sketched in figure 3 as a function of the keyhole radius a. There are two intersection points where the system is in mechanical equilibrium. However, only the point B is stable, whereas A is an unstable point.

In figure 4 we show as an example a calculation for welding of steel at a speed of $1\,\mathrm{m/min}$. There is now a set of curves for p_{abl} corresponding to different values of the parameter P_{abs}/d, where d is the workpiece thickness. There is a threshold of $0.78\,\mathrm{kW/mm}$ below which there is no intersection, i.e., no stable keyhole exists. For P_{abs}/d exceeding this value, we recover the qualitative behavior of figure 3.

Another result emerging from this investigation is that the keyhole radius is at least 1.7 times the laser focus radius, and increases quite rapidly with increasing laser power. However, for exceedingly high laser powers our model will overestimate the keyhole radius mainly because of the inherent assumption that all absorbed energy actually reaches the keyhole walls. This assumption will break down for high laser power when strong ablation leads to an optically thick plasma which essentially shields the material under irradiation.

4 Heat flow and weld-pool geometry

The temperature distribution within the workpiece in welding is of obvious importance. Not only does it directly determine the welding depth and the extension of the heat affected zone; it also plays a decisive role in mass and energy transport during the welding process. The evaporation rate, for example, is particularly sensitive to the surface temperature. Other important phenomena include a possible surface-tension-gradient driven convection (Marangoni effect) and instabilities related to the weld pool geometry (see below).

It is not the place here to give appropriate account of the many heat transport calculations published in the present context. We only mention [11] as a general reference and [12] as simple applications in welding using point and line sources, and [13] for a cylindrical source. In calculations with an extended source there occurs as a scaling parameter the Péclet number $\mathrm{Pe} = r_0 v/2\kappa$ (r_0 laser focus radius, v travel speed, κ thermal diffusivity). For $\mathrm{Pe} \ll 1$ the isotherms are nearly concentric circles around the laser focus, while for $\mathrm{Pe} \gg 1$ the shape of the isotherms is more elongated at the side already welded. A recent calculation aimed particularly at the modeling of high-speed welding ($\mathrm{Pe} \approx 10$) [14] assumes a distributed heat source acting on a moving sheet of finite thickness. Figure 5 shows the momentary temperature profile along the feed direction at the bottom surface of a metal, in comparison with a simple estimate which ignores the finite thickness. The two curves differ by a factor of two.

The validity of the calculation is demonstrated by the comparison to experimental data [15] in figure 6. Here, the weld depth is plotted as a function of travel speed. The agreement is satisfactory.

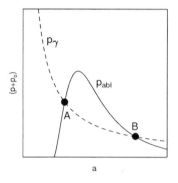

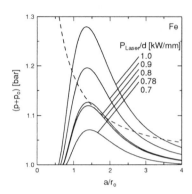

Fig. 3: Schematic behavior of the pressure from ablation p_{abl} and surface tension p_γ as a function of the keyhole radius a. p_0 is the atmospheric pressure.

Fig. 4: Pressure balance for the welding of iron. The keyhole radius a is normalized to the laser focus radius r_0.

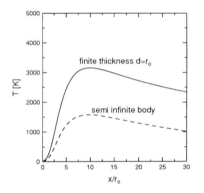

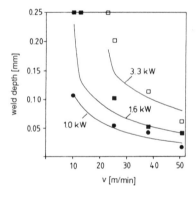

Fig. 5: Temperature profile at depth d from the surface for two different models.

Fig. 6: Comparison of weld depths measured under different conditions with respective theoretical predictions.

5 Time scales and pulsed-mode behavior

In pulsed-laser applications it is of primary interest to know the various time scales of the system in order to predict how much the process will be influenced by the imposed modulation as compared to an average, stationary power supply. Characteristic time scales are also important for an appropriate design in online diagnostics and control.

Whereas phenomena associated with fluctuations of the plasma are generally fast ($1\,\mu s$) [16], the thermal and hydrodynamical response of the system may be many orders of magnitude slower [17, 18]. A recent analysis of heat conduction with a time modulated laser beam [19] reveals that temperature changes penetrate only into a thin layer around the keyhole; at distances around one keyhole radius away the fluctuations rapidly decrease leaving a temperature distribution which is determined by the *average* laser power. The appropriate scaling parameter for this problem is $\Omega = \omega r_0^2 / \kappa$, with ω denoting the laser modulation frequency. In the kHz region, it is typically $\Omega \gg 1$.

An important question in this context concerns the dynamic properties of the melt, and, in particular, the response of the keyhole to fluctuations—whatever wanted or unwanted—of the laser energy supply. Recently, a model study was performed [18] where the collapse of a keyhole due to an abrupt beam shut-off was investigated. The time necessary for a complete collapse is of the order of $\sqrt{a^3 \rho / \gamma} \approx 0.1\,\text{ms}$, where a is the initial keyhole radius, ρ is the density of the melt and γ is its surface tension. Current research is being pursued which aims at calculating free and forced oscillations of a keyhole under cw and modulated beam conditions.

6 Capillary instability and humping

For economical reasons, industrial applications of welding processes demand progressively higher travel speeds. However, the maximum speed is limited in general by various defects of the weld seam, which occur above some threshold value of the travel speed. One such defect, which consists in the formation of an almost periodic chain of humps along the weld seam, has been considered in more detail recently [20].

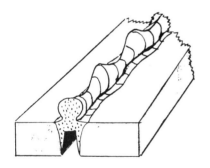

Fig. 7: Sketch of a typical humping pattern.

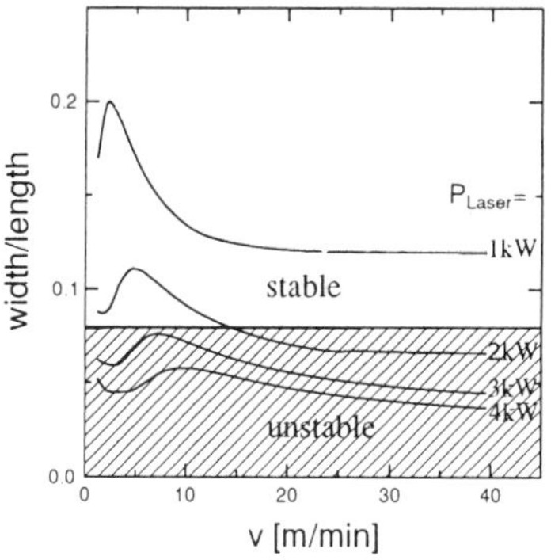

Fig. 8: Weld pool geometry and corresponding stability behavior for various operation parameters.

There has been—and still is—some debate as to which physical phenomenon is responsible for the occurrence of humping. In view of the almost spherical shape of the individual humps it is quite evident that surface tension must play a central role in any attempt of explanation. However, the role of surface tension *gradients* due to surface temperature gradients (Marangoni convection) is much less clear. Whereas some authors [21] propose that thermocapillary flow may be sufficient to entirely explain the humping phenomenon, the present authors attribute humping to a Rayleigh type instability and consider thermocapillarity at most a prerequisite to achieve certain conditions at which this instability might occur.

The basic idea is sketched in figure 7. The molten region is modeled as a bounded cylinder, where the convex shape may be caused by the increase of volume of the melt, and possibly by hydrodynamic pressure build-up due to thermocapillary or otherwise driven [22] melt flow. The curved surface may be shown to be unstable with respect to spatially periodic deformations of long wavelengths [20]. It is argued that such deformations can only exist if the weld pool is sufficiently long compared to its width, which is typically the case at high travel speeds. Figure 8 gives quantitative account of this argument, predicting that the critical travel speed in this case is around $15\,\mathrm{m/min}$ for a $2\,\mathrm{kW}$ laser. This compares well with the value of $10\,\mathrm{m/min}$ found experimentally.

References

(1) J. Dowden, P.D. Kapadia and M. Davis: "Encyclopedia of Fluid Mechanics", vol. 6, ch. 16, ed. N.P. Cheremisinoff (Huston, Gulf 1987).
(2) G. Herziger, E.W. Kreutz and K. Wissenbach: "SPIE Laser Processing: Fundamentals, Applications and Systems Engineering", vol. 668 (1986) 2.
(3) M. Beck and F. Dausinger: "European Scientific Laser Workshop on Mathematical Simulation", ed. H. Bergmann (Coburg, Sprechsaal 1989).
(4) A. Poueyo, L. Sabatier, G. Deshors, R. Fabro, A.M. de Frutos, D. Bermejo and J.M. Orza: J. Physique, Colloque C7 (1991) 183.
(5) G. Chita, M. Dell'Erba, M. Martano and P. Sforza: "Quantum Electronics and Plasma Physics", vol. 29, ed. G.C. Righini (Bologna, SIF 1991).
(6) J. Dowden, P.D. Kapadia and N. Postacioglu: J. Phys. D: Applied Physics 22 (1989) 741.
(7) B.R. Finke, P.D. Kapadia and J. Dowden: J. Phys. D: Applied Physics 23 (1990) 643.
(8) K.-U. Riemann: J. Phys. D: Applied Physics 24 (1991) 493.
(9) H.-B. Valentini: Contr. Plasma Phys. 31 (1991) 311.
(10) J. Kroos, U. Gratzke and G. Simon: "Mechanical stability of the keyhole in laser welding", submitted for publication in J. Phys. D: Applied Physics.
(11) H.S. Carslaw and J.C. Jaeger: "Conduction of Heat in Solids", (Oxford, OUP 1959).
(12) W.M. Steen, J. Dowden, M. Davis and P.D. Kapadia: J. Phys. D: Applied Physics 21 (1988) 1255.
(13) F. Noller: "3rd international conference on welding and melt, electrons and laser beams" Lyon, (1983), 89.
(14) U. Gratzke, P.D. Kapadia and J. Dowden: J. Phys. D: Applied Physics 24 (1991) 2125.
(15) C.E. Albright and S. Chiang: J. Laser Appl. (1988) 18.
(16) E. Beyer: "Einfluß des laserinduzierten Plasmas beim Schweißen mit CO_2-Lasern", Thesis, Darmstadt (1985).
(17) M. Vicanek, G. Simon, H.M. Urbassek and I. Decker: J. Phys. D: Applied Physics 20 (1987) 140.
(18) J. Kroos, U. Gratzke and G. Simon: "Collapse time of the keyhole in laser welding", submitted for publication in J. Phys. D: Applied Physics.
(19) G. Simon, U. Gratzke and J. Kroos: "Heat conduction analysis in welding with a time modulated laser beam", submitted for publication in J. Phys. D: Applied Physics.
(20) U. Gratzke, P.D. Kapadia, J. Dowden, J. Kroos and G. Simon: "Theoretical approach to the humping phenomenon in welding processes", submitted for publication in J. Phys. D: Applied Physics.
(21) K.C. Mills and B.J. Keene: International Materials Reviews 35 (1990) 185.
(22) M. Beck, P. Berger, P. Nagendra and J.A. Dantzig: "Aspekte der Schmelzbaddynamik beim Laserschweißen mit hoher Bearbeitungsgeschwindigkeit", Laser'91, München (1991) in press.

Cladding with Laser Radiation: Properties and Analysis

B. Ollier, N. Pirch, E.W. Kreutz, H. Schlüter, Lehrstuhl für Lasertechnik, RWTH Aachen
A. Gasser, K. Wissenbach, Fraunhofer-Institut für Lasertechnik, Aachen

Introduction

The cladding process employing laser radiation involves the application of a coating to the workpiece via the addition of supplementary materials. The object of this process is to create a surface layer which is adapted to the stress to be encountered under operational conditions. The process is controlled in such a manner as to ensure that the purest possible cladding layer is applied, with good adhesion to the substrate.

In the case of the single-stage laser cladding process, the pulverised supplementary material is transported via the powder supply system into the zone of interaction between laser radiation and workpiece. The supplementary material is solidified via the conduction of heat into the substrate, to the already solidified cladding track and via the release of heat (radiation and convection) into the environment. Between the cladding material and the substrate a metallurgical fusion bond zone develops, which is defined via the mixing level D.

Various models have been published relating to numerical simulation of the physical processes involved in laser cladding. Weerasinghe and Steen (1) use a three-dimensional model, taking due consideration of the interaction between the laser radiation and the powder particles. The energy transportation in the cladding and in the substrate is defined by diffusive thermal conduction. Investigations of the absorption and scattering of CO_2 laser radiation on powder particles based on Mie's scattering theory are described in Lagain (2). A two-dimensional model for single-stage powder cladding is described by Hoadley et al. (3). This model simulates the two-dimensional temperature field in quasi-stationary state, and the heating and melting process of the supplementary material is defined via a uniform heat sink which is distributed in the molten bath. The influence of the molten bath convection on the temperature distribution is not taken into account. Picasso and Hoadley (4) have recently published an extended two-dimensional model for laser cladding. In addition to the diffusive thermal conduction, the convective thermal transportation in the molten bath is also taken into account. The influence of the powder-gas flow and self-consistent determination of the liquid-gaseous phase boundary are also discussed.

This paper presents a two-dimensional model for determining the temperature and the distribution of velocities in the workpiece, taking due account of the results obtained in experiments. The interaction between hydrodynamics, thermal conduction and powder-gas flow determine the physical processes involved in cladding operations and, subsequently, the processing results (layer height, mixing level). The thermal conductivity and Navier-Stokes equations with the relevant boundary conditions are solved according to the finite element method. The development of the free fusible surface is established from high-speed video recordings and duly taken into account in the numerical model. A two-dimensional model for determining the local absorption takes due account of the angle- and polarisation-related absorption of the incident laser beam.

Numerical Model

Stationary-state solution of the conservation equations for mass, energy and momentum is carried out via an iterative process. The Navier-Stokes equations and the thermal conduction equation are separated for each iterative step by solving them separately for each respective temperature and velocity field. The iterative process is continued until the changes in the velocity and temperature fields, respectively, are within specified limits

Thermal conduction

The stationary thermal conduction equation in Eulerian coordinates is represented by

$$\rho c \vec{v} \nabla T = \nabla(\lambda \nabla T) \ , \qquad\qquad \text{Equ. 1}$$

where T is the temperature, ρ the density, c the specific heat capacity, λ the thermal conductivity and v the local velocity.

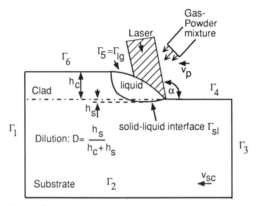

Fig. 1: Schematic representation of laser cladding (longitudinal section).

Fig. 1 shows a schematic representation of the powder-cladding process. The workpiece is moved in a negative x-direction at a constant feed rate v_{sc}. At an adequate distance from the laser-powder interaction zone the ambient temperature is specified at the edge Γ_3. Ignoring the release of heat into the environment, the following equation applies for the edges $\Gamma_{1,2,4,6}$

$$\frac{\partial T}{\partial n} \ | \ \Gamma_{1,2,4,6} = 0 \ . \qquad\qquad \text{Equ. 2}$$

The energy content of the introduced powder mass flow is taken into account at the solid-liquid phase boundary, where it is assumed that the cold powder (temperature To) is heated instantaneously along edge Γ_5 to the local surface temperature T_Γ. Usually, part of the laser power will also preheat the substrate on Γ_4. This results in the following equation

$$- \lambda \frac{\partial T}{\partial n} \ | \ \Gamma_{4,5} = q_l - \rho_p c_p v_{np} (T_\Gamma - T_0) \ , \qquad\qquad \text{Equ. 3}$$

where q_l is the absorbed local laser intensity, ρ_p is the powder density, c_p is the specific heat capacity of the powder and v_{np} is the normal component of the powder velocity at the phase boundary. The local heat flow is calculated on the basis of the polarisation- and angle-related absorption. The powder velocity in normal direction at the phase boundary is determined via the integral mass balance

$$\dot{m}_p = \rho_p \int_{\Gamma_{lg}} v_{np} \, d\Gamma = \rho_s v_{sc} h_c \quad , \qquad \text{Equ. 4}$$

where ρ_s is the density of the liquid phase and h_c is the cladding height.

Discretisation of the thermal conduction equation is carried out in accordance with Galerkin's projection method (5). Phase transitions are ignored here. Cross-linkage of the area is carried out with bilinear elements. In areas of dominant convective terms an upwind process (6) is employed.

Hydrodynamics

The motion equations for an incompressible medium are described in stationary state, ignoring volume forces, considering the following equation

$$(u_j \partial_j) u_i - \partial_j \left(\frac{1}{Re} (\partial_j u_i + \partial_i u_j) \right) + \partial_i p = 0 \qquad \text{Equ. 5}$$

$$\partial_i u_i = 0 \quad ,$$

where u_i represents the velocity components, Re Reynolds' number and p the liquid pressure.

The solid-liquid phase boundary is established by solving the thermal conduction equation. At the solid-liquid phase boundary the velocities are reduced to the feed velocity, and the following adhesion condition applies

$$\vec{u} \mid_{\Gamma_{sl}} = -\vec{v}_{sc} \quad . \qquad \text{Equ. 6}$$

At the liquid-gaseous phase boundary, the boundary conditions are formulated in tangential and normal direction via a suitable coordinate transformation process. In normal direction the velocity is established on the basis of the powder mass flow

$$u_n \mid_{\Gamma_{lg}} = \frac{\rho_l}{\rho_p} v_{np} \quad . \qquad \text{Equ. 7}$$

In a tangential direction Marangoni convection is induced due to the temperature-dependent surface tension. Additionally, shear stress can be transmitted via the powder-gas flow ($\mu_g = 0$). The impulse balance equation in a tangential direction at the liquid-gas phase boundary is as follows

$$\rho_g v_{ng} v_{tg} = \mu \left(\frac{\partial u_n}{\partial t} + \frac{\partial u_t}{\partial n} \right) + \rho_l v_{nl} v_{tl} + \frac{\partial \delta}{\partial T} \frac{\partial T}{\partial t} \quad , \qquad \text{Equ. 8}$$

where δ is the surface tension, v_n / v_t the normal and tangential components, respectively, of the velocity and μ the dynamic viscosity of the liquid.

The Uzawa algorithm (7) is used to solve the Navier-Stokes equations. The non-linear terms are linearised via Newton's method. Biquadratic form functions are employed for the velocity and bilinear form functions for the pressure. The resultant linear equation

690

system is transformed via a rotation matrix in such a manner that the normal and tangential components of the velocity occur at the nodes of the liquid-gaseous phase boundary as unknown quantities (9).

Results

The local absorbed intensity along the cladding bead, where the p-polarised laser beam is inclined at an angle of α in relation to the surface of the workpiece is considered. The elliptical molten bath surface was established from high-speed video recordings of the cladding process (Fig. 2). Fig. 3 shows both the calculated integral absorptance A_{int} and the experimental ones employing calorimetry as a function of angle of incidence α.

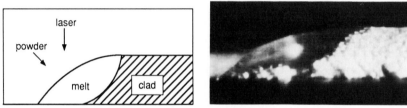

Fig. 2: Schematic representation and high-speed-video recording from the clad (longitudinal section).

The angle of incidence is limited by the finite contact angle of the cladding bead. When the angle of incidence is too large, the molten bath causes shading in the front area of the cladding, which results in process instabilities, as the necessary preheating of the substrate is no longer guaranteed.

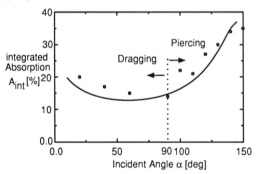

Fig. 3: Integrated absorption (o experiment, - simulation) as a function of the incident angle, p-polarized laser radiation, constant clad geometry.

The numerical simulations (Fig. 4-5) are carried out with p-polarised rectangular laser radiation of constant intensity in the feed direction (length L). For the normal component of the powder velocity v_{np} a Gaussian power density distribution along the fusible surface is assumed. The flow conditions (flow geometry and flow velocity) are such that the impulse transfer of the powder-gas flow in a tangential direction is negligible in relation to the surface tension gradients. As a result of the powder supply configuration and the powder velocities measured in the free beam (2), the shear stress transfer of the powder-gas flow is substantially lower than the surface tension gradients.

It is further assumed that the powder particles are heated sufficiently quickly in the molten bath to guarantee the condition of a heat sink along the liquid-gas phase boundary and the development of Marangoni convection. The estimation of the heating-up time for the powder particles ($d=50\mu m$) shows that the melting time of the particles is two orders of magnitude below the mean retention time of the particles on the surface, where the surface temperature of the spherical particles is maintained at the temperature of the molten bath (9). The particles which hit the solid substrate or the solidified track are not taken into account in the balance equations.

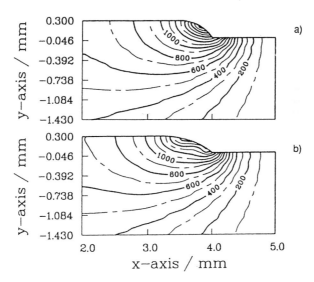

Fig. 4: Calculated temperature field, (stellit 6 on Ck 45, v_{sc}=250 mm/min, P_l=1,5 kW/cm, T_m=1300 $^\circ$C, L=0.6 mm) without a) and with b) Marangoni-convection.

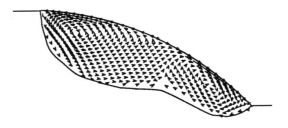

Fig. 5: Calculated velocity field (v_{max}≈90 cm/sec), (parameters see fig. 4b).

Fig. 4 shows the calculated temperature field including and excluding the molten bath convection. As a result of the minimal temperature conductivity of the supplementary material, the contour of the molten bath alters significantly when the convective energy transport in the molten bath is taken into account. In the area of the front vortex (Fig. 5) the mixing level is increased, and in the area of the rear vortex the molten bath is extended. The temperature gradients along the solidification contour and the solidification speeds are influenced by the molten bath convection.

The maximum velocities are attained at the surface, and are in the range of 0.9 m/sec. Due to the powder-molten bath density shift, the normal component of the surface velocity is very small, and becomes visible only in the area of low velocities between the two vortices. There is a good level of correspondence between the calculated velocities and the experimental results (10).

Conclusion

By utilising the angle- and polarisation-related absorption, the process efficiency and, consequently, the coating speed can be improved, particularly when employing a piercing process. To obtain usable information on the temperature field and the mixing level, the numerical simulation must take due account of the hydrodynamic aspects.

The powder-laser-molten bath interactions require more detailed investigation with regard to improved process approximation. The two-dimensional simulation enables qualitative assessment of the influence of the process parameters (powder quantity, beam geometry, distribution of intensity, feed rate) on the results of the cladding process.

References

(1) V.M. Weerasinghe, W.M. Steen: Transport phenomena in materials processing, ASME (1983) 15.
(2) P. Lagain: "Contribution exprimentale aux traitements de surface par laser avec apport de poudres", These de docteur de l'universite d'Aix-Marseille II, (1989).
(3) A.F.A. Hoadley, C. Mardsen, M. Rappaz: "A thermal model of the laser cladding process", International Symposium on Manufacturing and Materials Processing, Dubrovnik (1990).
(4) M. Picasso, A.F.A. Hoadley: Numerical Methods in Thermal Problems 7 (1991) 199.
(5) C. Cuvelier, A. Segal, A. v. Stennhoven: "Finite element methods and Navier-Stokes equations", in Mathematics and its applications, D. Reidel Publishing Company, Dordrecht (1986) 74.
(6) F. Thomasset: "Implementation of finite element methods for Navier-Stokes equations", in Springer series in computational physics, Springer Publishing Company, New York (1981) 37.
(7) M. Fortin, R. Glowinski: Studies in Mathematics and its Applications 15 (1983) 75.
(8) M.S. Engelman, R.L. Sani, P.M. Gresho: International Journal for Numerical Methods in Fluids 2 (1982) 225.
(9) H.S. Carslaw, J.C. Jaeger: "Conduction of heat in solids", Oxford University Press, London (1959) 233.
(10) H.W. Bieler: private communication.

Modelling of Keyhole/Melt Interaction in Laser Deep Penetration Welding

M.Beck, P.Berger, H.Hügel
Institut für Strahlwerkzeuge, IFSW, Universität Stuttgart, FRG

ABSTRACT

Above a critical intensity, the backpressure of evaporating material forms a keyhole, which allows the laserbeam to deeply penetrate into the workpiece. The evaporation process and the flow of the vapour inside this keyhole is investigated, theoretically. For a stable keyhole, the geometry and the resulting flow of vapour is described self-consistently. The dependence on process parameters, such as beam quality, F-number of the focusing optic, and the position of the focal plane is discussed. Additionally, the effect of a sudden change of absorbed laser power on evaporation and pressure distribution is described.

Model of keyhole formation

The keyhole formed in deep penetration welding is totally surrounded by liquid melt. It is stabilized by a continuous evaporation on the surface of the keyhole and by the outflow of the metallic vapour. The equilibrium shape of the keyhole is calculated by a local balance between the energy flux, absorbed on its surface, and the energy losses due to heat conduction and evaporation. The losses due to heat conduction can be calculated for a given keyhole temperature. In contrast, the absorbed energy flux on the keyhole surface cannot be directly calculated from the energy distribution of the laser beam, because of its redistribution by multiple reflection and due to partial absorption in the plasma. As the absorption, the evaporation is dependent on the keyhole geometry. To account for this feedback, a numerical iteration procedure is used (1). Its flow diagram is given in Fig.1. To allow a numerical solution, the following simplifications are introduced:

Evaporation temperature at the wall of a circular keyhole and 2-dimensional heat flow perpendicular to the keyhole axis are assumed. The heat flow is calculated following Dowden's model (2), which takes the thermophysical properties of the material and the welding speed into account. The absorption in the plasma is calculated according to Lambert'Beers law and that on the keyhole walls according to the Fresnel formulas. Complete coupling of the energy absorbed in the plasma to the adjacent keyhole walls is hypothetically assumed. The propagation of a focused Gaussian beam is taken into account up to its first reflection on the keyhole surface. Geometric optics is used to calculate the reflections. The calculation of the evaporation and the related

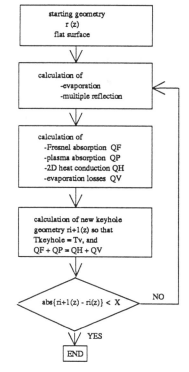

Fig. 1: Flow diagram

losses (3) are described in the following chapter. It has to be pointed out, that the vapour is assumed to be a monoatomic ideal gas at local evaporation temperature. Therefore, the evaporation model is not valid, when plasma absorption dominates the process.

Model of evaporation

A keyhole is stable, when the pressure due to evaporation p_{evap} balances all forces, tending to obliterate the cavity. The following equilibrium condition, therefore, can be described for the keyhole surface:

$$p_{evap}(\alpha,z) = p_{stens}(\alpha,z) + p_{hdyn}(\alpha,z) + p_{hstat}(z) \qquad (1) \; .$$

The pressure due to **surface tension** p_{stens} can be calculated by

$$p_{stens}(\alpha,z) = \sigma \left(\frac{1}{r_1(\alpha,z)} + \frac{1}{r_2(\alpha,z)} \right) \qquad (2) \; ,$$

with the surface tension σ and the two main curvatures r_1 and r_2 of the keyhole surface. For a keyhole radius of 0.1 mm, p_{stens} is in the order of 10^4 Pa.

The **hydrostatic pressure** p_{hstat} depends on the gravity g, on the density of the liquid ϱ_{liquid} and on the depth in the keyhole z and is given by

$$p_{hstat}(z) = \rho_{liquid} \, g \, z \qquad (3) \; .$$

For welding depths below 10 mm, the hydrostatic pressure is less than 10^3 Pa, what is low compared to the pressure due to surface tension.

To describe the **hydrodynamic pressure** p_{hdyn} due to the flow of melt around the keyhole, the Navier Stokes equations have to be solved. If viscous forces are negligible, which is a reasonable assumption only for low welding speeds, the maximum pressure variation Δp_{hdyn} around the keyhole can be estimated by

$$\Delta p_{hdyn}(z) = \frac{1}{2} \, \rho_{liquid} \left(v_{min}(z)^2 - v_{max}(z)^2 \right) \qquad (4) \; .$$

The maximum flow velocity v_{max} appears at the side of the keyhole (4). It is dependent on the material properties such as thermal diffusivity $\varkappa$, heat capacity c_p, evaporation temperature T_{vap}, melting temperature T_{melt}, room temperature T_0, the latent heat of melting L_m, the welding speed v_0 and can be approximated by

$$v_{max}(z) = v_0 \cdot \left[1 + \sqrt{ \frac{v_0 \, r(z)}{\varkappa_{liquid}} \cdot \left(1 + \frac{2 \, c_{p,liquid} \, (T_{vap}(z) - T_{melt})}{c_{p,liquid} \, (T_{melt} - T_0) + L_m} \right) } \; \right] \qquad (5) \; .$$

Since there is a stagnation point at the front of the keyhole, v_{min} is equal to 0. Δp_{hdyn} depends on the welding speed in the power of three, approximately. As the welding speed rises from 5m/min to 15 m/min, Δp_{hdyn} grows from 50 Pa to 10^4 Pa.

As the model concentrates on low welding speeds, Δp_{hdyn} is neglected and the keyhole is assumed to be circular. For that reason, the pressure of the fluid acting on the surface of the keyhole is calculated dependent on r(z), only.

Since p_{stens} and p_{hstat} increase towards the tip of the keyhole, the pressure in the keyhole volume has to increase as well. This vertical pressure variation inside the keyhole results in a flow of vapour. For the long and narrow keyhole, this flow is similar to a flow through a tube where mass flux and flow diameter are not constant (5). With this approximation and for laminar flow, the volume flow V of the vapour can be derived from the pressure gradient dp_{vap}/dz, the keyhole geometry $r(z)$, and from the viscosity η (eq.6), and the evaporation velocity perpendicular to the keyhole surface u_{vap} is given by eq.7:

$$\frac{dp_{vap}(z)}{dz} = \frac{8}{\pi}\,\eta(z)\,(\frac{1}{r(z)})^4\,\dot{V}(z) \quad (6)\,, \qquad u_{vap} = \frac{1}{16\,r(z)}\,\frac{d}{dz}\,(\,\frac{r(z)^4}{\eta(z)}\,\frac{dp_{vap}(z)}{dz}\,) \quad (7)\,.$$

For an ideal gas in thermodynamic equilibrium, the viscosity η is dependent on the temperature T_{vap}, the atomic mass m_a, and the collision diameter σ_a (eq.8). The local evaporation temperature T_{surf} is calculated by the Clausius-Clapeyron relationship and is dependent on the latent heat of vaporization L_{vap}, the molecular weight M^*, and on the known evaporation temperature T_{vap0} at pressure p_{vap0} (eq.9):

$$\eta(z) = \frac{5}{16}\,\frac{1}{\sigma_a^2}\sqrt{\frac{k\,T_{vap}(z)\,m_a}{\pi}} \quad (8)\,, \qquad T_{vap}(z) = T_{vap0}\,/\,(1 - \frac{R^*\,T_{vap0}}{M^*\,L_{vap}}\,\ln\frac{p_{surf}(z)}{p_{vap0}}) \quad (9)\,.$$

On the ablating keyhole surface exists a Knudsen layer of thickness of a few molecular free paths, which can be represented by discontinuities in temperature, density, and pressure. The structure of this layer can only be determined by consideration of molecular behaviour (6.7). The discontinuities across the Knudsen layer depend on the local Mach number in the exit plane of the Knudsen layer and on the ratio of specific heats γ of the vapour. Temperature T_{vap} and pressure p_{vap} in the keyhole volume are, therefore, lower than on the keyhole surface. Following Knight (7), the relationship between the conditions on the evaporating surface and the exit plane of the Knudsen layer are described by eq.10 - eq.14:

$$\frac{p_{vap}(z)}{p_{surf}(z)} = \frac{T_{vap}(z)}{T_{surf}(z)}\,\frac{\rho_{vap}(z)}{\rho_{surf}(z)} \quad (10)\,, \qquad \frac{T_{vap}(z)}{T_{surf}(z)} = (\sqrt{1+\pi(\frac{\gamma-1}{\gamma+1}\frac{M(z)}{2})^2} - \sqrt{\pi}\,\frac{\gamma-1}{\gamma+1}\frac{M(z)}{2})^2 \quad (11)\,,$$

$$\frac{\rho_{vap}(z)}{\rho_{surf}(z)} = \sqrt{\frac{T_{surf}(z)}{T_{vap}(z)}}\,(\,(M^2(z)+\frac{1}{2})\,C(z) - \frac{M(z)}{\sqrt{\pi}}\,) + \frac{1}{2}\,\frac{T_{surf}(z)}{T_{vap}(z)}\,(\,1 - \sqrt{\pi}\,M(z)\,C(z)\,) \quad (12)\,,$$

$$C(z) = e^{M(z)^2}\,erfc(\,M(z)\,) \quad (13)\,, \qquad M(z) = u(z)\sqrt{\frac{k}{2\,m_a\,T_{vap}(z)}} \quad (14)\,.$$

The pressure drop in the Knudsen layer depends on the evaporation rate, which in turn is a function of the pressure in the keyhole volume. Combining eq.6 to eq.14 leads to a differential equation of second order, which has to be solved for p_{vap} numerically. The flow velocities, mass flow- and evaporation rate and, subsequently, the evaporation losses are part of the solution. Since the model is based on the flow in a tube, the solution fails for the tip and for the exit of the keyhole. Whereas the keyhole tip can be simply excluded in the calculation without affecting the flow in the rest of the keyhole, the conditions at the exit show a strong affect on the results. Here, the flow diameter and, subsequently, the mass flow and evaporation rate tend to infinity, because the detachment of the flow, which occurs in reality, is not described by the Poiselle equation (eq.6). This flow detachment is artificially introduced by a separation criterion,

As the pressure inside the keyhole is less than $2*10^5$Pa in the tip, and typically 10^4Pa in the main part of the keyhole, the variation of surface temperature (eq.9) can be neglected.

Beam Quality and focusing optics

Fig. 4 shows for a given optic (F=7) the dependence of the welding depth on the position of the focal plane z_f and on the beam mode. Optimal welding depth is achieved when the focal position is below the surface of the workpiece. Compared to a $TEM_{01}*$ mode, the maximum welding depth for a TEM_{20} mode is reduced by 50%. In contrast, an optimal beam (TEM_{00}) ensures an increase in welding depth by about 20%, because of its smaller beam diameter and its extended Rayleigh lenght.

For every beam mode exists a single focusing optic, which ensures an optimum compromise between diameter and Rayleigh length of the focused beam. The welding depth, dependent on the F number and the focal position is exhibited in Fig.5.

Absorption mechanisms

Fig. 6 shows the influence of the absorption mechanisms on the welding depth. Taking Fresnel absorption into account, the self-focusing effect of multiple reflection leads to an increase in welding depth, when the focal position is close to the surface. Thus, the absorption rate for every single absorption is only about 12%, the total coupling at the optimum focal position is more than 85%. If the focal position is far from optimum, the total coupling is reduced and the process is interrupted. The additional introduction of an absorbing plasma with an absorbing length of 2/cm decreases the welding depth by reducing the self-focusing of the beam and by broadening the keyhole. In the situation described, the laser is absorbed in the plasma by 81% and by 17% at the keyhole wall (reflections = 2%).

Stability

So far, only stable situations have been investigated. Let us now describe an unstable situation, where the absorbed laser power has increased so rapidly that the keyhole geometry has not changed yet. This could be the case, for example, due to a

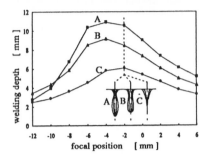

Fig. 4: Welding depth dependent on the beam mode, A: TEM_{00}, B: $TEM_{01}*$, C: TEM_{20}.

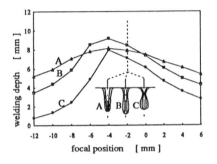

Fig. 5: Welding depth dependent on the F-number of the focusing optic, A: F=12, B: F=7, C: F=4.

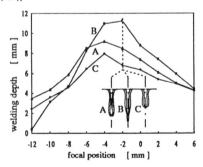

Fig. 6: Welding depth dependent on absorption mechanism, A: 100% wall absorption, B: Fresnel absorption, C: Fresnel absorption and plasma absorption β=2/cm.

allowing a maximum flow divergence of 10° to the flow axis. Nevertheless, the region where the flow leaves the keyhole, needs some further investigation.

Results

All calculations were performed for a laser power of P_L=4000W and a welding velocity of v_0 = 2 m/min. If not mentioned otherwise, a TEM_{01}* mode is assumed. The material properties of pure iron are used.

Evaporation and vapour flow

Fig. 2 compares the calculated keyhole- and melt pool geometries, as well as the flow velocities for different positions of the focal plane. To exclude the effect of different absorption mechanisms, plasma absorption is neglected and 100% absorption at the keyhole wall is assumed.

For a focal position 5mm below the surface (z_f=-5mm), the keyhole radius decreases continuously towards the tip of the keyhole. Since the pressure inside the keyhole is mainly dependent on the keyhole radius, the pressure rises continuously as well (Fig.3a). This results in a continuously accelerating flow out of the keyhole (Fig.2). For a focal position on the surface, the keyhole broadens slightly below the surface before it closes towards the tip. This results in a relativ pressure maximum below the surface (Fig.3b), which decouples the flow in the lower part of the keyhole, where recondensation occurs, from the flow near the keyhole exit (Fig.2).

Apart from the pressure distribution on the surface of the keyhole, Fig.3 shows the pressure drop in the Knudsen layer and its effect on the pressure distribution in the keyhole volume. Whereas the pressure drop due to the ablating material is small in most parts of the keyhole, reflecting the low evaporation rate, the pressure drop in the Knudsen layer becomes dominant at the exit

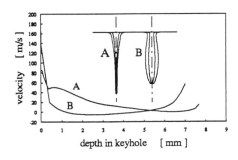

Fig. 2: Keyhole geometries and the related melt pool geometries and flow velocities, A: z_f = -6mm, B: z_f = 0mm.

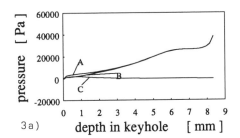

3a)

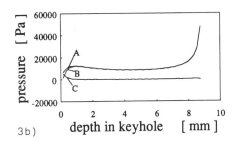

3b)

Fig. 3: Pressure in the keyhole. A: pressure on the keyhole surface, B: pressure in the keyhole volume, C: pressure drop in the knudsen layer. 3a: z_f = -6 mm, 3b: z_f = 0mm.

of the keyhole, where the evaporation rate is high. This reduces the pressure gradient in the keyhole volume, the resulting flow rate, and the evaporation rate.

The energy losses due to evaporation are 1% of P_L for z_f=0 mm and 3% of P_L for z_f=-5 mm and have no significant effect on the welding depth.

698

sudden peak in laser power. Since the evaporation temperature is already reached, the additional absorbed energy leads to an additional evaporation. For two different focal positions, Fig.7 shows the pressure distribution in the keyhole compared to the stable situation. Here, an additional energy input of 5% of the original laser power has led to an increase in pressure of more than 1000%. For this reason, the keyhole changes its geometry immediately, and subsequently alters the conditions for energy absorption and vapor flow. Whether this mechanism only corrects the keyhole shape or leads to a melt ejection, has to be further investigated.

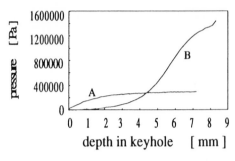

Fig. 7: Pressure in the keyhole for the unstable situation due to an increase in laser power of 5%, A: z_f = 0 mm, B: z_f = -6 mm.

Conclusions

The calculations demonstrate the importance of a good beam quality to achieve high welding depths effectively. Furthermore, the optimum focal position and F-number of the focusing optics can be estimated, dependent on the material properties, the welding speed and laser beam mode. The modelling of absorption mechanisms show the self-focusing effect of multiple reflections, which leads to enhanced absorption in the lower part of the keyhole, and its reduction due to an assumed plasma absorption in the upper part of the keyhole. The discussion of evaporation shows that a small evaporation rate is sufficient to counteract all forces that tend to obliterate the keyhole. As a consequence, forced evaporation due to low volatile alloying constituents, changes in laser power or in absorption conditions leads to high pressures inside the keyhole and subsequently alters the keyhole geometry.

References:

(1)   M.Beck, F.Dausinger : "Modelling of Laser Deep Welding Processes", European Scientific Laser Workshop on Mathematical Simulation, Lisbon 1989, ed. H.W.Bergmann, Coburg: Sprechsaal Publishing Group, 1990.

(2)   J.Dowden, M.Davis, P.Kapadia: "Some Aspects of the Fluid Dynamics of Laser Welding", J. Fluid Mech., Vol 126, 1983.

(3)   H.Treulieb;,"Modellierung der Verdampfung beim Lasertiefschweißen", Universität Stuttgart, Studienarbeit,1992(Inst. f. Strahlwerkzeuge).

(4)   M.Beck, F.Dausinger, P.Berger, H.Hügel : "Aspects of Keyhole-Melt Interaction in High Speed Laser Welding", 8th Int. Symp. on Gas Flow and Chemical Lasers, GCL, Madrid 1990, ed. J.M. Orza, C. Domingo. Bellingham (Wa): SPIE, 1991, (Proc. SPIE 1397).

(5)   P.G.Klemenz : "Heat balance and flow conditions for electron beam and laser welding", J.Appl.Pys, Vol. 47, 1976.

(6)   T. Ytrehus : "THEORY AND EXPERIMENTS ON GAS KINETICS IN EVAPORATIOMN", 10th International Symposium on Rarefied Gas Dynamics , Aspen, Colorado, 1976, ed. J.L.Potter, New York: AIAA, 1977.

(7)   C.J.Knight : "Theoretical Modelling of Rapid Surface Vaporization with Back Pressure", AIAA Journal, Vol.17., No.5, May 1979.

Melting-Solidification Phenomena in Pulsed-Laser Treatment of Metallic Materials. Two Dimensional Simulations.

A. Lashin, M. Poli - Istituto di Tecnologie Industriali e Automazione, Milano, Italy
I. Smurov - Ecole Nationale d'Ingénieurs de Saint Etienne, Saint Etienne, France

Introduction

Concentrated energy flow such as lasers are used widely to melt surface layers of metallic materials (1). By means of subsequent rapid cooling a modified structure of treated material can be produced. Many microstructural parameters can be related to the macroscopic temperature field. For instance growth rate and microstructure of columnar or equi-axed grains depend primarily on the speed at which the isotherms are moving (2).
The most important problem is the determination of the relationships between the laser machining parameters and the temperature distribution. A direct experimental investigation of the temperature distribution field is a difficult task due to the short pulse duration and the small treated zone. Therefore there are many papers concerning both the analytical calculation and numerical simulation of the evolution temperature field in laser treatment of materials. The purpose of this report is to present results of numerical simulation of the distribution of temperature field and melt geometry both for the process melting of metallic slab during pulse laser action and the process solidification after the end of pulse.

Scope of the model and governing equations

The mathematical model proposed includes the process of heating, melting, surface evaporation, cooling and solidification under the irradiation of a laser with different cylindrical symmetry space shape on a metal slab. Is is assumed that energy is absorbed on the irradiated surface, radiation mechanisms of heat losses from both sides of the slab are taking into account.
In the present model is neglected the convective heat transfer in the melting pool in comparison with the conduction one. It is also neglected the deformation of free surface melt due to the action of evaporation reactive pressures, i.e., suppose that surface of the melt is a plane.
The mathematical model used can be written in the following form:

$$\frac{\partial H}{\partial t} = \frac{1}{r}\frac{\partial}{\partial r}\left(\lambda r \frac{\partial T}{\partial r}\right) + \frac{\partial}{\partial z}\left(\lambda \frac{\partial T}{\partial z}\right) \quad , \quad t>0, \, 0<r<\infty, \, 0<z<L; \qquad \text{Equ. 1}$$

$$H(t,r,z) = \int_{T_0}^{T} \rho \, c(T') \, dT' + \rho \, L_m \, F(T), \qquad T(t=0,r,z)=T_0\text{-const;}$$

$$-\lambda\frac{\partial T}{\partial z}\bigg|_{z=0} = q_0(t,r) - q_1\left[T(t,r,z=0)\right], \quad -\lambda\frac{\partial T}{\partial z}\bigg|_{z=L} = q_1\left[T(t,r,z=L)\right], \quad \frac{\partial T}{\partial r}\bigg|_{r=0} = 0 \, ; \qquad \text{Equ. 2}$$

$$q_0(t,r) = \begin{cases} I(t) \, e^{-kr^2}, & 0\leq r\leq R \\ 0 \, , & r>R \end{cases} \, ; \quad I(t) = \begin{cases} I\text{-const}, & 0\leq t\leq \tau \\ 0 \, , & t>\tau \end{cases} \, ;$$

$$q_1(T) = \begin{cases} \rho L_v \dfrac{v_*}{\sqrt{T}} e^{-T_*/T} + \; \varepsilon \, \sigma T^4 \, , & T > T_m \\ \varepsilon \, \sigma T^4 \, , & T \leq T_m \end{cases} \, ;$$

$$v_* = \frac{P_v}{2\rho(2\pi k/m)^{1/2}} \exp\left[\frac{L_v}{T_v \, (k/m)}\right], \quad T_* = \frac{L_v}{(k/m)}$$

where: H- enthalpy per unit volume; T - temperature; c - specific heat per unit mass; L_m - latent heat of melting per unit mass; λ - thermal conductivity; q_0 - density of absorbed surface energy flux; k - concentration coefficient; q_1 - density of surface energy flux due to radiation and evaporation heat losses; ε - emissivity of metal surface; v_*, T_* - constants determined by Herz-Knudsen law of evaporation; L_v - latent heat of evaporation per unit mass; T_v - boiling temperature corresponding to pressure P_v; F(T) - liquid fraction temperature relationship, in the case of an isothermal phase change (Stefan boundary condition) the liquid fraction is given by Heaviside step function, in other cases F(T) can have more elaborate forms (e.g., the Scheil equation in metallurgical solidification).

Numerical solution

The time discretization of Equ.1 is based on the two level backward Euler (fully implicit) scheme. For spatial discretization was employed the control - volume finite - difference method (3) The general iterative source based method on the fixed grid was used for the treatment of latent heat evolution (4). To solve the linear algebraic equations the fridiagonal matrix algoritm was used. The convergence at a given time step is declared when the residuals of enthalpy become less than 10^{-8}.

Result and discussion

Results of numerical simulation of processes melting solidification under the pulse laser action with parameters $I= 1.5 \, 10^9 \, Wm^{-2}$, $k=5 10^6 m^{-2}$, $\tau = 7$ ms on the titanium slab 1 mm thick are presented (Fig. 1,2). The energy flux q_0- q_1 spent on process heating and melting of the slab decrease during the time, (Fig.1a) due to the loss of evaporation and as the result the surface temperature reaches the value of saturation $\sim 4000 \, °K$ (Fig 1b). After the end of the laser pulse (t>7ms), the process of solidification begins on edges of the pool while the process of melting continues on the bottom of the pool (Fig. 2c, curves 0,1). The radiation heat loss from the free surface of the melt lead to the solidification of the pool from the surface at later time (Fig 2c, curve 5).

References

(1) C.W. Draper, P. Mazzoldi: "Laser Surface Treatment of Metals", Martinus Nijhoff Publishers (1988).
(2) M. Rappaz, B. Carrupt, M. Zimmermann, W. Kurz: Helv. Phys. Acta. 60 (1987) 924.
(3) S.V. Patankar: "Numerical heat transfer and fluid flow", Hamisphere (1980).
(4) V.R. Voller, C.R. Swaninathan: Num. Heat Trans. 19B (1991) 175

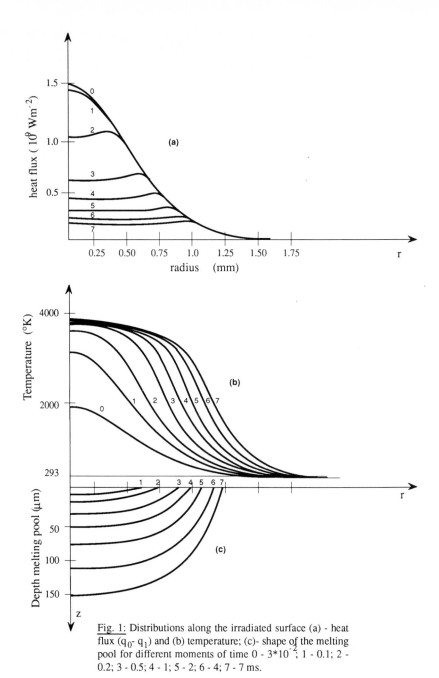

Fig. 1: Distributions along the irradiated surface (a) - heat flux (q_0- q_1) and (b) temperature; (c)- shape of the melting pool for different moments of time 0 - 3*10^{-2}; 1 - 0.1; 2 - 0.2; 3 - 0.5; 4 - 1; 5 - 2; 6 - 4; 7 - 7 ms.

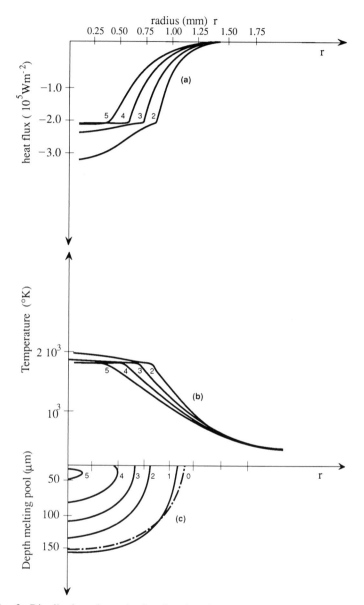

Fig. 2: Distribution along the irradiated surface (a) - heat flux (q_0 -q_1) and (b) - temperature; (c) - shape of the melting pool for different moments of time: 0-7; 1-8; 2-12; 3-14; 4-16; 5-18 ms.

Approches in Modelling of Evaporative and Melt Removal Processes in Micro Machining

E. Meiners, A. Kessler, F. Dausinger, H. Hügel, Universität Stuttgart, Institut für Strahlwerkzeuge (IFSW), Paffenwaldring 43, D- 7000 Stuttgart 80, FRG

Introduction

Laser micro machining is a new technology offering new possibilities in 3- dimensional shaping of materials. The energy of the laserbeam heats the material's surface. The ablation is driven by the pressure gradient between the working zone and the surrounding atmosphere. This pressure gradient can be built by an external gas jet driving molten material out of the interaction zone, by vaporizing material or a combination of both processes.

Evaporative processes have been theoretically investigated in various drilling and cutting models. In laser micro machining these models have to consider 3- dimensional geometries. Ablation experiments with pulsed Nd:YAG- lasers on vaporizing materials, an example is the ceramic Silicon- Nitride (Si_3N_4), show a strong dependency of ablation rate and surface quality not only on the laser parameters but also on the machining parameters. A model thus has to calculate ablation rate and surface structure taking into account physical parameters as well as machining parameters.

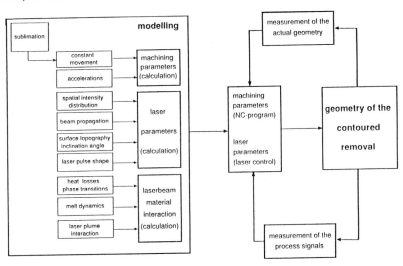

Fig. 1:   Scheme of the strategy to achieve a contoured removal

On the left side of figure 1 the modelling parameters for a general ablation model are summarized. Founded by a national project Läßiger et al [1] and Meiners et al [2] have investigated modelling programs including machining and laser parameters for a sublimation based process calculating 3 dimensional geometries. Based on the results of modelling calculations machining and laser parameters for a NC-program can be developed. Controlling the process is possible by measuring laser parameters or process-signals (e.g. plasma) [1] or directly measuring the contoured removal. Modelling and geometrical controlling render possible a drastic reduction in the number of experiments needed to achieve new workpiece geometries as can be shown [3][4].

Experiments show that drilling (the geometrically seen easiest machining process) of metals is partly achieved by melt expulsion, even in the case of very high intensity. Calcu-lations of the melt expulsion rate and the vaporization rate as function of the intensity for various metals will be shown for the drilling process. Experimental investigations indicate that superheating effects or boundary layer effects must -besides other effects- be considered in micro machining metals as well but are not yet fully understood.

Modelling of an evaporative machining process

For most laser processes the interaction of the laserbeam with the workpiece can be calculated as a surface heat source. The energy transport is basically heat conduction. Neglecting heat losses the energy needed for vaporization equals the absorbed laser energy. Regarded as a stationary, 1- dimensional problem the velocity of the vaporization front then is given by:

$$v = \frac{A*I}{E_{Vap}} \tag{1}$$

and therefore the ablated depth within a certain pulse duration τ is

$$s = \frac{A*I}{E_{Vap}} * \tau \tag{2}$$

A is the coupling rate, I the laser beam intensity and E_{Vap} the energy needed to vaporize the material. The specific energy needed to vaporize the material is given by formula 3:

$$E_{Vap} = \int_{T_0}^{T_z} \Delta C_p \, dT + \Delta H + \Delta Q_Z \ . \tag{3}$$

The terms of formula 3 give the energy needed to heat the ablated volume to the vapori-zation temperature, the energy needed for phase transitions ΔH and the energy of chemi-cal reactions ΔQ_Z.

The discretisation of the workpiece can be a 2- dimensional mesh describing a surface, not a volume, assuming the 1- dimensional vaporization front velocity of formula 1. An ortogonal, äquidistant mesh has been tested satisfying the model demands. Dependend on the intensity interacting on a discret element and the pulse duration the depth $s_{x,y}$ is calculated and stored as the new depth value within the mesh.

For the calculation of the geometry the machining parameters and the laser parameters must be considered. The machining parameters, in most applications rectangular scanning patterns [2], are defined by the feed per pulse and the displacement between the scanning traces. The laser parameters involved are:
- laser beam intensity distribution,
- beam quality,
- focussing conditions and
- pulse duration.

Modelling adaption to experiments on Si_3N_4

In micro machining Si_3N_4 with lasers the ceramic decomposes similar to a vaporisation process. The resulting pressure gradient drives the ablated material out of the interaction zone. The energy to vaporise Si_3N_4 can be calculated to [5]:

$$\Delta E_{Vap\ Si_3N_4} = 27\frac{kJ}{cm^3} \tag{4}$$

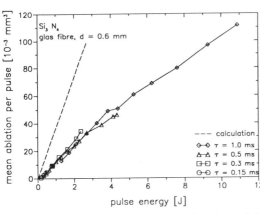

Figure 2 gives a comparison of the experimental results of the mean ablation per pulse as function of the pulse power for various pulse durations but at constant scanning parameters with calculated results from formula 4. In contrast to the model calculations the experiments show a threshold of the pulse power to start the ablation process and slight differences in the gradient of the curves. Formula 1 therefore must be modified to:

Fig. 2: Ablation per pulse as function of the pulse power for various pulse durations

$$v_{x,y,z} = C_1 * \frac{I_{x,y,z}}{E_{Vap}} + C_2 \tag{5}$$

The intensity threshold to start the ablation process C_2 can easily be measured experimentally. The total absorption corresponding to C_1 is much more difficult to measure exactly. Resulting from beam propagation effects caused by the variation of the focal position due to the nature of the ablation process the part of the laser beam intensity distribution not reaching the threshold value of C_2 does not cause ablation effects. Based on experimental data, e.g. from fig. 2, C_1 is calculated within the modelling itself.

The ablation rate as well as the surface roughness therefore depend not only on the laser parameters but on the scanning parameters as shown in figure 3. The modelling (data for C_1 and C_2 from fig. 3) is in very good coincidence with the experimental results. The effect of different ablation rates for different scanning parameters thus turns out to be

708

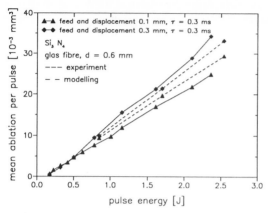

Fig. 3: Ablation per pulse as function of the pulse power for various machining parameters

dominated by beam propagation effects which determine the energy loss caused by the intensity threshold. Not only the ablation rate but also the surface geometry, e.g. surface roughness, edge angels can be calculated according to the scanning strategy and the laser parameters. This is shown within the figures 4 and 5 where experimental results given by SEM- pictures are compared to corresponding calculations.

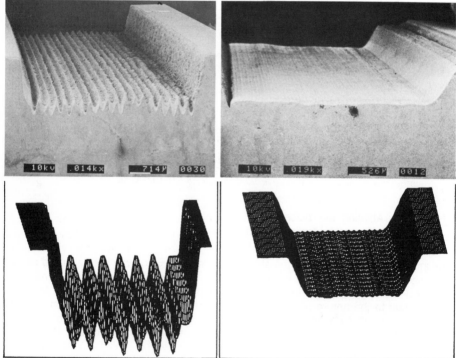

Fig. 4: SEM- pictures and corresponding calculations for 0.3 mm feed

Fig. 5: SEM- pictures and corresponding calculations for 0.1 mm feed

The presented modell thus for a vaporisation based process enables the calculation of - according to figure 1- as well the machining parameters as the laser parameters to achieve a defined geometry.

Modelling of a machining process with melt expulsion

In contrast to the described vaporization process of Si_3N_4, for all metals investigated in various caving processes melt expulsion could not be avoided. To enhance the efficiency of the process several methods support melt expulsion by an external, mostly reactive gas jet [6]. The combined interaction of e.g. vapour pressure, external gas jet, exothermal reactions on molten material within the interaction zone is not jet fully understood, especially in the case of 3- dimensional geometries.

Provided the ablated material is partly vaporized, with j_V being the vaporisation rate and L_V the specific energy for vaporization, and partly expulsed as melt, with j_M and L_M for the liquid phase, the total energy flux for the absorbed laserbeam energy is given by:

$$\dot{Q} = j_M * L_M + j_V * L_V \qquad (6)$$

The 1- dimensional, stationary theory presented in [7] was modified [8] iteratively calculating the interdependent parameters vaporisation rate and liquid expulsion rate as function of the absorbed intensity adjusting the vaporisation temperature und cavity pressure. Results for Fe and Al give figure 6.

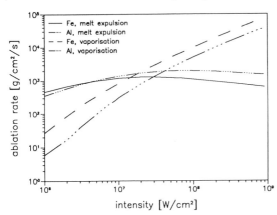

At lower intensity ablation in the melt regime is dominant. At higher intensity the ablation of vaporized material exceeds the expulsion rate of the molten material. The ablation rate of the vaporized material is continously growing with intensity, the expulsion rate of the molten material running a maximum at about 10^7 W/cm². It must be stated, however, that even at an intensity of 10^9 W/cm² ablation of molten material cannot be prevented.

Fig. 6: Calculated ablation rates for Fe and Al as funtion of the absorbed intensity

In case of sufficiently high intensity and short pulse duration (to avoid 3- dimensional heat conduction effects) which can be delivered for example from a Q- Switch lasersystem the ablation process can be calculated as a vaporization process [1]. Figure 6 gives the calculated efficiency as function of the intensity. The lower the intensity and thus the higher the liquid part of the removal rate the higher efficiencies are achieved. It has to be considered, however, that

710

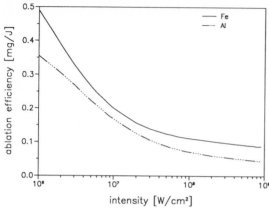

the process parameters yielding acceptable surface qualities (< 20 μm) are limited to a small parameter field [9]. Contrasting to the advantages of a vaporisation based process -according to e.g. surface roughness, melt deposition as known from Excimer laser processing or vaporizing materials like Si_3N_4- the efficiency then is very poor.

Fig. 7: Calculated ablation efficiency for Fe and Al as functions of the absorbed intensity

Conclusions

Based on calculations the machinig and laser parameters can be defined for a vaporization process, which has been demonstrated for Si_3N_4 as an example. The problems of melt expulsion, typical for the maching process of metals up to now can not be calculated sufficiently.

References

[1]   Lässiger, B. et al: *Contoured Material Removal with Cw-Q-Switch Nd:YAG-Laserradiation*, proc. ECLAT 92, Göttingen, 1992

[2]   Meiners, E. et al: *Micro Machining of Ceramics by Pulsed Nd:YAG-Laser*, proc. ICALEO 91, San Jose, USA, 1991

[3]   Wiedmaier, M. et al: *Integration of Materials Processing with YAG- Lasers in a Turning Center*, proc. ECLAT 92, Göttingen, 1992

[4]   Meiners, E: *Dreidimensionale Keramikbearbeitung mit Festkörperlaser*, Werkstoff und Innovation 3/92, Giesel Verlag

[5]   Salmang, H.: *Keramik*, Teil 1: Allgemeine Grundlagen und wichtige Eigenschaften, Springer, Berlin, 1982

[6]   Meiners, E. et al: *Micro Machining of Hard Materials using Nd:YAG- Laser*, proc. LASER 91, Springer, 1992

[7]   Allmen, M.: *Laser drilling velocity in metals*, J. of Appl. Phys., Vol. 47, No. 12, 1976, p. 5460-5463

[8]   Meiners, E. et al: *Phänomenologische Untersuchungen zum Laserbohren von Metallen*, IFSW internal report, to be published

[9]   Eberl, G.: *Abtragen von Freiformflächen mit dem LASERCAV*, Handbuch zum Seminar "Feinbearbeitung mit dem Laserstrahl" in Stuttgart 1992, VDI- Bildungswerk

Acknowledgements

The project this report is based on is supported by the BMFT of the Federal Republic of Germany and recorded under code number 13N5747. The authors are responsible for the content of this publication.

Knowledge Based System for Laser Hardening

Schlebeck, D.; Bachmann, M.
Technische Hochschule Ilmenau
Institut für Fertigung
Ilmenau, Germany

Laser hardening with CO_2 high power lasers is used in production for 20 years. Considerable progress had been made in the characterisation of laser hardening processes by means of the mathematical analysis during the last years.

We wanted to connect the results of experiments, further work in industry and thermal field calculations and turned them into a knowlegde based system. A designer shall get suitable help or advise in material selection and designing for close limit production with the help of that.

Thermal field calculations and classification

Calculation of temperature fields in the workpiece caused by the laser beam is the first point of treatise. That urgently requires to find out the relationship between power density distribution, scanning velocity, interaction time, the shape of the hardened or annealed tracks in the workpiece and local material properties after hardening. That calculations are based on thermal diffusion equations and are expressed by means of regression functions in solution space (Fig. 1). It is possible to calculate the technological parameters from inputs like depth or width of the hardened or annealed areas with the help of these. The calculations are possible for seven kinds of different shapes of the workpieces or classes.

The differences between the classes consist in the position of the treated tracks related to the workpiece's boundaries and different kinds of hindered heat conduction caused by the workpiece's boundaries (Fig. 2).

The seven classes represent: thick and thin plates (classes 1,2) with additional lateral boundaries (classes 3,4,6), small bodies (where the whole area to be hardened is overlapped by the laser beam without any relative velocity between them) and cylindrical ones.

This key may serve as a descriptor like the Opitz- classification in design to find repetition or similar parts with the pertinent technologiy in data banks.

Although is seems very simple more than 80% of our treated parts from machine and device building are settled down into this classification with some simplifications.

The occurance of additional elements like holes near the tracks or changing cross sections who handicap the conduction of the heat is most problematic. They cause transient processes. The wear properties of the parts may be impaired significantly.

A series of design- technological rules and semi-qualitative relationships are allocated to every class . Mostly they deal with (Fig. 3):

- rules to avoid transient processes
- rules to minimize thermal induced warpings and distorsions after the treatment
- rules to avoid annealed areas at high loaded workpiece sections
- material properties in dependence of the technological parameter set or
- non perpendicular irradiation

Exact information about the materials and the behavior of the coating is needed for a successful simulation reaction.

Characterisation of the coating

The absorption coefficent is mostly considered as constant during the simulation of laser hardening processes and the transfer of the results into practise. The values are often won by calorimetric and reflective methods.
Test calculations have shown that this is commonly false and may cause considerable differences between computed and practically realised shapes of the tracks.
Graphite coating is often used because of its really convienient handling. The thermal conductivity of the coating is small. A temperature diffence is induced between the surface of the coating and that of the specimen. The difference depends on the thickness of the coating and the scanning velocity (Fig. 4).
Thin coatings are permeable for the infrared radiation of a CO_2-Laser. These effects superimpose to that absorption behavior.
Graphite coatings are burning up at temperatures more than 1.100°C. The speed depends on the oxygen concentration in the inert gas shield.
Thats why we have local differences of the absorption coefficient within the fokus. The absorption coefficient may decrease down to the value of the substrat material. The temperature field in the treated bodies is different to the calculated one and so the practical found hardened track is smaller than the computed.
As we know the use of mixed Graphite/ Silicone-dioxide/ Titanium-dioxide coatings is profitable. Thermal conductivity and resistance of these coatings are better.

Classification of the material properties

Boundary conditions may be derived in order to classify the hardened strucure of different materials to 4 groups from the master plots. If

-the temperature field in the body is self consistent and
-the maximum working temperature at the bodies surface is the solidus point of the material

the classification of a material to a class for some materials is dependent on the interaction time (Fig. 6).

Class 1 is characterised by a soft and continouos decrease of the hardness with increasing hardening depth. It is profitly in use for bearings or guidances with a high percentage of solid friction . Typical materials are hypoeutectoide steels or such with an high difference between the austeniting temperature AC1 and AC3.

In Class 2 we find a plateau of the hardness with abrupt decrease. This is good for cutting tools which are often to be maintained by grinding. Typical materials are nearly eutectoide steels.

If the hardness is increasing from surface to depth z class 3 is applicable. This reaction is to be found for example at hypereutectoide low alloyed steels. Class 3 is suitable for tools, which will be finised after hardening by grindig.

Class 4 is characterized by a small hardness on the surface, caused by retained austenite. We find this at higher carbon steels and ledeburitic cold work steels. These layers usually with a thickness from 100 to 200 micron have to be removed by finishing the hardened specimen (Fig. 6).

All 4 states are adjustable by changing the interaction time for some steels . The example shows a diagram of a ledeburitic cold work steel. That is legal for thick bodies (Class 1 in the body's classification) and gaussian intensity distribution. The typical interaction time areas may be derease up to 50% because of the handicaped thermal conduction while treating thin bodies .
This statement is not always valid in the case of unalloyed steels. To harden this steels, a high quenching rate is required. A handicapped thermal conductivity may cause complete new mechanical properties.

Example

A short practical example shall demonstrate the efficency of today's simulation methods. This cutter for agricultural use had to be hardened without damaging the finished cutting edge. An austeniting temperature was given in the range of 900-1100 ° C. The interaction time is around 4 sec (Fig. 5).

The mathematical simulation regarded

-irradiation under an angle of 55 degrees
-the workpieces annual thickness
-the lateral cutting edge as boundary
-laser power reflected and reabsorbed in the V- grooves
-properties of this material depending on interaction time

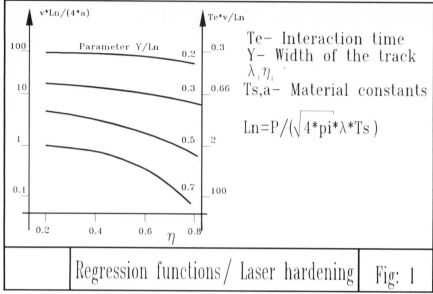

Regression functions / Laser hardening | Fig: 1

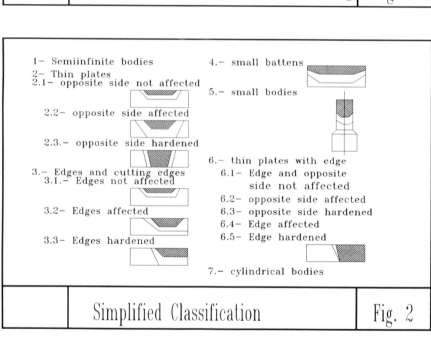

Simplified Classification | Fig. 2

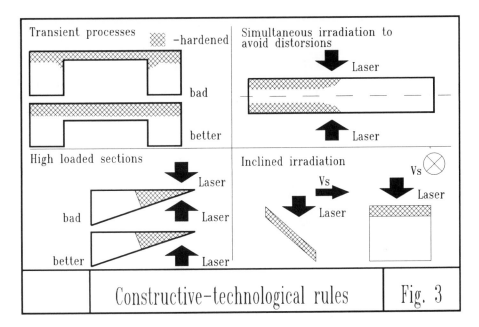

Constructive-technological rules | Fig. 3

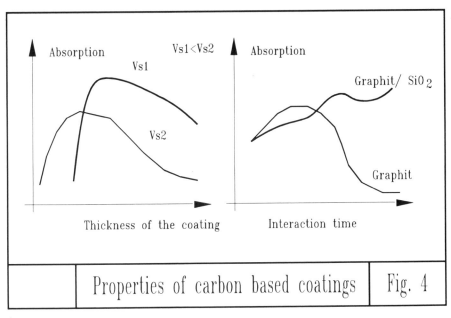

Properties of carbon based coatings | Fig. 4

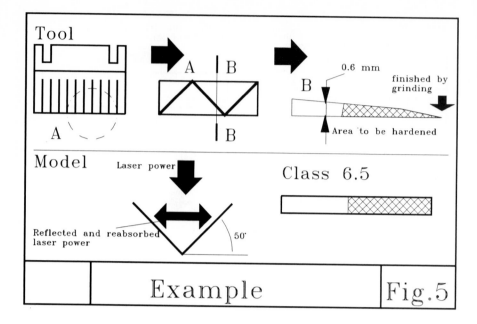

Tool

A

A | B

B

0.6 mm

B

finished by grinding

Area to be hardened

Model

Laser power

Class 6.5

Reflected and reabsorbed laser power

50°

| | Example | Fig.5 |

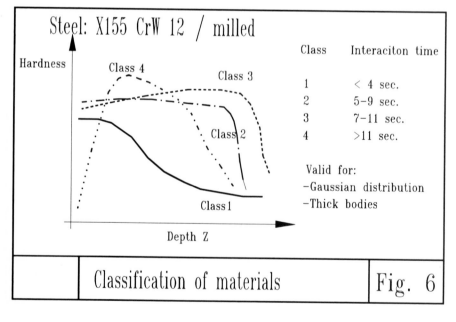

Steel: X155 CrW 12 / milled

Hardness

Class 4

Class 3

Class 2

Class 1

Depth Z

Class	Interaciton time
1	< 4 sec.
2	5–9 sec.
3	7–11 sec.
4	>11 sec.

Valid for:
−Gaussian distribution
−Thick bodies

| | Classification of materials | Fig. 6 |

X.
Thin Powder Production

Modelling of Laser Heating of the Fine Ceramic Particles

A. I. Bushma, I. V. Krivtsun, E. O. Paton Electric Welding Institute of the Ukrainian Academy of Sciences, Kiev, Ukraine

Introduction

Many works are devoted to investigation of heating and evaporation of the spherical particles in the electromagnetic radiation field (1–3). But the assumptions, used in them (stationary surface heating source or evenly distributed one) don't take into account some peculiarities of space–time distribution of the electromagnetic energy, absorbed by the particles at laser heating. Let's consider these peculiarities at the example of CO_2–laser radiation $(\lambda = 10.6\mu m)$ absorbing by ceramic particles Al_2O_3, SiO_2 with radius a=5...50μm. In contrast to the metals, absorbing IR–radiation in the thin surface layer, the ceramic materials under consideration, being nonideal dielectrics, are transparent enough for radiation with wavelength chosen and under condition $a/\lambda \approx 1$ absorb its energy in the whole volume of the particle (4). Besides, the heating nonuniformity connected with the interference structure of the electromagnetic field, excited in the particle, may give rise to the essential nonuniformity of material optical properties distribution. Consequently, at the heating process the whole absorbtion cross section and space distribution of the absorbed power can change.

Absorbtion of Electromagnetic Radiation by Ceramic Particles

Let's suppose, that the plane electromagnetic wave is incident in the dielectric particle with radius a in the negative direction of OZ axis of the spherical coordinate system with the referent point at the centre of the particle. Supposing the particle being rotated around its own axis, which is perpendicular to OZ axis, and considering the radiation being unpolarized, one can introduce the angle-averaged power, absorbed in the volume unit of the particle substance (5)

$$D(r) = -\frac{1}{4\pi} \cdot \int_0^{2\pi} d\varphi \int_0^\pi \sin\theta \; \frac{1}{r^2} \frac{\partial}{\partial r} \; [r^2 P_r^d(r,\theta,\varphi)] d\theta. \qquad \text{Equ. 1}$$

If the rotation period is less then the time of temperature space distribution changing, the temperature field in the particle is considered to be the spheric–simmetrical and nonuniformity of its dielectric permittivity to be considerable only in the radial direction. To find the fields, exited by electromagnetic wave in the radial–nonuniform spherical particle and to determine the pure form of D(r) let's use the layer–approximation method. Let's divide the particle for N spherical layers, where dielectric permittivity is considered to be constant

$$\varepsilon(r) = \varepsilon_s \equiv \varepsilon_s' + i\varepsilon_s'' \quad at \quad a_{s-1} < r \le a_s \quad (s=1,2,...,N), \qquad \text{Equ. 2}$$

where a_s and a_{s-1} – outer and inner radii of s layer ($a_0 = 0$, $a_N = a$). Increasing the number of layers is certain to give more precise approximation of the optical properties of the nonuniformly heated particle. From the other side, this method gives the opportunity to calculate the distributions of the absorbed power in the stratified particles, consisting of different materials. Using the solution of the problem of the plane wave diffraction by the stratified sphere (6), one can find

$$D(r) = P^{inc} \cdot \frac{\varepsilon_s''}{8k} \cdot \sum_{m=1}^\infty (2m+1) \cdot \sum_{\gamma=1,2} \left\{ |d_\gamma^{(s)}|^2 \cdot F_\gamma^{(s)}(r) \; + \; 2Re\left[d_\gamma^{(s)} \tilde{d}_\gamma^{(s)*} \cdot H_\gamma^{(s)}(r) \right] + |\tilde{d}_\gamma^{(s)}|^2 \cdot G_\gamma^{(s)}(r) \right\}$$

$$at \quad a_{s-1} < r \le a_s \quad (s = 1,2,...,N). \qquad \text{Equ. 3}$$

Here

$$F_1^{(s)}(r) = k_s \cdot j_m(k_s r) \cdot k_s^* \cdot \tilde{j}_m^*(k_s r);$$

$$F_2^{(s)}(r) = \frac{1}{r^2} \cdot \frac{\partial}{\partial r}[r j_m(k_s r)] \cdot \frac{\partial}{\partial r}[r \tilde{j}_m^*(k_s r)] \; + \; \frac{m(m+1)}{r^2} j_m(k_s r) \tilde{j}_m^*(k_s r);$$

$$H_\gamma^{(s)}(r) = F_\gamma^{(s)}(r) \qquad at \qquad \overset{*}{j}_m, \overset{'*}{j}_m \to \overset{*}{n}_m, \overset{'*}{n}_m \; ; \; (\gamma = 1,2)$$

$$G_\gamma^{(s)}(r) = F_\gamma^{(s)}(r) \qquad at \qquad j_m, \overset{'}{j}_m, \overset{*}{j}_m, \overset{'*}{j}_m \to n_m, \overset{'}{n}_m, \overset{*}{n}_m, \overset{'*}{n}_m;$$

$$k = 2\frac{\pi}{\lambda}; \quad k_s = k\sqrt{\varepsilon_s}; \qquad\qquad\qquad \text{Equ. 4}$$

p^{inc} – density of the incident power; λ – wavelength at the outer medium ($\varepsilon_{ext} = 1$);

$$d_\gamma^{(s)} = \frac{2i}{a\frac{\partial}{\partial a}[ah_m^{(1)}(ka)]} \cdot \frac{p_\gamma^{(s)}}{\pi_\gamma p_\gamma^{(N)} + \tilde{\pi}_\gamma \tilde{p}_\gamma^{(N)}};$$

$$\tilde{d}_\gamma^{(s)} = \frac{2i}{a\frac{\partial}{\partial a}[ah_m^{(1)}(ka)]} \cdot \frac{\tilde{p}_\gamma^{(s)}}{\pi_\gamma p_\gamma^{(N)} + \tilde{\pi}_\gamma \tilde{p}_\gamma^{(N)}}, \qquad \text{Equ. 5}$$

where $j_m(z)$, $n_m(z)$, $h_m^{(1)}(z) = j_m(z) + in_m(z)$ – spherical m–order Bessel functions of the first, second and third kind. The quantities $p_\gamma^{(s)}$ and $\tilde{p}_\gamma^{(s)}$ are determined by the correlations:

$$p_\gamma^{(s+1)} = \mu_\gamma^{(s+1)} p_\gamma^{(s)} + v_\gamma^{(s+1)} \tilde{p}_\gamma^{(s)};$$

$$\tilde{p}_\gamma^{(s+1)} = \tilde{\mu}_\gamma^{(s+1)} p_\gamma^{(s)} + \tilde{v}_\gamma^{(s+1)} \tilde{p}_\gamma^{(s)};$$

$$p_\gamma^{(1)} = 1; \quad \tilde{p}_\gamma^{(1)} = 0, \qquad\qquad \text{Equ. 6}$$

where

$$\mu_1^{(s+1)} = -a_s k_s \left\{ \frac{\partial}{\partial r}[r j_m(k_s r)] n_m(k_{s+1} r) - j_m(k_s r) \cdot \frac{\partial}{\partial r}[r n_m(k_{s+1} r)] \right\} \Big|_{r=a_s};$$

$$\mu_2^{(s+1)} = -a_s k_{s+1} \left\{ \frac{\partial}{\partial r}[r j_m(k_s r)] n_m(k_{s+1} r) - \frac{k_s^2}{k_{s+1}^2} j_m(k_s r) \cdot \frac{\partial}{\partial r}[r n_m(k_{s+1} r)] \right\} \Big|_{r=a_s};$$

$$\tilde{\mu}_\gamma^{(s+1)} = -\mu_\gamma^{(s+1)} \qquad at \quad n_m, \overset{'}{n}_m \to j_m, \overset{'}{j}_m;$$

$$v_\gamma^{(s+1)} = \mu_\gamma^{(s+1)} \qquad at \quad j_m, \overset{'}{j}_m \to n_m, \overset{'}{n}_m;$$

$$\tilde{v}_\gamma^{(s+1)} = -\mu_\gamma^{(s+1)} \qquad at \quad j_m, \overset{'}{j}_m \to n_m, \overset{'}{n}_m; \; n_m, \overset{'}{n}_m \to j_m, \overset{'}{j}_m, \qquad \text{Equ. 7}$$

and quantities π_γ and $\tilde{\pi}_\gamma$ in Equ. 5 are

$$\pi_1 = k_N \cdot \left\{ j_m(k_N r) - \frac{\partial}{\partial r}[r j_m(k_N r)] \cdot \frac{h_m^{(1)}(kr)}{\frac{\partial}{\partial r}[r h_m^{(1)}(kr)]} \right\} \Big|_{r=a};$$

$$\pi_2 = k \cdot \left\{ \frac{k_N^2}{k^2} j_m(k_N r) - \frac{\partial}{\partial r}[r j_m(k_N r)] \cdot \frac{h_m^{(1)}(kr)}{\frac{\partial}{\partial r}[r h_m^{(1)}(kr)]} \right\} \Big|_{r=a};$$

$$\tilde{\pi}_\gamma = \pi_\gamma \qquad at \quad j_m, \overset{'}{j}_m \to n_m, \overset{'}{n}_m. \qquad\qquad \text{Equ. 8}$$

Parallel with quantity $D(r)$, the solution of the diffraction problem (6) gives the opportunity to find the whole absorbtion Q^d and scattering Q^{sc} cross sections of the plane electromagnetic wave by radial–nonuniform dielectric particle. In the limit of the uniform spherical particle one can receive the well–known formulas of Mie theory (7) by putting in the received correlations $N=1$. At Fig. 1 the calculation results of the absorbed power distributions in the uniform ceramic particles Al_2O_3 and SiO_2 are shown. In the case of Al_2O_3 the angle-averaged power, absorbed in the volume unit of the spherical particle, decreases from its surface to the centre. The heating inhomogeneity is intensified with a increasing (comp. curves 1 and 3 at Fig. 1,a). For the SiO_2 particles the opposite effect is observed. The maximum of the quantity $D(r)$, located inside the particle, decreases with its radius growth (Fig. 1,b), that results in more homogeneous heating of the SiO_2 particles when a increases.

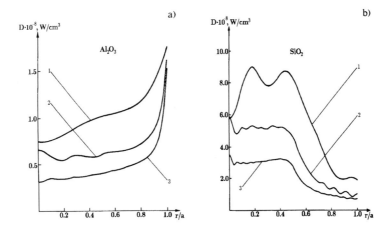

Fig. 1: The distributions of the absorbed power in the particle volume ($P^{inc}=10^6$ W/cm^2, $\lambda=10.6\,\mu$m): a - Al_2O_3; b - SiO_2; 1 - a=10 μm; 2 - 30 μm; 3 - 50 μm.

Particles Heating by Electromagnetic Radiation

To calculate the temperature fields in the ceramic particles, being heated by laser radiation, let's use the nonstationary thermoconductivity equation with distributed heat source. Taking into account the suggestion about the spherical simmetry of the thermal field inside the particle and considering the dependence of the material optical properties on the temperature, this equation can be written as

$$\rho C \frac{\partial T}{\partial t} = \frac{1}{r^2} \cdot \frac{\partial}{\partial r}(r^2 \chi \frac{\partial T}{\partial r}) + D(r,t). \qquad \text{Equ. 9}$$

Here $T(r,t)$ – temperature, $\rho(T)$ – mass density, $\chi(T)$ – thermoconductivity; $C(T)$ – effective thermal heat capacity of the particle substance, determined with taking into consideration the specific melting heat W_m and vaporization heat W_b

$$C(T)=c(T)+W_m\delta(T-T_m)+W_b\delta(T-T_b), \qquad \text{Equ. 10}$$

where $c(T)$ – specific thermal heat capacity of the material, T_m and T_b – melting and boiling temperatures correspondently. The quantity $D(r,t)$ is calculated from the Equ. 3–8 with taking into account time alteration of space distribution $\varepsilon(T(r,t))$, connected with nonuniform particle heating. To describe the heat emission from the particle surface we can use the Newton heat transfer model and Stephan-Boltzmann radiation law. Then, the boundary conditions to the Equ. 9 can be written as

$$-(\chi \frac{\partial T}{\partial r})|_{r=a}=\alpha(T_{sur}-T_{ext})+\beta\sigma(T_{sur}^4-T_{ext}^4);$$

$$\frac{\partial T}{\partial r}|_{r=0}=0, \qquad \text{Equ. 11}$$

where T_{sur}–surface temperature, T_{ext}–temperature of the external gas; β–the integral blackness degree of the particle material; σ–Stephan–Boltzmann constant; α–heat exchange coefficient. Let's write the starting condition for Equ. 9 as

$$T(r)\big|_{t=0} = T_{ext}. \qquad \text{Equ. 12}$$

Equ. 9 was being solved numerically, using the finite differences method. The vaporized material was supposed to be moved away from the particle surface, not weakening the incident radiation and not influencing the conditions of the heat exchange between the particle and environment. If the boiling temperature was reached in the inner point r_k of the calculated domain, the particle radius was spasmodicly diminished and solving the Equ. 9 was continued within the domain $0 \leq r \leq r_k$. At $r_k = 0$, the thermal explosion of the particle was considered to take place, and solving procedure was stopped. On the basis of the suggested mathematical model, the numerical investigation of the CO_2-laser radiation thermal influence on the Al_2O_3 and SiO_2 particles in argon flow (T_{ext}=293 K) was held. The starting radius of the particles is a^0 =30 μm; the intensity of the incident radiation is P^{inc}=10^6 W/cm^2 . For comparison, the cases when the optical properties of the material were changing with temperature and when they remained constant, being equal to their values at T=T_{ext} were considered. In the first case the real part of the dielectric permittivity was supposed not to depend on the temperature and the imaginary part, being proportional to the phonons free path length (8), to be calculated with the aid of the correlation

$$\varepsilon'' = A\, \frac{c(T)}{\chi(T)}, \qquad \text{Equ. 13}$$

where the constant A was determined from the known value ε'' at T=293 K. Within the temperature domain above melting temperature ε'' was constant, determined from Equ. 13 at T=T_m. Time changing of the temperature inside and at the surface and time changing of the radius of the Al_2O_3

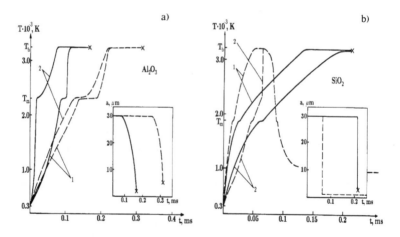

Fig. 2: Time changing of the temperature and radius of the Al_2O_3 (a) and SiO_2 (b) particles: solid curves - $\varepsilon = \varepsilon' + i\varepsilon''(T)$, dashed curves - ε=const: 1 - T(0,t); 2 - T(a,t); × - thermal explosion of the particle.

and SiO_2 particles at electromagnetic radiation heating are shown at Fig.2. Nonuniform distribution of the heat sources in the particles under consideration (see Fig. 1,a;b) gives rise to the fact, that even in optically uniform case the surface temperature of Al_2O_3 particle grows more quickly, then its center temperature (dashed curves 1,2 at Fig. 2,a), and in the case of SiO$_2$ the reverse effect is observed (Fig. 2,b). According to this, Al$_2$O$_3$ particle is vaporized from the surface up to the radius value a $\approx$ 5 μm and then the thermal explosion of the remained material takes place (dashed curve at Fig. 2,a). SiO$_2$ particle is heated so, that the boiling temperature is reached at r$\approx$1.5 μm and, according to the model in use, its radius is spasmodicly diminished up to the pointed value (dashed curve at Fig. 2,b). The radiation power, being absorbed by the remained particle, is so small, that the particle begins to get cold uniformly and its radius remains constant (Fig. 2,b). The heating nonuniformity of the Al$_2$O$_3$ particles is intensified (solid curves 1,2 at Fig. 2,a), and of the SiO$_2$ particles is sligtly weakened (Fig 2,b), when the temperature dependence of the optical properties of the materials is taken into account. It's connected with the correspondent redistribution of $\varepsilon''(r)$, and consequently D(r) in the

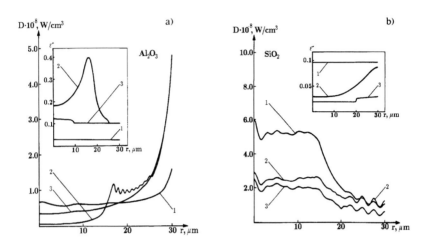

Fig. 3: Redistribution of D(r) and $\varepsilon''(r)$ in the particle volume at laser heating: a - Al$_2$O$_3$; 1 - t=0; 2 - 38 μs; 3 - 88 μs; b - SiO$_2$; 1 - t=0; 2 - 20 μs; 3 - 50 μs.

particle volume, the considerable changes of the pointed quantities distributions are observed at the begining of heating (comp. curves 1,2 at Fig. 3). Taking into consideration the temperature dependences of the dissipative characteristics of Al$_2$O$_3$ and SiO$_2$ also gives rise to the considerable increasing of the heating rate of the Al$_2$O$_3$ particles in comparison with the optically uniform particles approach (Fig. 2,a) and to its decreasing in the case of SiO$_2$ (Fig. 2,b). The reason of this phenomenon is the correspondent changes of the whole absorbtion cross sections, observed at the heating of the considered particles.

References
(1)   A. V. Kuzikovsky: Sov. J. Izv. Vuz. Phys. 5 (1970) 89.
(2)   Yu. I. Yalamov, N. A. Silin, A.I.Sidorov etc.: Sov. J. Tech. Phys. 2 (1980) 380.
(3)   V. I. Igoshin, S. Yu. Pichugin: Sov. J. Quant. Electr. 10 (1985) 2187.
(4)   N. N. Belov: Sov. J. of Appl. Spectr. 6 (1986) 948.
(5)   A. I. Bushma, I. V. Krivtsun: Sov. J. Phys. and Chem. Treat. of Mater. 2 (1992) 40.
(6)   V. S. Gvozdetsky, A. G. Zagorodny, I. V. Krivtsun etc.: Preprint ITP-83-167P., Kiev (1983) 26.
(7)   M. Born, E. Wolf: Principles of Optics, Pergamon (1973) 720.
(8)   C. Kittel: Introduction to Solid State Physics, Wiley (1978) 792.

List of Authors

Author Index*

A
Aksenov, L. 235
Alunovic, M. 451
Amon, M. 325
Anjos, M.A. 263
Arsenault, B. 621

B
Bachmann, M. 711
Basting, D. 3, 609
Bea, M. 63
Beck, M. 693
Becker, R. 565
Behler, K. 151
Berger, P. 63, 111, 693
Bergmann, H.W. 157, 163, 293, 325, 363, 521, 527, 533, 547, 591, 649, 673
Berkmanns, J. 33, 151
Betz, J. 93
Beyer, E. 27, 33, 143, 151, 585
Beylat, L. 257
Bharti, A. 299
Biermann, S. 51
Bignonnet, A. 635
Binroth, C. 39
Blank, E. 193
Blanpain, B. 539
Böhm, S. 485
Bolender, H. 369
Boufoussi, M. 635
Bransden, A.S. 117
Brenner, B. 199, 337, 343
Buerhop, C. 521, 603
Burchards, D. 223
Burghardt, B. 609
Burmester, S. 565
Bushma, A.I. 719

C
Celis, J.P. 539
Chevrier, J.Ch. 635
Com-Nougue, J. 181
Costa, A.R. 263
Coulon, P.A. 181

Covelli, L. 235, 251

D
Dausinger, F. 99, 125, 137, 313, 411, 559, 705
De Paris, A. 131
Denis, S. 635
Denney, P.E. 71
Deshors, G. 331
Dezert, D. 211
Dietsch, R. 479
Dubé, D. 621

E
Ebert, R. 491
Ebner, R. 187
Edler, R. 111
Endres, T. 117, 157, 293, 325, 363
Engst, P. 509
Erkens, G. 451
Ermolaev, A. 15

F
Fantozzi, G. 131
Fischer, A. 491, 497
Fouguet, F. 211
Franck, M. 539
Freisleben, B. 363
Frenk, A. 375
Froben, F.W. 515
Funken, J. 451
Fux, V. 205, 387

G
Gaensicke, F. 491
Gasser, A. 687
Gassmann, R. 205, 405
Geiger, M. 87, 577, 603
Geith, A. 521
Gélinas, C. 621
Giesen, A. 63
Glumann, C. 63, 99

*The page numbers refer to the 1st page of the relevant article

Author Index*

Goller, M. 577
Gratzke, U. 681
Gropp, A. 87
Grünenwald, B. 411

H
Haddenhorst, H. 245
Haferkamp, H. 461
Heider, T. 293
Helebrant, A. 521
Hensel, F. 39
Herrmann, G. 533
Hesse, D. 105
Hinse-Stern, A. 223
Hofmann, J. 591
Holste, C. 343
Hontzopoulos, E. 553
Hopfe, S. 479
Hopfe, V. 485
Hornbogen, E. 245
Hourdakis, G. 553
Hügel, H. 99, 111, 137, 559, 693, 705
Hunziker, O. 193

I
Ignatiev, M. 15, 241
Ihlemann, J. 615
Illmann, U. 491
Ishide, T. 57, 81
Ito, S. 57

J
Jagiella, M. 51
Janhofer, K. 143
Jaschek, R. 521, 533, 673
Jolys, P. 257
Juch, K. 205, 281

K
Kabasawa, M. 667
Kahlert, H.-J. 609
Kalla, G. 143

Kar, A. 655
Kechemair, D. 257
Keiper, B. 497
Kerrand, E. 181
Kessler, A. 705
Keßler, G. 473
Kirmse, W. 419
Kirner, P. 217, 393
Klopper, W. 591
König, W. 217, 393
Kosuge, S. 667
Kovalev, E. 241
Krappitz, H. 369
Krause, V. 27, 33
Krepulat, W. 63
Kreutz, E.W. 269, 451, 539, 687
Kriszt, B. 187
Krivtsun, I.V. 719
Kroos, J. 681
Kubát, P. 509
Kukla, H. 527
Kupfer, R. 547, 591

L
Lagain, P. 257
Lambert, P. 621
Lang, A. 163, 673
Lashin, A. 45, 699
Lässiger, B. 585
Laurens, P. 257
Lemmer, O. 451
Lepski, D. 349
Leyendecker, T. 451
Linke, J. 661
Liu, Y. 381
Loosen, P. 27
Löschau, W. 205, 281
Luft, A. 199, 205, 387, 405
Lugscheider, E. 369
Lutz, N. 577, 603

M
Macherauch, E. 629
Mai, H. 479

*The page numbers refer to the 1st page of the relevant article

Author Index*

Mann, K. 503
Mariaux, S. 193
Marsden, C.F. 375
Matsumoto, O. 57, 81
Mazumder, J. 381, 655
Mega, M. 57
Meiners, E. 559, 705
Menzel, S. 467, 661
Merlin, J. 635
Metev, S. 565
Militzer, B. 473
Mituhashi, T. 57
Möhlmann, C. 461
Mommsen, J. 597
Mongis, J. 331
Mordike, B.L. 171, 223, 229, 629
Müller, D. 157, 293, 325, 363
Müller, F. 503

N

Nagashima, T. 81
Nagura, Y. 81
Nakada, K. 667
Naser, J. 527
Niessen, M. 585
Nizery, F. 331
Nowotny, S. 387, 411

O

Ollier, B. 687
Ono, M. 667
Ott, P. 585
Overmeyer, L. 21

P

Pelletier, J.M. 211
Peyre, J.-P. 331
Pirch, N. 269, 687
Poli, M. 699
Pollack, D. 387
Pompe, W. 445, 479
Puig, T. 257

Q

Queitsch, R. 527, 649

R

Rapp, J. 99
Rebhan, T. 577
Reisse, G. 491, 497
Reiter, H.-T. 199, 343
Reitzenstein, W. 199, 205, 281, 349
Richter, A. 473
Ritz, M. 515
Robin, M. 131, 211
Rockstroh, T. 655
Roos, J.R. 539
Roy, M. 299
Rozsnoki, L. 217
Rozsnoki, M. 269
Rudlaff, T. 313, 559
Rund, M. 319

S

Sabatier, L. 257
Sadovsky, V.D. 307
Sakamoto, H. 125
Schädlich, S. 199, 337, 405
Schastlivtsev, V.M. 307
Scheibe, H.-J. 445
Scheller, D. 393
Schlebeck, D. 711
Schlüter, H. 687
Schmidt, H. 615
Schoeneich, B. 479
Scholtes, B. 629
Scholz, R. 479
Schultrich, B. 445
Schüßler, A. 287
Schutte, K. 527, 649, 673
Schwager, K.-D. 629
Schwarz, W. 387
Sepold, G. 39
Shen, J. 411
Shibata, K. 125, 381, 433
Siemroth, P. 445
Simon, A. 635

*The page numbers refer to the 1st page of the relevant article

Author Index*

Simon, G. 681
Smurov, I. 15, 45, 235, 241, 251, 699
Specht, B. 681
Spur, G. 93
Stephan, D. 199
Stöhr, M. 643
Storch, W. 199
Stournaras, C.J. 553
Strulese, S. 15, 241
Stürmer, M. 571
Sunderarajan, G. 299
Sung, H. 451
Surry, C. 45

T

Tabatchikova, T.I. 307, 355
Tagirov, K. 235
Taube, R. 673
Techel, A. 405
Tehel, A. 485
Thauvin, G. 181
Titov, I. 15, 241
Tönshoff, H.K. 21, 105, 319, 597
Topkaya, A. 51
Treusch, H.-G. 27, 33, 585
Tsetsekou, A. 553

U

Uglov, A. 241

V

Vannes, A.B. 211
Vicanek, M. 681
Vilar, R. 263
Völlmar, S. 479
Volz, R. 399
von Alvensleben, F. 21
von Bergmann, H. 591
Voss, A. 451

W

Wagnière, J.-D. 375

Wahl, T. 137
Wehner, B. 479
Wehner, M. 539
Weisheit, A. 229
Weiß, H.-J. 445
Weißbrodt, P. 479
Weißmann, R. 521, 03
Wetzig, K. 467, 661
Wiedemann, G. 199
Wiedmaier, M. 559
Winderlich, B. 199, 337, 343
Wissenbach, K. 539, 687
Wolff-Rottke, B. 615

Y

Yakovleva, I.L. 307, 355
Yokoyama, A. 81
Yu, H. 515

Z

Zambetakis, T. 553
Zeyfang, E. 643

Subject Index

Subject Index*

A

Ablation 445, 473, 479, 577,
      609

Abrasive wear 281, 411

Absorption 313, 375, 673, 693,
      719

Adaptive optics 63

Aerodynamical window 63

Ageing treatment 355

Alloying 15, 117, 187, 205,
      229, 235, 241, 251,
      281, 293

Aluminium

- alloys 33, 105, 117, 125,
      193, 245, 399

- Lithium alloys 163

- Magnesium alloys 163

Alumino-silicate 131

Automotive industry 433

B

Beam

- absorptivity 57, 257

- characterisation 643

- delivery 63

- diagnostic system 349

- homogenizer 3

- oscillation 71, 349

- quality 63, 105, 693

- shaping 331, 349, 363

Brewster-effect 313

C

Carbide coatings 369

Carbon films 497

Ceramic particles 245, 719

Ceramics 21, 131, 411, 485,
      515, 553, 577,

Cladding 71, 205, 211, 223,
      369, 375, 381, 387,
      399, 405, 411

CO2-laser 57, 111, 131, 313,
      399, 405, 521, 533,
      621, 643, 667, 673

- welding 667

Coat bonding 405

Coating 241, 405, 621

Cobalt-Tungsten hardfacing 71

Composite layers 287

Copper vapour laser 547, 591

Cutting 15, 21, 71, 51, 111,
      117, 433, 553, 591,
      643

D

Defects in LPVD produced layers 467

Deposition 445, 509

Diamond-like-carbon 503

Dielectric permittivity 719

Diffusion zone 319

Drill hole geometry 547

Drilling of semiconductor materials 547

Dual wavelength processing 577

Duplex treatment 319

* The page numbers refer to the 1st page of the relevant article

Subject Index*

E

Electron microscopy 479, 515

Emission spectroscopy 515

Emissivity 15

Erosion behaviour 199, 661

Evaporation 45, 591, 693, 699, 705

Excimer laser 3, 105, 533, 553, 597,
609

- treatment 527, 577, 649

F

Fatigue 199, 257, 337, 343,
629

Fiber optics 71

Fibre reinforced composites 485

Filler wire 223

Fire-polishing 521

Fusion zone microstructure 105

G

Green`s function method 349

H

Hard coatings 205, 387, 485

Hardening 57, 199, 313, 319,
331, 337, 343, 349,
553, 629, 635, 673

Hardness 199, 241, 245, 281,
307, 331, 355, 411

Heat

- affected zone 547, 591

- conduction 105

- flow analysis 681

Heating nonuniformity 719

Homogenizer 609

Humping 681

I

Image processing 479

Industrial robots 87

Influence of laser parameters 565

Intensity distribution 349

J

Japanese Laser Industry 3

K

Keyhole phenomena 681

L

Laser

- coupled SEM 467, 661

- hard particle injection 281

- heating 719

- induced deposition 461

- induced plasma 461, 565, 667

- PVD 461, 467

- quenching 307, 355

- track geometry 349

- vaporization 515

- wire welding 223

- CVD 485, 491

Layer-approximation method 719

* The page numbers refer to the 1st page of the relevant article

Subject Index*

M

Magnesium evaporation 105

Marking 603

Material transfer 621

Materials processing 3, 27, 105, 577

Mean stress sensitivity 343

Mechanical properties 163, 211, 565, 577

Melt ejection 111, 705

Melting 45, 205, 699, 719

Micromachining 3, 105, 609, 705

Microstructure 187, 199, 245, 281,
      369, 411

Microstructuring 603

Modelling 45, 521, 635, 681,
      693, 705

Modulated laser beam 681

Molybdenum films 461

Monitoring 81

Multi variate analysis 81

Multilayers 287

Multiphoton induced deposition 491

Multiplexing 27

N

Nd: YAG 51, 87, 313, 533

Neutral network system 81

Ni-base alloys 399

Ni-Cr-B-Si (Colmonoy 6) 369

Ni/C-multilayer structures 479

Nickel substrates 621

Niobium carbide 187

Nitriding 257

O

ODS B15

Optical

- distortion 63

- emission spectroscopy 577

- fiber 57, 81, 235, 251

- films 461

Optics 15, 63, 363

P

Paste bound materials 369

Penetration depth 667

Phase transformation 635

Photoluminescsnce spectroscopy 667

Photolytic deposition 491, 509

Physical vapor deposition 445

Plasma

- diagnostic 673

- effects 649, 681

- nitriding 319

- plume diagnostics 577

- sources 479

- temperature 667

Polymeres 597

Powder 411

Power attenuation 375

Precision drilling 547

Preheated wire 223

Pulse periodic action 235, 251

Pulsed laser deposition 497

Pyrolytic deposition 491

Pyrometer 15, 621

* The page numbers refer to the 1st page of the relevant article

Subject Index*

R

Radiation 15
Rapid solidification 187, 229
Reflectometry 533
Residual stresses 319, 331, 629, 635

S

Screen printing 281, 293
Sensors 51
Shielding gas 667
Short pulse laser 445, 547, 533, 673
Silicate glasses 603
Silicon nitride 705
Solid state laser 33
Solidification 45, 205, 699
Spatial
- distribution 235, 251
- thickness variation 497
Spike 45
Steel 33, 87, 111, 187, 205,
      281, 293, 307, 319,
      331, 337, 355
Stellite 205, 369, 375
Stripping 527, 533, 597
Surface treatment 193, 245, 257, 293,
      363, 381, 399, 433

T

Temperature field 349, 491, 635
Texturing 621
Thermal
- explosion 719
- load 565
- shock behavior 661
Thermoconductivity 719
Thermocycle 15
Thin films 445, 461, 473, 479,
      491, 515
Three-dimensional laser 87
Titanium
- alloys 229, 491, 509
- nitride 527
Transverse hot cracking 105
Tungsten carbide 211, 223
Turbodrill 241

U

UV
- optical systems 609
- detection 21
- excimer laser 461

V

Vanadium carbide 187

W

Wavefront deformation 63
Wear 199, 205, 229, 241,
      245, 257, 287, 369,
      399
Weld seam shape 105
Welding 15, 33, 51, 71, 105,
      117, 125, 131, 163,
      211, 433, 553, 643,
      681

* The page numbers refer to the 1st page of the relevant article

Subject Index*

Wetting 287
Wire feed cladding A10

X
X-ray mirrors 479
XeCl-excimer laser 603

Y
YAG
- laser 3, 57,63, 559
- laser welding 81

Z
Zinc coated sheets 33
Zirconia 553

* The page numbers refer to the 1st page of the relevant article